# AI and Blockchain Applications in Industrial Robotics

Rajashekhar C. Biradar
*Reva University, India*

Geetha D.
*Reva University, India*

Nikhath Tabassum
*Reva University, India*

Nayana Hegde
*Reva University, India*

Mihai Lazarescu
*Politecnico di Torino, Italy*

A volume in the Advances in Computational Intelligence and Robotics (ACIR) Book Series

IGI Global
PUBLISHER of TIMELY KNOWLEDGE

Published in the United States of America by
IGI Global
Engineering Science Reference (an imprint of IGI Global)
701 E. Chocolate Avenue
Hershey PA, USA 17033
Tel: 717-533-8845
Fax: 717-533-8661
E-mail: cust@igi-global.com
Web site: http://www.igi-global.com



Library of Congress Cataloging-in-Publication Data

Names: Biradar, Rajashekhar C., 1966- editor. | Devanagavi, Geetha D., 1971- editor. | Tabassum, Nikhath, 1983- editor. | Hegde, Nayana, 1978- editor. | Lazarescu, Mihai Teodor, 1967- editor.
Title: AI and blockchain applications in industrial robotics / edited by Rajashekhar Biradar, Geetha Devanagavi, Nikhath Tabassum, Nayana Hegde, Mihai Lazarescu.
Description: Hershey PA: Engineering Science Reference, 2024. | Includes bibliographical references. | Summary: "The combination of AI and blockchain is a relatively new and exciting field, with the potential to revolutionize industries and create new opportunities for businesses. This book on AI and blockchain serves as an educational resource for those looking to learn more about the technologies and their potential applications"-- Provided by publisher.
Identifiers: LCCN 2023034540 (print) | LCCN 2023034541 (ebook) | ISBN 9798369306598 (hardcover) | ISBN 9798369306604 (paperback) | ISBN 9798369306611 (ebook)
Subjects: LCSH: Robots, Industrial. | Artificial intelligence--Industrial applications. | Blockchains (Databases)--Industrial applications. | Internet of things--Industrial applications.
Classification: LCC TS191.8 .A43 2024 (print) | LCC TS191.8 (ebook) | DDC 629.8/9263--dc23/eng/20231226
LC record available at https://lccn.loc.gov/2023034540
LC ebook record available at https://lccn.loc.gov/2023034541

This book is published in the IGI Global book series Advances in Computational Intelligence and Robotics (ACIR) (ISSN: 2327-0411; eISSN: 2327-042X)

British Cataloguing in Publication Data
A Cataloguing in Publication record for this book is available from the British Library.

# Advances in Computational Intelligence and Robotics (ACIR) Book Series

Ivan Giannoccaro
University of Salento, Italy

ISSN:2327-0411
EISSN:2327-042X

Mission

While intelligence is traditionally a term applied to humans and human cognition, technology has progressed in such a way to allow for the development of intelligent systems able to simulate many human traits. With this new era of simulated and artificial intelligence, much research is needed in order to continue to advance the field and also to evaluate the ethical and societal concerns of the existence of artificial life and machine learning.

The **Advances in Computational Intelligence and Robotics (ACIR) Book Series** encourages scholarly discourse on all topics pertaining to evolutionary computing, artificial life, computational intelligence, machine learning, and robotics. ACIR presents the latest research being conducted on diverse topics in intelligence technologies with the goal of advancing knowledge and applications in this rapidly evolving field.

Coverage

- Machine Learning
- Intelligent Control
- Evolutionary Computing
- Artificial Intelligence
- Artificial Life
- Natural language processing
- Heuristics
- Brain Simulation
- Neural Networks
- Adaptive and Complex Systems

IGI Global is currently accepting manuscripts for publication within this series. To submit a proposal for a volume in this series, please contact our Acquisition Editors at Acquisitions@igi-global.com or visit: http://www.igi-global.com/publish/.

The Advances in Computational Intelligence and Robotics (ACIR) Book Series (ISSN 2327-0411) is published by IGI Global, 701 E. Chocolate Avenue, Hershey, PA 17033-1240, USA, www.igi-global.com. This series is composed of titles available for purchase individually; each title is edited to be contextually exclusive from any other title within the series. For pricing and ordering information please visit http://www.igi-global.com/book-series/advances-computational-intelligence-robotics/73674. Postmaster: Send all address changes to above address.

## Titles in this Series

*For a list of additional titles in this series, please visit:* www.igi-global.com/book-series/advances-computational-intelligence-robotics/73674

*Handbook of Research on AI and ML for Intelligent Machines and Systems*
Brij B. Gupta (Asia University, Taichung, Taiwan & Lebanese American University, Beirut, Lebanon) and Francesco Colace (University of Salerno, Italy)
Engineering Science Reference • copyright 2024 • 503pp • H/C (ISBN: 9781668499993) • US $380.00 (our price)

*Advances in Explainable AI Applications for Smart Cities*
Mangesh M. Ghonge (Sandip Institute of Technology and Research Centre, India) Nijalingappa Pradeep (Bapuji Institute of Engineering and Technology, India) Noor Zaman Jhanjhi (School of Computer Science and Engineering, Faculty of Innovation and Technology, Taylor's University, Malaysia) and Praveen M Kulkarni (KLS Gogte Institute of Technology, India)
Engineering Science Reference • copyright 2024 • 300pp • H/C (ISBN: 9781668463611) • US $270.00 (our price)

*Perspectives on Artificial Intelligence in Times of Turbulence Theoretical Background to Applications*
Nuno Geada (ISCTE, University Institute of Lisboa, Portugal) and George Leal Jamil (Informações em Rede C e T Ltda., Brazil)
Engineering Science Reference • copyright 2024 • 232pp • H/C (ISBN: 9781668498149) • US $300.00 (our price)

*Artificial Intelligence in the Age of Nanotechnology*
Wassim Jaber (ESPCI Paris - PSL, France)
Engineering Science Reference • copyright 2024 • 330pp • H/C (ISBN: 9798369303689) • US $300.00 (our price)

*Application and Adoption of Robotic Process Automation for Smart Cities*
R.K. Tailor (Manipal University Jaipur, India)
Engineering Science Reference • copyright 2023 • 226pp • H/C (ISBN: 9781668471937) • US $270.00 (our price)

*Deterministic and Stochastic Approaches in Computer Modeling and Simulation*
Radi Petrov Romansky (Technical University of Sofia, Bulgaria) and Nikolay Lyuboslavov Hinov (Technical University of Sofia, Bulgaria)
Engineering Science Reference • copyright 2023 • 513pp • H/C (ISBN: 9781668489475) • US $265.00 (our price)

*Technological Tools for Predicting Pregnancy Complications*
D. Satishkumar (Nehru Institute of Technology, India) and P. Maniiarasan (Nehru Institute of Engineering and Technology, India)
Engineering Science Reference • copyright 2023 • 392pp • H/C (ISBN: 9798369317181) • US $365.00 (our price)

IGI Global
PUBLISHER of TIMELY KNOWLEDGE

701 East Chocolate Avenue, Hershey, PA 17033, USA
Tel: 717-533-8845 x100 • Fax: 717-533-8661
E-Mail: cust@igi-global.com • www.igi-global.com

# Table of Contents

# Detailed Table of Contents

bolstering patient identity verification. Smart contracts are also deployed across layers to delineate rules and permissions for healthcare data access, enhancing data integrity and security. This architecture is built on a general, non-specific blockchain platform chosen for its robust security features.



The internet of things (IoT) has changed the comforts of every individual. Sensors and communication models play an important role in the effective implementation of IoT systems for any applications. Sensors can be classified based on the parameters to be measured and on the working principles. The second significant component in IoT-based systems is communication models, which help to exchange the data between the two IoT devices. Basically, there are four types of communication models that are commonly used: device to device, device to cloud, device to the gateway, and backend data sharing communication models. This chapter explains various sensors that can be connected to the internet for transferring data between IoT devices, application areas and communication models. The classification of IoT application areas is based on the scope of functionality, use, and adaptation. Sensors are classified on the basis of types of measurement such as vision, position, physical, and others. Basic communication models and open research challenges are explained at the end.



As the internet of things (IoT) and industrial internet of things (IIoT) continue to expand, the need for robust cyber security solutions becomes increasingly critical. The convergence of blockchain technology and artificial intelligence (AI) offers promising opportunities to address the security challenges posed by IoT and IIoT applications. This chapter provides an overview of the potential synergies between blockchain and AI in the context of cyber security for IoT and IIoT. Blockchain technology, with its decentralized and immutable nature, can provide enhanced security and trust in IoT and IIoT networks. It offers features such as data integrity, transparency, and tamper resistance, making it well-suited for securing critical data and transactions. Additionally, blockchain can facilitate secure device identity management, access control, and secure communication among IoT and IIoT devices.

The rapid advancements in artificial intelligence (AI) and blockchain technologies have brought about significant transformations in various sectors, including healthcare. This proposal outlines a research paper that aims to investigate the different approaches and applications of AI and blockchain technologies in the medical domain. By exploring the integration of these technologies, the authors seek to analyze their potential benefits, challenges, and ethical implications in revolutionizing healthcare practices. The study will involve a comprehensive review of relevant literature, case studies, and interviews with experts in the field. The findings will contribute to a deeper understanding of the synergistic effects of AI and blockchain in healthcare, paving the way for enhanced decision-making, data security, interoperability, and patient-centric care.



A country's growth is gauged with reference to its power consumption and energy use, both of which are fast expanding. Energy consumption management improves energy generation and delivery. The smart grid is a tremendous improvement over the power system of the 20th century, using two-way electrical and communication exchanges to create a sophisticated, distributed, computerized energy delivery network. The fields of artificial intelligence (AI) and block chain distributed ledger technology (BDLT) are the most fascinating areas of study in the field of green energy and related power automation. An in-depth analysis of the most advanced automated planning, governance, optimization, confidentiality, and safety methods for the distribution of power and smart grid using integrated artificial intelligence and block chain is presented in this chapter.



Intelligent internet of vehicles is an emerging technology and has an impact on all smart vehicles on the road due to rapid development in satellite communication integrated with internet technology and enables the integration of smart vehicles with system components like inbuilt vehicle sensors and internet, surrounding vehicles, and smart devices. This integration creates a safe and better platform for vehicles to travel on the road and at the same time be connected to the surrounding environment and vehicles for better performance. Integrating blockchain technology with vehicular communication and smart vehicles is the latest emerging concept. In this chapter, the authors present the survey and latest advancement in block chain technology for intelligent transport systems. This chapter's main highlight is on the application scenario of the internet of vehicles using block chain after a decent review of the literature survey. Also, the authors focus on key challenges and explore further research paths in the intelligent internet of vehicles using block chain technology.

This chapter explores the integration of artificial intelligence (AI) and blockchain technology to create a secure and transparent online conversation platform. Advancements in AI have given rise to virtual assistants capable of human-like interactions and versatile task execution. By harnessing blockchain technology, these virtual assistants enhance security and response times, making them invaluable in applications ranging from healthcare to gaming. The chapter presents a project that enhances virtual assistant capabilities through blockchain integration, featuring a self-learning module and a Solana-based blockchain network with smart contracts. Implementation employs Vanilla JavaScript, Express.js, and Node.js, offering end-to-end customer support and adapting to modern lifestyles. The fusion of AI and blockchain showcases the potential for a robust and secure online conversation platform.



Money laundering is a worldwide issue that jeopardizes the stability and integrity of financial institutions. Many countries have implemented anti-money laundering laws and regulations to combat this. The basics of money laundering and its influence on the financial system, as well as existing strategies for detecting and combating it, are covered in this chapter. K- nearest neighbors, random forest, naive bayes, deep neural networks, and evolution metrics are examples of machine learning techniques and algorithms used to identify suspicious transactions. Financial institutions and regulatory bodies can strengthen their ability to detect and prevent money laundering activities and help to protect the integrity of the financial system by utilizing a variety of measures. In this work, the authors have presented a deep comparative analysis among various machine learning algorithms that are used in money laundering detection.

# Preface

Welcome to this edited reference book on "AI and Blockchain Applications in Industrial Robotics." As editors, we are delighted to present a comprehensive exploration of the synergies between artificial intelligence (AI) and blockchain technologies, and their transformative impact on industrial robotics.

The combination of blockchain technology unchanging and distributed ledger abilities, along with AI's predictive and pattern recognition skills, has brought about a new era of advanced security, clarity, and effectiveness. This book aims to clarify the complexities of this innovative merger, highlighting its potential to transform industries and open up new possibilities for businesses.

AI and blockchain can be used together to solve important social and industrial challenges. Socially, they can predict epidemics, detect cyberbullying and hate speech, and analyze complex behavioral patterns to prevent unwanted outcomes. On the industrial front, these technologies can help with inventory management, automation, and traceability by providing intelligent solutions.

This volume is a helpful guide for a wide range of readers, including graduate students, researchers, industry professionals, undergraduates, and freelancers. It covers different aspects of the topic and gives an in-depth understanding of how AI and blockchain can be implemented in industrial robotics. The book explores essential topics for understanding and implementing advanced technologies, like improving performance and security and analyzing ethical, legal, and social consequences.

The book includes many inspiring chapters discussing different aspects of AI and blockchain applied in industrial robotics. These chapters cover smart manufacturing, the convergence of AI and blockchain, ethical considerations, and future trends. Real-world case studies and practical examples are provided throughout the book to give the readers a better understanding of how these technologies can be applied in different contexts.

We believe that this book is an invaluable asset for anyone seeking to deepen their understanding of AI and blockchain technologies, and how they impact industrial robotics. Whether you are a student, researcher, industry professional, or developer, the diverse range of topics covered in this book is designed to cater to your interests and needs.

Embark on an intellectual journey with us to discover how AI and blockchain intersect in the realm of industrial robotics. We will uncover the potential benefits, challenges, as well as future directions.

## CHAPTER OVERVIEW

### Chapter 1: AI and Blockchain Fundamental: Their convergence in the Realm of Industrial Robotics

Authored by Rashmi Rani Samantaray, Zahira Tabassum, Abdul Azeez

This chapter explains how AI and blockchain technologies work together in industrial robotics. The authors discuss their ability to change manufacturing and examine important uses, difficulties, and future prospects. Energy-efficient results and more productive automation processes are possible when AI and blockchain are integrated.

### Chapter 2: AI and Blockchain for Industrial Robotics: Ethical, Legal, and Social Implications

Authored by Sanjana Prasad, Deepashree RajendraPrasad

Addressing the profound transformation AI and blockchain offer in industrial robotics, this chapter explores the promising era of heightened automation and transparency. However, it also acknowledges the ethical, legal, and social concerns accompanying these advancements, emphasizing the need for comprehensive consideration of societal implications.

### Chapter 3: Enhancing Industrial Robotics Performance and Security with AI and Blockchain Technologies

Authored by Deepak Varadam, Sahana P. Shankar, Aryan Bharadwaj, Tanvi Saxena, Sarthak Agrawal, Shraddha Dayananda

Focusing on making industrial robotic systems better, this chapter explains how AI and blockchain can improve performance and security. AI helps robots learn, while blockchain provides a secure infrastructure. These improvements allow for greater precision, adaptability, and security of industrial robots, reducing downtime and increasing production efficiency.

### Chapter 4: Intelligent Resource Allocation and Optimization for Industrial Robotics using AI and Blockchain

Authored by Tarun Kumar Vashishth, Vikas Sharma, Kewal Krishan Sharma, Bhupendra Kumar, Sachin Chaudhary, Rajneesh Panwar

Efficient resource allocation is crucial in getting the most out of industrial robotics. This chapter explores how AI and blockchain technologies can be integrated to develop dynamic resource allocation strategies. By highlighting different AI algorithms and how blockchain ensures transparency, it provides a framework for optimizing industrial robotics.

## Chapter 5: Multi-Layered Blockchain-Based Security Model for Electronic Health Record Systems

Authored by Mrutyunjaya S. Hiremath, Rajashekhar C. Biradar

Addressing healthcare data security concerns, this chapter presents the Multi-Layered Blockchain-Based Electronic Health Records (ML-BbEHR) framework. Segmenting EHR data into distinct blockchain layers, this model enhances security, ensures patient identity verification, and implements smart contracts to regulate data access, bolstering data integrity in healthcare systems.

## Chapter 6: IoT Sensors for Smart Automation: A Systematic Review

Authored by Chayalakshmi C. L., Mahabaleshwar S. Kakkasageri, Rajani S. Pujar, Nayana Hegde

This chapter provides a comprehensive overview of IoT sensors and communication models necessary for effective IoT system implementation. It categorizes sensors based on measurement types and analyzes communication models essential for data exchange among IoT devices. Additionally, it delves into application areas and emerging challenges in this domain.

## Chapter 7: Blockchain and Artificial Intelligence for Cyber Security in the Era of Internet of Things and Industrial Internet of Things Applications

Authored by Amit Kumar Tyagi

Highlighting the need for strong cybersecurity in the growing fields of IoT and IIoT, this chapter explores the synergy between blockchain and AI. it discusses how blockchain's decentralized system and AI's features can secure critical data, and examines ways to control access and secure communication between IoT devices.

## Chapter 8: Application of AI and Blockchain Technologies in Medical Domain

Authored by Deepak Varadam, Sahana P. Shankar, Anirudh Shankar, Ashwath Narayan JR, Nidhi N. P., Tushar Narendra Kumar

This proposal aims to explore how AI and blockchain can be utilized in healthcare. Investigating their potential benefits and ethical implications, the study involves a comprehensive review, case studies, and expert interviews. It seeks to improve decision-making, data security, and patient-centric care in the medical domain.

## Chapter 9: Artificial Intelligence and Block Chain Technology for Secure Smart Grid and Power Distribution Automation

Authored by Zahira Tabassum, Rashmi Rani Samantaray, Syeda Husna Mohammadi, Anees Fathima

Focusing on the growth of the energy industry, this chapter explores how AI and blockchain can integrate into power distribution and smart grid systems. Delving into automated planning, governance, optimization, and safety methods, it offers insights on how to effectively manage energy through the use of AI and blockchain.

## Chapter 10: Blockchain Technology: A Study Applied on Intelligent Transport System

Authored by Shridevi Kamble, Manjunath Kounte, Saujanya M. S.

Exploring the integration of blockchain into intelligent vehicular communication systems, this chapter surveys the advancements in Blockchain Technology for Intelligent Transport Systems. It investigates applications, challenges, and research avenues, emphasizing a secure and connected environment for smart vehicles.

## Chapter 11: Building a Secure and Transparent Online Conversation Platform: A Blockchain-Based Website with Integrated Virtual Assistant

Authored by Md Tauseef, Sayak Jana, Aashish R. Gouda, Ashutosh Mishra, Utkarsh Raj Srivastava

Describing the merging of AI and blockchain for a secure online conversation platform, this chapter showcases a project augmenting virtual assistant capabilities through blockchain integration. Employing JavaScript frameworks and smart contracts, it emphasizes end-to-end customer support and enhanced security measures.

## Chapter 12: Enhancing Money Laundering Detection Through Machine Learning: A Comparative Study of Algorithms and Feature Selection Techniques

Authored by Soumya Ranjan Mishra, Pranati, Anika, Hitesh Mohapatra

This chapter comparatively analyze different machine learning algorithms used to detect suspicious transactions and combat money laundering. It explores K-nearest neighbors, random forest, naïve Bayes, deep neural networks, and evolutionary metrics to help financial institutions fight illicit activities.

## Chapter 13: Apple Leaf Disease Identification and Segmentation Using Enhanced Learning Driven Feature Representation Model

Authored by Harsha Raju, Veena Kalludi Narasimhaiah, Mukil Alagirisamy

Focusing on identifying plant disease from leaf images, this study presents an algorithmic approach using enhanced learning-driven feature representation. It aims to improve disease classification accuracy by isolating diseased leaf segments, utilizing a fitness function to optimize feature representation, and enhancing deep learning generalization.

## Chapter 14: Machine Learning and Blockchain Integration in Industrial Robotics Challenges and Opportunities

Authored by Dankan Gowda V., Pullela S. V. V. S. R. Kumar, Manojkumar S. B., Aparna Atul Junnarkar, Sheetalrani Rukmaji Kawale

Exploring the potential of using machine learning and blockchain in industrial robotics, this chapter outlines the benefits and challenges. It covers decentralized control systems, enhanced security measures, and transparent decision-making, emphasizing collaboration between academia, businesses, and government for mainstream integration.

Each chapter presents a unique perspective on the integration of AI, blockchain, and robotics, contributing to a deeper understanding of their applications and implications across diverse domains.

## IN SUMMARY

As we bring this edited reference book to a close, we reflect on the remarkable exploration of "AI and Blockchain Applications in Industrial Robotics." The diverse perspectives, insightful analyses, and practical examples shared within these pages collectively underscore the transformative potential of integrating artificial intelligence and blockchain technologies in the field of industrial robotics.

The exploration of enhanced performance and security through the convergence of AI and blockchain has illuminated the path toward a more efficient and intelligent future. The case studies presented, from smart manufacturing to secure data sharing in IoT-based smart industries, serve as beacons, guiding us towards innovative solutions and new horizons.

Our intention in compiling this collection was to provide a comprehensive resource for a broad audience—from graduate students to industry professionals—offering both theoretical frameworks and real-world applications. Aiming to bridge the gap between academic knowledge and practical implementation, we recognize the importance of fostering a nuanced understanding of the ethical, legal, and social implications that accompany these technologies.

As we visualize the future of work in the age of AI and blockchain, we recognize that the landscape of industrial robotics is evolving rapidly. This book outlines trends, research gaps, and potential benefits that indicate areas ripe for exploration and innovation where we should direct our attention.

We express our gratitude to the experts who enriched this volume and extend our appreciation to the readers who embark on this intellectual journey. Our hope is that the information collected here will inspire research, spark new ideas, and catalyze advancements within rapidly evolving fields of AI, blockchain, and industrial robotics.

In conclusion, as editors, we invite you to carry forward the spirit of inquiry and innovation fostered by this book. The future is shaped by those who engage with the possibilities of today, and we trust that the knowledge shared here will contribute to a future where the integration of AI and blockchain technologies becomes an indispensable and transformative force in the realm of industrial robotics.

Warmest regards,

*Rajashekhar C. Biradar*
*Reva University, India*

*Geetha D.*
*Reva University, India*

*Nikhath Tabassum*
*Reva University, India*

*Nayana Hegde*
*Reva University, India*

*Mihai Lazarescu*
*Politecnico di Torino, Italy*

# Chapter 1
# AI and Blockchain Fundamental: Their Convergence in the Realm of Industrial Robotics

**Rashmi Rani Samantaray**
https://orcid.org/0000-0002-9696-3404
*HKBK College of Engineering, India*

**Zahira Tabassum**
*HKBK College of Engineering, India*

**Abdul Azeez**
https://orcid.org/0000-0002-4009-0958
*HKBK College of Engineering, India*

## ABSTRACT

*This chapter explores the intersection of artificial intelligence (AI) and blockchain technologies in the realm of industrial robotics. The integration of AI and blockchain has the potential to revolutionize various aspects of the manufacturing sector, from enhancing robotic capabilities to optimizing supply chain management. This chapter delves into the key applications, benefits, challenges, and future prospects of combining AI and blockchain in industrial robotics. Both blockchain and artificial intelligence have generated a lot of excitement in the technology industry for a number of years. AI and blockchain applications for robotics applications thus provides productive, energy-efficient outcome and thus ease the process of automation and maintenance.*

## 1. INTRODUCTION

Industrial robotics has undergone significant advancements in recent years, transforming manufacturing processes across various industries. The integration of Artificial Intelligence (AI) and Block chain technologies into industrial robotics has further revolutionized the field. AI offers the ability to enhance robot perception, decision-making, and adaptability, while Block chain ensures secure and transparent

DOI: 10.4018/979-8-3693-0659-8.ch001

data management. Understanding the background and potential of these technologies is crucial to appreciate their applications in industrial robotics.

The motivation behind exploring the applications of AI and Block chain in industrial robotics stems from several factors.

Enhanced Efficiency: AI can optimize robot operations, improving productivity and reducing errors.

Safety Improvements: AI can make robots safer to work alongside humans by enabling them to react to changing environments.

Data Security: Block chain can safeguard sensitive data and prevent tampering, critical in manufacturing settings.

Supply Chain Transparency: Block chain can provide transparency in the supply chain, ensuring quality control and traceability.

Industry 4.0 Adoption: As Industry 4.0 becomes more prevalent, the synergy of AI, Block chain, and robotics becomes imperative.

Objective: The primary objective of this study is to explore the applications of AI and Block chain in industrial robotics with a focus on achieving the following goals:

Assess the impact of AI on industrial robots in terms of automation, adaptive learning, and predictive maintenance.

Evaluate how Block chain technology can enhance data security, traceability, and trust in robotic applications.

Identify real-world use cases where AI and Block chain have been successfully integrated into industrial robotics.

Analyze the potential challenges and limitations of implementing these technologies in industrial settings.

Provide insights into the future trends and innovations in AI and Blockchain within the field of industrial robotics.

Scope: The scope of this study encompasses a broad range of topics related to the applications of AI and Block chain in industrial robotics, including but not limited to:

AI in Industrial Robotics:

Machine learning for robot perception and object recognition, AI-driven decision-making and autonomous robotic systems, Predictive maintenance and failure detection using AI,

Human-robot collaboration and safety improvements through AI, Blockchain in Industrial Robotics: Securing robot control systems and data using Blockchain .Supply chain management and provenance tracking through Blockchain, Ensuring the integrity of robotic process data, Decentralized control and coordination of robotic networks using Blockchain.

Thus, the integration of AI and Blockchain in industrial robotics holds immense promise for transforming manufacturing and related industries (Turjman F.,2021). This study aims to provide a comprehensive understanding of their applications, benefits, and challenges, paving the way for more informed decisions in adopting these technologies for improved industrial automation and efficiency.

# 2. BLOCKCHAIN AND ARTIFICIAL INTELLIGENCE

## 2.1 Blockchain

The distributed database technology known as block chain has a significant impact on a number of industries, including banking and finance, healthcare, retail, and real estate, among others. Among the most important benefits of block chain are the following:

A block chain database delivers a highly decentralised database, is completely tamper-proof and unchangeable, ensures maximum data security and privacy, permits data access through a ledger system, and assures maximum data transparency.

Despite the benefits and promises of block chain, there are still a number of demerits to its use in contemporary business models:

Speed and efficiency: It does signature verification .Each node is verified for redundancy.

- Exorbitant implementation fees. Block chain is more expensive than a traditional database. Additionally, businesses must correctly plan and implement the block chain integration into their workflow.
- Data modification. It is not so easy to alter data that has already been stored using block chain technology; for this, all of the block's codes must be rebuilt, which is expensive and time consuming.

## 2.2 Artificial Intelligence

Artificial intelligence (AI) technology seeks to imitate human intelligence or offer a logical approach to problem-solving so that it may independently make decisions and carry out activities. The ability of machines to accomplish things cognitively has led to a paradigm-shifting approach in the commercial world (Al-Fuqaha, A.,2019).

AI's salient characteristics include the following:

Enhancements in data management capabilities, data personalization, data-centric predictive analysis, reduction of development errors, real-time or immediate support, and data mining.

Despite all the excitement around AI, a number of significant obstacles prohibit firms from deploying AI widely. These difficulties include:

- The lack of computing or data processing capacity necessary for an AI implementation;
- The lack of talent to work on AI projects; and
- The unreliability of AI outputs.

## 2.3 Swarm Robotics, Blockchain, and AI All Together

It is believed that the use of block chain and AI together can advance the field of swarm robots. In this field, a "swarm" of physical robots cooperate to carry out operations or tasks. Each robot in this sector uses artificial intelligence (AI) to interact with its surroundings according to predefined rules.These robots become more resilient and more scalable in terms of their collective behaviour and interaction capacity when they are networked (Sharma, R.,2022).

Applications in the entertainment business, precision farming, and medical transportation are all using this technology more and more. Security for these complex apps is a persistent problem, despite their rising popularity. Security is a major problem, nevertheless, due to the dynamic informational landscape and the peculiarities of swarm robotics. Blockchain technology may provide a solid solution in this case.

Blockchain technology uses cutting-edge encryption methods like cryptographic digital signatures and cryptographically secure public-key cryptography to ensure the highest level of data security via shared channels. The specific private key that is accessible to a robot regulates access to the data. Blockchain provides robotics with the best security solution, while AI-powered robotics has emerged as a cutting-edge technology.

## 3. SIGNIFICANT BENEFITS OF AI AND BLOCKCHAIN

We must also comprehend the benefits that these technologies provide if we are to comprehend how blockchain and AI meet in the field of robotics (Kumar, A., 2022). AI can address many blockchain difficulties in a variety of ways, including:

a) Energy conservation: Data mining demands a lot of computational power, and AI can be quite good at reducing energy usage.
b) Scalability: Blockchain's ability to grow at an exponential rate can be used by AI to enable decentralised learning and other activities.
c) Improving security: While blockchain is incredibly safe and impenetrable, operating alongside machine learning and deep learning technologies can make apps even more secure.
d) Efficiency: AI can offer efficient resource utilisation to reduce blockchain-related costs.
e) Data hardware
f) Meeting talent needs: There are still very few experts and professionals working in the space of blockchain technology. AI-powered virtual agents can be extremely useful for tasks like independently establishing new ledgers as the talent gap widens and the need for blockchain implementation rises.
g) Data screening can help regulate access to private data while also making personal data and privacy more exposed due to the growing volume of data being used in blockchain applications.

### 3.1 Blockchain Enhances AI

AI solutions can benefit from the addition of blockchain. Blockchain can significantly contribute to the augmentation of an AI system by providing a decentralised ledger system, such as:

Explaining AI decisions: AI-based solutions frequently struggle to articulate the reasoning behind a choice or course of action. The development of the argument can be explained using the data records stored in a blockchain database.

Increasing the effectiveness of AI: Blockchain can help create more effective data models, data-centric activities, and data-driven insights by protecting the shared data flow.

A market for all data: A data marketplace that is accessible might be made possible by the decentralised database system that securely, transparently, and easily gathers diverse data from all different sources. The increased amount data from AI-powered systems with this democratization of the database will give more data power to organizations.

Reducing catastrophic risks: Many are looking to blockchain for safe access to mission-critical data regarding specific activities, from weather reports to military data. Mission-critical data can be further secured using AI and blockchain in order to send alarms about potential disasters.

It will be a lot of fun. Many new technology, like quantum computers or robots in a swarm, as well as models, like distributed ledgers or smart contracts, must make a broad and sustained case for themselves. There is still much to be tested and many topics to be accelerated.

Robot programming and parameterization are now even simpler for everyone thanks to AI. Perhaps software won't be required at all in the future.

With the blockchain, a new, large database could arise that is not dominated by, and thus not managed and controlled by, a small number of businesses.

In this chapter, we have discussed the of need of Artificial Intelligence and benefits of Block Chain on Robotics Application. Chapter implements many case studies on this area.

## 3.2 AI in Industrial Robotics

### 3.2.1 Evolution of Industrial Robotics

The evolution of industrial robotics is a fascinating journey that has spanned over several decades, reshaping industries and revolutionizing manufacturing processes worldwide. This detailed description will provide an in-depth exploration of how industrial robotics has evolved, including key milestones, technological advancements, and its profound impact on various sectors Unuvar, M. (2020).

1. The Birth of Industrial Robotics (1950s-1960s):

The origins of industrial robotics can be traced back to the mid-20th century. The pivotal moment came in 1954 when George Devol and Joseph Engelberger introduced the Unimate, the world's first programmable robot. The Unimate was designed primarily for the automotive industry and could perform tasks like die-casting and welding.

Case Study: Unimate in General Motors

In 1961, General Motors became the first company to deploy the Unimate robot in their die-casting operations. This marked a significant shift in manufacturing as robots took on dangerous and repetitive tasks.

Result: Improved worker safety, increased productivity, and reduced labor costs.

2. Technological Advancements (1970s-1980s):

The 1970s and 1980s witnessed substantial advancements in industrial robotics technology. These developments laid the foundation for more sophisticated and versatile robots.

Case Study: PUMA (Programmable Universal Machine for Assembly) Robot

Victor Scheinman developed the PUMA robot in the late 1970s, and it became popular in industries such as electronics and automotive manufacturing.

Result: Precision assembly, increased production speed, and greater flexibility in manufacturing processes.

Case Study: ABB's RobotWare

In the 1980s, ABB introduced the RobotWare system, which allowed robots to be easily programmed and reconfigured for various tasks.

Result: Enhanced adaptability and minimized downtime in manufacturing facilities.

3. The Emergence of Collaborative Robots (Cobots):

The 21st century brought a new era of human-robot collaboration, with the introduction of collaborative robots or cobots. These robots are designed to work alongside humans, fostering cooperation and enhancing productivity.

Case Study: Universal Robots UR5

Universal Robots launched the UR5 cobot in 2008, known for its user-friendly interface and versatility.

Result: UR5 cobots found applications in industries ranging from logistics to healthcare, enabling automation in various sectors.

Case Study: Rethink Robotics' Baxter

Rethink Robotics introduced Baxter in 2012, a versatile cobot capable of performing a wide range of tasks.

Result: Baxter became popular in small and medium-sized enterprises, automating tasks previously done manually.

4. Integration of Artificial Intelligence (AI) and Machine Learning (ML):

The integration of AI and ML technologies has made industrial robots smarter, more adaptive, and capable of handling complex tasks with precision.

Case Study: Fanuc's Zero Downtime (ZDT)

Fanuc utilized AI and IoT technologies to develop the Zero Downtime (ZDT) system, which predicts maintenance needs in real-time.

Result: Reduced unplanned downtime, improved maintenance efficiency, and significant cost savings.

Case Study: ABB's YuMi

ABB launched YuMi in 2015, a dual-arm collaborative robot with advanced vision and control capabilities.

Result: YuMi found applications in delicate tasks like electronics assembly and even collaborated with tenor Andrea Bocelli in a live concert performance.

5. Robotics in the Age of Industry 4.0:

The concept of Industry 4.0 ushered in the integration of robotics, the Internet of Things (IoT), and data analytics, creating smart factories where automation and connectivity are central.

Case Study: Tesla's Gigafactories

Tesla's Gigafactories employ a vast array of industrial robots for tasks such as battery production and vehicle assembly.

Result: Enhanced efficiency, reduced production costs, and rapid scaling of electric vehicle production.

Case Study: Siemens' MindSphere

Siemens' MindSphere platform combines robotics, IoT, and data analytics to optimize manufacturing processes.

Result: Increased productivity, reduced energy consumption, and improved product quality in various industries.

6. The Rise of Mobile Robots:

Mobile robots have become essential in sectors like logistics, warehousing, and healthcare, enabling automation in previously unautomated environments.

Case Study: Amazon Robotics

Amazon utilizes thousands of mobile robots in its fulfillment centers for automated order picking and transportation.

Result: Drastically improved order fulfillment speed, reduced error rates, and increased warehouse efficiency.

Case Study: Boston Dynamics' Spot

Boston Dynamics' Spot robot, initially designed for research and exploration, found applications in inspection, security, and logistics.

Result: Enhanced inspection capabilities in industries such as construction and oil & gas.

7. Robotics in Healthcare:

The healthcare sector has witnessed substantial advancements in the use of robots for surgery, rehabilitation, and patient care.

Case Study: Da Vinci Surgical System

Intuitive Surgical's Da Vinci Surgical System has revolutionized minimally invasive surgery with its precision and dexterity.

Result: Reduced patient trauma, shorter recovery times, and improved surgical outcomes.

Case Study: PARO Therapeutic Robot

PARO, a therapeutic robot in the shape of a seal, assists in emotional support and therapy for patients, particularly in elderly care facilities.

Result: Improved mental well-being and reduced stress among patients.

8. Future Directions in Industrial Robotics:

As we look ahead, several trends and emerging technologies are shaping the future of industrial robotics:

AI-driven Automation: Further integration of AI and ML for autonomous decision-making and adaptive learning.

Human-Robot Collaboration: Enhanced safety features and closer human-robot collaboration.

3D Printing and Additive Manufacturing: Robots capable of 3D printing complex structures and components.

Quantum Computing: Potential applications in optimizing robotic algorithms and simulations.

Biologically-Inspired Robotics: Mimicking natural organisms for increased agility and adaptability.

In conclusion, the evolution of industrial robotics has been a remarkable journey filled with innovation, technological advancements, and real-world applications. From the humble beginnings of the Unimate to the present era of smart, adaptive robots, industrial robotics has reshaped industries, improved efficiency, and redefined the possibilities of automation in the modern world. As technology continues to advance, industrial robotics will play a pivotal role in shaping the future of work and industry, ushering in an era of increased productivity, safety, and innovation.

## 3.3 AI Techniques for Robotics

### 3.3.1 Machine Learning for Robotics

Machine learning is integral to robotics, revolutionizing the way robots perceive and interact with their environment. In the realm of computer vision, ML algorithms empower robots to recognize and classify objects, track their movements, and navigate through complex spaces. Sensor fusion techniques leverage ML to integrate data from diverse sensors, enhancing the robot's overall perception. Path planning benefits from ML optimization, enabling robots to find efficient and safe routes through dynamic environments. Manipulation and grasping tasks are refined through ML-driven strategies, allowing robots to adapt their grip based on object characteristics.

Simultaneous Localization and Mapping (SLAM) is improved using ML, facilitating accurate mapping and navigation. Human-robot interaction becomes more natural as ML aids in gesture and speech recognition. Learning from Demonstrations (LfD) enables robots to acquire new skills by observing and imitating human actions. Anomaly detection and predictive maintenance are enhanced through ML, ensuring optimal performance and minimizing downtime. Adaptive control mechanisms, driven by online learning, enable robots to dynamically adjust to changing conditions. In task planning and scheduling, ML optimizes resource allocation, enhancing the efficiency of robotic systems. As ML continues to advance, the synergy between machine learning and robotics propels the development of increasingly autonomous and capable robotic systems.

### 3.3.2 Computer Vision for Robotics

In robotics, computer vision plays a pivotal role in object recognition and classification, allowing robots to identify and understand the nature of the objects in their surroundings. This capability extends to tracking objects over time, facilitating tasks such as following moving targets. Moreover, computer vision contributes significantly to navigation and obstacle avoidance, empowering robots to plan paths and navigate through complex environments.

Simultaneous Localization and Mapping (SLAM) is another critical application where computer vision excels. It enables robots to create maps of their surroundings while concurrently determining their own position, a crucial feature for autonomous exploration and navigation in unfamiliar spaces.

Human-robot interaction is greatly enhanced by computer vision through gesture and facial recognition. Robots can respond to human gestures and expressions, making interactions more intuitive and natural. Additionally, computer vision aids in augmented reality interfaces for teleoperation, allowing remote control with visual feedback.

In industrial settings, computer vision is employed for quality control and inspection tasks. Robots equipped with vision systems can identify defects and ensure the precision and quality of manufacturing processes. Moreover, computer vision assists in 3D reconstruction, providing robots with detailed spatial information about their environment.

### 3.3.3 Natural Language Processing for Robotics

Natural Language Processing (NLP) is a transformative technology in the realm of robotics, bridging the communication gap between humans and machines. By enabling robots to understand, interpret, and generate human language, NLP facilitates more intuitive and user-friendly interactions in various applications.

One crucial application of NLP in robotics is human-robot collaboration. NLP allows robots to comprehend and respond to natural language commands, making it easier for users to instruct and control robots without the need for complex programming or manual input devices. This is particularly valuable in scenarios where seamless human-robot teamwork is essential, such as manufacturing, healthcare, and domestic assistance.

Moreover, NLP contributes to enhancing the adaptability and flexibility of robots. Robots equipped with NLP capabilities can receive and process verbal instructions for task execution, allowing for dynamic task reprogramming and quick adjustments to changing environments. This adaptability is vital in environments where tasks are not predefined or may evolve over time.

In human-robot interaction, NLP enables robots to understand and respond to spoken language, making interactions more natural and user-friendly. This is particularly relevant in service robotics, where robots may assist with information retrieval, answer queries, or engage in dialogue with users.

## 3.4 Benefits of AI in Industrial Robotics

The integration of Artificial Intelligence (AI) in industrial robotics brings about transformative benefits, enhancing efficiency, flexibility, and overall performance in manufacturing processes. One significant advantage is improved automation and optimization of tasks. AI algorithms enable robots to adapt and learn from data, optimizing production workflows and minimizing downtime. This adaptability is crucial in handling diverse tasks within dynamic manufacturing environments.

AI enhances predictive maintenance, reducing the likelihood of equipment failures. By analyzing sensor data, AI can predict when machinery requires maintenance, allowing for proactive interventions and minimizing unplanned downtime. This predictive capability extends to quality control, as AI-driven vision systems can identify defects in real-time, ensuring the production of high-quality products.

Machine learning algorithms in industrial robotics improve precision and accuracy. Robots equipped with AI can learn to perform intricate tasks with high levels of precision, contributing to increased product quality and consistency. Furthermore, collaborative robots, or cobots, enhanced by AI, can work alongside human workers, improving overall productivity and workplace safety.

The data-driven insights generated by AI in industrial robotics facilitate better decision-making. Manufacturers can analyze production data to identify bottlenecks, optimize resource allocation, and improve overall operational efficiency. As industries continue to embrace Industry 4.0 principles, the amalgamation of AI and industrial robotics is pivotal in creating smarter, more adaptive, and efficient manufacturing ecosystems.

## 4. UNDERSTANDING BLOCKCHAIN TECHNOLOGY

Blockchain technology is a revolutionary innovation that has disrupted various industries and holds immense promise for the future. Originally designed as the underlying technology for cryptocurrencies like Bitcoin, blockchain has evolved far beyond its original use case. In this essay, we will delve into the fundamentals of blockchain technology, its key features, applications across diverse sectors, and the potential it offers for shaping our future.(Kesavaraja, 2021)

### 4.1 What Is Blockchain Technology?

At its core, a blockchain is a decentralized and distributed ledger that records transactions across a network of computers. These transactions are grouped into blocks, which are linked together in chronological order, forming a chain. Each block contains a list of transactions and a unique cryptographic hash of the previous block, creating a secure and immutable record of all transactions. This design ensures transparency, security, and trust among participants in the network.

### 4.2. Key Features of Blockchain Technology

Decentralization: Unlike traditional centralized systems, blockchain operates on a decentralized network of nodes (computers) that validate and record transactions. This eliminates the need for intermediaries, such as banks or payment processors, reducing the risk of fraud and censorship.

Transparency: Transactions recorded on a blockchain are visible to all participants in the network. This transparency fosters trust and accountability, as anyone can verify the authenticity of transactions and data.

Security: Blockchain employs strong cryptographic techniques to secure data. Once a block is added to the chain, altering it becomes computationally infeasible, making the blockchain highly resistant to tampering and fraud.

Immutability: Once information is recorded on the blockchain, it cannot be altered or deleted. This immutability ensures the integrity of historical data, making it suitable for applications requiring a tamper-proof record.

Consensus Mechanisms: Blockchains rely on consensus mechanisms to validate transactions and maintain the integrity of the ledger. Common mechanisms include Proof of Work (PoW) and Proof of Stake (PoS), each with its own strengths and weaknesses.

### 4.3 Applications of Blockchain Technology

Blockchain technology has a wide range of applications across various industries, including (Kalla,2021).

Cryptocurrencies: Bitcoin, the first and most well-known cryptocurrency, uses blockchain technology to enable peer-to-peer digital transactions without the need for intermediaries like banks.

Supply Chain Management: Blockchain can be used to track the movement of goods throughout the supply chain, ensuring transparency and authenticity. This is particularly valuable in industries like food and pharmaceuticals to trace the origin and quality of products.

Financial Services: Blockchain has the potential to revolutionize the financial sector by reducing transaction costs, increasing efficiency, and enabling quicker cross-border payments and settlements.

Smart Contracts: Smart contracts are self-executing contracts with the terms of the agreement directly written into code. They automatically execute when predefined conditions are met, reducing the need for intermediaries in legal and business processes.

Healthcare: Blockchain can securely store and share patient records, ensuring data privacy and interoperability among healthcare providers. This can improve patient care and reduce administrative overhead.

Voting Systems: Blockchain-based voting systems can enhance the security and transparency of elections, making it difficult for unauthorized parties to manipulate results.

Intellectual Property: Blockchain can be used to register and protect intellectual property rights, including patents, copyrights, and trademarks, by creating an immutable record of ownership.

Identity Verification: Blockchain offers a secure and decentralized way to manage digital identities, reducing the risk of identity theft and providing individuals with more control over their personal information.

Real Estate: Property transactions involve a multitude of intermediaries and paperwork. Blockchain can streamline this process, reducing fraud and the time it takes to transfer property ownership.

Gaming and Collectibles: Blockchain-based tokens and digital assets are being used in the gaming industry for in-game purchases, unique collectibles, and provable ownership of virtual items.

## 4.4. Future Potential of Blockchain Technology

The future of blockchain technology holds numerous possibilities and challenges. Some of the key areas where blockchain is expected to make significant contributions include:

Scalability: Addressing the scalability issues of current blockchain networks is crucial to their wider adoption. Innovations like sharding and layer-2 solutions aim to increase the transaction throughput of blockchain networks.

Interoperability: Achieving interoperability between different blockchain platforms and networks will be vital for seamless data and value transfer across various applications and industries.

Regulation and Compliance: As blockchain technology matures, regulatory frameworks will need to evolve to ensure consumer protection, prevent illicit activities, and promote responsible innovation.

Environmental Concerns: Proof of Work (PoW) consensus mechanisms, used by some blockchains, consume significant energy. Transitioning to more energy-efficient consensus mechanisms like Proof of Stake (PoS) will be essential to address environmental concerns.

Integration with Legacy Systems: Integrating blockchain technology with existing legacy systems in various industries will be a complex process, requiring careful planning and investment.

Blockchain technology has evolved from a niche concept into a transformative force with the potential to reshape industries and enhance transparency, security, and efficiency. Its decentralized nature, transparency, and immutability make it a powerful tool for applications ranging from cryptocurrencies to supply chain management, healthcare, and more.

As blockchain technology continues to mature, addressing scalability, interoperability, and regulatory challenges will be crucial for its widespread adoption. However, the potential benefits, including reduced fraud, increased efficiency, and enhanced trust, make the future of blockchain technology promising and exciting.

In conclusion, blockchain technology represents a paradigm shift in how we record, share, and trust data, and its impact on society is likely to grow significantly in the years to come. As we navigate this evolving landscape, it is essential to explore its possibilities while addressing its challenges responsibly and ethically.

## 4.5 Role of Blockchain in Robotics

Blockchain technology has the potential to play a significant role in enhancing the capabilities and trustworthiness of robotics systems. One primary application is in the realm of secure and transparent data management. Blockchain's decentralized and tamper-resistant ledger ensures the integrity and traceability of data generated by robotic systems. This is particularly crucial in scenarios where data accuracy is paramount, such as in industrial automation and critical infrastructure.

In robotics, where secure communication and coordination are vital, blockchain provides a decentralized and secure framework. It can enable secure peer-to-peer communication and transactions between robots, ensuring that commands and information are transmitted and received with a high level of trust.

Moreover, blockchain can be leveraged to establish and verify the authenticity of robotic identities. Each robotic entity can have a unique identity recorded on the blockchain, preventing unauthorized access and enhancing the overall security of robotic networks.

Smart contracts, programmable self-executing contracts stored on the blockchain, can automate and enforce agreements between robotic entities. This can streamline transactions, facilitate cooperation between robots, and ensure that predefined rules and conditions are met without the need for intermediaries.

By providing a secure, transparent, and decentralized framework, blockchain technology can enhance the reliability, security, and efficiency of robotic systems, fostering greater trust in the deployment of robots in various applications, from industrial automation to collaborative robotics.

# 5. BLOCKCHAIN FRAMEWORK

Four core building blocks of blockchain frameworks (Alexandre,2018).

Blockchain frameworks typically include the following four building blocks:

A shared ledger: The distributed transaction record is the only addition to the shared ledger. The blockchain behind Bitcoin was created with the goal of democratising visibility; nevertheless, while using blockchain, consumer data regulations must also be taken into account. Immutability or append-only semantics can be achieved with a correctly setup distributed SQL or noSQL database.

Cryptography: In a blockchain, cryptography guarantees transaction authentication and verifiability. This requirement is part of blockchain architecture due to the emphasis on assuming computational hardness and making encryption more difficult for an attacker to crack. The economic motivation and system design of the bitcoin blockchain make this an intriguing problem. Considerations regarding cryptography vary while operating in a less democratic or permissioned company ledger network.

Trust systems or consensus: Consensus or trust systems refer to the use of the network's strength to validate transactions. In my opinion, trust systems are at the core of blockchain systems. Since not all validation occurs through consensus, we believe the phrase "trust system" is preferable to "consensus system." The general layout and financial investment in a blockchain infrastructure are determined by this fundamental component of trust. The trust system changes with every new player in the blockchain industry, giving rise to variants that are tailored for certain blockchain use cases. Blockchain technology is built on the principles of ownership, trade, and trust. The trust system regulates trade between member

*Figure 1.*

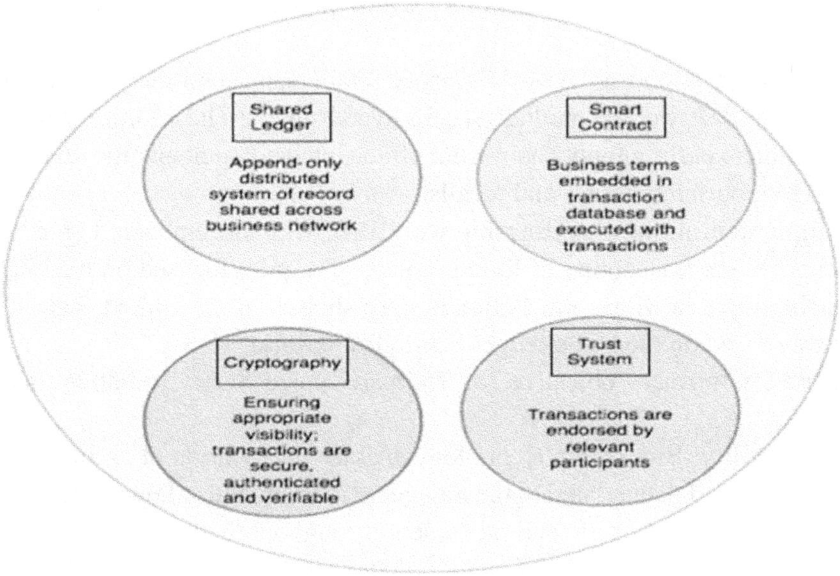


companies for inter-company transactions. There is still a lot of work to be done to define the best trust system for specific use cases, such as P2P and sharing economy models with B2B models.

d) Business rules or smart contracts: Smart contracts are commercial agreements that are integrated into a blockchain transaction database and carried out in conjunction with transactions. This is also a part of a blockchain solution's ruleset. The value flow and state of each transaction must be specified.

The following use diagram as shown in Figure 1 gives a good idea of these concepts:

The four building blocks are generally accepted and well understood. They have existed for decades prior to blockchain. Shared ledgers are an evolutionary change, similar to the move to computer-based spreadsheets, but the underlying business rules have stayed the same.

## 6. CASE STUDIES OF BLOCKCHAIN

### 6.1 Case Study One: Supply Chain Optimization Through Blockchain

Supply Chain Optimization through Blockchain case studies- (Al-Fuqaha,2019).

Blockchain technology has been touted as a game-changer for supply chain optimization. Its ability to provide transparency, traceability, and security in a decentralized manner has the potential to revolutionize various industries. Here are some case studies that showcase how blockchain is being used to optimize supply chains:

Walmart's Food Safety Initiative: Walmart, one of the world's largest retailers, has been experimenting with blockchain to enhance food traceability. In a pilot project, they tracked the journey of mangoes from farms in Mexico to store shelves in the United States. By using blockchain, they reduced the time it took to trace the origin of a mango from days to seconds, which enhances food safety and allows for quicker recalls in case of contamination.

Walmart's Food Safety Initiative is a notable case study demonstrating the application of blockchain technology to enhance food safety and traceability within its supply chain. Here are the key details of this initiative:

**Background**: Walmart, as one of the world's largest retailers, places a strong emphasis on food safety to ensure the quality and safety of the products it sells to consumers. The traditional methods of tracking and tracing food products in the supply chain were often slow and cumbersome, making it challenging to quickly respond to food safety issues and recalls.

**Blockchain Implementation**: In partnership with IBM, Walmart embarked on a blockchain-based pilot project to improve the traceability of food products. The pilot focused on tracking the journey of mangoes from their source in Mexico to Walmart store shelves in the United States. The blockchain technology used was based on the Hyperledger Fabric framework.

**Key Objectives**: The primary objectives of Walmart's Food Safety Initiative through blockchain included:

**Enhanced Traceability**: Blockchain technology enables the creation of an immutable, transparent ledger of all transactions and data related to the mangoes' journey. This allows for real-time monitoring and traceability of each mango's origin and movement throughout the supply chain.

**Faster Recall Response**: In the event of a food safety issue or product recall, the blockchain system enables rapid identification of affected batches or products, significantly reducing the time it takes to pinpoint the source of contamination and remove tainted products from store shelves.

**Increased Transparency**: Consumers can access information about the origin of the mangoes by scanning a QR code on the product's packaging, promoting transparency and trust in the supply chain.

**Results and Impact**: Walmart's pilot project demonstrated several key benefits of using blockchain in the food supply chain:

**Reduced Traceability Time**: The time required to trace the origin of a mango was reduced from days or even weeks to a matter of seconds.

**Improved Food Safety**: The ability to quickly trace the source of a food safety issue allows for more targeted and efficient recalls, minimizing the potential harm to consumers.

**Enhanced Consumer Trust**: The transparency provided by blockchain technology instills confidence in consumers, as they can easily access information about the product's journey and origin.

**Efficiency Gains**: By streamlining the traceability process and reducing paperwork, the supply chain becomes more efficient, potentially reducing costs.

**Expansion**: Following the success of the mango pilot project, Walmart expanded its use of blockchain technology to other products within its supply chain, including poultry and leafy greens. The goal is to implement blockchain for a broader range of products to further enhance food safety and traceability.

In summary, Walmart's Food Safety Initiative is a pioneering case study that showcases how blockchain technology can significantly improve food safety, traceability, and transparency in the food supply chain, benefiting both retailers and consumers.

## 6.2 Case Study Two: Collaborative Robotics Network (Pereira, 2019)

Robotics using AI in healthcare applications is a rapidly evolving field with the potential to revolutionize patient care, diagnostics, surgery, and more. Here are some key applications and examples of how AI-powered robotics is making a significant impact in healthcare:

Robotic Surgery:

da Vinci Surgical System: The da Vinci surgical system is one of the most well-known examples. It assists surgeons in performing minimally invasive procedures with high precision. AI algorithms enhance the surgeon's movements and provide real-time feedback as in Figure 2.

Rehabilitation and Physical Therapy:

Robotic Exoskeletons: Devices like the EksoGT or ReWalk provide mobility assistance to patients with spinal cord injuries. AI algorithms adapt to the patient's movements and provide support accordingly.

Physical Therapy Robots: Robots like the Kinova Jaco assist in physical therapy exercises and rehabilitation routines for patients recovering from injuries or surgeries.

Telemedicine and Remote Monitoring:

Telepresence Robots: AI-powered telepresence robots, like the RP-VITA, enable doctors to remotely examine patients and consult with other healthcare professionals.

Home Healthcare Robots: These robots can assist with medication management, check vital signs, and communicate with healthcare providers, allowing patients to receive care at home as in Figure 2.

Medication Management:

Robotic Dispensing Systems: Robots like the Yuyama are used in pharmacies and hospitals to automate the dispensing of medication, reducing errors and improving efficiency.

Diagnostic Imaging:

AI-Enhanced Imaging: AI algorithms can analyze medical images (e.g., X-rays, MRIs, CT scans) to assist radiologists in detecting and diagnosing diseases like cancer or neurological conditions as shown in Figure 3.

Laboratory Automation:Like sample processing,store in proper place,keep record of sample data safely robots are trained and utilised .To name a few of additional robotic responsibilities are Robotic Sample Processing: Robots can automate the handling and processing of laboratory samples, speeding up testing and reducing human error.

Patient Care and Assistance: Robotic Nurses and Caregivers: Robots like Pepper are used in healthcare settings to provide companionship, answer questions, and remind patients to take medication as shown in Figure 4.

Disease Monitoring: Robotic Devices for Continuous Monitoring: Robots equipped with sensors can continuously monitor patients' vital signs and report any deviations to healthcare providers in real-time.

Personalized Medicine: AI algorithms analyze patient data to personalize treatment plans and predict drug responses.

Emergency Response:

Robots for Disaster Response: AI-powered robots can be deployed in disaster-stricken areas to assess damage, search for survivors, and deliver supplies.

Mental Health Support:

Mental Health Robots: AI-powered robots and chatbots are used to provide emotional support and therapy for patients with mental health conditions.

Patient Education and Engagement:

Interactive Robots: Robots equipped with AI can educate patients about their conditions, treatment options, and recovery plans.

Benefits of AI-Powered Robotics in Healthcare:

- Enhanced precision and accuracy in surgical procedures.
- Reduced risk of human error in diagnostics and medication management.

*Figure 2.*

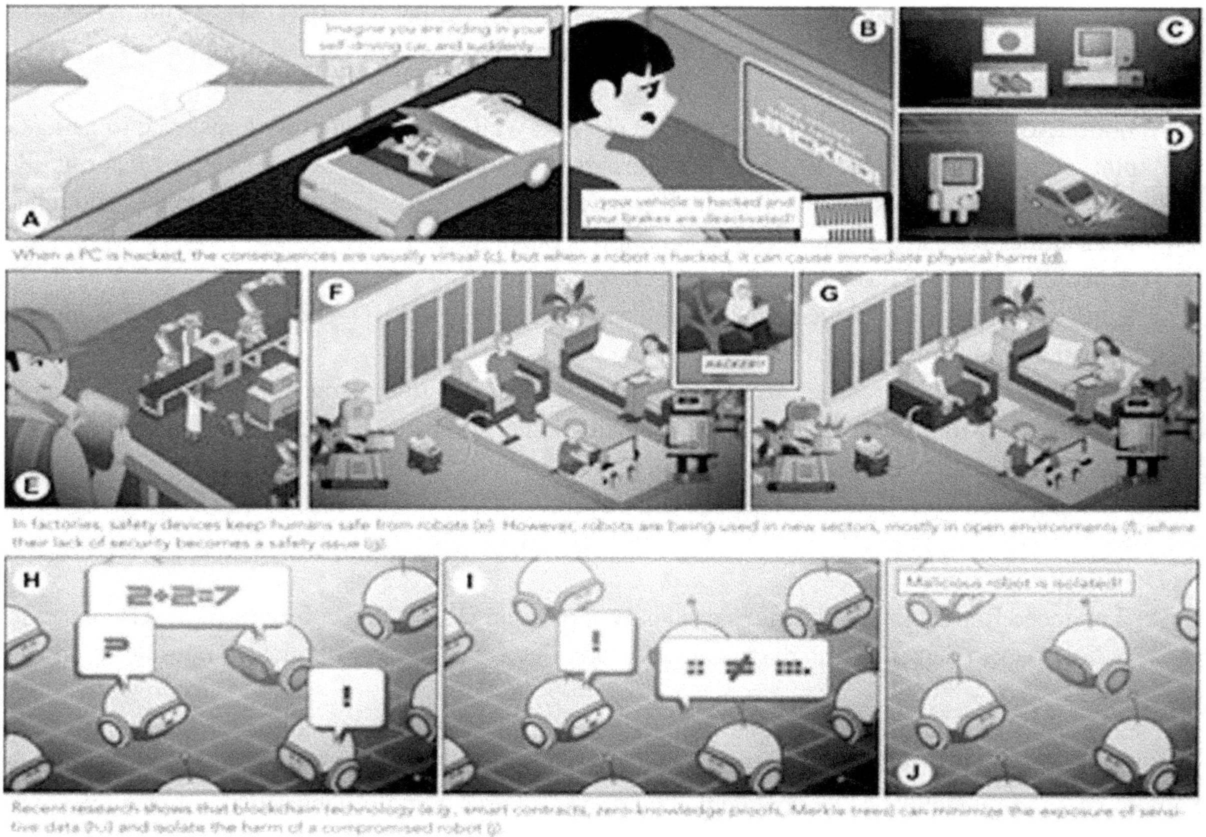


*Figure 3.*

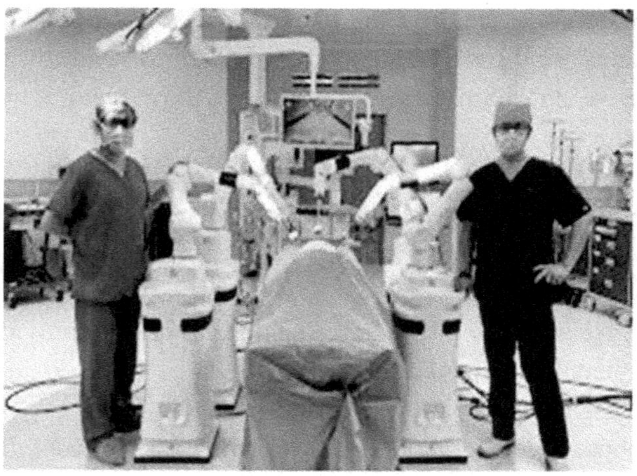

*Figure 4.*

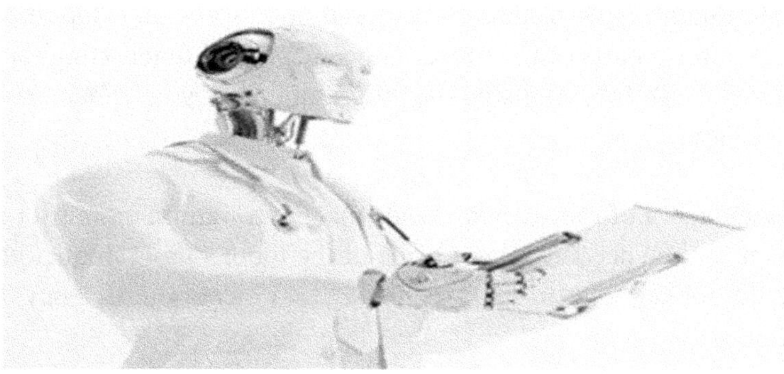

- Improved patient outcomes and reduced recovery times.
- Increased accessibility to healthcare services, especially in remote or underserved areas.
- Enhanced patient engagement and education.
- Efficient use of healthcare resources.

While AI-powered robotics holds great promise in healthcare, it is essential to ensure regulatory compliance, data security, and ethical considerations to fully harness its potential while maintaining patient safety and privacy.

## 6.3 Case Study Three: Energy-Efficient Robotic Manufacturing

Increasing Robotic Energy Efficiency: A Road to Sustainable Industrialization in (Al-Turjman,2021).

Energy efficiency in robots has become a key trend in the developing field of industrial automation, serving as a beacon of innovation and sustainability. Robotics adoption is proving to be a crucial solution as global companies struggle with rising energy costs and the demand to lessen their carbon impact, bringing both energy and cost benefits.

Industrial facilities have long contributed significantly to the world's energy use, with old assembly lines using a huge amount of energy. A new era of energy efficiency has, however, begun with the development of robotics. Because modern robots are made particularly to use less energy, they have reduced operating expenses and are more competitive in a period of rising energy prices.

Robots offer a dual advantage over conventional assembly lines. First off, they can save a lot of energy by using less heating. Second, because they can work quickly, robots can increase production rates and improve the time and energy efficiency of manufacturing.

The automotive sector serves as a perfect illustration of this twofold benefit. Auto assembly lines frequently employ robots to complete a variety of activities, from final assembly to welding and painting. Robots, which can work effectively in a variety of temperatures, can considerably minimise the need for heating, which lowers energy consumption. These processes require precise temperature control.

Luckily, there are several strategies for energy conservation while deploying industrial robots. Techniques for minimising energy consumption include improving programming trajectories, enhancing speed and acceleration optimisation, and decreasing friction.

Here are 10 practical methods for using industrial robots to save energy and money:

1. Employ advanced programming tools: Robots move a lot without need, which is one of the major energy wasters. Using the appropriate robot programming software is the first step towards saving energy. You can optimise parts of the robots' movements and functioning that will have a variety of effects on energy usage by optimising the programming of your robots.
2. Accelerate Cycle Time

Cycle time is a key indicator of robot performance that has a significant impact on energy consumption. This is the amount of time needed for one cycle of the robot's programme to finish.

You can decrease the robot's idle time between tasks and increase efficiency, which will decrease energy consumption.

3. Introduce a maintenance plan

To keep them operating at their optimum, the majority of robotic systems need routine maintenance. Make sure you have a maintenance programme in place for your robots if you want to prevent energy waste. When performing maintenance, make sure to examine and replace any worn parts, clean any moving parts, and test the robot to make sure it is functioning properly.

4. Turn on Sleep Mode.

The use of dynamic power modes, which make sure the robot only consumes energy when executing a task, is a relatively recent development in robotics.

The robot can use less energy when switching dynamically between modes like operation, standby, and hibernation.

5. Employ power management techniques

Additionally, some contemporary robots include cutting-edge power management technologies that can drastically lower energy use.

These enable the robot to monitor its energy use and modify its operation to stay as efficient as possible, much like smart home energy systems.

6. Boost Robotics Programming

You can change your robot's operation in a number of ways to save energy by optimising the programming of its movements. Optimising trajectories so the robot only moves the minimal amount necessary to reach where it needs to go and reducing speed during slow-moving job segments are two examples of improvements you may make to increase the robot's efficiency.

7. Select the Best Robot

It's crucial to pick the appropriate robot for the task at hand. For instance, a dedicated palletizing robot can consume around five times less energy than a similar-spec industrial robot that serves many purposes. While multipurpose robots have their uses, there are occasions when specialised or limited-purpose robots utilise less energy.

8 Lower Friction

Energy loss during the robot's motions can also be minimised by lowering friction and resistance in the joints. This saves energy and money by lowering the amount of energy required for each motion. Proper lubrication, routine maintenance, and awareness of how the environment affects friction are all ways to lessen friction in a robot's joints.

9 Use brakes with regeneration

Regenerative brakes are a technique that is gaining popularity for a variety of uses, including your home automobiles. To save energy during braking, they operate by transforming kinetic energy into electricity.

10.Reduce acceleration

Reduced acceleration in robotic systems can use 30% less energy, according to researchers at Chalmers University of Technology. Only employ the accelerations and decelerations necessary for the specific robotic task at hand. Your robot may be able to move more quickly, but that doesn't mean it always should.

## 6.4 Challenges of Blockchain

A blockchain is a digital ledger where several blocks are used to preserve digital records of certain acts, such as transactions. The transaction cannot be altered or deleted after it has been verified and added to a block. Additionally, it is a very transparent technology because every user may observe the history of records (Zheng,2022).

Distributed ledger technology (DLT), a decentralised digital records database, powers the blockchain. As was already indicated, any user of this database has the ability to examine every transaction or piece of data that was recorded on the distributed ledger. It is practically impossible to make unauthorised modifications to the database because every member of the blockchain has an identical copy of the information on the shared network.

Every industry that needs intricate logistics or a high degree of reliability might employ blockchain. But why hasn't it yet been utilised, for example, to track containers in docks that are frequently overloaded or during worldwide government elections?

Let's quickly go over the blockchain's primary problems.

- **High Prices**: Blockchain operations are incredibly expensive. Blockchains are powered by the equipment required to mine cryptocurrencies, but this equipment is extremely expensive. As a result, a large fleet of cryptocurrency mining equipment is often under the ownership of a major cryptocurrency corporation. Due to high costs, it is very challenging for independent researchers to use the blockchain.
- **Burdensome Hardware:** As was already noted, a facility will often have thousands of miners installed to maintain a blockchain's functionality. This takes up a lot of room in your office and

makes it challenging to move the equipment. Even independently constructed mining rigs are cumbersome and challenging to shift from one location to another in the event of mechanical problems.

- **Potential Vulnerability:** The number of blocks on each decentralised infrastructure is frightening. Each one of them has its own data, codes, documents, receipts, and a wealth of intricate information. On Earth, a new project using a decentralised ledger is started every minute. Alternatively, a new cryptocurrency is produced and added on a new blockchain for circulation. Despite being generally secure, this data explosion is still susceptible to security problems because there aren't enough resources to individually protect each piece of sensitive information. It's not implausible to believe that hackers could occasionally tamper with the transaction receipts or ledgers given the waves upon waves of data flowing over the blockchain.
- **Excessive Energy Use:** Mining equipment for cryptocurrencies has a significant carbon footprint. Even while many businesses try to power the system using hydroelectric energy or windmill facilities, it still costs a lot of money.
- **Skills Gap:** Even though cryptocurrency has been around for more than ten years, it is still not widely used. As a result, there is a skills gap in the area because there are far fewer IT employees who are knowledgeable with blockchain than there is need for them.

## 7. CHALLENGES OF ARTIFICIAL INTELLIGENCE

Artificial intelligence is creating machines that can mimic human intelligence. Its goal is to develop products as smart as, or even much smarter than humans so that the issues that are beyond the realm of human abilities could be solved. To understand how blockchain can help AI, let's focus on the main challenges in this area (Razavi,2020).

### 7.1 Main Challenges of AI

a) Same as with blockchain, the budgets for implementing AI technologies can be immense. Even the Big Tech companies are sometimes struggling to pay for it. In addition to that, let's not forget the extremely costly (and not that numerous) workforce that is required to deal with AI.The major factor that sets humans apart from animals (and robots) is our capacity for moral judgement, understanding of sophisticated abstract concepts, and adherence to moral principles. that is, to think logically. This manner of thinking cannot be duplicated by artificial intelligence. Particularly given that AI is actively used in warfare technologies, this can result in enormous issues.

b) Unaffordable Workforce and Equipment

Similar to bitcoin, integrating AI technologies might come with a hefty price tag. Even the big tech corporations occasionally struggle to make ends meet. Additionally, let's not forget the very expensive (and sparse) staff that is needed to handle AI.

c) Recurring Errors

The fact that there is a constant possibility of trailing is one of the most fundamental problems in machines, and AI may be partially to blame for this. It might be challenging to determine whether an incident was caused by a user, developer, or manufacturer when it occurs.

d) Remote Navigation

Unmanned navigation, like the autopilot mode in cars, is less accurate than human navigation. Since it occasionally ignores obstacles and causes accidents, AI navigation is currently unreliable.

## 8. CAN BLOCKCHAIN, ROBOTICS, AND AI CAN FIT TOGETHER?

AI simulates human intelligence, and robotics imitates human actions. When combined, the two can significantly decrease the amount of human labor. In particular, it relates to the time spent on simple tasks, like taping up boxes or bolting details together. Swarm robots could deal with those tasks effectively.

One of blockchain's main goals is to reduce time and costs, too. Same with AI: as it evolves, it can perform any type of boring job, whereas the human workforce can be used for more complex tasks. Three different tech industries, therefore, share a similar aim. Let's have a look at how merging blockchain, swarm robotics, and AI technologies can solve the challenges in these fields to give us a better net product.

Enhanced Decentralization

So far blockchain technology provides us with a great ecosystem to deal with finance (crypto and tokens). As we discussed above, one of its key advantages is secure transaction approval within the blockchain. Therefore, one possible use case is to delegate approving the blocks to the robots. Or the robots can communicate one with another during the work process using public and private keys to evade human control and mistakes.

More Security

The blockchain is already doing great with transparency and security, as information contained in blocks is really hard to alter. However, could it be even more efficient with the help of AI? Imagine a pattern that detects hackers. Or a neural network that detects double spending automatically.

Virtual Workforce

Going even further, AI and robotics could help blockchain by replacing the limited workforce in the area. Just for a moment, think of robotic Bitcoin miners that approve transactions way faster than humans. Could that be a solution to speed up the transactions and reduce energy usage, too?

Solving Tasks

Blockchain can help AI with difficult tasks, such as moral choices that represent a huge issue for robots. Mining rigs can solve complex computational issues, which, combined with the speed of AI, can create self-thinking robots and further decrease the amount of human verification in robotics.

Smarter, Cost-Effective Systems

To sum up, blockchain can enhance the functions of robotics by providing decentralization to create a self-regulating ecosystem. In the fields where computation is required, AI and blockchain mixed could bring to life robots that solve complex mathematical equations efficiently. This will significantly lower the overall costs and bring in more profit (Tanwar,2020).

## 9. KEY METRICS OF BLOCKCHAIN

Blockchain technology has gained widespread attention and adoption across various industries due to its unique features and capabilities. When assessing the performance and effectiveness of a blockchain network, several key metrics are commonly used to gauge its health, efficiency, and security. Here are some of the key metrics of blockchain:

1. **Transaction Throughput**: Transaction throughput measures the number of transactions a blockchain can process per unit of time, usually measured in transactions per second (TPS). It is a critical metric for evaluating the scalability of a blockchain network. High throughput is essential for blockchain applications with a large user base, such as cryptocurrencies and payment networks.
2. **Confirmation Time**: Confirmation time refers to the average time it takes for a transaction to be added to the blockchain and considered final. For cryptocurrencies, shorter confirmation times are preferred as they allow for faster transactions. Longer confirmation times can lead to delays and hinder user experience.
3. **Consensus Mechanism**: The consensus mechanism employed by a blockchain network is a crucial metric. Common consensus mechanisms include Proof of Work (PoW), Proof of Stake (PoS), Delegated Proof of Stake (DPoS), and others. Each has its own advantages and disadvantages in terms of security, energy efficiency, and decentralization.
4. **Network Security**: Blockchain security is paramount. Metrics to assess security include the hash power (for PoW networks), the number of validating nodes (for PoS and DPoS networks), and the historical record of successful security breaches or attacks on the network.
5. **Decentralization**: Decentralization is a fundamental principle of blockchain technology. Metrics to evaluate decentralization include the distribution of nodes, the level of control by a small group of participants, and the degree of censorship resistance. High levels of decentralization are generally desirable for maintaining trust and security.
6. **Immutability**: Immutability refers to the inability to alter or delete data once it's recorded on the blockchain. The number of confirmations required for a transaction to be considered immutable is an important metric for assessing the level of security and trustworthiness of a blockchain.
7. **Transaction Fees**: Transaction fees are the costs associated with sending transactions on a blockchain. These fees can vary significantly depending on network congestion and the blockchain's design. Low transaction fees are preferable for users and businesses.
8. **Block Size and Interval**: The block size and block interval determine the capacity and speed of a blockchain network. A larger block size can accommodate more transactions per block, while a shorter block interval reduces confirmation times. However, these parameters should strike a balance to prevent centralization and ensure network stability.
9. **Network Uptime**: Network uptime measures the percentage of time a blockchain network is operational and available. High uptime is critical for business-critical applications. Downtime can result from network upgrades, maintenance, or unexpected issues.
10. **Interoperability**: Interoperability measures a blockchain's ability to communicate and transact with other blockchains and external systems. The ease of integrating a blockchain into existing infrastructure is an important consideration for enterprise adoption.

11. **Energy Efficiency**: Energy efficiency is becoming increasingly important due to concerns about the environmental impact of blockchain networks, especially those using PoW consensus. Metrics may include the energy consumption per transaction or block.
12. **Developer Activity**: The level of developer activity, including the number of contributors, code commits, and community support, can indicate the health and growth potential of a blockchain project.
13. **User Adoption**: The number of active users, dApps (decentralized applications), and use cases built on a blockchain network can provide insights into its real-world applicability and success.
14. **Token Metrics**: For blockchain networks with native tokens, token metrics such as market capitalization, trading volume, and token distribution are essential for evaluating the network's economic stability and potential for growth.

These key metrics collectively help stakeholders, including users, developers, and investors, assess the performance, security, and viability of a blockchain network. Depending on the specific use case and goals of a blockchain project, different metrics may hold varying degrees of importance. Therefore, it's essential to consider these metrics in the context of the blockchain's intended purpose and ecosystem.

## 10. SECURITY CHALLENGES OF BLOCKCHAIN

Blockchain technology offers numerous security advantages, such as immutability, transparency, and cryptographic protection. However, it is not immune to security challenges and risks. Understanding these challenges is essential for mitigating potential threats and ensuring the security of blockchain-based systems (Culmone,2023). Here are some of the key security challenges of blockchain:

1. **51% Attacks**: In Proof of Work (PoW) blockchains, a 51% attack occurs when an entity or group of miners gains control of over 50% of the network's mining power. This allows them to manipulate transactions, double-spend coins, or halt transaction processing. While 51% attacks are difficult and costly to execute on large networks like Bitcoin, smaller networks with lower mining power are more vulnerable.
2. **Double Spending**: Double spending is a risk in all blockchain networks, where a user can spend the same digital asset more than once. Blockchain mitigates this through consensus mechanisms, but small, less-secure networks may still be susceptible.
3. **Smart Contract Vulnerabilities**: Smart contracts, self-executing code on a blockchain, can contain vulnerabilities that are difficult to spot during development. These vulnerabilities can be exploited to drain funds or disrupt operations. Notable examples include the DAO hack and various DeFi exploits.
4. **Private Key Management**: Users are responsible for securing their private keys, which provide access to their blockchain assets. Loss or theft of private keys can result in irreversible loss of funds. The challenge lies in creating user-friendly, secure methods for private key management.
5. **Sybil Attacks**: Sybil attacks involve an attacker creating multiple fake nodes to control a significant portion of the network. This can be used to manipulate consensus and disrupt network operations. PoW blockchains are particularly susceptible to this if miners can easily create multiple mining nodes.

6. **Quantum Computing Threat**: Quantum computers have the potential to break the cryptographic algorithms used in blockchain technology, such as SHA-256 (used in Bitcoin). This poses a long-term threat to blockchain security, and research into quantum-resistant cryptographic solutions is ongoing.
7. **Regulatory and Compliance Challenges**: Blockchain's pseudonymous nature can pose challenges for regulatory compliance, especially in the context of anti-money laundering (AML) and know-your-customer (KYC) requirements. Some governments are concerned about the use of cryptocurrencies for illegal activities.
8. **Privacy Concerns**: While blockchain offers transparency, privacy is a concern. Public blockchains expose transaction details to anyone, potentially revealing sensitive information. Privacy-focused blockchains like Monero and Zcash aim to address this challenge.
9. **Smart Contract Bugs**: Smart contracts are prone to coding errors or vulnerabilities that can be exploited by attackers. Due to their immutability, once deployed, fixing these issues can be difficult or impossible.
10. **Distributed Denial of Service (DDoS) Attacks**: Blockchain nodes and networks can be targeted by DDoS attacks to disrupt services and transactions, affecting availability and performance.
11. **Interoperability Risks**: Interoperability between different blockchain networks can introduce security risks if not implemented carefully. Vulnerabilities in one blockchain may affect connected networks.
12. **Phishing and Social Engineering**: Attackers can use phishing attacks and social engineering techniques to trick users into revealing private keys or login credentials for wallets and exchanges.
13. **Governance Challenges**: Blockchain networks often require decision-making processes for upgrades and changes. Governance disputes can lead to network splits (forks) and affect security and user trust.
14. **Economic Attacks**: Attackers can manipulate the economics of a blockchain network by exploiting vulnerabilities in incentive mechanisms, such as staking and voting systems.

To address these security challenges, blockchain developers, users, and stakeholders must continuously improve security practices, perform code audits, stay updated with the latest security developments, and collaborate with the broader blockchain community to enhance the overall security posture of blockchain networks. Additionally, regulatory compliance and risk management strategies are essential for businesses and organizations utilizing blockchain technology.

## 11. INDUSTRIAL ROBOTICS

Industrial robots are automated machines designed and programmed to perform various tasks and operations within industrial settings. These robots are used across a wide range of industries, from manufacturing and automotive to electronics, healthcare, and logistics. They are characterized by their ability to perform repetitive, precise, and often dangerous tasks with speed and accuracy. Here are some key aspects of industrial robots:

## 11.1 Types of Industrial Robots

- **Articulated Robots**: These robots have multiple joints and are designed to mimic the movement of a human arm. They are highly versatile and are commonly used for tasks such as welding, painting, and material handling.
- **SCARA Robots (Selective Compliance Articulated Robot Arm)**: SCARA robots have two parallel joints, allowing them to move in a planar fashion. They excel in tasks requiring precise horizontal movements, such as assembly and pick-and-place operations.
- **Delta Robots**: Delta robots are known for their high-speed, precision, and agility. They are often used in applications like packaging, sorting, and high-speed assembly.
- **Cartesian Robots**: Cartesian robots, also known as gantry robots, move in linear paths along three axes (X, Y, and Z). They are suitable for tasks that require accurate positioning and heavy lifting.
- **Cylindrical Robots**: These robots have a rotating base and an extendable arm, making them suitable for tasks like material handling and machining.
- **Parallel Robots**: Parallel robots use a system of interconnected arms and joints to achieve high-speed and precision in applications like 3D printing and surgery.

## 11.2 Applications

- **Manufacturing**: Industrial robots are extensively used in manufacturing industries for tasks like welding, painting, assembling products, and quality control.
- **Automotive**: The automotive industry relies heavily on robots for tasks such as welding car bodies, painting, and assembling components.
- **Electronics**: Robots are used in electronics manufacturing for soldering, pick-and-place operations, and circuit board assembly.
- **Healthcare**: In healthcare, robots are used in surgery (robotic-assisted surgery), medication dispensing, and patient care.
- **Logistics and Warehousing**: Robots are employed in logistics and warehousing for tasks like sorting, packing, and moving goods within warehouses.
- **Agriculture**: Agriculture robots assist with tasks like planting, harvesting, and monitoring crop health.
- **Construction**: Some construction tasks, such as bricklaying and concrete pouring, can be performed by robotic systems.

## 11.3 Advantages of Industrial Robots

- **Precision and Accuracy**: Robots can perform tasks with high precision and repeatability, reducing errors and improving product quality.
- **Efficiency**: They can work continuously without fatigue, leading to increased production efficiency.
- **Safety**: Robots can handle dangerous and hazardous tasks, protecting human workers from harm.
- **Cost Reduction**: Over time, the use of industrial robots can lead to cost savings through reduced labor costs and improved productivity.

- **Flexibility**: Many industrial robots are programmable and can be easily reconfigured for different tasks, making them versatile.

## 11.4 Challenges

- **Initial Cost**: The upfront cost of acquiring and installing industrial robots can be substantial.
- **Programming Complexity**: Programming robots can be complex and may require skilled technicians.
- **Maintenance and Downtime**: Robots require regular maintenance, and downtime for repairs can impact production schedules.
- **Integration**: Integrating robots into existing manufacturing processes may require modifications and adjustments.
- **Job Displacement**: The automation of tasks can lead to concerns about job displacement for human workers.

## 11.5 Future Trends

- **Collaborative Robots (Cobots)**: The use of collaborative robots designed to work alongside humans is on the rise, particularly in tasks that require human-robot cooperation.
- **Advanced Sensors and AI**: Industrial robots are increasingly equipped with advanced sensors and artificial intelligence to enhance their perception and decision-making capabilities.
- **IoT Connectivity**: Industrial robots are being integrated into the Internet of Things (IoT) for remote monitoring, predictive maintenance, and data analytics.

Industrial robots continue to evolve, offering greater capabilities and playing a vital role in modern manufacturing and various other industries. As technology advances, they are expected to become even more versatile, efficient, and integrated into various industrial processes.

## 12. SWARM ROBOTICS

Swarm robotics is a field of robotics that draws inspiration from the collective behaviour of social insects, such as ants, bees, and termites, to design and control groups of relatively simple robots, known as swarm robots or agents. Unlike traditional robotics, where a single complex robot is programmed to perform specific tasks, swarm robotics focuses on the coordination and collaboration of multiple robots to accomplish tasks collectively (Tyagi,2022). Here's a detailed description of swarm robotics:

## 12.1 Key Characteristics of Swarm Robotics

1. **Decentralization**: In swarm robotics, there is typically no centralized control or leader. Instead, each robot operates autonomously based on local information and interactions with neighboring robots.

2. **Local Communication**: Swarm robots communicate with nearby robots to share information and coordinate their actions. Communication can be either direct (e.g., through wireless communication) or indirect (e.g., through sensing the behavior of nearby robots).
3. **Emergent Behavior**: Swarm robotics relies on the principle of emergent behavior, where complex and intelligent group behaviors emerge from the simple interactions and rules followed by individual robots. There is no need for complex centralized algorithms to achieve coordination.
4. **Redundancy**: Swarm robotics benefits from redundancy. If one robot fails or is removed from the group, the swarm can adapt and continue its task, enhancing robustness and fault tolerance.
5. **Scalability**: Swarm robotic systems are often scalable, meaning that the number of robots in the swarm can be adjusted to suit the task requirements. This scalability allows for flexibility in various applications.

## 12.2 Applications of Swarm Robotics

Swarm robotics has a wide range of applications across various fields (Calandra, 2022):

1. **Search and Rescue**: Swarm robots can be deployed in disaster-stricken areas to search for survivors, map environments, and deliver supplies. Their ability to cover large areas and work collaboratively makes them valuable in such scenarios.
2. **Agriculture**: In agriculture, swarm robots can be used for tasks like planting, harvesting, and monitoring crops. They can work efficiently in large fields and adapt to changing conditions.
3. **Environmental Monitoring**: Swarm robots are employed to collect data on environmental conditions, such as water quality, air pollution, and wildlife tracking. They can provide real-time data for research and conservation efforts.
4. **Manufacturing**: Swarm robots can be used in manufacturing for tasks like material handling, assembly, and quality control. They offer flexibility in adapting to changes in production demands.
5. **Exploration**: In space exploration, underwater exploration, and planetary research, swarm robots can explore unknown and hazardous environments, map terrain, and collect scientific data.
6. **Security and Surveillance**: Swarm robots can be used for perimeter security, surveillance, and monitoring of critical infrastructure. They can collaborate to track intruders or investigate suspicious activities.
7. **Transportation and Delivery**: Swarm robots can be employed in logistics and delivery services to transport goods efficiently within warehouses or for last-mile deliveries.
8. **Education and Research**: Swarm robotics is used in educational settings to teach robotics concepts and algorithms. It also serves as a platform for research in distributed robotics and collective intelligence.

## 12.3 Challenges in Swarm Robotics

Despite its promise, swarm robotics presents several challenges:

1. **Coordination Algorithms**: Designing effective coordination algorithms that allow robots to work together without centralized control can be complex, especially in dynamic and unpredictable environments.

2. **Communication**: Ensuring reliable communication among robots, especially in environments with obstacles or interference, is a challenge.
3. **Scalability**: While swarm robotics is scalable, managing a large number of robots and ensuring efficient coordination can be challenging.
4. **Robustness**: Ensuring that the swarm remains robust in the face of robot failures, sensor errors, or environmental changes is critical.
5. **Task Allocation**: Allocating tasks efficiently among swarm robots and avoiding conflicts or duplications can be non-trivial.
6. **Energy Efficiency**: Managing the energy resources of individual robots in a swarm to maximize their operational time is essential, especially for battery-powered robots.

In conclusion, swarm robotics leverages the principles of decentralization, emergent behavior, and local interactions to create coordinated and adaptable robotic systems. Its applications span a wide range of fields, and ongoing research aims to address the challenges associated with scalability, coordination, and robustness. Swarm robotics has the potential to revolutionize various industries by enabling cost-effective and efficient solutions for complex tasks.

## 12.4 Implementation of Swarm Robotics

The implementation of swarm robotics involves designing and programming a group of autonomous robots to work collaboratively and achieve a specific task or set of tasks (Darko, 2022; Makori, 2020; Suman, 2021).

Here are the steps and considerations for implementing swarm robotics:

1. **Define the Task(s):** Clearly define the task(s) that the swarm of robots will be expected to perform. The task(s) should be decomposed into individual subtasks that can be accomplished by the robots working together.
2. **Select Robot Hardware:** Choose the appropriate hardware for your swarm robots. This includes selecting the robot platforms, sensors, actuators, and communication modules. The hardware should be suitable for the task requirements and environmental conditions.
3. **Develop Robot Software:** Write the software that controls the behavior of each individual robot. This involves programming the robots to perform their designated roles and interact with other robots based on predefined rules or algorithms.
4. **Communication Protocols:** Implement communication protocols that allow the robots to exchange information and coordinate their actions. This may involve both direct communication (e.g., through wireless communication) and indirect communication (e.g., through sensing the behavior of neighboring robots).
5. **Coordination Algorithms:** Design coordination algorithms that enable the robots to work together effectively. These algorithms can be based on local interactions, emergent behavior, and decentralized decision-making. Common coordination approaches include swarm intelligence, flocking, and consensus algorithms.
6. **Simulation and Testing:** Before deploying the swarm in the real world, it's often beneficial to simulate the behavior of the swarm in a controlled virtual environment. This allows for testing

and refining the coordination algorithms and assessing the swarm's performance under various conditions.

7. **Physical Implementation:** Deploy the swarm of robots in the physical environment where the task will be carried out. Ensure that the robots are properly calibrated, and sensors are functioning correctly.
8. **Task Allocation and Distribution:** Develop mechanisms for allocating tasks among the robots. Determine how the robots will distribute tasks among themselves based on their capabilities and the current state of the environment.
9. **Monitoring and Feedback:** Implement monitoring systems to collect data and feedback from the swarm. This information can be used for real-time decision-making, performance evaluation, and troubleshooting.
10. **Adaptability and Robustness:** Design the swarm to be adaptable and robust. The robots should be capable of responding to changes in the environment or to failures within the swarm. This may involve redundancy and fault-tolerance mechanisms.
11. **Evaluation and Optimization:** Continuously evaluate the performance of the swarm in achieving the defined tasks. Use this feedback to optimize the coordination algorithms and improve the efficiency and effectiveness of the swarm.
12. **Scaling and Deployment:** Depending on the requirements of the task, you may need to scale up or down the number of robots in the swarm. Deploy the swarm in the target environment and closely monitor its performance.
13. **Maintenance and Upkeep:** Regularly maintain and update the robots and their software to ensure they remain operational and adapt to changing conditions. This includes addressing hardware wear and tear and updating coordination algorithms.
14. **Data Analysis and Reporting:** Analyze the data collected during swarm operations to gain insights into performance, identify areas for improvement, and generate reports for stakeholders.
15. **Safety and Ethical Considerations:** Ensure that safety protocols are in place to prevent accidents or harm to humans or the environment. Consider ethical implications, especially when deploying swarm robots in sensitive areas or applications.

    Implementing swarm robotics requires a multidisciplinary approach, involving expertise in robotics, artificial intelligence, communication, and domain-specific knowledge related to the task at hand. Successful implementation depends on careful planning, robust algorithms, and ongoing maintenance to ensure the swarm operates effectively and efficiently as per Figure 5.

    Scenes from an animation video describing the key takeaways of this article. The video is available here: https://youtu.be/HCyomzw4tdU.

15. **Challenges in Implementing Blockchain in Robotics:** While the potential benefits of integrating blockchain technology into robotics are substantial, there are several challenges and considerations that must be carefully addressed: (Guizani, 2020), (Tawalbeh, 2021), (Sharma, 2022), (Trichina, 2022).

1. **Scalability:** Blockchain networks, especially public ones like Bitcoin and Ethereum, can suffer from scalability issues. The slow transaction processing times as the network grows can be problematic in applications where robots need to operate in real-time.
2. **Latency:** Blockchain transactions require confirmation from multiple nodes, leading to latency. In situations where robots must respond rapidly, delays caused by blockchain transactions can hinder their performance.

*Figure 5.*

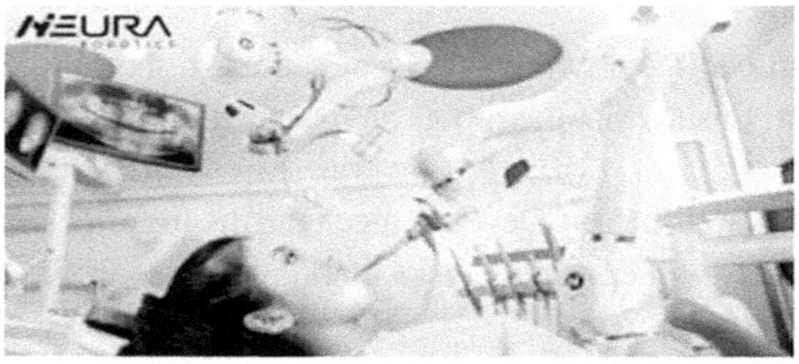

3. **Energy Efficiency:** Many blockchain networks, particularly those using proof-of-work consensus, are energy-intensive. Robots often rely on limited power sources, making energy efficiency a concern.
4. **Cost:** Using blockchain can be costly due to transaction fees, especially on public blockchains. This can impact the economic viability of robotic applications, particularly for smaller businesses or projects.
5. **Data Privacy:** While blockchain provides transparency, it also exposes data to all participants in the network. In scenarios where sensitive data is involved, ensuring privacy becomes challenging. Striking a balance between transparency and data protection is crucial.
6. **Integration Complexity:** Integrating blockchain technology into existing robotic systems can be complex and require significant development effort and expertise. The existing infrastructure must be adapted to accommodate blockchain, which can be a challenging task.
7. **Smart Contract Security:** Errors or vulnerabilities in smart contracts can lead to significant security breaches. Ensuring the security of smart contracts used in robotics applications is crucial and often requires specialized knowledge.
8. **Regulatory and Legal Challenges:** The legal and regulatory landscape for blockchain and robotics is still evolving. Compliance with data protection laws and other regulations can be complex and require ongoing monitoring.
9. **Oracles and External Data:** Robots often require access to external data sources for decision-making. Integrating these data sources with blockchain (oracles) can introduce vulnerabilities if not done securely.
10. **Private vs. Public Blockchains:** Choosing between public and private blockchains depends on the use case but involves trade-offs. Private blockchains offer more control but may sacrifice some decentralization benefits.
11. **Interoperability:** Robots in different environments may use different blockchain platforms or technologies, making interoperability between disparate systems a challenge.
12. **Security Risks:** While blockchain can enhance security, it is not immune to attacks. Smart contract vulnerabilities, 51% attacks (in the case of proof-of-work blockchains), and other threats can still pose risks to robotic systems.
13. **Adoption Barriers:** Implementing blockchain technology in robotics requires not only technical knowledge but also changes in business processes and organizational culture. Resistance to change can be a significant hurdle.

14. **Environmental Concerns:** The environmental impact of energy-intensive blockchain networks, especially those using proof-of-work consensus, may raise ethical and sustainability concerns.
15. **Reliability:** Blockchain networks may not always be available, leading to concerns about system reliability. This is especially critical in applications where robots must operate continuously.

## 13. CONCLUSION

In conclusion, while blockchain technology holds promise for enhancing the security and transparency of robotic systems, its integration presents a complex set of challenges and considerations. The decision to adopt blockchain in robotics should be guided by a careful assessment of the specific use case, taking into account factors such as scalability, latency, data privacy, and cost.

Addressing these challenges requires a multidisciplinary approach that combines expertise in robotics, blockchain technology, cybersecurity, and compliance. Solutions may involve selecting the right blockchain platform, optimizing system architecture, and establishing robust security protocols.

As blockchain technology continues to evolve and mature, it holds the potential to revolutionize the way robotic systems operate and communicate securely. However, successful integration will require a clear understanding.

## REFERENCES

Aditya, U. S., Singh, R., Singh, P. K., & Kalla, A. (2021). A survey on blockchain in robotics: Issues, opportunities, challenges and future directions. *Journal of Network and Computer Applications*, *196*, 103245. doi:10.1016/j.jnca.2021.103245

Alladi, T., Chamola, V., Sahu, N., & Guizani, M. (2020). Applications of blockchain in unmanned aerial vehicles: A review. *Vehicular Communications*, *23*, 100249. doi:10.1016/j.vehcom.2020.100249

Chakraborti, T., Isahagian, V., Khalaf, R., Khazaeni, Y., Muthusamy, V., Rizk, Y., & Unuvar, M. (2020). From Robotic Process Automation to Intelligent Process Automation: –Emerging Trends–.*Business Process Management: Blockchain and Robotic Process Automation Forum: BPM 2020 Blockchain and RPA Forum, Seville, Spain, September 13–18, 2020 Proceedings*, *18*, 215–228.

Debrah, C., Chan, A. P., & Darko, A. (2022). Artificial intelligence in green building. *Automation in Construction*, *137*, 104192. doi:10.1016/j.autcon.2022.104192

Gupta, R., Shukla, A., & Tanwar, S. (2020). BATS: A blockchain and AI-empowered drone-assisted telesurgery system towards 6G. *IEEE Transactions on Network Science and Engineering*, *8*(4), 2958–2967. doi:10.1109/TNSE.2020.3043262

Hussain, A. A., & Al-Turjman, F. (2021). Artificial intelligence and blockchain: A review. *Transactions on Emerging Telecommunications Technologies*, *32*(9), e4268. doi:10.1002/ett.4268

Javaid, M., Haleem, A., Singh, R. P., Khan, S., & Suman, R. (2021). Blockchain technology applications for Industry 4.0: A literature-based review. *Blockchain: Research and Applications*, *2*(4), 100027.

Kumar, A. (2022, October). A Broad Survey on AI Integration in Blockchain: A Forward-Looking Approach. In *Proceedings of the National Conference on Recent Trends of Engineering & Technologies,(RTET-2022) Ramgovind Group of Colleges, Koderma, Jharkhand, India* (pp. 5-7). Academic Press.

Lopes, V., & Alexandre, L. A. (2018). *An overview of blockchain integration with robotics and artificial intelligence.* arXiv preprint arXiv:1810.00329

Lopes, V., Alexandre, L. A., & Pereira, N. (2019). *Controlling robots using artificial intelligence and a consortium blockchain.* arXiv preprint arXiv:1903.00660

Makori, E. O. (2020). Blockchain applications and trends that promote information management. In *Emerging trends and impacts of the internet of things in libraries* (pp. 34–51). IGI Global. doi:10.4018/978-1-7998-4742-7.ch002

Muheidat, F., & Tawalbeh, L. A. (2021). Artificial intelligence and blockchain for cybersecurity applications. In *Artificial Intelligence and Blockchain for Future Cybersecurity Applications* (pp. 3–29). Springer International Publishing. doi:10.1007/978-3-030-74575-2_1

Porkodi, S., & Kesavaraja, D. (2021). Healthcare robots enabled with IoT and artificial intelligence for elderly patients. *AI and IoT-Based Intelligent Automation in Robotics*, (pp. 87-108).

Salah, K., Rehman, M. H. U., Nizamuddin, N., & Al-Fuqaha, A. (2019). Blockchain for AI: Review and open research challenges. *IEEE Access : Practical Innovations, Open Solutions*, *7*, 10127–10149. doi:10.1109/ACCESS.2018.2890507

Salama, R., Al-Turjman, F., & Culmone, R. (2023, March). AI-Powered Drone to Address Smart City Security Issues. In *International Conference on Advanced Information Networking and Applications* (pp. 292-300). Cham: Springer International Publishing.

Saldamli, G., & Razavi, A. (2020, December). Swarm robotics meets blockchain to deploy surveillance missions. In *2020 32nd International Conference on Microelectronics (ICM)* (pp. 1-6). IEEE. 10.1109/ICM50269.2020.9331824

Sousa, M. J., Dal Mas, F., Gonçalves, S. P., & Calandra, D. (2022). AI and blockchain as new triggers in the education arena. *European Journal of Investigation in Health, Psychology and Education*, *12*(4), 445–447. doi:10.3390/ejihpe12040032 PMID:35447750

Verma, A., Bhattacharya, P., Madhani, N., Trivedi, C., Bhushan, B., Tanwar, S., Sharma, G., Bokoro, P. N., & Sharma, R. (2022). Blockchain for industry 5.0: Vision, opportunities, key enablers, and future directions. *IEEE Access : Practical Innovations, Open Solutions*, *10*, 69160–69199. doi:10.1109/ACCESS.2022.3186892

Vrontis, D., Christofi, M., Pereira, V., Tarba, S., Makrides, A., & Trichina, E. (2022). Artificial intelligence, robotics, advanced technologies and human resource management: A systematic review. *International Journal of Human Resource Management*, *33*(6), 1237–1266. doi:10.1080/09585192.2020.1871398

Yang, Q., Zhao, Y., Huang, H., Xiong, Z., Kang, J., & Zheng, Z. (2022). Fusing blockchain and AI with metaverse: A survey. *IEEE Open Journal of the Computer Society*, *3*, 122–136. doi:10.1109/OJCS.2022.3188249

# Chapter 2
# AI and Blockchain for Industrial Robotics:
## Ethical, Legal, and Social Implications

**Sanjana Prasad**
*HKBK College of Engineering, India*

**Deepashree RajendraPrasad**
*Boston University, USA*

## ABSTRACT

*The fusion of artificial intelligence (AI) and blockchain technology holds the promise of a profound transformation in the realm of industrial robotics. This amalgamation has the potential to usher in a new era characterized by heightened automation, transparency, and trust within production and supply chain processes. Nevertheless, as is often the case with technological advancements of this magnitude, it is imperative to recognize and address several ethical, legal, and social concerns that accompany this paradigm shift.*

## 1. INTRODUCTION

Robots are designed to perform repetitive tasks and relieve human workers from strenuous physical work. Industrial robots are developed for automating intensive production tasks which requires constantly moving assembly lines. Industrial robots in manufacturing are used for material handling, pick & place functions and inspection to assembly, packaging & palletizing and finishing applications. Advantages of Industrial Robots include increased efficiency, higher quality, improved working environment, increased profitability, longer working hours, and prestige. Limitations includes Capital cost, expertise, limitations in the tasks it can perform. Equipping robots with AI tools allow them to learn and make decisions autonomously and in real time through the use of algorithms and techniques that enable them to process information from sensors that connect them to their environment. Blockchain can help robotics to take a step further towards complete autonomy, where they can operate without much human intervention.

DOI: 10.4018/979-8-3693-0659-8.ch002

Roboethics analyses the ethical, legal and social aspects of robotics with respect to advanced robotics applications. Issues related to this include liability, protection of privacy, human dignity defence, dignity of work and distributive justice. Some of the social impacts of industrial robots include job displacement, changes in social norms and relationships, distribution of wealth and power, and mitigating the impact of robots on society. In this chapter, we will be discussing about the Role of AI and Blockchain technology in Industrial robotics and the ethical, legal and social implications when AI and Blockchain technology in Industrial robots. We will be discussing about how to overcome such challenges (Adithya et al.,2021).

Artificial Intelligence (AI) has empowered robots with the ability to learn from data, adaptability to changing environments, and making decisions, making them more versatile and efficient. In industrial settings, AI-driven robots enhance productivity, reduce errors, and contribute to safer work environments. Blockchain, on the other hand, provides a distributed and immutable ledger for data storage and transactions. It ensures data transparency, security, and traceability, making it invaluable for recording and verifying information in industrial robotics.

Robots can be equipped with machine vision and artificial intelligence systems that enable them to respond to a variety of situations and provide feedback on system performance in real-time. Some of the other functionalities done by robots includes Assembly & Dispensing, Handling & Picking, Machining & Cutting, Welding & Soldering, Casting & Moulding, Finishing & Sanding, Painting & Coating, Cleaning & Hygiene, Logistics & Storage, Packing & Palletizing, Inspection & Quality Control and Harvesting.

In the realm of industrial robotics, the convergence of artificial intelligence (AI) and blockchain technology has ushered a new era of automation and data management. These technologies, individually potent, join forces to revolutionize the way industries operate. While the potential benefits are remarkable, the ethical, legal, and social implications accompanying this synergy cannot be overlooked.

Industries currently employing Robots are Agriculture, Robot picking, Crop harvesting/wedding, Manufacturing, Damage control and quick maintenance, Automatic control, nuclear waste management, assembly line quality inspection, Smart home appliances, aerospace, Transportation and Healthcare. Figure 1 shows the picture of Industrial Robot.

*Figure 1. Industrial robot [Telephonica]*

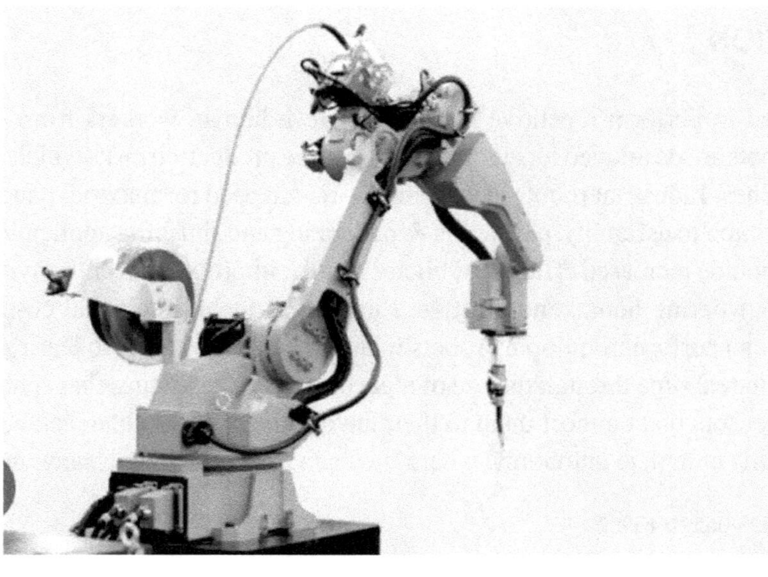

In the context of industrial robotics, artificial intelligence (AI) refers to a set of advanced computer algorithms and machine learning techniques which enables robotic systems to perform tasks with a degree of autonomy and adaptability. AI equips these robots with the ability to analyse data from various sensors, make informed decisions, and optimize their operations without constant human intervention. For industrial robotics, AI enhances efficiency, precision, and flexibility, allowing robots to excel in tasks ranging from manufacturing and quality control to logistics and predictive maintenance.

Blockchain technology is a decentralized and distributed ledger system that records transactions and data in a secure, transparent, and tamper-resistant manner. In industrial robotics, blockchain serves as a data management and authentication tool. It helps track the production process, verify the authenticity of components, and ensure the integrity of data generated by robotic systems. This technology offers heightened data security and transparency, making it invaluable in maintaining the quality and traceability of products in industrial settings, while also providing a means to securely manage transactions, such as payments or contracts, within a network of robotic systems and stakeholders (Kumar et al.,2022). Together, AI and blockchain are transforming industrial robotics by infusing them with intelligence, data integrity, and automation capabilities, ushering in a new era of efficiency and reliability.

AI and Blockchain play pivotal roles in enhancing industrial robotics systems by bringing intelligence, data integrity, and advanced capabilities to the forefront of automation.

The whole structure of this chapter is as follows. In Section 2, AI and Blockchain in Industrial Robotics: Ethical Implications, Legal implications and social implications are discussed. Section 3 briefs about the various Technology Risks involved in the technology. Section 4 briefs detail Legal Framework and Regulations. Section 5 and 6 outlines the future scope and Conclusion in this field is discussed.

## 1.1 Advantages of Artificial Intelligence (AI) in Industrial Robotics

AI in robotics has seen vast success across multiple industries and gained a significant market over the last few years. The AI robotics market stood at US $6.9 Billion in 2021 and is forecasted to reach US $35.5 Billion by 2026 at a CAGR of 38.6% (Bhatt et al., 2022). The global pandemic forced workers to find ways to work remotely, paving the way for intelligent robots that can be programmed and controlled from anywhere.

Many organizations integrate AI robots into their routine procedures for increased productivity, efficiency, and better customer experience. Dominating both tech and non-tech industries, robots can be seen greeting customers at stores, waiting tables at restaurants, harvesting crops, or lifting heavy loads at manufacturing plants. In industrial settings, AI-enabled robots keep workers safe by operating in shared spaces (Kumar et al.,2021). They also perform complex tasks such as cutting, grinding, welding, and inspection autonomously. A research study concluded that the introduction of 1.34 robots per 1000 workers (one standard deviation) reduced workplace injuries by approximately 1.2 injuries per worker.

- Enhanced Autonomy: AI equips industrial robots with the ability to learn from data and adapt to changing conditions. This enhances their autonomy and allows them to make real-time decisions, optimizing operations and improving efficiency.
- Predictive Maintenance: AI-powered robotics systems can predict when maintenance is needed, reducing downtime and preventing costly breakdowns. They analyze sensor data to detect anomalies and recommend proactive maintenance actions.

- Quality Control: AI-driven robots excel in quality control tasks by inspecting products with precision, identifying defects, and ensuring consistent product quality (Hans et al., 2023). This minimizes errors and waste.

Optimized Workflows: AI algorithms can optimize workflows, such as production scheduling and material handling, by dynamically adjusting processes based on real-time data, ultimately saving time and resources. Figure 2 shows the usage of robots in industrial setup.

*Figure 2. Usage of robots in industrial setup [CIO]*

## 1.2 Advantages of Blockchain Technology in Industrial Robotics

- Data Integrity: Blockchain provides a secure and tamper-proof ledger for recording data generated by industrial robots. This ensures data integrity, which is critical for maintaining product quality and traceability.
- Supply Chain Transparency: Blockchain enables end-to-end visibility in the supply chain. Robots can record data about the origin, production, and transportation of components or products, enhancing transparency and trust.
- Smart Contracts: Blockchain facilitates the use of smart contracts, self-executing agreements that automate transactions when predefined conditions are met. This is beneficial for managing contracts, payments, and compliance in robotic systems.
- Security: Blockchain's cryptographic security features protect sensitive data and prevent unauthorized access, making it a robust solution for safeguarding information in industrial robotics.

In combination, AI and blockchain empower industrial robotics systems to operate with greater efficiency, reliability, and transparency. They enable businesses to streamline operations, reduce costs, improve product quality, and ensure compliance with stringent industry standards and regulations.

*Table 1. Advantages of AI and blockchain technology in industrial robotics*

| Artificial Intelligence (AI) | Blockchain Technology |
|---|---|
| Enhanced Autonomy | Data Integrity |
| Predictive Maintenance | Supply Chain Transparency |
| Quality Control | Smart Contracts |
| Optimized Workflows | Security |

Section 2 discusses about the various real time case studies done in the area of AI and Blockchain technology in Industrial Robotics. Section 3 discusses about the various ethical, legal and social implications when AI and Blockchain technology is applied in Industrial robotics (Nikolinakos,2023). Section 4 describes about the gaps identified. Section 5 discusses about the conclusion and future scope.

## 1.3 Case Studies

Certainly, here are some case studies and use cases that illustrate the ethical, legal, and social implications of AI and blockchain in industrial robotics:

### 1.3.1 Case Study One: Amazon Warehouse Robotics and Worker Safety

*Ethical Implications:* Amazon's extensive use of AI-driven robots in its warehouses has raised concerns about the well-being of its workers. Reports of employees facing physically demanding conditions alongside robotic coworkers highlight ethical concerns related to worker safety, job displacement, and the responsibility of companies in ensuring the welfare of their workforce. In Figure 3 the usage of robots in Amazon warehouse is shown picture of Industrial Robot.

*Legal Implications:* Various lawsuits and regulatory scrutiny have emerged around Amazon's use of robots. Legal questions about liability for workplace accidents involving robots have been raised. Additionally, labor unions and workers' rights advocates have called for better regulations to protect human workers in such automated environments.

### 1.3.2 Case Study Two: IBM's AI and Blockchain for Supply Chain Transparency

*Social Implications:* IBM's Food Trust blockchain platform is used to enhance transparency in the food supply chain. It allows consumers to trace the origin and journey of food products. This use case has positive social implications by ensuring food safety and reducing the risk of counterfeit goods entering the market. Figure 4 shows the structure of Supply chain transparency in IBM.

*Legal Implications:* The adoption of blockchain in supply chains requires adherence to various regulations related to food safety and traceability. Compliance with data privacy laws, especially in cross-border operations, adds to the legal complexities in such applications.

*Figure 3. Amazon warehouse robotics [NDTV-Amazon]*

*Figure 4. Private smart contracts [IBM]*

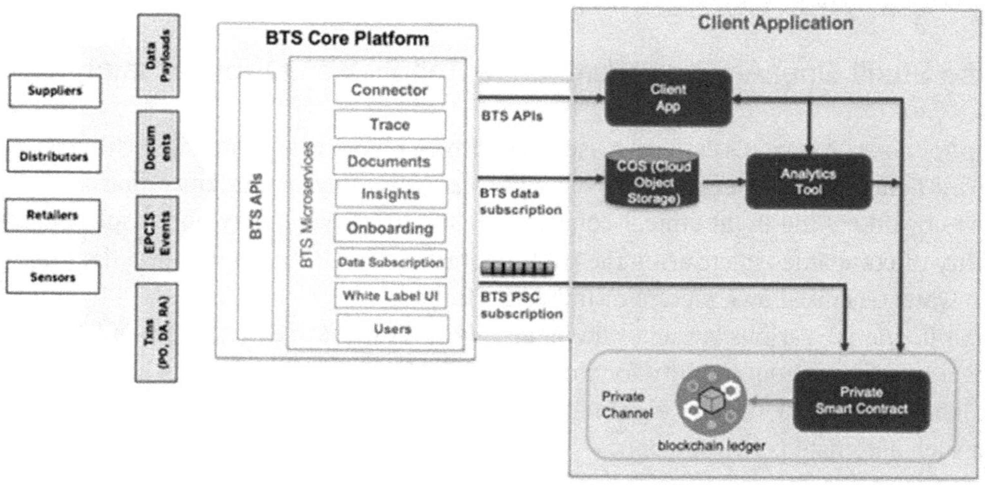


## 1.3.3 Case Study Three: Tesla's Autopilot and Autonomous Vehicle Ethics

*Ethical Implications*: Tesla's Autopilot, an advanced driver-assistance system, has faced ethical dilemmas. Incidents of drivers misusing Autopilot have raised questions about driver responsibility and overreliance on AI systems, with potential safety consequences (Lopes & Alexandre,2018). Figure 5 shows the Autonomous vehicle used in Tesla.

*Legal Implications:* The legal landscape for autonomous vehicles is complex. Regulatory frameworks are still evolving, and the assignment of liability in the event of accidents involving autonomous vehicles is a legal challenge that has yet to be fully addressed.

*Figure 5. Tesla's autopilot [LinkedIn]*

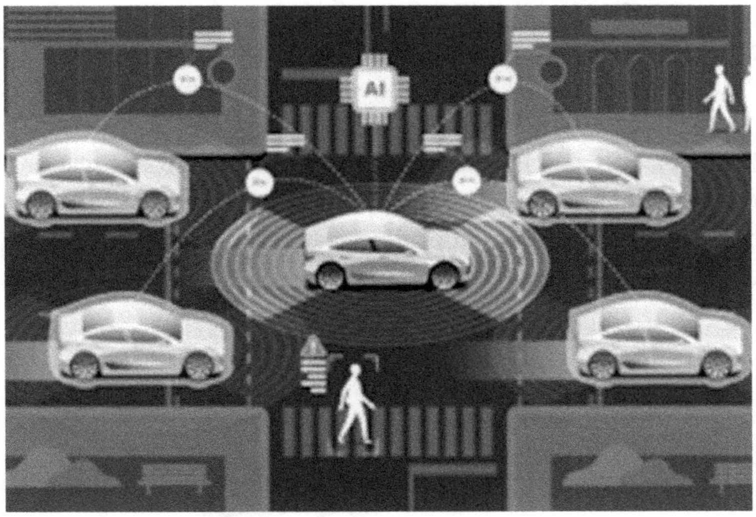


### 1.3.4 Case Study Four: Siemens' Industrial Robots and Human-Robot Collaboration

*Social Implications*: Siemens employs collaborative robots (cobots) that work alongside human employees. This approach enhances productivity but also raises social questions about the future of work and the need for reskilling and adapting to a human-robot collaborative environment. Figure 6 depicts the Human-Robot collaboration .

*Ethical Implications*: Ethical considerations include ensuring the safety of human workers, addressing concerns about job displacement, and fostering a workplace culture that encourages harmonious collaboration between humans and robots.

These case studies highlight the real-world complexities and implications of AI and blockchain integration in industrial robotics, emphasizing the need for a balanced approach that considers ethical, legal, and social dimensions while harnessing the potential benefits of these technologies.

Certainly, to provide a balanced perspective, let's examine both positive and negative examples of AI and blockchain in industrial robotics:

*Figure 6.*

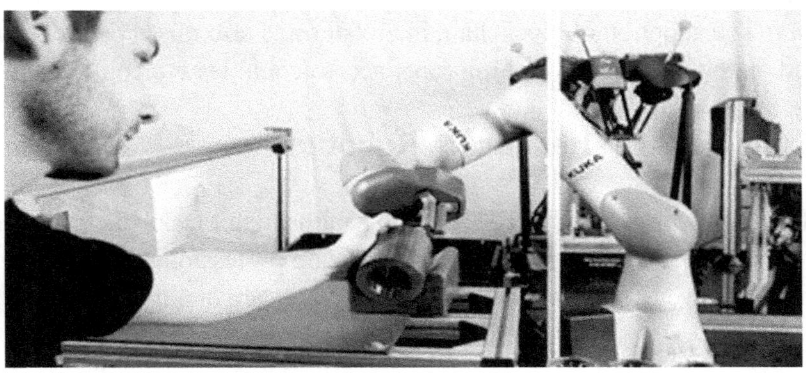

## 1.4 Positive Example: Boston Dynamics Robotics and Manufacturing

*Positive Impact:* Boston Dynamics has created advanced robots like the Spot robot, which can navigate industrial settings and perform tasks such as inspection and data collection. These robots enhance safety by taking on tasks in hazardous environments, reducing human exposure to risks. Workers can focus on higher-level tasks, leading to job enrichment and improved workplace safety.

*Negative Impact*: Concerns have arisen about the potential misuse of advanced robots. If not properly regulated, these robots could be weaponized or used for surveillance purposes, infringing on privacy and security. Figure 7 shows the Boston Dynamics .

*Figure7.*

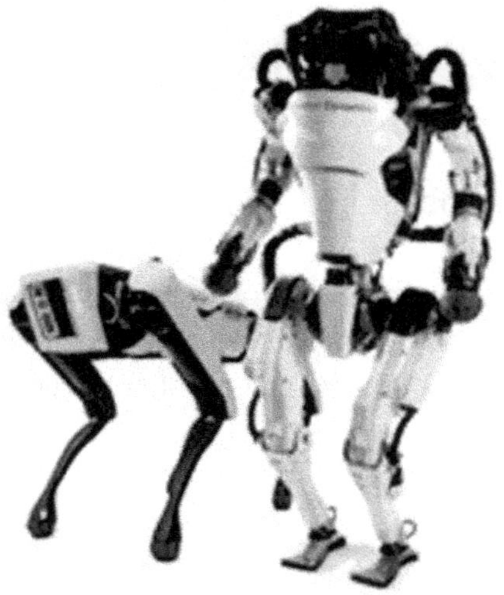

## 1.5 Positive Example: Maersk and IBM's Blockchain in Shipping

*Positive Impact:* Maersk and IBM have collaborated on TradeLens, a blockchain platform for the shipping industry. It improves transparency, reduces fraud, and enhances the efficiency of global trade logistics. This positively impacts companies and communities by reducing delays and enabling better supply chain management. Figure 8 shows the working of Maersk line and IBM in different sectors .

*Negative Impact:* The adoption of blockchain in global trade also raises privacy concerns. Sensitive trade data is shared on the blockchain, and improper access could lead to confidentiality breaches.

## 1.6 Positive Example: AI in Healthcare Robotics

Positive Impact: Robots integrated with AI are used in healthcare for tasks like patient care and surgery. These robots can enhance patient outcomes, assist healthcare professionals, and perform precision surgeries with minimal invasiveness, ultimately improving healthcare quality.

*Figure 8. Working of room container management (RCM) [Supplychain]*

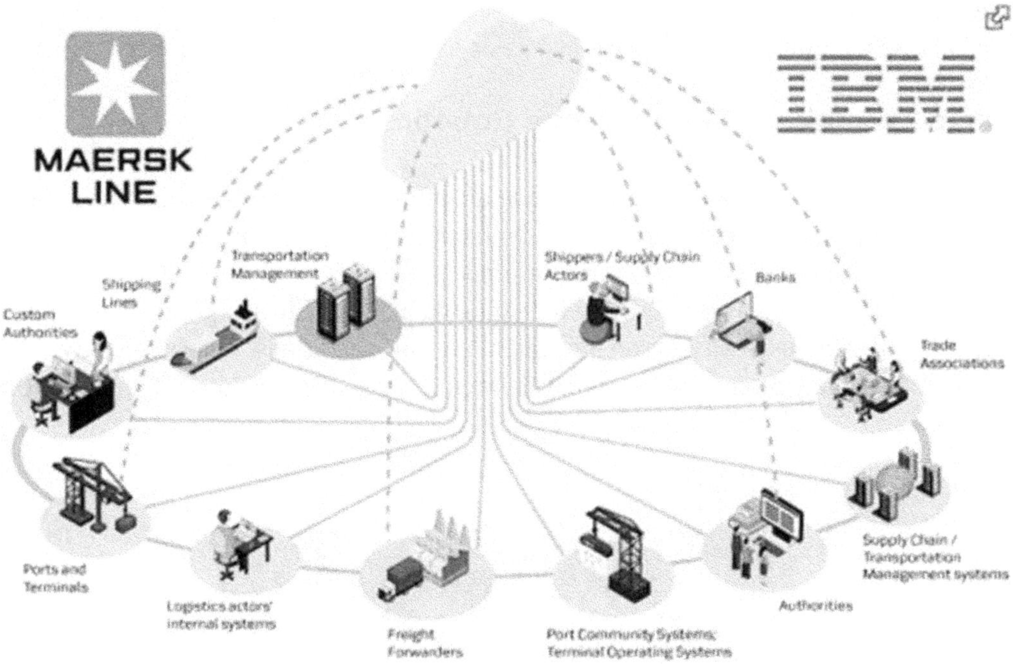


Negative Impact: Ethical concerns arise around the potential for errors or malfunctions in medical robots, which could lead to serious harm to patients. Ensuring the reliability and safety of these systems is critical. Figure 9 shows the usage of AI in Healthcare robotics.

*Figure 9. AI in healthcare robotics [ExpressHealthcare]*

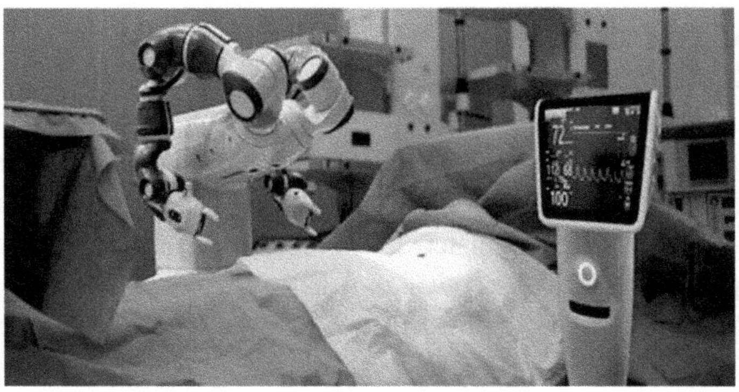

## 1.7 Positive Example: Tesla's Autopilot

Positive Impact: Tesla's Autopilot and Full Self-Driving (FSD) features aim to increase road safety by reducing accidents caused by human error. These features have the potential to save lives and reduce traffic congestion.

Negative Impact: The negative side involves ethical dilemmas. Misuse of Autopilot and overreliance on its capabilities have led to accidents, sparking debates about the level of human supervision required and the ethical responsibility of both drivers and Tesla as the technology provider.

## 1.8 Positive Example: AI in Agriculture Robotics

Positive Impact: AI-powered robots in agriculture help optimize crop management, reduce resource usage, and increase yields. This positively impacts both companies and communities by enhancing food production efficiency and reducing environmental impact.

Negative Impact: Ethical concerns emerge regarding data ownership and privacy in agriculture. The data collected by AI systems, such as crop information or soil conditions, may be sensitive and subject to misuse if not adequately protected. Figure 10 shows the usage of AI in Agriculture robotics.

*Figure 10. AI in agriculture robotics [FreeThink]*

# 2. AI AND BLOCKCHAIN TECHNOLOGY IN INDUSTRIAL ROBOTICS: ETHICAL IMPLICATIONS

As AI-driven robots become more autonomous, ethical dilemmas emerge. One key concern is job displacement. While AI and robotics can boost productivity, they also raise questions about the impact on human employment. As jobs become automated, there's a need for reskilling and upskilling programs to address workforce displacement. Another ethical dimension revolves around privacy. AI-driven robots often process vast amounts of data, including sensitive information. Ensuring that this data is used responsibly and that individuals' privacy rights are respected is paramount. Unauthorized access or misuse of this data can lead to severe consequences. Transparency is another ethical concern. As AI algorithms make decisions, understanding their decision-making processes becomes challenging (Taherdoost,2022). This "black box" problem can lead to distrust, particularly when the outcomes affect individuals' lives, such as in healthcare or autonomous vehicles. The integration of AI and blockchain in industrial robotics brings forth a multitude of ethical considerations that must be carefully addressed to ensure responsible development and deployment of these technologies:

1. **Job Displacement**: One of the foremost ethical concerns is the potential displacement of human workers. As AI-driven robots become more autonomous and capable, there's a risk of job loss in certain industries. Ethical solutions involve providing opportunities for reskilling and upskilling the workforce to transition into new roles that complement AI-powered automation. People may become unemployed because of automation. This has been a fear for decades, but experience shows that the introduction of information technology and automation creates far more jobs than those which are lost (Economist, 2016). Further, many will argue that jobs now are more interesting than the repetitive routine jobs that were common in earlier manufacturing companies. Artificial intelligence systems and robots help industry to provide more cost-efficient production especially in high cost countries. Thus, the need for outsourcing and replacing all employees can be reduced. Still, recent reports have argued that in the near future, we will see overall loss of jobs. However, other researchers mistrust these predictions. Fewer jobs and working hours for employees could tend to benefit a small elite and not all members of our society. One proposal to meet this challenge is that of a universal basic income. Further, current social security and government services rely on the taxation of human labor—pressure on this system could have major social and political consequences. Thus, we must find mechanisms to support social security in the future, these may be similar to the "robot tax" that was recently considered but rejected by the European Parliament.

How much and in what way are we going to work with increased automation? If machines do everything for us, life could, in theory, become quite dull. Normally, we expect that automating tasks will result in shorter working hours. However, what we see is that the distinction between work and leisure becomes gradually less evident, and we can do the job almost from anywhere. Mobile phones and wireless broadband give us the opportunity to work around the clock. Requirements for being competitive with others result in many today working more than before although with less physical effort than in jobs of the past. Although artificial intelligence contributes to the continued development of technology and this trend, we can simultaneously hope that automated agents might take over some of our tasks and thus also provide us some leisure time.

2. **Data Privacy**: The use of AI and blockchain involves the collection and processing of vast amounts of data. Ensuring the privacy of this data is paramount. Ethical considerations demand robust data protection measures, clear consent mechanisms, and transparency about how data is utilized to prevent privacy infringements.
3. **Algorithmic Bias:** AI algorithms can inadvertently perpetuate biases present in historical data, leading to unfair or discriminatory outcomes. Ethical AI in industrial robotics necessitates regular audits, bias mitigation strategies, and ongoing monitoring to ensure fairness in decision-making processes.
4. **Transparency and Accountability**: AI systems can often operate as "black boxes," making it challenging to understand their decision-making logic. Ethical considerations require transparency in AI algorithms, along with mechanisms for accountability in cases where AI systems make incorrect or harmful decisions.
5. **Safety and Security:** Industrial robots equipped with AI may raise concerns about safety, particularly when operating in close proximity to humans. Ensuring that robots are programmed to prioritize safety and follow ethical guidelines is essential to prevent accidents or harm.

6. **Blockchain Security and Data Ownership**: In the context of blockchain, ethical questions revolve around data ownership and control. Users and organizations need to understand their rights and responsibilities in decentralized networks, and mechanisms should be in place to address disputes or unauthorized access to data.
7. Inequality and Access: The benefits of AI and blockchain technology should be accessible to all, rather than exacerbating societal inequalities. Ethical considerations require measures to ensure equitable access and opportunities for all stakeholders, regardless of their economic or social status.
8. Environmental Impact: The energy-intensive nature of blockchain technology raises concerns about its environmental impact. Ethical practices involve exploring energy-efficient blockchain solutions and minimizing the carbon footprint of blockchain networks.

## 3. TECHNOLOGY RISK

Losing human skills due to technological excellence. The foundation for our society for hundreds of years has been training humans to make things, function, work in and understand our increasingly complex society. However, with the introduction of robots, and information and communication technology, the need for human knowledge and skills is gradually decreased with robots making products faster and more accurately than humans. Further, we can seek knowledge and be advised by computers. This lessens our need to train and utilize our cognitive capabilities regarding memory, reasoning, decision making etc. This could have a major impact on how we interact with the world around us (Salah et al.,2023). It would be hard to take over if the technology fails and challenging to make sure we get the best solution if only depending on information available on the web. The latter is already today a challenge with the blurred distinction between expert knowledge and alternative sources on the web. Thus, there seems to be a need for training humans also in the future to make sure that the technology works in the most effective way and that we have competence to make our own judgments about automatic decision making.

Artificial intelligence can be used for destructive and unwanted tasks. Although mostly remotely controlled today, artificial intelligence is expected to be much applicable for future military unmanned aircrafts (drones) in air and for robots on to the ground. It saves lives in the military forces, but can, by miscalculations, kill innocent civilians. Similarly, surveillance cameras are useful for many purposes, but many are skeptical of advanced tracking of people using artificial intelligence. It might become possible to track the movement and behavior of a person moving in a range of interconnected surveillance camera and position information from the user's smartphone. The British author George Orwell (1903–1950) published in 1949 the novel "1984," where a not-so-nice future society is described: Continuous audio and video monitoring are conducted by a dictatorial government, led by "Big Brother." Today's technology is not far away from making this possible, but few fear that it will be used as in "1984" in our democratic societies. Nevertheless, disclosures (e.g., by Edward Snowden in 2013) have shown that governments can leverage technology in the fight against crime and terror at the risk of the innocent being monitored.

Successful AI can lead to the extinction of mankind? Almost any technology can be misused and cause severe damage if it gets into the wrong hands. As discussed in the introduction, a number of writers and filmmakers have addressed this issue through dramatic scenes where technology gets out of control. However, the development of technology has not so far led to a global catastrophe. Nuclear power plants have gotten out of control, but the largest nuclear power plant accidents at Chernobyl in

Russia (1986) and Fukushima in Japan (2011) were due to human and mechanical failure, not the failure of control systems. At Chernobyl, the reactor exploded because too many control rods were removed by experimentation. In Fukushima cooling pumps failed and reactors melted as a result of the earthquake and subsequent tsunami. The lesson of these disasters must be that it is important that systems have built in mechanisms to prevent human errors and help to predict risk of mechanical failure to the extent possible. new technology brings many benefits, and damage is often in a different form than we first would think of. Misuse of technology is always a danger, and it is probably a far greater danger than the technology itself getting out of control. An example of this is computer software which today is very useful for us in many ways, while we are also vulnerable from those who abuse the technology to create malicious software in the form of infecting and damaging virus programs. In 1999, the Melissa virus spread through e-mails leading to the failures of the e-mail systems in several large companies such as Intel and Microsoft due to overload. There are currently a number of people sharing their concerns regarding lethal autonomous weapons systems. Others argue that such systems could be better than human soldiers in some situations, if they are programmed to never break agreed laws of war representing the legal requirements and responsibilities of a civilized nation.

Ethical perspectives of AI and robotics should be addressed in at least two ways.

- First, the engineers developing systems need to be aware of possible ethical challenges that should be considered including avoiding misuse and allowing for human inspection of the functionality of the algorithms and systems (Bostrom and Yudkowsky, 2014).
- Second, when moving toward advanced autonomous systems, the systems should themselves be able to do ethical decision making to reduce the risk of unwanted behaviour (Wallach and Allen, 2009).

Professor and science fiction writer Isaac Asimov (1920–1992) was already in 1942 foresighted to see the need for ethical rules for robot behaviour. Subsequently, his three rules have often been referenced in the science fiction literature and among researchers who discuss robot morality:

1. A robot may not harm a human being, or through inaction, allow a human to be injured.
2. A robot must obey orders given by human beings except where such orders would conflict with the first law.
3. A robot must protect its own existence as long as such protection does not conflict with the first or second law.

## 4. LEGAL FRAMEWORK AND REGULATIONS

The legal landscape in the AI and blockchain-powered industrial robotics domain is still evolving. Existing regulations may not adequately address the unique challenges posed by these technologies. Ensuring that legal frameworks keep pace with technological advancements is crucial. Issues such as liability for accidents involving autonomous robots and the establishment of standards for data security and privacy must be addressed. In some cases, industry-specific regulations may be necessary to ensure the Risk & Liability (Zhamshid et al., 2023).

One key issue is who is liable if something goes wrong with the robot. Traditionally, when something went wrong with manufacturing machinery it could generally be traced to either a defect in the machine itself or to the machine being incorrectly operated. However, in a robotic environment, faults might stem from the machine, or from the software which forms a part of the machine, or the telecommunications which allow machines within the factory to communicate with one another, or from an intervention by a human operator.

This mean that reviewing the contracts under which robotic lines are procured is more important than ever to ensure that manufacturers can apportion liability when it arises, and recover any losses or costs which they incur as a result of any failures from the relevant supplier. Manufacturers will also need to make sure that they have insurance available to cover relevant additional risks.

**Data Protection:** The increased use of robotics will also have data protection implications. This is because information a robot captures about employees (e.g. through a camera, microphone or sensor) is likely to be personal data. Manufacturers will therefore need to ensure that any personal data captured by a robot is processed in a way which accords with relevant privacy rules and their own policies that tell employees how the employer will deal with their personal data. Manufacturers will also need to be confident that the data and information captured by robotics is not being passed back to the supplier without authorisation.

**Health & Safety:** Deploying robots in amongst a human workforce will also engage health and safety issues. Recent changes to sentencing guidelines mean that penalties for serious health and safety breaches can exceed £10 million in some cases. Manufacturers must therefore engage with their health and safety advisers on this point and put appropriate policies and procedures in place.

**Maintenance:** The increased use of robotics also creates issues around obsolescence. Traditional machinery manufacturers are sometimes required by customers (who have sufficient bargaining power) to make spare parts available for perhaps 10 years, but this is very difficult when dealing with robotics where a key element may be third party software, which may become (if not necessarily obsolete) at least unsupported more quickly than that.

**Intellectual Property:** In terms of Intellectual Property, manufacturers must remember that while they may own the robot, they are likely to only get a licence to use the associated software. In the case of robots utilising AI, it is the user who is providing the data and stimuli which is facilitating the robot's learning on site, but the user is unlikely to get any rights in respect of that.

**Employment Law:** Finally, if manufacturers seek to replace human workers with robots, they will of course need to give appropriate notice and appropriate consultation on any redundancies, and on a practical level will need to manage the transition in a way that limits any (perhaps inevitable) negative publicity. safe and ethical use of AI and blockchain in industrial robotics.

## 5. AI AND BLOCKCHAIN TECHNOLOGY IN INDUSTRIAL ROBOTICS: SOCIAL IMPLICATIONS

The social implications of AI and blockchain in industrial robotics are multifaceted. While these technologies promise improved efficiency and quality, they also introduce concerns. Firstly, there's the issue of inequality. The benefits of AI and blockchain may not be evenly distributed, potentially exacerbating socioeconomic disparities. Access to these technologies and the opportunities they create must be equitable (Tyagi & Bhatnagar,2023). Secondly, the role of workers may change. With automation, workers may find themselves in roles that require different skills. Ensuring that individuals have the means to acquire these skills is essential to prevent job displacement. Lastly, the integration of AI and blockchain in industrial robotics can lead to a cultural shift in the workplace. Employees may need to adapt to working alongside robots and AI systems. Managing this transition effectively is crucial for maintaining a harmonious work environment (Fleres et al.,2023). In the industrial robotics landscape, the ethical, legal, and social implications of AI and blockchain are complex and intertwined. As these technologies continue to advance, stakeholders must collaborate to develop strategies that maximize the benefits while addressing these challenges responsibly. In the chapters to follow, we will delve deeper into case studies, future trends, and recommendations to navigate this evolving landscape.

The integration of AI and blockchain in industrial robotics has far-reaching societal impacts that span various dimensions, both positive and potentially challenging:

1. **Improved Efficiency and Productivity**:

AI-powered industrial robots can execute tasks with precision and consistency, leading to increased efficiency and productivity. This translates into cost savings for industries, which can subsequently result in economic growth and job opportunities.

2. **Job Transformation:**

While automation can lead to job displacement in some areas, it often results in job transformation rather than complete elimination. New roles emerge to oversee and maintain AI-driven systems, design algorithms, and manage data. However, ensuring a smooth transition and providing reskilling opportunities are essential to minimize the negative effects on the workforce.

3. **Safety Enhancement**:

Industrial robots equipped with AI can work in hazardous environments, reducing the risk to human workers. This enhances workplace safety and reduces workplace-related injuries and fatalities.

4. **Quality Improvement:**

AI and blockchain can be utilized for quality control and assurance in manufacturing processes. This results in products of higher quality and reliability, positively impacting consumers and industries alike. Figure 11 shows the various advantages of Blockchain and AI for industrial automation.

5. **Supply Chain Transparency:**

Blockchain in supply chains increases transparency and traceability. Consumers can gain insights into the origin and journey of products, enhancing trust and reducing the likelihood of counterfeit goods.

*Figure 11. Advantages of blockchain and AI for industrial automation [Insights2techinfo]*

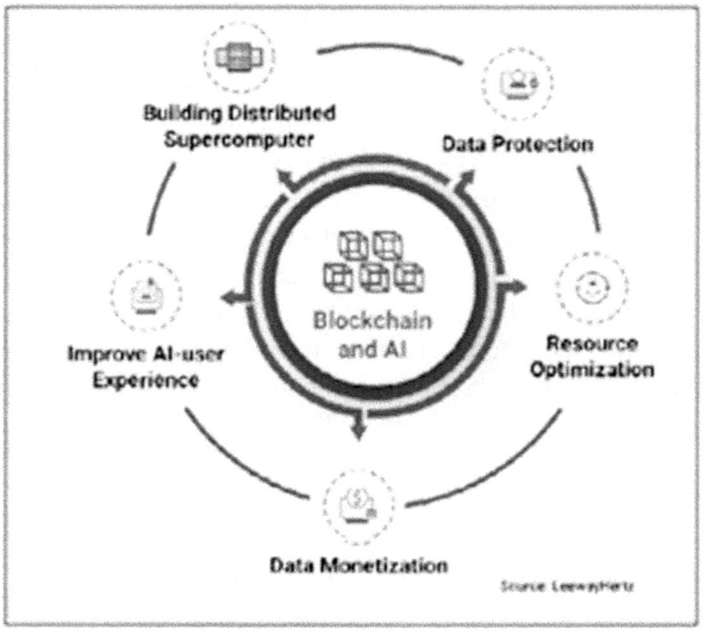


6. **Economic Disparities:**

The benefits of AI and blockchain may not be distributed equally across society, potentially exacerbating economic disparities. Access to these technologies and the skills required to work with them can create a digital divide (Alja'afreh et al., 2022).

7. **Ethical Concerns**:

The ethical implications of AI in industrial robotics are significant. These technologies can raise concerns related to privacy, bias, accountability, and fairness. Ethical considerations need to be addressed to ensure responsible use.

8. **Environmental Impact:**

The energy consumption of blockchain networks, especially in proof-of-work systems, can have a significant environmental footprint. Efforts are underway to develop more energy-efficient consensus mechanisms.

9. **Cultural Shift:**

The integration of AI and robotics in the workplace can lead to a cultural shift. Employees may need to adapt to working alongside machines and AI systems, which can require adjustments in organizational culture and work practices (Verma et al.,2022).

In conclusion, the societal impact of AI and blockchain in industrial robotics is multifaceted. While these technologies offer opportunities for increased efficiency, safety, and transparency, they also pose challenges related to workforce disruption, ethical considerations, and environmental concerns (Dhirani et al.,2023). It is imperative for stakeholders, including governments, industries, and communities, to work collaboratively to maximize the benefits while addressing the challenges to ensure that the integration of AI and blockchain in industrial robotics serves the broader interests of society.

## 5.1 Real-World Examples of Ethical Challenges

Real-world examples highlight the pressing ethical challenges that arise in the integration of AI and blockchain in industrial robotics:

## 5.2 Job Displacement: Automotive Manufacturing

In the automotive industry, AI-driven robots have been increasingly used for tasks like welding and assembly. While this has improved efficiency, it has also led to job displacement. For instance, a major car manufacturer implemented robotic systems that significantly reduced the need for human welders. This resulted in layoffs, sparking ethical concerns about the impact on workers' livelihoods. The challenge here is to find ways to balance automation's benefits with a commitment to reskilling and providing alternative employment opportunities for affected workers.

## 5.3 Privacy: Healthcare Robotics

In healthcare, robots equipped with AI are employed for various tasks, including patient care and data analysis. The challenge is to protect patient privacy while utilizing these technologies effectively. For instance, in hospitals using AI-powered robots for patient monitoring, there are concerns about the collection and transmission of sensitive patient data. Ensuring strict adherence to privacy regulations and robust encryption techniques is essential to safeguard patient confidentiality.

## 5.4 Transparency: Supply Chain Management

Blockchain is often used in supply chain management for tracking products and ensuring authenticity. However, transparency challenges can arise. For example, a food supply chain may implement blockchain to trace the origin of products. The challenge here is to ensure that the data entered into the blockchain is accurate and trustworthy. If any party within the supply chain falsifies data, it could lead to a false sense of transparency and reliability. This ethical issue calls for mechanisms to verify the authenticity of data at various points in the supply chain.

## 5.5 Security: Autonomous Vehicles

In the development of autonomous vehicles, AI and blockchain are utilized for data sharing and vehicle-to-vehicle communication. The ethical challenge emerges in the event of a cybersecurity breach. If hackers gain access to the blockchain system governing vehicle communication, they could potentially manipulate traffic data, leading to accidents or other safety issues. Ensuring robust security measures within the blockchain network becomes crucial to prevent such breaches.

These real-world examples underscore the importance of addressing ethical challenges in the integration of AI and blockchain in industrial robotics. Balancing technological advancements with ethical considerations is essential to harness the benefits of these technologies while safeguarding human interests, privacy, and safety in various industries.

Updated regulations to address emerging challenges in the realm of AI and blockchain in industrial robotics is imperative for several reasons:

1. *Rapid Technological Advancements:* AI and blockchain technologies are evolving at an unprecedented pace. The capabilities and applications of these technologies continue to expand, presenting new challenges and risks. Outdated regulations may not adequately cover these emerging issues, leaving a regulatory gap that can be exploited or cause confusion.
2. *Ethical Concerns:* Ethical considerations surrounding AI and blockchain are complex and evolving. Issues such as algorithmic bias, job displacement, and privacy violations require new or updated regulations to ensure that technology developers and users adhere to ethical principles and responsible practices.
3. *Privacy and Data Security:* The collection and utilization of vast amounts of data in AI and blockchain systems raise profound privacy and data security concerns. Regulations need to adapt to protect individuals' rights and provide clear guidance on data handling, encryption, and disclosure in the context of industrial robotics.
4. *Safety and Liability:* As AI-driven robots become more autonomous, determining liability in the event of accidents or malfunctions becomes challenging. Regulations must establish clear guidelines for safety standards, risk assessment, and liability allocation to protect both manufacturers and end-users.
5. *International Harmonization:* In an increasingly globalized world, harmonized regulations facilitate cross-border collaboration and trade. The absence of consistent regulations can hinder the adoption and growth of AI and blockchain technologies in international markets.
6. *Transparency and Accountability:* Ensuring transparency in AI decision-making and holding AI systems accountable for their actions are essential for trust and ethical operation. Regulations should promote transparency and mechanisms for auditing and oversight.
7. *Cybersecurity:* The integration of blockchain in industrial robotics introduces new cybersecurity challenges. Regulations should address these challenges by setting standards for secure blockchain implementations and incident response protocols.
8. *Inclusivity and Accessibility:* Regulations should promote inclusivity and accessibility, ensuring that the benefits of AI and blockchain are accessible to all segments of society and that potential biases or discrimination are mitigated.

In conclusion, the dynamic nature of AI and blockchain technologies necessitates regulatory agility. Updating and creating new regulations is essential to address emerging challenges, protect individuals and organizations, and promote responsible and ethical use of these transformative technologies in the context of industrial robotics. Collaboration between governments, industry stakeholders, and technology experts is key to striking the right balance between innovation and regulation.

International standards and initiatives related to ethics and legality in the area of AI and blockchain in industrial robotics.

These standards and initiatives aim to provide guidelines, best practices, and a framework for responsible development and deployment of these technologies:

ISO Standards (International Organization for Standardization):

ISO has developed a range of standards relevant to AI and blockchain in industrial robotics. For example:

*ISO 13482:* Specifies safety requirements for personal care robots, addressing safety concerns related to human-robot interactions.

*ISO 18646 series*: Focuses on the safe design, testing, and operation of industrial robots, taking into account aspects like safety, interoperability, and risk assessment.

IEC Standards (International Electrotechnical Commission):

*IEC 80601-2-77:* Concerns the safety and essential performance of autonomous robots for use in healthcare settings, emphasizing safety for patients and users.

IEEE Standards (Institute of Electrical and Electronics Engineers):

*IEEE P7000*: A series of standards projects related to the development of model processes for addressing ethical concerns during system design using autonomous and intelligent systems.

*EU's AI Regulation:* The European Union introduced the Artificial Intelligence Act in April 2021, which aims to provide a comprehensive framework for the regulation of AI systems. It addresses AI ethics, transparency, accountability, and conformity assessments.

*UNESCO's Recommendation on the Ethics of AI:* UNESCO has been actively working on developing a global standard-setting instrument on the ethics of AI, with the goal of promoting human rights, dignity, and diversity.

*Partnership on AI:* The Partnership on AI is a consortium of organizations, including leading tech companies, that collaborate on various aspects of AI ethics and policy. It promotes responsible AI practices across industries.

*World Economic Forum's Global AI Council:* The World Economic Forum established the Global AI Council to work on developing governance frameworks and policy guidelines for AI, with a focus on ethics, safety, and transparency.

*IEEE Global Initiative on Ethics of Autonomous and Intelligent Systems:* This initiative focuses on advancing ethical considerations in AI and autonomous systems, including robots, through workshops, publications, and guidelines.

AI and blockchain integration in industrial robotics offers a range of potential benefits and drawbacks for various stakeholders, including workers, companies, and communities:

## 5.6 Benefits for Workers

1. Increased Safety: AI-driven robots can take on dangerous tasks, reducing the risk of workplace injuries and fatalities. Workers can be relieved from hazardous roles, enhancing their safety.
2. Skill Enhancement: The adoption of AI and blockchain may necessitate upskilling and reskilling of workers. This can lead to a more skilled and adaptable workforce, potentially opening up higher-paying job opportunities.
3. Reduced Monotony: Repetitive and mundane tasks can be automated, allowing workers to focus on more intellectually stimulating and creative aspects of their jobs.

## 5.7 Drawbacks for Workers

1. Job Displacement: Automation can lead to the displacement of certain jobs, especially those involving routine tasks. Workers in these roles may face unemployment or need to transition to new job profiles.
2. Skills Gap: Not all workers may have the necessary skills to work alongside AI and robots. This can result in a skills gap, making it challenging for some individuals to secure new employment.
3. Job Insecurity: The fear of job displacement can create job insecurity among workers, impacting their overall well-being and job satisfaction.

### Benefits for Companies:

1. Increased Productivity: AI and blockchain can significantly enhance productivity, leading to cost savings and improved competitiveness for companies.
2. Quality Improvement: Automation and data-driven decision-making can lead to higher product quality and reduced defects, boosting customer satisfaction.
3. Cost Reduction: Automated processes can lead to lower labor costs, as well as reductions in errors and inefficiencies.

### Drawbacks for Companies:

1. Implementation Costs: The initial investment in AI and blockchain technology can be substantial, and companies may need to allocate resources for training and maintenance.
2. Resistance to Change: Employees may resist the adoption of new technology, leading to organizational challenges during the transition phase.

### Benefits for Communities:

1. Economic Growth: Companies benefiting from AI and blockchain integration may expand, creating new job opportunities in the community and contributing to economic growth.
2. Safer Work Environments: With AI-driven robots taking on dangerous tasks, workplace safety can improve, reducing the burden on healthcare and social services in the community.

### Drawbacks for Communities:

1. Economic Disparities: Communities may experience economic disparities if certain segments of the population do not have access to the necessary training and education to participate in the changing job market.
2. Environmental Impact: Depending on the energy source and efficiency of blockchain networks, there may be environmental impacts in the community due to increased energy consumption.

## 5.8 Concerns Related to Inequality and Access to Technology

Concerns related to inequality and access to technology in the domain of AI and blockchain in industrial robotics are pressing and must be addressed to ensure equitable benefits (Chaka,2023):

1. Digital Divide:
    - Access to Technology: Unequal access to AI and blockchain technology can exacerbate existing disparities. Some individuals and communities may lack the infrastructure and resources needed to leverage these technologies fully.
    - Skills Gap: The rapid evolution of AI and blockchain requires continuous learning. Those with limited access to education and training opportunities may fall behind in acquiring the skills necessary to work with these technologies.
2. Economic Disparities:
    - Job Market: AI and robotics can disrupt traditional job markets, and individuals with limited access to technology may face job displacement without the means to transition to new roles.
    - Ownership and Entrepreneurship: Unequal access can hinder entrepreneurship and ownership in technology-driven industries. Limited access to funding, education, and networks can prevent individuals from starting AI and blockchain ventures.
3. Ethical Considerations:
    - Bias and Discrimination: AI algorithms can perpetuate biases present in training data. If not addressed, this can lead to discriminatory outcomes that disproportionately affect marginalized communities.
    - Privacy: Vulnerable communities may be at greater risk of privacy violations, as they often have less control over their data and limited recourse in the event of misuse.
4. Access to Benefits:
    - Healthcare: Innovations in AI-driven healthcare can offer significant benefits, but unequal access to these technologies may result in disparities in healthcare outcomes and treatment options.
    - Education: AI-powered educational tools can enhance learning, but students without access to such tools may face disadvantages in their education.

### Addressing Inequality and Access:

1. Equitable Education: Ensuring that quality education and training programs are accessible to all, including underserved communities, is crucial. This includes providing opportunities for digital literacy and technology education.
2. Affordable Access: Policies should aim to make technology more affordable and accessible. Government initiatives, subsidies, or community programs can help bridge the digital divide.
3. Diversity and Inclusion: Encouraging diversity in the technology sector and addressing bias in AI algorithms can contribute to more equitable outcomes.
4. Community Involvement: Engaging communities and stakeholders in technology adoption can foster understanding, collaboration, and inclusive technology development.
5. Policy and Regulation: Governments and international bodies should implement policies that promote equitable access, protect privacy, and address discrimination in technology use.

Efforts to reduce inequality and increase access to AI and blockchain technology in industrial robotics are essential for ensuring that the benefits of these innovations are distributed fairly and that marginalized communities are not left behind in the rapidly evolving digital age.

## 5.9 Legal Framework and Regulations

The existing legal framework and regulations governing AI and blockchain in industrial robotics are in a state of development and evolution due to the rapidly advancing nature of these technologies. As of my last knowledge update in September 2021, the legal landscape can be characterized as follows:

1. General Data Protection Regulations (GDPR) and Privacy Laws:
   - GDPR in the European Union sets strict rules for data protection, including data generated by AI and blockchain systems in industrial settings (Kory-Westlund, 2023). It emphasizes the rights of individuals over their personal data, requiring clear consent for data collection and stipulating severe penalties for data breaches. Similar privacy laws exist in various countries worldwide.
2. Product Liability Laws:
   - Product liability laws are crucial for determining responsibility in case of accidents or damage caused by AI-driven robots. These laws hold manufacturers, operators, and other parties accountable for any harm resulting from a malfunctioning or misbehaving robot. They help ensure safety in industrial settings.
3. Intellectual Property Rights:

Intellectual property laws cover inventions, patents, and copyrights related to AI algorithms and blockchain solutions used in industrial robotics. Companies must navigate these laws to protect their innovations and avoid infringing on others' intellectual property.

4. Safety Standards and Regulations:
   - Safety standards for industrial robots are established by organizations such as the International Organization for Standardization (ISO). These standards set guidelines for the design, operation, and maintenance of robotic systems, including those employing AI (Fox & Griffy,2023). Adherence to these standards is often legally mandated.
5. Trade and Export Controls:
   - In certain cases, the export of AI or blockchain technology may be subject to trade restrictions and export controls, particularly when it has military or dual-use applications. Legal frameworks such as the Wassenaar Arrangement aim to prevent the proliferation of advanced technologies.
6. Contractual Agreements and Smart Contracts:
   - Contracts formed on blockchain networks are subject to contract law, which includes considerations of enforceability and legality. The development and use of smart contracts on blockchains need to adhere to existing contract law principles.
7. Anti-Trust and Competition Laws:

- As AI and blockchain integration can impact markets and competition, antitrust and competition laws may come into play. These regulations aim to prevent anti-competitive behavior and promote fair competition.

8. AI-Specific Regulations:
    - Some regions and countries have started exploring or implementing AI-specific regulations. These may include requirements for transparency, accountability, and bias mitigation in AI systems (Luan et al., 2022).

## 6. FUTURE SCOPE

It's important to note that the legal landscape for AI and blockchain in industrial robotics is dynamic. Governments and regulatory bodies are actively working on updating and crafting new regulations to address emerging challenges and ensure the responsible development and deployment of these technologies. Businesses involved in AI and blockchain in industrial robotics must stay informed about evolving legal requirements and engage with relevant authorities to ensure compliance.

## 7. CONCLUSION

In this chapter, an overview of the role of Artificial Intelligence and Blockchain technology in Industrial Robotics is provided. In conclusion, the integration of AI and blockchain technologies in industrial robotics holds significant promise for enhancing efficiency, transparency, and security in manufacturing processes. However, the convergence of these powerful technologies also brings forth a range of ethical, legal, and social implications that must be carefully considered and addressed

From an ethical standpoint, the use of AI in industrial robotics raises concerns related to job displacement, as automation may lead to a reduction in human employment opportunities. It is crucial for businesses and policymakers to prioritize strategies for reskilling and upskilling the workforce to ensure a smooth transition to an automated landscape.

Additionally, the ethical use of AI in robotics involves addressing bias in algorithms and ensuring fair and accountable decision-making processes. As AI systems become more complex and autonomous, there is a need for clear ethical guidelines and standards to govern their behavior, with an emphasis on transparency and accountability.

On the legal front, the deployment of AI and blockchain technologies in industrial robotics necessitates the development and adaptation of existing regulations. Legal frameworks must keep pace with technological advancements to address issues such as liability, data ownership, and intellectual property in the context of autonomous systems.

Moreover, the integration of blockchain in industrial robotics introduces new legal considerations, particularly regarding data privacy and security. Smart contracts and decentralized ledgers can enhance transparency and traceability, but they also require careful consideration of compliance with data protection regulations.

Socially, the widespread adoption of AI and blockchain in industrial robotics may contribute to a more efficient and productive economy, but it also requires careful navigation of potential societal challenges. Public awareness and education about the benefits and risks of these technologies are essential to foster informed discussions and mitigate concerns.

In conclusion, the successful and responsible implementation of AI and blockchain in industrial robotics demands a collaborative effort from technology developers, businesses, policymakers, and society at large. By prioritizing ethical considerations, adapting legal frameworks, and engaging in open dialogue, we can harness the transformative potential of these technologies while safeguarding the well-being of individuals and communities in the rapidly evolving landscape of industrial automation.

## REFERENCES

Aditya, U. S., Singh, R., Singh, P. K., & Kalla, A. (2021). A survey on blockchain in robotics: Issues, opportunities, challenges and future directions. *Journal of Network and Computer Applications, 196*, 103245. doi:10.1016/j.jnca.2021.103245

Alja'afreh, M., El Ahmar, W., Hoda, M., & Karime, A. (2022), A Review on Industrial Blockchain. In *2022 Fourth International Conference on Blockchain Computing and Applications (BCCA)*. IEEE. 10.1109/BCCA55292.2022.9922219

Bhatt, H., Bahuguna, R., Singh, R., Gehlot, A., Akram, S. V., Priyadarshi, N., & Twala, B. (2022). Artificial Intelligence and Robotics Led Technological Tremors: A Seismic Shift towards Digitizing the Legal Ecosystem. *Applied Sciences (Basel, Switzerland), 12*(22), 11687. doi:10.3390/app122211687

Chaka, C. (2023). Fourth industrial revolution—A review of applications, prospects, and challenges for artificial intelligence, robotics and blockchain in higher education. *Research and Practice in Technology Enhanced Learning, 18.*

Dhirani, L. L., Mukhtiar, N., Chowdhry, B. S., & Newe, T. (2023). Ethical dilemmas and privacy issues in emerging technologies: A review. *Sensors (Basel), 23*(3), 1151. doi:10.339023031151 PMID:36772190

Fox, S., & Griffy-Brown, C. (2023). Robotics in society: Technology in Society Briefing. *Technology in Society*, *72*, 102174. doi:10.1016/j.techsoc.2022.102174

Hans, S., Chohan, J. S., Kumar, R., & Pareek, S. 2023, March, Artificial Intelligence with Robotics-A Bibliometric Study, In *2023 10th International Conference on Computing for Sustainable Global Development (INDIACom)*. IEEE.

Kory-Westlund, J. M. (2023). Implications of children's social, emotional, and relational interactions with robots for human–robot empathy. *Conversations on Empathy: Interdisciplinary Perspectives on Empathy, Imagination and Othering*, (pp. 256-279). Taylor & Francis.

Kumar, R. L., Khan, F., Kadry, S., & Rho, S. (2022). A survey on blockchain for industrial internet of things. *Alexandria Engineering Journal*, *61*(8), 6001–6022. doi:10.1016/j.aej.2021.11.023

Kumar, R. L., Wang, Y., Poongodi, T., & Imoize, A. L. (Eds.). (2021). *Internet of Things, artificial intelligence and blockchain technology*. Springer. doi:10.1007/978-3-030-74150-1

Lopes, V., & Alexandre, L. A. (2018). *An overview of blockchain integration with robotics and artificial intelligence.* arXiv preprint arXiv:1810.00329

Luan, F., Yang, X., Chen, Y., & Regis, P. J. (2022). Industrial robots and air environment: A moderated mediation model of population density and energy consumption. *Sustainable Production and Consumption, 30*, 870–888. doi:10.1016/j.spc.2022.01.015

Nikolinakos, N. T. (2023). EU Policy and Legal Framework for Artificial Intelligence, Robotics and Related Technologies-The AI Act. *Law, Governance and Technology Series, 53.*

Salah, K., Rehman, M. H. U., Nizamuddin, N., & Al-Fuqaha, A. (2019). Blockchain for AI: Review and open research challenges. *IEEE Access : Practical Innovations, Open Solutions*, *7*, 10127–10149. doi:10.1109/ACCESS.2018.2890507

Taherdoost, H. (2022). Blockchain technology and artificial intelligence together: A critical review on applications. *Applied Sciences (Basel, Switzerland)*, *12*(24), 12948. doi:10.3390/app122412948

Tyagi, S., & Bhatnagar, P. (2023). Roboethics: A Bibliometric Review [2005–2021]. *International Journal of Business Analytics & Intelligence (IJBAI), 11*(1).

Verma, A., Bhattacharya, P., Madhani, N., Trivedi, C., Bhushan, B., Tanwar, S., Sharma, R., Bokoro, P. N., & Sharma, R. (2022). Blockchain for industry 5.0: Vision, opportunities, key enablers, and future directions. *IEEE Access : Practical Innovations, Open Solutions*, *10*, 69160–69199. doi:10.1109/ACCESS.2022.3186892

Zhamshid, V., Nusratilloevich, Y. A., Shahida, K., & Azamat, K. (2023). Rise of the Machines: The Legal Implications of Robotics and Automation for the Digital Workforce. *International Journal of Cyber Law*, *1*(4).

# Chapter 3
# Enhancing Industrial Robotics Performance and Security With AI and Blockchain Technologies

**Deepak Varadam**
*Ramaiah University of Applied Sciences, India*

**Sahana P. Shankar**
https://orcid.org/0000-0001-8977-9898
*Ramaiah University of Applied Sciences, India*

**Aryan Bharadwaj**
*Ramaiah University of Applied Sciences, India*

**Tanvi Saxena**
*Ramaiah University of Applied Sciences, India*

**Sarthak Agrawal**
*Ramaiah University of Applied Sciences, India*

**Shraddha Dayananda**
*Ramaiah University of Applied Sciences, India*

## ABSTRACT

*Industrial robotics are becoming more widely used, but their performance and security must be urgently enhanced to satisfy the needs of contemporary industrial contexts. This chapter focuses on how AI and blockchain technology might improve industrial robotic systems' performance while guaranteeing strong security precautions. The capabilities of industrial robots are greatly enhanced by AI technologies. Robots may improve their performance, gain new abilities, and adapt to changing circumstances by utilising cutting-edge machine learning techniques. Robots may learn from their experiences thanks to the incorporation of AI, which improves their operational effectiveness, precision, and decision-making abilities. AI enables robots to optimise their performance, spot anomalies, and proactively resolve potential difficulties, resulting in increased production and less downtime. This is done through real-time data analysis and predictive analytics. Incorporating blockchain technology also provides an industrial robotics system with a safe and open framework.*

DOI: 10.4018/979-8-3693-0659-8.ch003

## INTRODUCTION

Industrial robotics are becoming more widely used, but their performance and security must be urgently enhanced to satisfy the needs of contemporary industrial contexts. The chapter mainly focuses on how AI and blockchain technology might improve industrial robots systems performance while guaranteeing strong security precautions. The capabilities of industrial robots are greatly enhanced by AI technologies. Robots may improve their performance, gain new abilities, and adapt to changing circumstances by utilising cutting-edge machine learning techniques. Robots may learn from their experiences thanks to the incorporation of AI, which improves their operational effectiveness, precision, and decision-making abilities. AI enables robots to optimise their performance, spot anomalies, and proactively resolve potential difficulties, resulting in increased production and less downtime. This is done through real-time data analysis and predictive analytics.

Incorporating blockchain technology also provides an industrial robotics system with a safe and open framework. Data integrity, immutability, and traceability are guaranteed by blockchains distributed ledger system, which is essential for preserving the integrity of industrial operations. Industrial robots can securely store and distribute data, such as performance measurements, maintenance records, and supply chain information, among several stakeholders by utilising blockchain. This decentralised strategy enhances the security of industrial robots systems by preventing unauthorised access, tampering, or harmful activities. Industrial robotics gains from the synergistic effects of AI and blockchain technology. AI algorithms are capable of analysing data gathered from sensors, cameras, and other sources to uncover insightful trends. The security and integrity of the data is guaranteed by using blockchain technology to store and exchange these insights securely. Additionally, blockchains decentralised structure offers resilience against single points of failure, enabling fault tolerance and guaranteeing continuity in industrial operations. However, there are obstacles to successfully integrating AI and blockchain into systems for industrial robotics. For robots with limited resources, the intricacy of AI algorithms and the resource-intensive nature of blockchain can provide computational difficulties. Scalable architectures, optimised algorithms, effective hardware and software design are required to meet these problems.

Real-time applications and their data uses the blockchain technology by making a chain like structure of secured data blocks that are used to store various information. In the present world, robotics is a multi-disciplinary domain that is thriving, and growing its roots deep into various fields of research, manufacturing industries, healthcare, cyber security and even in our day-to-day lives. It has the potential to revolutionize production processes by enhancing productivity, precision, and efficiency, industrial robots has attracted a lot of interest. There are also worries about data security, transparency, and trust as industrial robots become more linked and incorporated into intricate supply chain networks. Blockchain technology, which is renowned for being decentralised and unchangeable, has the ability to solve these problems and improve industrial robots performance. Robotics is made up of vast fields of studies which make it as the root to many challenges in the scenario of performance.

## EMERGENCE OF BLOCKCHAIN TECHNOLOGY IN INDUSTRIAL ROBOTICS

Blockchain technology is examined to promote operational efficiency in industrial robots systems by facilitating transparent and auditable transactions, facilitating cooperation without relying on trust, and facilitating data integrity and security. The technology enables producers to check the quality and maintenance history of robotic systems, optimise logistics and inventory management, and secure the validity and traceability of raw materials and componentry. Blockchain is known as a distributed ledger technology for its secure and openly record transactions. It is therefore a potential technique for boosting the security of systems for industrial robots. Blockchain can be utilised, for instance:

**Track the origin of robotics parts:** This can assist in preventing the use of fake components in industrial robots, which could result in safety risks.

**Note every interaction with a robot:** The technology can assist in locating and looking into security events, such as assaults using malicious software.

**Control who has access to private robot data:** In this case, the technology aids in limiting unauthorised access to information that might be used to damage robots or their operators.

**Create a safe environment where robots and other systems can transact:** Safe environment results in ensuring that transactions are exact and unchangeable.

In addition to these security advantages, blockchain can also give industrial robots systems a variety of additional advantages, such as:

**Increased transparency:** Making it simpler to trace the origin of components and record all robot interactions, blockchain can contribute to greater transparency in industrial robotics systems. This can enhance the systems overall security and assist stakeholders develop a sense of trust with one another.

**Decreased expenses:** By doing away with the need for central servers and middlemen, blockchain can aid in cost reduction. Businesses may save money in this way on transaction costs and IT infrastructure.

**Increased effectiveness:** Simpler to track the state of robots and their components, blockchain can aid in enhancing the effectiveness of industrial robotics systems. Systems overall productivity is tremendously increased and assists to reduce downtime.

This promising technology improves the overall security of industrial robots systems by using blockchain. It provides several security advantages in addition to other advantages like improved transparency and lower costs. Blockchain is anticipated to become a more crucial technology for assuring the security of these systems as the usage of industrial robotics keeps expanding. Additionally, machine learning algorithms can learn from patterns and historical data, allowing the system to adapt and improve its security measures over time. Moreover, AI-powered security systems can continuously monitor and assess the behavior of industrial robots, identifying any abnormal activities that may indicate a security breach. The potential challenges and limitations associated with implementing AI-based security solutions for industrial robotics is addressed with clear explanation. These challenges include the need for robust data collection and training sets, the potential for false positives or negatives in anomaly detection algorithms, and the requirement for regular updates to address evolving cyber threats.

In summary, combining blockchain with AI technology has enormous promise for improving the efficiency and security of industrial robotics systems. Industrial robots can attain greater levels of productivity, flexibility, and dependability by utilising AI's learning capabilities and blockchains safe and transparent framework. To fully realise the potential of AI and blockchain in industrial robots, future research and development efforts should concentrate on resolving technical issues and investigating cutting-edge solutions.

## EVOLVING OF INDUSTRY 4.0

Industries regularly blend several technologies to achieve the best results. The market is fiercely competitive, and consumers have higher expectations of brands now more than ever. Businesses must therefore adjust beyond their original plans. Technologies are integrating more easily than ever before, making it simpler for businesses to adopt them and create systems that benefit both them and the end user. Additionally, businesses have access to enough data to test out unique combinations. Due to the integration of blockchain technology with robotic process automation, all of this is possible. The main goal of robotic process automation is to assume control over the numerous facets of the organisation. It is a consensual entity that can communicate with people and the system alike. Its architecture enables real-time data interaction and interpretation. Here, finding processes that can be automated and educating coworkers are tasks that the entire team must complete. These blockchains offer a distributed shared ledger to manage all of the data and information, as well as an intelligent, reliable, and simple trading platform for all parties involved (Tailor et al., 2022). The industry is continually changing to suit modern customer demands with greater efficiency and sustainability as technology continues to advance at a breakneck pace. Industry 3.0 also saw the development of automation and robotics, but today they are connected through networks to make it simpler to manage, watch over, and manage production facilities. The fourth industrial revolution has created countless possibilities, including machine to machine connections and online/remote location-based control. Instead of focusing on a single technology or process, a summary of significant industry 4.0 technologies is combined that can help with data management difficulties and safe data management (Rane et al., 2022). These technologies are autonomous mobile robots (AMR), internet of things (IoRT), block chains, artificial intelligence (AI), cyber security, big data analytics, and machine learning (ML) to effectively improve the security and privacy of the services (Islam et al., 2021).

Blockchain technology is evolving quickly every day, and it may be integrated with a variety of systems, including robotics and AI applications. However, as this discipline is still relatively new, it is unclear what it might eventually develop into. There is a wide range of approaches and platforms that aim to integrate blockchain technology into robotic systems, enhance AI services, or address issues that exist in the major blockchains, with the goal of developing more capable and secure robotic systems (Lopes & Alexandre, 2018). The industrial Internet of Things (IIoT) is essential for putting safe, scalable, and quickly accepted technologies into practice for the smart industry. Traditional IIoT systems frequently rely on centralized designs that might be a standalone point of failure and are open to various threats. Blockchain is widely employed in recent business because to its security and decentralization. This research offers blockchain-based architectural styles that ensure dependable and secure plant operations. Cutting edge contemporary concepts and technologies like big data, AI, cloud computing, and cyber-physical systems are a few instances of how they have revolutionized the industrial sector (Alsamhi et al., 2021). IIoT and smart industry integration have elevated industrial processes to a new level. The IIoT considerably modifies a variety of industrial processes by providing facilities and bringing new technology and economic models. Modern industries can benefit from a number of services provided by IIoT platforms, including big data analysis, cloud computing, high-speed networking, and development of applications. By utilizing novel and cutting-edge approaches, these services improve on already-existing industrial processes by streamlining production procedures, enhancing client experiences, and lowering manufacturing costs (Latif, Idrees, Ahmad et al, 2021). Industry 4.0 largely refers to manufacturing facilities with automated, intelligent machinery that is backed by cyber-physical systems, AI, ML, and IoT. According to most definitions, blockchain technology is not a primary driving element behind Industry 4.0. The main areas of focus include the IoT, AI, ML, and cyber-physical systems (CPS). It is evident that blockchain offers a solution to the bulk of Industry's issues. Security and protection are important factors for Industry applications (Aoun et al., 2021). Sustainability and a good equilibrium between environmental, social, and commercial consequences are increasingly necessities for the industrial industry. Data is exchanged over a peer-to-peer, open-access network, and authorized transactions. Blockchain ensures the transaction's security and transparency. Blockchain was initially applied in the banking industry, in order to replace manual transaction authentication with digital for cryptocurrencies. Blockchain has generated interest from both academics and business professionals because it may enhance data security, immutability, trust, traceability, and transparency. All parties agree on established procedures that are used to communicate and validate data (Latif, Idrees, Huma et al, 2021). Direct stakeholder-to-stakeholder transaction verification is made possible by the decentralized structure of blockchain, eliminating the need for middlemen. The risk of information loss or modification, as well as any transactional human mistake, are all eliminated by modern encryption. Blockchain technology has a favorable influence on the effectiveness of decision-making, the pace at which supply chain operations take place, and the dependability and integrity of supply chain transactions (Khanfar et al., 2021). The current change in contemporary production systems known as Industry 5.0 is moving from mass production to mass customization. By integrating machinery, robots, people, goods, and consumers into interconnected networks, Industry 5.0 seeks to advance conventional manufacturing techniques. The goal of this integration is to combine the real and virtual worlds and build intelligent hybrid workspaces where humans and robots work together to improve industrial sectors' efficiency and safety. Cobots or collaborative robots are used to mix the flexibility and expertise of traditional labor with the efficiency of technology that is automated. Maintaining safety and efficiency in collaborative situations requires precise and seamless communication between human operators and robots. These technologies provide seamless data connectivity and flow (Raffik et al., 2023).

The development of digital machinery prototypes, the use of twin designs, the appearance of hyper cognitive systems, and the fusion of extended reality and virtual reality are credited with this transition. Collaborative robots, safe machine boundaries, and AI-driven supply networks have all contributed to this paradigm shift. The future phase, dubbed as Industry 5.0, is expected to extend beyond the reach of Industry 4.0 ecosystems by incorporating enormous manufacturing with user-centered customization. A number of industrial sectors, including supply chain management systems, smart agriculture, healthcare, and aircraft manufacture, are anticipated to benefit from Industry 5.0. However, it might be challenging to safely share information across networks that span different authority domains. For maintaining synergy and security inside industrial boundaries, a trusted and secure data link is necessary (Hemdan et al., 2023). The Industrial Internet of Things (IIoT), autonomous robots, soft robotics, and collaborative robots (COBOTs) provide potential benefits for the automobile industry. These technological advancements have the potential to boost efficiency, production, safety, and cost-savings. Automakers are embracing these innovations in order to maintain their competitiveness in a market that is evolving swiftly. By addressing these issues, researchers want to inspire innovation and provide solutions that will advance these fields (Thanki & Joshi, 2023). The banking industry has also seen significant changes as a result of the digital transformation sparked by technologies like AI and Blockchain. These technologies have transformed banking by introducing Chatbots and Robo advisors, which provide virtual customer service and investment advice. The main goal is to analyze how AI and blockchain are now used in banking, to highlight the challenges banks have had incorporating these technologies, and to provide some perspectives on how banking will change in the digital era. Enhancing platform performance, resolving security, privacy, and regulatory concerns, and partnering with humans and robots are key strategies that banks may use to get beyond challenges and capture opportunities given by AI and blockchain (Dewasiri et al., 2023).

One of the many sectors where blockchain technology is being employed as a safe and practical solution is healthcare. Despite not being a universally applicable solution, blockchain has the potential to improve the healthcare sector. By employing blockchain networking, patient data may be securely stored and transferred across hospitals, laboratories, pharmacies, and doctors. This technology enhances the efficiency, security, and openness of the interchange of medical data by accurately identifying critical diagnostic errors (Shankar et al., 2022). Combining blockchain technology with artificial intelligence models enables the creation of an open network for the exchange and authorization of information in the healthcare industry. Blockchain technology allows for access to medical information by healthcare practitioners, while AI makes better use of data and decision-making processes to promote better healthcare delivery (Hannan, 2023). Cyber-Physical Systems (CPS) provide the foundation for a number of applications, such as smart cars, smart cities, and defensive systems. CPS link physical parts and computing algorithms. It is challenging to defend CPS against hostile attacks, and in this dynamic context, resilience building is crucial. This chapter covers the characteristics of CPS, the existing circumstance, the security issues that CPS faces, and potential solutions. By merging blockchain with CPS, processes across many industries can be made more efficient, and data can be transmitted securely to improve security applications (Vignesh Saravanan et al., 2023). The healthcare industry is focused on raising life expectancy since it can contribute to economic growth. It is impossible to put off technological advancements because they are necessary for providing high-quality healthcare services. When using new technologies in healthcare, especially when managing sensitive medical records, privacy, security, interoperability, and dependability are crucial factors to take into account. Blockchain technology has become a tool for resolving issues with complexity and privacy in a variety of industries, including healthcare. Patients are at risk from counterfeit pharmaceuticals and medical equipment, which also erodes public confidence in healthcare systems (Uma, 2023). Blockchain ensures the privacy, accuracy, and security of patient records, while AI improves the safety and accuracy of detection and diagnosis. The vulnerability of existing healthcare systems is caused by out-of-date data storage methods. But for securely storing and monitoring patient medical data, blockchain offers a decentralized solution.

Fourth Industrial Revolution (4IR) are being met through education while also transforming our lives and places of employment. To succeed in the 4IR era, innovative and sustainable teaching and learning technologies are crucial. With potential advantages like personalized learning, secure credentialing, and decentralized learning networks, AI and blockchain technologies hold great promise in this aspect. This study evaluates previous research on the use of blockchain and AI in education, looking at examples and discussing the potential benefits and drawbacks of these technologies. The study puts out a novel approach to incorporating blockchain and AI into ethical methods of education. Further empirical research is also discussed, as well as the necessity to take ethical and societal ramifications into account. The most significant finding is that blockchain and AI have the potential to revolutionize education by enhancing accessibility, effectiveness, and security (Islam, 2023). Internet of Things (IoT) has received a lot of attention as computational resources keep growing, according to the authors of (Falayi et al., 2023). Many industries and professions are incorporating IoT technologies to offer smart applications. However, due to environmental risks and criminal activity, IoT network security continues to be a major problem. Access control and other traditional security methods are insufficient to protect IoT devices from dangers on their own. In order to address these issues, deep learning and blockchain technology have emerged as critical technology that offers unique and secure approaches to IoT network security.

The convergence of artificial intelligence (AI) and blockchain technology has ushered in a new era of possibilities in the realm of industrial robotics. In the year 2020, researchers delved into the potential synergies between these two transformative technologies, exploring how their integration could propel industrial robotics towards enhanced performance and heightened security (Benotsmane et al., 2020).

Industrial robotics is a rapidly growing field with the potential to revolutionize manufacturing and other industries. In order to fully realize the potential of industrial robotics, there are a number of challenges that need to be addressed. These challenges include:

**Performance:** Industrial robots need to be able to perform tasks accurately and reliably, even in challenging environments.

**Security:** Industrial robots need to be secure from cyberattacks, which could disrupt production or even cause physical harm.

**Cost:** Industrial robots are expensive, which can make them out of reach for some businesses.

AI and blockchain technology can be combined to enhance the performance and security of industrial robotics in a number of ways (Kumar & Saha, 2020).

To understand the use of AI and Blockchain more profoundly in the field of industrial robotics, the below topics will explain in detail.

Blockchain's Contributions to Secure Data Management: The pursuit of heightened security in industrial robotics led researchers to explore the integration of blockchain technology. With its inherent attributes of immutability and transparency, blockchain emerged as a robust solution for secure data sharing and storage. The research of 2020 highlighted blockchain's capacity to safeguard data integrity, thwart unauthorized access, and bolster traceability in the domain of robotics. By ensuring that all data transactions are tamper-proof and readily auditable, blockchain technology establishes a level of trust among stakeholders, reinforcing the security framework of industrial processes (Singh & Singh, 2020).

Decentralized Control Architectures: The confluence of AI and blockchain sparked the emergence of novel decentralized control architectures for industrial robots. These architectures distribute decision-making processes across interconnected robotic entities, imbuing the system with heightened fault tolerance and resilience. The research of 2020 introduced frameworks that leverage AI algorithms for localized decision-making while utilizing blockchain for global consensus establishment. This innovative approach not only amplifies system reliability but also enhances adaptability to dynamic operational scenarios (Rahman et al., 2020).

Resource Optimization and Energy Efficiency: The challenge of resource optimization and energy efficiency in industrial robotics was met head-on by the amalgamation of AI and blockchain. AI algorithms scrutinized real-time data to make informed decisions regarding energy utilization, leading to heightened efficiency and diminished operational costs. In parallel, blockchain integration brought transparency to energy consumption and resource allocation records, promoting accountability and sustainability practices within the industrial domain (Kumar et al., 2020).

## ENHANCING INDUSTRIAL ROBOTICS SECURITY WITH AI

Enhancing industrial robots has emerged as a crucial endeavor in the current industrial landscape, which is continually changing. Previously limited to repetitive and predetermined tasks, industrial robots are rapidly becoming flexible and intelligent devices. A new era of automation is beginning to emerge as a result of the introduction of cutting-edge technologies and creative business models. In this investigation, analyzing into the complex world of improving industrial robotics, revealing its importance, applications, and potential in the future is clearly elucidated. There are a wide variety of ways where the performance can be improved. The incorporation of robotics has ushered in a dramatic revolution in the constantly changing field of industrial automation, redefining the landscape of production, manufacturing, and beyond. Robots have established themselves as priceless assets in a variety of industries because to their exceptional potential for accuracy and efficiency. However, the constant growth of robotic systems within our industrial ecosystems, a compelling demand for strong security measures to protect these devices develops. It is impossible to emphasize how crucial their security is when robots interact with sensitive data, carry out complex jobs, and oversee crucial operations. The incorporation of AI into industrial robots stands out as a strong solution in this age of networked systems and digital vulnerabilities. With its transformational potential, AI not only improves the performance and capacities of robots but also acts as a security sentinel. A thorough investigation into the complex interactions between industrial robotics security and AI, revealing the many ways that AI technologies strengthen security precautions, reduce risks, and serve as the cornerstone of reliable automation in our industries rapid technological change (Aditya, Singh, Singh, & Kalla, 2021). AI assumes the guardian position for industrial robots, ushering in an era marked by both innovation and safety, where machines are not just tools but trusted sentinels of progress in navigating the intersection of technology and security.

*Figure 1. Commonality between industrial robots, AI, and blockchain technologies*

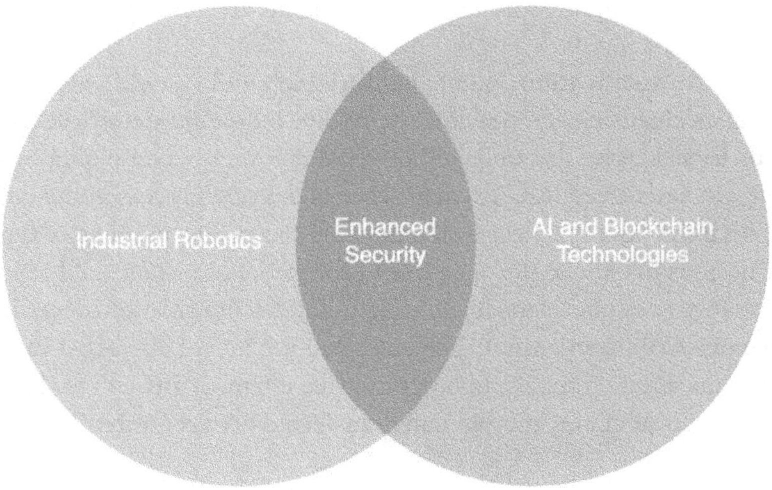


Figure 1 shows a Venn diagram which shows how industrial robots, AI and blockchain is collaborated for a better system.

**Imperative for Industrial Robotics Security:** Understanding the urgent need for security measures in the field of industrial robots is crucial before exploring how AI may improve security. Industrial robots have evolved from their original uses on manufacturing floors to become adaptable, intelligent beings who work alongside people, carry out complex jobs, and handle private information. Due to their development, they are now vulnerable to a variety of dangers, including physical tampering, hacking, and unauthorized access. Industrial robotics security breaches may result in everything from production interruptions to financial losses. Securing the security of these autonomous systems becomes essential as the stakes increase. Although still necessary, traditional security methods frequently fall short of keeping up with the constantly changing threat environment. This is where AI comes into play, acting both as a tool to improve performance and as a keeper of industrial robot's security (Park & Li, 2021).

## INDUSTRIAL ROBOTICS SECURITY MEASURES POWERED BY AI

Industrial robots have the capacity to detect abnormalities and possible dangers quickly because to AIs ability to handle enormous volumes of data in real-time. Artificial intelligence (AI) systems can spot odd patterns in a robots behavior or departures from anticipated standards by analyzing data from numerous sensors and cameras. This capacity covers both physical risks like tampering or unauthorized entry as well as cyber dangers like malware or hacking attempts. Detecting and preventing intrusions AI-driven intrusion detection systems can strengthen the security of industrial robotics even more. These systems constantly scan the robots network for any attempts at unauthorized access or other odd network activity. Artificial intelligence (AI) algorithms have the power to spot suspicious trends and prompt rapid actions, such isolating the compromised system or notifying security staff (Mahyuni et al., 2020).

Few of the anomalies and threats are identified below:

**Enhancing cybersecurity:** As industrial robots become more plugged into networks, hacks are a possibility. Robot systems may be actively shielded from malware, viruses, and other cyber threats with the help of AI. Machine learning algorithms offer a dynamic defense by developing and adapting to counter new and changing threats.

**Access Control and Authentication:** Within industrial robotics systems, AI can improve authentication and access control mechanisms. It is possible to combine biometric identification, facial recognition, and voice recognition to make sure that only authorized employees may use or access robot systems. In addition to preventing unauthorized use, this also protects delicate processes and data.

**Security Based on Behavior:** For industrial robots, AI can build a baseline of typical behavior. Any departure from this norm is marked as posing a risk to security. For instance, AI can initiate alarms and actions to reduce the danger if a robot starts reading unauthorized information or moving in unusual ways.

**Protocols for Secure Communication:** For industrial robots, AI can assist in creating and implementing secure communication protocols. In order to make it almost difficult for bad actors to intercept or change crucial information, it can encrypt data transferred between robots, controllers, and central systems.

**Challenges and Considerations:** Even while AI has a lot to offer in terms of improving industrial robots security, there are still certain issues to be aware of. As weaknesses in AI systems might be exploited to undermine security, it is imperative to ensure that AI itself is protected from such assaults. To find a balance between innovation and responsible usage, it is also important to carefully consider the ethical and legal ramifications of deploying AI in security measures.

As a result, the incorporation of AI into industrial robots is a crucial step towards strengthening security in a setting where the stakes are always growing. Industrial robots are becoming essential parts of complex networks that power our economies, rather than being separate entities. It is essential to protect these systems against a variety of hazards, both physical and digital. Industrial robots are now equipped with an array of tools and skills thanks to AI, including the ability to recognize hazards, adjust to changing conditions, and strengthen security measures. Industrial robots can confidently function in a world that is becoming more complicated and linked by utilizing AI-powered threat detection, anomaly recognition, intrusion prevention, and behavior analysis. The symbiotic link between AI and industrial robotics security will stay at the forefront of technical developments as we continue to push the limits of automation and robots. AI ensures our industrial future and unleash new levels of efficiency and production if the correct policies, tools, and ethical guidelines are put in place (Stadtler, 2014).

## ENHANCING INDUSTRIAL ROBOTICS SECURITY WITH BLOCKCHAIN

The flourishing phase of a smart factory has begun thanks to the Industrial Internet of Things. The old IIoT architecture can no longer effectively handle such massive systems as the number of nodes and network size increase. In order to redesign the conventional IIoT architecture, the blockchain architecture was introduced. A newly developed method for building distributed networks. The main problems with the traditional IIoT architecture are first examined, after which the most recent developments are gathered. Industrial robotics is expanding quickly, and robots are becoming more independent and intelligent. There are also more security issues that need to be resolved as a result of attackers attempting to exploit these systems vulnerabilities. Blockchain technology offers a method to improve the effectiveness and transparency of industrial robotics operations in addition to resolving some of these security concerns (Bang & Choi, 2019).

Figure 2 shows the overview blockchain requirement.

*Figure 2. Use of blockchain*

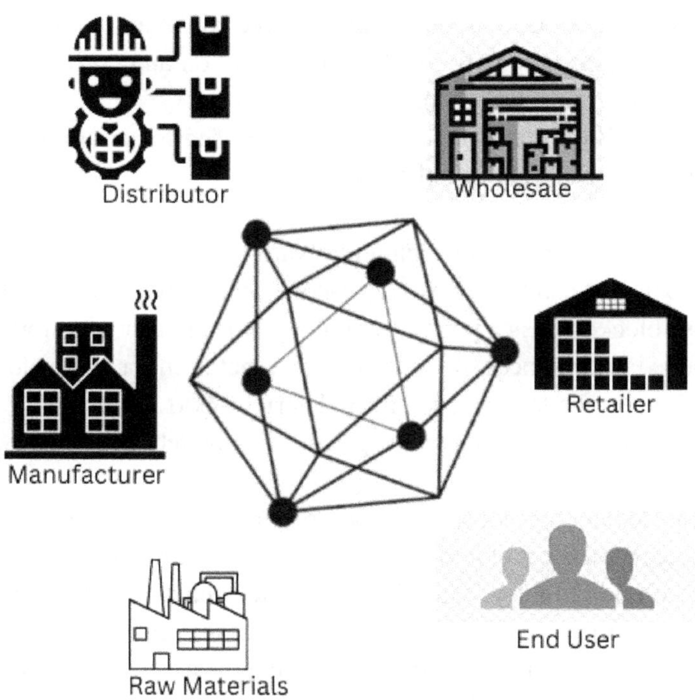

**Blockchain basics:** A distributed ledger technology called blockchain makes it possible to share and store data in a secure manner that cannot be altered. This is accomplished by utilizing a network of computers that keep track of transactions in a shared ledger. The network verifies each transaction before it is put to the ledger, making it incredibly challenging to alter or hack.

Benefits of blockchain for industrial robotics:

Industrial robots can gain a variety of advantages from blockchain, including:

**Data security:** Blockchain may be used to tamper proof store the data of industrial robots. This can aid in limiting the ability of bad actors to access or change critical data, such as customer or production schedules.

**Identity and authentication:** Blockchain may be used to securely and independently verify the identities of industrial robots. This can aid in preventing hostile actors from posing as industrial robots and getting access to systems or facilities without authorization.

**Authorization:** Blockchain may be used to regulate how industrial robots can access resources. This can assist in limiting the ability of bad actors to obtain unauthorized access to information or physical assets.

**Communication Security:** Blockchain may be used to encrypt conversations between industrial robots and other equipment, improving communication security. This can lessen the likelihood of malicious parties listening in on or manipulating with conversations.

**Control security:** Blockchain may be used to store industrial robot control codes in a safe and unchangeable manner. By doing this, it may be possible to thwart hostile actors from commandeering industrial robots and harming or damaging them (Bilancia et al., 2023).

Blockchain may be used to increase the effectiveness and transparency of industrial robotics operations in addition to these security advantages:

**Tracking the movement of products and materials:** Blockchain technology may be used to monitor the flow of goods and resources inside a production, which can enhance inventory control and lower waste.

**Tracking the performance of industrial robots:** Blockchain technology may be used to track the performance of industrial robots, allowing for the early detection and correction of issues that would otherwise result in accidents or property damage.

**Automating payments:** Blockchain technology may be used to streamline and lower the cost of payments made between industrial robots and other devices.

Challenges and Future Outlook:

Industrial robots may benefit from blockchain in a variety of ways, but there are several obstacles that must be overcome before it can be extensively used. These difficulties include:

**Scalability:** Because blockchain is a young technology, it is unclear how it will develop to meet the needs of industrial robots. Performance: Blockchain has a reputation for being sluggish, and it is not yet apparent how to speed it up to the level needed for industrial robots.

**Cost:** Because blockchain technology may be pricey, it's not yet apparent how to make it more accessible to industrial robotics.

The advantages of blockchain for industrial robots are substantial, notwithstanding these difficulties. As the technology advances and the problems are solved, we can anticipate a growing adoption of blockchain in this field. One solution to these issues is to use a side chain or a hybrid blockchain-centralized system. However, these solutions are not without their own challenges. Blockchain is a secure technology, but assaults are still conceivable. Malicious parties may hack into the blockchain network or the authentication and authorization server. Legislation to control blockchain-based systems is still being created. Prior to putting a blockchain-based system into place, it is essential to follow all relevant rules and regulations. Blockchain is a new technology, thus it's possible that corporate players won't like it. It is critical to establish stakeholder trust in the technology and to inform them of the benefits of blockchain. Industrial robotics security might be revolutionized by the formidable technology of blockchain.

## ENHANCING INDUSTRIAL ROBOTICS PERFORMANCE WITH BLOCKCHAIN

Industrial robotics refers to the use of robots in manufacturing and industrial settings to automate tasks and processes. These robots are designed to perform a wide range of functions, from simple repetitive tasks to complex, precise operations and Blockchains a decentralized and distributed ledger technology that underlies cryptocurrencies like Bitcoin.

**Articulated Robots:** These robots have multiple joints and resemble a human arm. They are highly versatile and commonly used in manufacturing, welding, and assembly.

**SCARA Robots:** Selective Compliance Assembly Robot Arm (SCARA) robots are known for their high speed and precision. They are often used in applications like pick-and-place, assembly, and packaging.

**Cartesian/Gantry Robots:** These robots move in straight lines along the X, Y, and Z axes. They are used in tasks that require precise positioning, such as CNC machining and 3D printing.

**Delta Robots:** Delta robots are designed for high-speed tasks like pick-and-place in the food and electronics industries.

**Collaborative Robots (Cobots):** These robots are designed to work alongside humans safely. They have sensors and safety features that enable close collaboration without risk to human workers (Xiao et al., 2023).

There are various sector of industrial robots described in Table 1.

The performance of all the sector can have a boost if it is combined by the properties of blockchain, which are mentioned below:

**Decentralization:** One of the fundamental features of blockchain is its decentralized nature. Instead of relying on a central authority or intermediary, transactions and data are stored on a network of computers that work together to validate and record transactions. This method can enhance security and reduce the risk of a single point of failure.

**Transparency:** Transactions on a blockchain are typically public and can be viewed by anyone with access to the blockchain. This transparency can help to prevent fraud and ensure the integrity of the ledger.

**Security:** Blockchain uses cryptographic techniques to secure data and transactions. Once a transaction is recorded on the blockchain, it is extremely difficult to alter or delete, making it tamper-resistant.

**Immutability:** Blockchain's design makes it resistant to modification of data once it has been added to the ledger. This immutability is a key feature for applications where data integrity is critical.

**Smart Contracts:** Blockchain can support programmable contracts called smart contracts. These self-executing contracts automatically enforce the terms and conditions of an agreement when predefined conditions are met. Smart contracts have applications in various industries, including finance, supply chain management, and legal services.

Application of Blockchain on industrial robotics:

Enhancing the performance of industrial robotics using blockchain technology can provide several benefits, including improved security, transparency, and efficiency in manufacturing and supply chain processes. Here are some ways blockchain can enhance industrial robotics:

*Table 1. Types of industrial robots*

| Robot Type | Description | AI Algorithms/Techniques |
|---|---|---|
| Robotic Arms | Traditional industrial robots with multiple axes for precise positioning and movement. | - Inverse Kinematics<br>- Vision-based guidance<br>- Reinforcement Learning (for control)<br>- Path planning algorithms (e.g., A* or RRT*) |
| Cobots (Collaborative Robots) | Robots designed to work safely alongside humans without safety barriers. | - Force/torque sensing and control<br>- Machine vision for object detection and tracking<br>- Natural language processing (for interaction)<br>- Safety algorithms |
| Mobile Robots | Autonomous robots capable of navigation and mobility in various environments. | - Simultaneous Localization and Mapping (SLAM)<br>- Deep Learning for object recognition<br>- Path planning and obstacle avoidance |
| AGVs (Automated Guided Vehicles) | Vehicles used for material transport within a controlled environment. | - Localization and navigation algorithms<br>- Collision avoidance |
| Drones (Unmanned Aerial Vehicles) | Aerial robots used in various industries for tasks like inspection and monitoring. | - Computer vision for object detection<br>- GPS and IMU-based navigation<br>- Path planning algorithms |
| Exoskeletons | Wearable robotic devices to enhance human strength and endurance for industrial tasks. | - Biomechanical modeling<br>- Control algorithms (e.g., PID controllers)<br>- Human-robot interaction interfaces |
| Automated Inspection Systems | Robots designed for quality control and inspection tasks. | - Machine vision for defect detection<br>- Neural networks for image analysis |
| AI-Enhanced Manufacturing | Systems that integrate AI into various aspects of manufacturing processes. | - Predictive maintenance<br>- Process optimization algorithms<br>- Quality control using machine learning |

## SUPPLY CHAIN MANAGEMENT

Supply chain management (SCM) is a critical component of modern business operations that focuses on the efficient planning, sourcing, production, distribution, and return of goods and services. It involves the coordination of various activities, processes, and stakeholders across the entire supply chain to deliver products to end consumers. Some of the major key points about SCM which can help us to improve performance of robotics are:

**Planning:** This involves forecasting demand, setting production schedules, and defining inventory levels to ensure products are available when and where they are needed.

**Sourcing:** Identifying and selecting suppliers, negotiating contracts, and managing relationships with suppliers to secure the necessary materials and components for production.

**Production:** Managing the manufacturing or assembly process efficiently to meet demand while controlling costs and maintaining quality.

**Cost Efficiency:** Reducing supply chain costs through improved processes, reduced waste, and efficient resource allocation.

**Quality Assurance:** Ensuring the quality and integrity of products throughout the supply chain to meet or exceed customer expectations.

**Lean Supply Chain:** It mainly focuses on minimizing waste and maximizing efficiency to reduce costs while maintaining product quality (Al-Jaroodi & Mohamed, 2019).

*Figure 3. Working of blockchain*

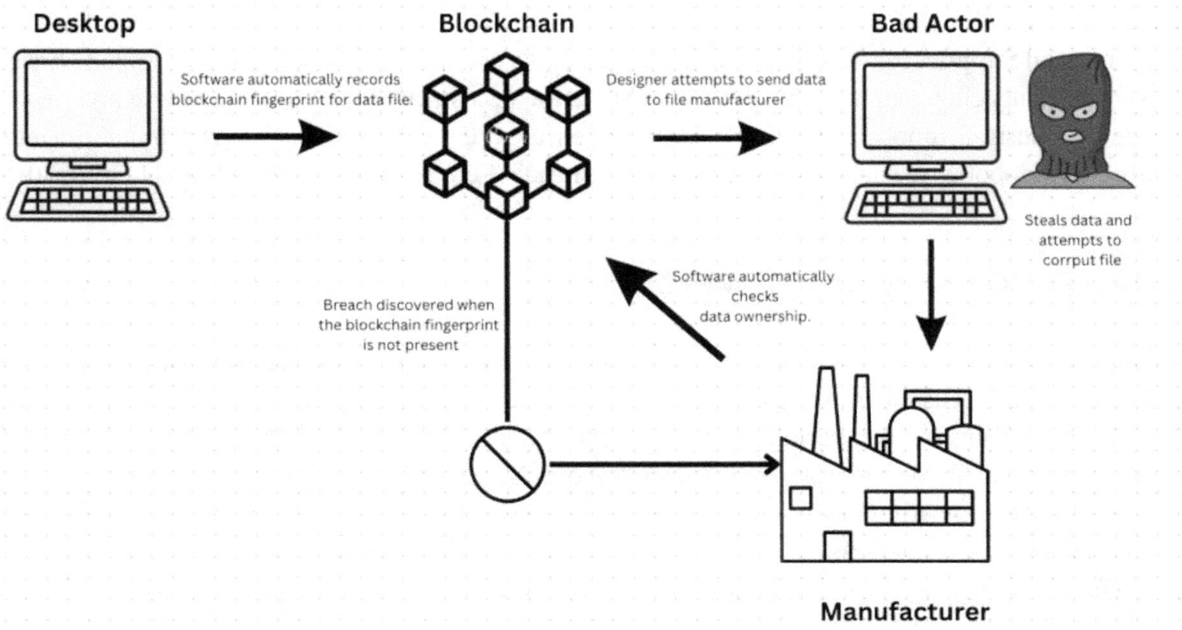


Figure 3 shows the overview of blockchain working.

Supply chain financing, also known as supplier financing or trade finance, refers to financial solutions and strategies that facilitate the funding of supply chain activities. It involves providing working capital or financing options to suppliers, manufacturers, or other participants within the supply chain. Both helps in their own way for improving the performance of the industrial robotics and some of the factors are shown below.

**Optimized Procurement and Sourcing:** SCM ensures the efficient procurement of essential components, parts, and materials required for manufacturing and maintaining industrial robots. Efficient sourcing and supplier management can lead to cost savings and the availability of high-quality components, which are critical for the performance and reliability of robotics systems.

**Inventory Management:** Effective inventory management, a key aspect of SCM, ensures that necessary spare parts for industrial robots are available when needed, reducing downtime. Supply chain financing can support inventory financing, allowing companies to maintain adequate stock levels without straining their working capital.

**Demand Forecasting:** SCM includes demand forecasting, which helps in predicting the need for robotic systems or components. Accurate forecasts enable companies to plan their production and inventory levels effectively. This forecasting can also support the planning of robot deployment to meet changing production requirements.

**Logistics and Distribution:** SCM optimizes logistics and distribution processes, ensuring that industrial robots, components, and spare parts are delivered to their destinations in a timely and cost-effective manner. Efficient logistics reduce lead times, enabling faster response to maintenance or production needs.

**Quality Control:** Quality control practices within SCM help ensure that components used in industrial robots meet the required quality standards. Robust quality control processes help maintain the overall quality and reliability of robotic systems.

**Financial Support for Robotics Investments:** Supply chain financing options, such as trade finance or supplier financing, can help organizations secure the necessary funds for the purchase and maintenance of industrial robots. This financial support ensures that companies can invest in and maintain cutting-edge robotics technology to enhance performance (Fernandez-Carames & Fraga-Lamas, 2019).

## QUALITY CONTROL AND COMPLIANCE

Quality control and compliance are essential aspects of product and service management across various industries. They focus on ensuring that products or services meet predefined standards, specifications, regulations, and customer expectations. Some of the major key points about Quality Control and Compliance which can help us to improve performance of robotics are:

**Consistency:** Ensuring that products or services consistently meet quality standards, specifications, and performance expectations.

**Defect Prevention:** Identifying and addressing potential issues in the production or service delivery process to prevent defects from occurring.

**Customer Satisfaction:** Meeting or exceeding customer expectations in terms of quality, reliability, and performance.

**Cost Reduction:** Minimizing costs associated with rework, repairs, and warranty claims by detecting and rectifying quality issues early in the process.

**Statistical Process Control**: The use of statistical techniques to monitor and control production processes to maintain quality.

**Quality Assurance:** A proactive approach that focuses on process improvement and prevention of defects rather than detection and correction.

**Six Sigma:** A data-driven methodology that aims to minimize defects and variations in processes.

There are several ways in which Quality Control and Compliance help in improving the performance of industrial robotics.

**Consistent Product Quality:** Quality control processes verify that the components used in the manufacturing of industrial robots meet quality standards. This ensures that the robots themselves are built to high-quality specifications. Industrial robots, when operating, can consistently produce products or perform tasks to the same high-quality standards. This is especially important in industries such as automotive manufacturing and electronics assembly.

**Reliability and Uptime:** Quality control practices extend to regular maintenance and inspection of industrial robots. Scheduled maintenance helps prevent unexpected breakdowns and downtime. Compliance with maintenance schedules and quality control standards ensures that robots operate reliably, contributing to higher uptime and improved overall performance.

**Data-Driven Optimization:** Quality control processes often involve collecting data from robotic operations. This data can be analyzed to identify areas for improvement and optimization. By complying with data collection and analysis procedures, organizations can fine-tune robotic operations to enhance performance, reduce errors, and increase efficiency.

**Compliance with Industry Standards:** Industrial robotics often need to adhere to industry-specific standards, such as ISO 9001 for quality management or ISO 13485 for medical devices. Compliance with these standards not only ensures quality but also opens opportunities for organizations to enter global markets and work with customers who require adherence to specific standards.

**Continuous Improvement:** Both quality control and compliance encourage a culture of continuous improvement. Organizations that routinely monitor and audit their robotic processes can identify areas where adjustments are needed to enhance performance. Compliance audits can uncover areas for improvement and lead to corrective actions that improve overall performance (Queralta et al., 2020).

## REAL-TIME MONITORING AND CONTROL

Real-time monitoring and control play a crucial role in providing performance support to industrial robotics by offering several benefits that enhance efficiency, reliability, and overall effectiveness. This also helps in various ways for enhancing the performance of the industrial robotics:

**Immediate Issue Detection:** Real-time monitoring allows for the continuous collection and analysis of data from industrial robots, including sensor readings, performance metrics, and operational parameters.

**Proactive Maintenance:** Real-time data enables predictive maintenance strategies. By analyzing the condition of robotic components and systems in real time, maintenance teams can anticipate when maintenance or repairs are needed.

**Performance Optimization:** Real-time monitoring provides insights into how industrial robots are performing during various tasks and processes.

In conclusion, industrial robotics, blockchain technology, supply chain management, quality control, and compliance are interconnected elements that play vital roles in modern business operations. When integrated effectively, they contribute to improved productivity, quality, efficiency, safety, and overall performance, ensuring that businesses remain competitive and adaptable in today's rapidly evolving industrial landscape. We also discussed how real-time monitoring and control enhance the performance of industrial robotics. Real-time monitoring detects issues, supports proactive maintenance, optimizes performance, ensures safety, and enables data-driven decision-making, all of which contribute to the efficiency and reliability of industrial robots.

## ENHANCING INDUSTRIAL ROBOTICS PERFORMANCE WITH AI

Industrial robots, formerly stationary and single-focused machines, have undergone a remarkable transformation. They are now endowed with the capabilities of AI, enabling them to adapt, learn, and optimize their performance in unprecedented ways. This fusion of AI and robotics is reshaping manufacturing, assembly, and various industries on a global scale. AI is elevating the performance of industrial robotics, delve into real-world applications, and assess the potential for future advancements in this captivating field. The convergence of AI and industrial robotics has ushered in a new era of productivity and efficiency. Traditional industrial robots were highly specialized, excelling at repetitive tasks with precision, yet confined to predetermined functions. However, AI has revolutionized the landscape by introducing adaptability, flexibility, and autonomy to these machines. Combining AI and Industrial Robotics helps to improve the efficiency and accuracy of the system (Al-Jaroodi & Mohamed, 2019).

*Figure 4. Security measures powered by AI and blockchain*

Figure 4 shows the collaboration of AI and Blockchain in industrial robotics with respect to security aspects.

## LEARNING AND ADAPTATION

AI's influence on industrial robotics primarily stems from its ability to acquire knowledge and adapt. Machine learning algorithms empower robots to continually enhance their performance. They can scrutinize data from diverse sensors, discern patterns, and make real-time adjustments to their actions. For example, in a manufacturing environment, an AI-equipped robot can identify defects in products and modify its processes on-the-fly to minimize defects, thereby elevating product quality.

**Predictive Maintenance:** AI also plays a pivotal role in predictive maintenance, a vital aspect of avoiding costly downtime in industrial settings. Robots can autonomously monitor their own state and performance, courtesy of sensors and AI algorithms. By scrutinizing this data, they can forecast when components are likely to malfunction and schedule maintenance in advance, thus curtailing disruptions in the production workflow.

**Autonomous Decision-Making:** AI empowers industrial robots with autonomous decision-making capabilities. This means that robots can make choices based on the information available to them, such as adapting to changes in the environment or choosing the most efficient path for completing a task. In logistics and warehousing, for instance, AI-driven robots can optimize routes for goods transportation, reducing travel time and energy consumption (Zhang et al., 2019).

## REAL-WORLD APPLICATIONS

The impact of AI-enhanced industrial robotics is not limited to theory; it has been put into practice across various industries, revolutionizing the way work is done.

**Automotive Manufacturing:** The automotive industry has been at the forefront of adopting AI in robotics. Car manufacturers use AI-driven robots for tasks such as welding, painting, and assembly. These robots can adapt to variations in car models and configurations, making production lines more flexible and efficient. This adaptability is particularly valuable in an industry where frequent model changes and customizations are common.

**E-commerce and Warehousing:** In e-commerce and warehousing, AI-powered robots have transformed logistics operations. Companies like Amazon use robots to automate the movement of goods within their fulfillment centers. These robots can autonomously navigate through the vast warehouses, locate items, and transport them to packing stations. This level of automation has significantly reduced the time and labor required for order fulfillment.

**Healthcare:** AI-enhanced robots have also found applications in the healthcare sector. Surgical robots, for instance, can assist surgeons with precision tasks, leading to safer and more efficient surgeries. Additionally, robots equipped with AI algorithms can be used for tasks such as medication delivery in hospitals or handling hazardous materials in research laboratories.

**Agriculture:** In agriculture, AI-driven robots are revolutionizing farming practices. Autonomous tractors and drones equipped with AI can analyze crop data and make decisions about planting, fertilizing, and harvesting. This not only increases efficiency but also allows for more sustainable and precise farming practices, reducing the use of resources and chemicals (Lopes & Alexandre, 2018).

## CHALLENGES OF AI AND ROBOTICS

While the integration of AI and industrial robotics holds immense promise, it is not without its challenges and considerations.

**Cost of Implementation:** One significant challenge is the initial cost of implementing AI-enhanced robotics systems. The integration of AI algorithms, sensors, and other advanced technologies can be expensive. Small and medium-sized enterprises may find it challenging to justify the upfront investment, even though the long-term benefits are substantial.

**Skilled Workforce:** AI-enhanced robotics systems require a skilled workforce to design, program, and maintain them. There is a growing demand for professionals with expertise in both robotics and AI, and industries must invest in training and education to meet this demand.

**Ethical and Safety Concerns:** As robots become more autonomous, ethical and safety concerns arise. There are questions about liability in the event of accidents involving autonomous robots and concerns about AI making biased decisions. Clear regulations and ethical guidelines are needed to address these issues (Alladi et al., 2020).

Table 2 discusses four case studies on applications of blockchain methodology.

*Table 2. Case studies on blockchain applications*

| Case Study | Method/Methodology | Description | Challenges Encountered | Key Takeaways |
|---|---|---|---|---|
| Proposing a low-cost self-testing and tracking system using AI and Blockchain for COVID-19 and other diseases, especially in resource-limited areas. | The project involves implementing Point-of-Care diagnostics using AI and Blockchain technologies, and developing a self-testing mobile application for local financial and supply chain issues. | The study suggests implementing a self-testing and tracking system to mitigate the impact of COVID-19 in sub-Saharan Africa, addressing the weaknesses in healthcare systems and surveillance. | Not explicitly mentioned in the provided content. | The proposed technology can effectively combat COVID-19 and similar epidemics in resource-limited areas. It emphasizes the importance of localized diagnostics. |
| Demonstrating the efficient management of the supply chain through the integration of blockchain technology with AI and Smart Contracts. Identifying challenges and discussing solutions. | The study utilizes a qualitative data analysis approach, integrating IPFS, blockchain, and smart contracts for data security, real-time data collection, and secure data exchange frameworks. | The study explores the integration of blockchain technology with AI and Smart Contracts to improve the security and transparency of healthcare supply chains. | Standardized protocols, blockchain management guidelines, and modern analytics techniques are needed for interoperability, data visibility, privacy, and real-time data collection, while integrating with IoT devices. | The study emphasizes the potential of blockchain and AI in enhancing supply chain transparency and security, emphasizing the need for strategic implementation and stakeholder collaboration. |
| Healthcare 4.0 aims to digitize healthcare using blockchain, AI, and IoT, improving reliability, convenience, and transparency through wearable sensors for real-time patient monitoring. | Wearable devices enable real-time health data tracking, utilizing blockchain for data security and smart contracts for data integrity. AI, ML, and machine learning algorithms aid in patient diagnosis and care. | The study explores Healthcare 4.0, a digital transformation utilizing wearable devices for remote patient monitoring and blockchain for data security, discussing its applications and methodology. | Challenges include scalability, latency, and interoperability. Solutions include open standards, efficiency-limiting validation participation, and patient data privacy through appropriate blockchain models. | The technology discussed has the potential to revolutionize healthcare by enabling accurate diagnosis and personalized treatment, requiring stakeholder collaboration for data-driven decision-making and improved patient outcomes. |

## SUPPLY CHAIN MANAGEMENT AND TRACEABILITY IN HEALTHCARE ROBOTICS USING BLOCKCHAIN

Robotic systems are being used more often in healthcare settings to perform tasks like patient care, surgery, and medication dispensing. These robots frequently need a constant supply of parts and consumables. For the sake of patient safety and operational effectiveness, it is imperative to guarantee the accessibility, verifiability, and legitimacy of these supplies. These problems can be solved with blockchain technology.

**Supply Chain Complexity:** A complicated supply chain including numerous suppliers, manufacturers, distributors, and healthcare facilities is frequently necessary for healthcare robotics. It is difficult to plan and monitor the flow of necessary supplies and parts.

**Traceability:** To guarantee that consumables (such as surgical instruments, drugs, and disposables) fulfill quality and safety requirements, traceability is crucial in the healthcare industry. For both patient safety and regulatory compliance, it is essential to be able to track every component back to its original source.

**Counterfeit Products:** In the healthcare industry, counterfeit medical supplies are a serious problem. By using blockchain technology, the risk of fake goods getting into the supply chain can be decreased by creating an unchangeable record of every product's journey from the manufacturer to the final consumer.

**Data Integrity:** A tamper-proof ledger, made possible by blockchain technology, can guarantee the security and integrity of product information and supply chain transaction records. This is especially crucial for the healthcare industry, as data integrity is paramount.

**Interoperability:** It is difficult to integrate blockchain technology with current robotics and healthcare systems. One of the main challenges is ensuring that various systems can securely share and communicate with one another.

**Regulatory Compliance:** Since the healthcare industry is heavily regulated, putting blockchain technology to use should comply with current laws and regulations, such HIPAA in the US. Using blockchain to ensure compliance can be difficult.

**User Privacy:** It is critical to strike a balance between the requirement for supply chain openness and patients' and healthcare professionals' privacy. Blockchain technology can facilitate selective information sharing, safeguarding private information and guaranteeing transparency when required.

**Smart Contracts:** Smart contracts on blockchain technology can be used to automate supply chain procedures like quality control inspections and payment upon delivery. However, certain knowledge is needed to create and manage smart contracts.

**Data Ownership:** It might be difficult to determine who owns and controls data on a blockchain. This entails negotiating agreements amongst different players in the healthcare industry.

**Scalability:** As a healthcare supply chain's volume of transactions and data increases, blockchain technology may run into scalability problems. It's crucial to make sure the blockchain network can support the demand.

**Education and Training:** Training is required for healthcare professionals and robotics operators to use blockchain technology safely and effectively. Investment in training and education initiatives is necessary for this.

## FUTURE WORK

The future of AI-enhanced industrial robotics is bright, with several exciting developments on the horizon. Some of the research areas are described below:

**Human-Robot Collaboration:** One of the most promising areas of development is human-robot collaboration. As robots become more intelligent and adaptable, they can work alongside humans more effectively. This collaboration can lead to safer and more productive work environments, particularly in industries like manufacturing and healthcare.

**Edge Computing:** Edge computing, which involves processing data closer to where it's generated, is poised to enhance AI in industrial robotics. Robots equipped with edge AI can make faster decisions without relying on distant data centers, which is critical for real-time applications.

**Swarm Robotics:** Swarm robotics involves coordinating multiple robots to work together in a decentralized manner. AI algorithms can enable swarm robotics, allowing a group of robots to collaborate on complex tasks, such as search and rescue operations or environmental monitoring (Aditya, Singh, Singh, & Kalla, 2021).

## CONCLUSION

The integration of AI and industrial robotics is transforming industries across the spectrum. These intelligent machines are no longer limited to repetitive tasks but have evolved into adaptable, autonomous, and versatile tools. Real-world applications in automotive manufacturing, e-commerce, healthcare, and agriculture showcase the tangible benefits of this synergy. However, challenges related to cost, workforce, ethics, and safety must be addressed as the adoption of AI-enhanced robotics continues to grow. Looking ahead, the future of industrial robotics holds promise in areas such as human-robot collaboration, edge computing, and swarm robotics. The evolving landscape of AI-enhanced industrial robotics is crucial to strike a balance between innovation and responsibility, harnessing the potential of these machines to improve productivity and quality of life while ensuring they operate successfully.

## REFERENCES

Aditya, U. S., Singh, R., Singh, P. K., & Kalla, A. (2021). A survey on blockchain in robotics: Issues, opportunities, challenges and future directions. *Journal of Network and Computer Applications*, *196*, 103245. doi:10.1016/j.jnca.2021.103245

Al-Jaroodi, J., & Mohamed, N. (2019a). Blockchain in industries: A survey. *IEEE Access : Practical Innovations, Open Solutions*, *7*, 36500–36515. doi:10.1109/ACCESS.2019.2903554

Al-Jaroodi, J., & Mohamed, N. (2019b, January). Industrial applications of blockchain. In *2019 IEEE 9th Annual Computing and Communication Workshop and Conference (CCWC)* (pp. 0550-0555). IEEE. 10.1109/CCWC.2019.8666530

Alladi, T., Chamola, V., Sahu, N., & Guizani, M. (2020). Applications of blockchain in unmanned aerial vehicles: A review. *Vehicular Communications*, *23*, 100249. doi:10.1016/j.vehcom.2020.100249

Alsamhi, S. H., Lee, B., Guizani, M., Kumar, N., Qiao, Y., & Liu, X. (2021). Blockchain for decentralized multi-drone to combat COVID-19 and future pandemics: Framework and proposed solutions. *Transactions on Emerging Telecommunications Technologies*, *32*(9), e4255. doi:10.1002/ett.4255

Aoun, A., Ilinca, A., Ghandour, M., & Ibrahim, H. (2021). A review of Industry 4.0 characteristics and challenges, with potential improvements using blockchain technology. *Computers & Industrial Engineering*, *162*, 107746. doi:10.1016/j.cie.2021.107746

Bang, J., & Choi, M. J. (2019, September). Design and implementation of storage system for real-time blockchain network monitoring system. In *2019 20th Asia-Pacific Network Operations and Management Symposium (APNOMS)* (pp. 1-4). IEEE. 10.23919/APNOMS.2019.8892967

Benotsmane, R., Dudás, L., & Kovács, G. (2020). Survey on artificial intelligence algorithms used in industrial robotics. *Multidiszciplináris Tudományok: A Miskolci Egyetem Közleménye, 10*(4), 194-205.

Bilancia, P., Schmidt, J., Raffaeli, R., Peruzzini, M., & Pellicciari, M. (2023). An Overview of Industrial Robots Control and Programming Approaches. *Applied Sciences (Basel, Switzerland), 13*(4), 2582. doi:10.3390/app13042582

Dewasiri, N. J., Karunarathne, K. S. S. N., Menon, S., Jayarathne, P. G. S. A., & Rathnasiri, M. S. H. (2023). Fusion of Artificial Intelligence and Blockchain in the Banking Industry: Current Application, Adoption, and Future Challenges. In Transformation for Sustainable Business and Management Practices: Exploring the Spectrum of Industry 5.0 (pp. 293-307). Emerald Publishing Limited.

Falayi, A., Wang, Q., Liao, W., & Yu, W. (2023). Survey of Distributed and Decentralized IoT Securities: Approaches Using Deep Learning and Blockchain Technology. *Future Internet, 15*(5), 178. doi:10.3390/fi15050178

Fernandez-Carames, T. M., & Fraga-Lamas, P. (2019). A review on the application of blockchain to the next generation of cybersecure industry 4.0 smart factories. *IEEE Access : Practical Innovations, Open Solutions, 7*, 45201–45218. doi:10.1109/ACCESS.2019.2908780

Hannan, S. A. (2023). Challenges of Blockchain Technology using Artificial Intelligence in Healthcare System. *International Journal of Innovative Research in Science, Engineering and Technology, 12*(01), 64–74.

Hemdan, E. E. D., El-Shafai, W., & Sayed, A. (2023). Integrating Digital Twins with IoT-Based Blockchain: Concept, Architecture, Challenges, and Future Scope. *Wireless Personal Communications, 131*(3), 1–24. doi:10.100711277-023-10538-6 PMID:37360142

Islam, M. A. (2023). *AI & Blockchain as sustainable teaching and learning tools to cope with the 4IR.* arXiv preprint arXiv:2305.01088.

Islam, M. J., Rahman, A., Kabir, S., Karim, M. R., Acharjee, U. K., Nasir, M. K., Band, S. S., Sookhak, M., & Wu, S. (2021). Blockchain-SDN-based energy-aware and distributed secure architecture for IoT in smart cities. *IEEE Internet of Things Journal, 9*(5), 3850–3864. doi:10.1109/JIOT.2021.3100797

Khanfar, A. A., Iranmanesh, M., Ghobakhloo, M., Senali, M. G., & Fathi, M. (2021). Applications of blockchain technology in sustainable manufacturing and supply chain management: A systematic review. *Sustainability (Basel), 13*(14), 7870. doi:10.3390u13147870

Kumar, G., Saha, R., Buchanan, W. J., Geetha, G., Thomas, R., Rai, M. K., Kim, T.-H., & Alazab, M. (2020). Decentralized accessibility of e-commerce products through blockchain technology. *Sustainable Cities and Society, 62*, 102361. doi:10.1016/j.scs.2020.102361

Kumar, S., & Saha, G. K. (2020). *Blockchain Technology Testing Challenges and Approaches*. Academic Press.

Latif, S., Idrees, Z., Ahmad, J., Zheng, L., & Zou, Z. (2021). A blockchain-based architecture for secure and trustworthy operations in the industrial Internet of Things. *Journal of Industrial Information Integration, 21*, 100190. doi:10.1016/j.jii.2020.100190

Latif, S., Idrees, Z., Huma, Z., & Ahmad, J. (2021). Blockchain technology for the industrial Internet of Things: A comprehensive survey on security challenges, architectures, applications, and future research directions. *Transactions on Emerging Telecommunications Technologies, 32*(11), e4337. doi:10.1002/ett.4337

Lopes, V., & Alexandre, L. A. (2018). *An overview of blockchain integration with robotics and artificial intelligence.* arXiv preprint arXiv:1810.00329.

Mahyuni, L. P., Adrian, R., Darma, G. S., Krisnawijaya, N. N. K., Dewi, I. G. A. A. P., & Permana, G. P. L. (2020). Mapping the potentials of blockchain in improving supply chain performance. *Cogent Business & Management, 7*(1), 1788329. doi:10.1080/23311975.2020.1788329

Park, A., & Li, H. (2021). The effect of blockchain technology on supply chain sustainability performances. *Sustainability (Basel), 13*(4), 1726. doi:10.3390u13041726

Queralta, J. P., Qingqing, L., Zou, Z., & Westerlund, T. (2020, April). Enhancing autonomy with blockchain and multi-access edge computing in distributed robotic systems. In *2020 Fifth International Conference on Fog and Mobile Edge Computing (FMEC)* (pp. 180-187). IEEE. 10.1109/FMEC49853.2020.9144809

Raffik, R., Sathya, R. R., Vaishali, V., & Balavedhaa, S. (2023, June). Industry 5.0: Enhancing Human-Robot Collaboration through Collaborative Robots–A Review. In *2023 2nd International Conference on Advancements in Electrical, Electronics, Communication, Computing and Automation (ICAECA)* (pp. 1-6). IEEE.

Rahman, A., Sara, U., Kundu, D., Islam, S., Islam, M. J., Hasan, M., & Nasir, M. K. (2020). *Distbsdoindustry: Enhancing security in industry 4.0 services based on distributed blockchain through software defined networking-iot enabled architecture.* arXiv preprint arXiv:2012.10011.

Rane, S., Shah, P., & Sekhar, R. (2022, August). Survey of technologies for industry 4.0. In *2022 6th International Conference on Computing, Communication, Control And Automation (ICCUBEA* (pp. 1-6). IEEE.

Shankar, S., Varadam, D., Agrawal, H. & E, N. (2022). Blockchain for IoT and Big Data Applications: A Comprehensive Survey on Security Issues. In M. Niranjanamurthy, S. Peng, E. Naresh, S. Jayasimha & V. Balas (Ed.), *Advances in Industry 4.0: Concepts and Applications* (pp. 65-86). De Gruyter. doi:10.1515/9783110725490-004

Singh, P., & Singh, N. (2020). Blockchain with IoT and AI: A review of agriculture and healthcare. *International Journal of Applied Evolutionary Computation, 11*(4), 13–27. doi:10.4018/IJAEC.2020100102

Stadtler, H. (2014). Supply chain management: An overview. *Supply chain management and advanced planning: Concepts, models, software, and case studies*, 3-28.

Tailor, R. K., Pareek, R., & Khang, A. (2022). Robot process automation in blockchain. *The data-driven blockchain ecosystem: Fundamentals, applications, and emerging technologies*, (pp. 149-164).

Thanki, R., & Joshi, P. (2023). Advanced Technologies for Industrial Applications. In *Advanced Technologies for Industrial Applications* (pp. 73–95). Springer International Publishing. doi:10.1007/978-3-031-33238-8_6

Uma, S. (2023). Blockchain and AI: Disruptive Digital Technologies in Designing the Potential Growth of Healthcare Industries. In *AI and Blockchain in Healthcare* (pp. 137–150). Springer Nature Singapore. doi:10.1007/978-981-99-0377-1_9

Vignesh Saravanan, K., Jothi Thilaga, P., Kavipriya, S., & Vijayalakshmi, K. (2023). Data Protection and Security Enhancement in Cyber-Physical Systems Using AI and Blockchain. In *AI Models for Blockchain-Based Intelligent Networks in IoT Systems: Concepts, Methodologies, Tools, and Applications* (pp. 285–325). Springer International Publishing. doi:10.1007/978-3-031-31952-5_13

Xiao, P., Salleh, M. I., Zaidan, B. B., & Xuelan, Y. (2023). Research on risk assessment of Blockchain-driven Supply Chain finance: A systematic review. *Computers & Industrial Engineering*, *176*, 108990. doi:10.1016/j.cie.2023.108990

Zhang, R., Xue, R., & Liu, L. (2019). Security and privacy on blockchain. *ACM Computing Surveys*, *52*(3), 1–34. doi:10.1145/3316481

# Chapter 4
# Intelligent Resource Allocation and Optimization for Industrial Robotics Using AI and Blockchain

**Tarun Kumar Vashishth**
https://orcid.org/0000-0001-9916-9575
*IIMT University, India*

**Vikas Sharma**
https://orcid.org/0000-0001-8173-4548
*IIMT University, India*

**Kewal Krishan Sharma**
*IIMT University, India*

**Bhupendra Kumar**
*IIMT University, India*

**Sachin Chaudhary**
https://orcid.org/0000-0002-8415-0043
*IIMT University, India*

**Rajneesh Panwar**
*IIMT University, India*

## ABSTRACT

*This chapter focuses on the application of intelligent resource allocation and optimization techniques for industrial robotics systems using the synergistic integration of artificial intelligence (AI) and blockchain technologies. Efficient resource allocation is crucial for maximizing the performance and productivity of industrial robotics, and AI-based approaches offer the ability to dynamically allocate resources based on real-time data and system requirements. Additionally, blockchain technology provides a decentralized and secure platform for recording and verifying resource allocation transactions, ensuring transparency and trust in the allocation process. The chapter explores various AI algorithms and models that can be employed for resource allocation and optimization in industrial robotics, including machine learning, evolutionary algorithms, and reinforcement learning. Furthermore, the chapter investigates how blockchain technology can enhance resource allocation and optimization by providing a distributed ledger for recording and verifying resource transactions.*

DOI: 10.4018/979-8-3693-0659-8.ch004

## 1. INTRODUCTION TO INDUSTRIAL ROBOTICS

Industrial robotics represents a transformative force in manufacturing and automation. Over the years, these sophisticated machines have evolved from simple mechanical arms to highly advanced systems capable of intricate tasks. This introduction serves as a gateway to understanding the world of industrial robotics, its historical evolution, current applications, and the profound impact it has on industries worldwide. As we delve into the depths of industrial robotics, we will explore their key components, working principles, and the pivotal role they play in optimizing production processes and enhancing efficiency. Moreover, we will discuss the latest trends and future prospects in the field, emphasizing the continuous innovation that drives this fascinating industry forward. So, let's embark on a journey to unravel the fascinating realm of industrial robotics.

## 2. SIGNIFICANCE OF RESOURCE ALLOCATION AND OPTIMIZATION IN INDUSTRIAL ROBOTICS

Efficient resource allocation and optimization are critical components of modern industrial robotics systems. In the rapidly evolving landscape of manufacturing and automation, these processes hold significant importance for several key reasons:

i. Enhanced Efficiency: Proper resource allocation ensures that industrial robots are used optimally, reducing downtime and increasing overall productivity. This efficiency is vital in meeting production targets and maintaining competitiveness in the market.
ii. Cost Savings: Optimized resource allocation can lead to cost savings by reducing unnecessary energy consumption, wear and tear on equipment, and the need for human intervention. This can have a direct impact on a company's bottom line.
iii. Improved Quality: By allocating resources effectively, industrial robots can consistently produce high-quality products, reducing defects and rework. This is crucial in industries where product quality is paramount, such as automotive and electronics manufacturing.
iv. Resource Scalability: As production demands fluctuate, flexible resource allocation allows for easy scalability. Industrial robots can adapt to changes in production volume, ensuring that resources are neither over utilized nor underutilized.
v. Data-Driven Decision Making: AI and blockchain technologies enable data collection and analysis on a granular level. This data can be used to make informed decisions about resource allocation and optimization, leading to continuous process improvement.
vi. Enhanced Safety: Efficient resource allocation can also improve workplace safety. Robots can be allocated to handle dangerous or repetitive tasks, reducing the risk of accidents and injuries to human workers.
vii. Sustainability: Optimized resource allocation can contribute to sustainability goals by minimizing energy consumption and waste. This aligns with the growing focus on environmentally friendly manufacturing practices.
viii. Competitive Advantage: Companies that excel in resource allocation and optimization are often more competitive in their respective industries. They can offer better prices, shorter lead times, and higher quality products, attracting more customers.

*Figure 1. Critical components of modern industrial robotics systems*

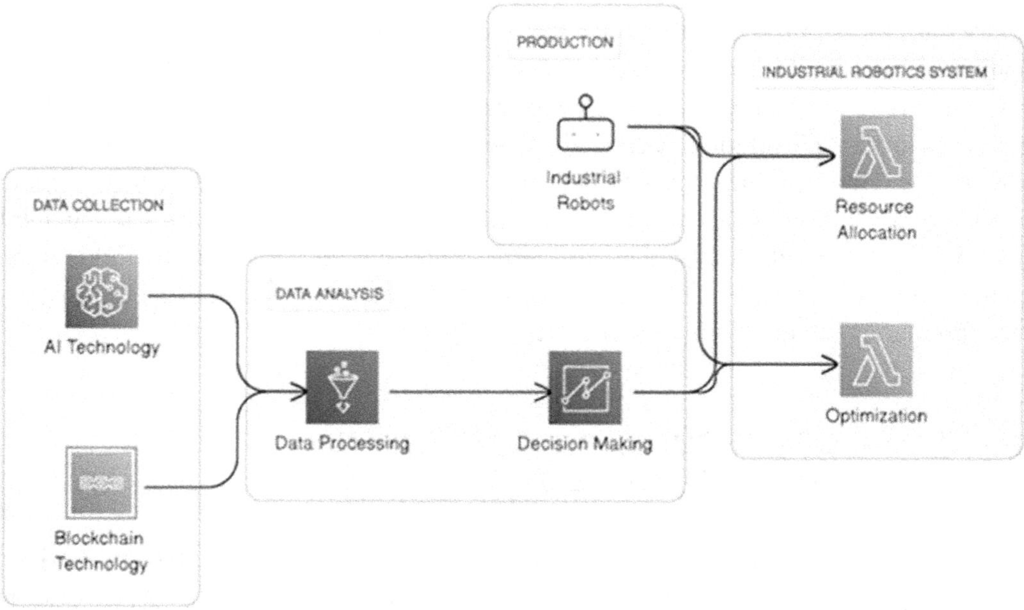


## 3. OVERVIEW OF AI AND BLOCKCHAIN TECHNOLOGIES

Artificial Intelligence (AI) and Blockchain are two transformative technologies that have gained immense prominence in recent years. Each technology has distinct characteristics and applications, and when combined, they can create powerful synergies. Here's an overview of both AI and Blockchain:

### 3.1 Artificial Intelligence (AI)

Artificial Intelligence (AI) plays a pivotal role in the context of intelligent resource allocation and optimization for industrial robotics, especially when combined with blockchain technology. This integrated approach leverages the power of AI to enhance the allocation and management of critical resources in industrial settings, offering an efficient, data-driven, and transparent ecosystem that ensures optimal performance and cost-effectiveness. At its core, AI in this context refers to the deployment of advanced algorithms, machine learning models, and data analytics to make intelligent decisions regarding resource allocation and optimization in industrial robotics. Industrial settings often involve complex and dynamic processes that require the efficient allocation of resources such as robotic assets, energy, labor, and materials. AI systems, armed with the ability to process vast amounts of data and adapt to changing conditions, can dynamically allocate these resources in real-time, optimizing operations and improving overall efficiency.

Intelligent resource allocation is a multifaceted process. It involves recognizing the available resources, understanding the current demands, and making decisions that ensure those resources are used optimally. In the case of industrial robotics, this may involve determining which robots should perform specific tasks, how much energy each robot should consume, and when to allocate additional labor or materials to the production line. AI achieves this by analyzing historical and real-time data, including production metrics, sensor data, and maintenance records, to make informed resource allocation decisions. One of the key benefits of integrating blockchain technology into this process is to ensure transparency, security, and traceability in resource allocation. Blockchain is essentially a distributed ledger that records all transactions in a secure, tamper-resistant manner. This technology is leveraged to provide a transparent and immutable record of resource allocation decisions, which can be crucial in industrial settings where trust and accountability are paramount. Stakeholders, including manufacturers, suppliers, and regulators, can have real-time visibility into resource allocation decisions and transactions, enhancing trust in the supply chain.

Blockchain's role in securing these transactions is especially significant when considering that resource allocation in industrial robotics involves valuable assets, data, and financial transactions. Ensuring the integrity and authenticity of these transactions is critical. Blockchain technology achieves this through its cryptographic mechanisms and decentralized consensus, making it extremely challenging for any unauthorized party to tamper with the records. In practice, AI-driven resource allocation and optimization in industrial robotics can take various forms. For example, machine learning models can be trained to predict equipment failures, allowing maintenance resources to be allocated preemptively, reducing downtime. AI algorithms can also optimize the routing and scheduling of robots within a manufacturing facility, reducing idle time and increasing overall efficiency. Moreover, the integration of blockchain ensures that the data used by AI for resource allocation is accurate and unmodifiable. This combination of AI and blockchain is a powerful tool in ensuring that the allocation of resources in industrial robotics is not only efficient but also secure and transparent, contributing to improved productivity and operational excellence.

AI in the context of intelligent resource allocation and optimization for industrial robotics using blockchain represents a technological synergy that has the potential to revolutionize industrial processes. It empowers organizations to make data-driven decisions, optimize their resource allocation strategies, and secure their transactions, ultimately enhancing efficiency, reducing costs, and building trust in the supply chain. This innovative approach serves as a cornerstone of Industry 4.0, where intelligent resource allocation and optimization are at the forefront of the manufacturing and industrial landscape.

## 3.2 Applications

a) Natural Language Processing (NLP): AI is used in chatbots, language translation, and sentiment analysis.
b) Computer Vision: AI powers facial recognition, object detection, and autonomous vehicles.
c) Healthcare: AI aids in medical diagnosis, drug discovery, and personalized medicine.
d) Finance: AI is employed in fraud detection, algorithmic trading, and credit scoring.
e) Manufacturing: AI-driven robotics and automation improve efficiency and quality.

## 3.3 Blockchain

Blockchain, in the context of intelligent resource allocation and optimization for industrial robotics using AI and blockchain, serves as a foundational technology that enhances transparency, security, and traceability in the management of critical resources within industrial settings. This innovative integra-

*Figure 2. Integration of artificial intelligence (AI) and blockchain*

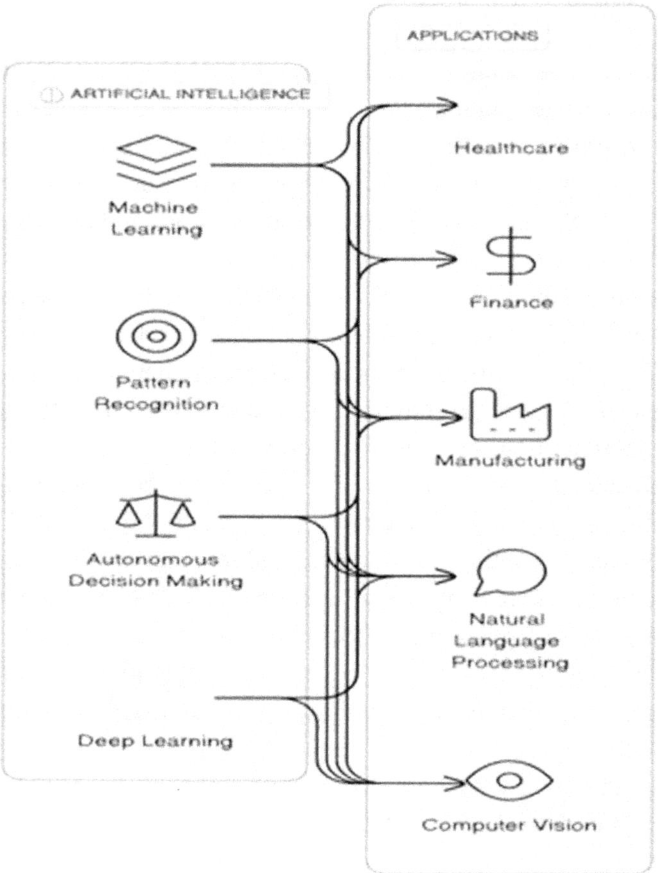


tion of blockchain technology plays a pivotal role in ensuring the reliability and efficiency of resource allocation while leveraging the power of AI to make data-driven decisions.

At its core, blockchain is a decentralized and distributed ledger technology that records transactions across a network of computers, creating a chain of immutable blocks. Each block contains a set of transactions, and once added to the chain, it cannot be altered, providing an unforgeable history of all actions. In the context of resource allocation for industrial robotics, blockchain serves several crucial purposes:

Transparency and Traceability: Blockchain ensures complete transparency in resource allocation decisions and their associated transactions. All stakeholders within the industrial ecosystem, from manufacturers to regulators, have real-time visibility into the allocation of robotic assets, energy, labor, and materials. This transparency enhances accountability and trust within the supply chain.

Security and Trust: Blockchain's security features, including cryptographic encryption and decentralized consensus mechanisms, make it extremely resilient to tampering and fraud. It is a secure and trust-building technology, which is especially important when dealing with valuable resources and sensitive data.

Immutable Records: Once a transaction is recorded on the blockchain, it is virtually impossible to alter or delete. This immutability ensures that resource allocation decisions and the records of their execution remain reliable and unchallengeable. It is a critical feature for maintaining the integrity of the supply chain.

Smart Contracts: Smart contracts are self-executing agreements with the terms of the contract directly written into code. In the context of industrial robotics, smart contracts can automate resource allocation based on predefined conditions, making the process more efficient and reducing the risk of human error.

*Figure 3. Blockchain technology distributed digital ledger technology*

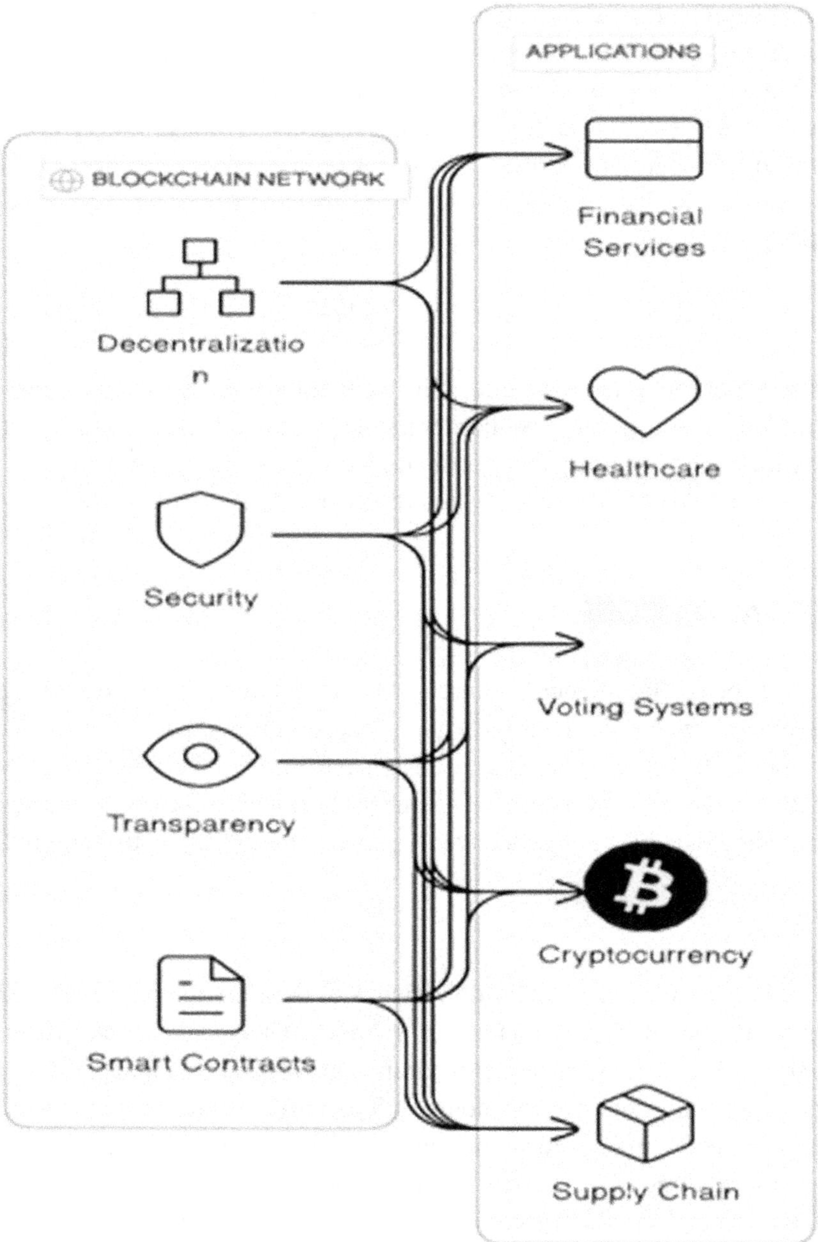

Decentralization: Blockchain operates on a decentralized network, eliminating the need for intermediaries and central authorities. This decentralization enhances the efficiency and reduces the costs associated with resource allocation, making it ideal for complex and distributed supply chain networks.

In the context of industrial robotics, where real-time resource allocation is essential for efficiency, blockchain's decentralized and secure nature enables dynamic decision-making based on real-time data. AI algorithms can utilize the secure and transparent data recorded on the blockchain to make intelligent resource allocation decisions, optimizing the use of robotic assets and other resources. This integrated approach ensures that the right resources are allocated at the right time, reducing waste and improving overall productivity.

In conclusion, blockchain in the context of intelligent resource allocation and optimization for industrial robotics using AI and blockchain is a transformative technology. It enhances transparency, security, and traceability in resource allocation decisions while enabling the power of AI to make data-driven and dynamic choices. This integration contributes to the efficiency, reliability, and trustworthiness of resource allocation in industrial settings, ultimately advancing the goals of Industry 4.0 and the optimization of manufacturing and supply chain processes.

### 3.4 Applications

a) Cryptocurrency: Blockchain is the underlying technology for cryptocurrencies like Bitcoin and Ethereum.
b) Supply Chain: Blockchain improves traceability and transparency in supply chain management.
c) Financial Services: It's used for cross-border payments, trade finance, and settlements.
d) Healthcare: Blockchain secures patient data and facilitates interoperability.
e) Voting Systems: It enhances the security and transparency of voting processes.

## 4. OBJECTIVES AND SCOPE

The objectives and scope of "Intelligent Resource Allocation and Optimization for Industrial Robotics using AI and Blockchain" encompass a range of goals and boundaries that define the focus and direction of this innovative approach. These objectives are designed to harness the synergies of AI and blockchain technologies to enhance the allocation and management of critical resources in industrial robotics, contributing to operational efficiency, cost-effectiveness, and transparency in the supply chain.

### 4.1 Objectives

Efficient Resource Allocation: The primary objective is to develop and implement AI-driven systems that can intelligently allocate resources in industrial robotics. This includes optimizing the allocation of robotic assets, energy, labor, and materials to maximize operational efficiency.

Cost Reduction: By making data-driven resource allocation decisions, the aim is to reduce operational costs. This includes minimizing energy consumption, reducing idle time, and streamlining labor allocation.

Real-time Decision Making: The system should enable real-time resource allocation and adapt to changing conditions, ensuring that resources are allocated where and when they are needed most.

Transparency and Traceability: Utilizing blockchain technology, the project aims to provide transparency in resource allocation decisions and transactions. Stakeholders within the industrial ecosystem should have real-time visibility into resource allocation, contributing to trust and accountability.

Security and Data Integrity: Ensure that resource allocation decisions and transactions are secure and tamper-resistant. Blockchain technology will be employed to safeguard the integrity of the data and transactions.

Smart Contract Automation: Implementing smart contracts to automate resource allocation processes based on predefined conditions. This can lead to more efficient and error-free resource allocation.

## 4.2 Scope

Resource Types: The project will focus on the allocation and optimization of a broad range of resources, including robotic assets, energy, labor, and materials. These are central to industrial operations and play a significant role in the supply chain.

Industrial Robotics: The scope encompasses resource allocation within the realm of industrial robotics, which includes manufacturing processes, warehouse automation, and other related applications.

AI and Blockchain Integration: The project will integrate AI algorithms for data-driven decision-making and blockchain technology for secure and transparent transaction recording.

Real-time Resource Allocation: The focus is on real-time resource allocation and optimization to respond to changing conditions and demands.

Transparency and Security: Ensuring that the allocation of resources is transparent and secure through the use of blockchain technology, enhancing trust within the supply chain.

Cost-Effective Solutions: Implementing solutions that reduce operational costs and improve overall efficiency, contributing to cost-effectiveness in industrial processes.

Smart Contracts: Leveraging smart contracts to automate resource allocation decisions and enhance the efficiency of the process.

The objectives and scope of "Intelligent Resource Allocation and Optimization for Industrial Robotics using AI and Blockchain" aim to create a comprehensive system that optimizes resource allocation, improves efficiency, reduces costs, and ensures transparency and security in the context of industrial robotics. This innovative approach aligns with the goals of Industry 4.0 and the evolution of smart manufacturing and supply chain management.

# 5. LITERATURE REVIEW

Singh, Rathore & Park (2020) introduced a Blockchain-enabled Intelligent IoT Architecture with Artificial Intelligence, offering an efficient convergence of blockchain and AI for IoT through cutting-edge techniques and applications.

He et al. (2020) present a comprehensive framework for blockchain-based edge computing in IoT scenarios, outlining the step-by-step process of a transaction between an IoT endpoint and an Edge Computing Node (ECN).

Du et al. (2017) address the computation offloading challenge in a mixed fog/cloud system. They propose a joint optimization approach that encompasses offloading decisions, computation resource allocation, transmit power, and radio bandwidth allocation, all while ensuring user fairness and meeting maximum tolerable delay requirements.

Zhang et al. (2017) introduce a collaborative optimization framework involving FNs, DSOs, and DSSs. This framework facilitates distributed decision-making to attain optimal resource allocation schemes.

Sharma et al. (2023) conducted an analysis of resource allocation problems within a blockchain-based MEC system architecture. This research explores solutions to address these challenges.

Wang et al. (2022) introduced a blockchain-enabled resource orchestration scheme for IoT using deep reinforcement learning (DRL). This innovative approach allows the IoT edge server and end users to collaboratively determine network resource allocations based on blockchain principles.

AlQerm and Pan (2022) introduced an innovative hierarchical resource facilitation framework called I-HARF, designed to adapt to dynamic Edge-IoT scenarios, including remote and mobile situations, application sensitivity, and varying resource demands based on time and location.

Sun, Peng, and Mao's (2018) work delves into the intricacies of resource management in F-RANs, addressing the challenges posed by network dynamics, resource diversity, and the integration of resource management with mode selection in a dynamic system.

Huang and colleagues (2021) introduce a blockchain system that achieves consensus with minimal energy consumption in edge devices, utilizing an innovative Proof of Stake mechanism.

Xu et al. (2020) explore the potential applications of blockchain in resource management and sharing for 6G across various scenarios, including the Internet of Things, device-to-device communications, network slicing, and inter-domain blockchain ecosystems.

Sharma et al (2021) investigate AIoT as a fusion of Mobile-Internet of Things (M-IoT) and AI, emphasizing the need for rapid responses, secondary support systems, and computational resources.

Duc et al. (2019) delve into the challenge of ensuring dependable resource allocation in collaborative edge-cloud settings. They conduct a comprehensive survey of technologies, mechanisms, and strategies designed to enhance the reliability of distributed applications operating in diverse and heterogeneous network environments.

## 6. DISCUSSION OF RESOURCE ALLOCATION METHODS

Resource allocation methods play a critical role in optimizing operations and achieving efficiency, especially in the context of industrial robotics using AI and blockchain. These methods are fundamental for distributing resources effectively, ensuring that the right resources are allocated to the right tasks at the right time. Here, we discuss various resource allocation methods and their relevance in this context:

Traditional Methods: Traditional resource allocation methods involve manual decision-making based on predetermined schedules and historical data. While these methods are straightforward, they lack adaptability and real-time responsiveness. In the context of AI and blockchain, there is an opportunity to replace or enhance traditional methods with data-driven and automated approaches.

AI-Driven Allocation: AI-based resource allocation methods leverage machine learning and predictive analytics to make data-driven decisions. Machine learning models can analyze historical data, sensor inputs, and various parameters to predict when and where resources are needed. This approach enables dynamic and real-time allocation, optimizing resource usage.

Dynamic Scheduling: Dynamic scheduling algorithms take into account real-time data and changing conditions. For instance, in industrial robotics, the allocation of robotic assets can be optimized by dynamically routing robots to tasks based on their availability, location, and the urgency of the task. AI can play a key role in optimizing dynamic scheduling, considering factors like energy consumption and task priorities.

Blockchain-Enhanced Security: While not a direct resource allocation method, blockchain technology ensures the security and transparency of resource allocation decisions. It records every transaction and resource allocation decision in an immutable ledger, providing a secure and tamper-resistant record of actions taken. This enhances trust among stakeholders and ensures the integrity of the allocation process.

Smart Contracts: Smart contracts are self-executing contracts with predefined rules. In the context of resource allocation, smart contracts can automate the process by executing resource allocation decisions when specific conditions are met. For example, a smart contract can automatically allocate robotic assets to specific tasks when certain criteria are fulfilled.

Optimization Algorithms: Various optimization algorithms can be employed for resource allocation. These algorithms aim to find the most efficient allocation of resources, taking into account constraints and objectives. AI-driven optimization algorithms can make decisions that result in resource allocation that minimizes costs, maximizes efficiency, or achieves other specified goals.

Optimization algorithms play a crucial role in intelligent resource allocation and optimization for industrial robotics using AI and blockchain. These algorithms help make informed decisions about how to allocate resources efficiently, reduce operational costs, and maximize productivity in real-time. Here is some optimization algorithms commonly used in this context:

a. Genetic Algorithms (GAs): Genetic algorithms are inspired by the process of natural selection. They use a population of potential solutions and apply genetic operators like mutation and crossover to evolve and find optimal resource allocation strategies. GAs are highly adaptable and can explore a wide range of solutions, making them suitable for complex and dynamic resource allocation problems. Here's a high-level mathematical description of how GAs work:

    1. Initialization:
        - Let P be the population of potential solutions, each represented by a chromosome.
        - Let n be the number of chromosomes in the population.
        - Each chromosome is encoded as a vector of genes.
    2. Evaluation:
        - Let f(x) be the fitness function that quantifies the quality of a solution x.
        - The fitness function maps a chromosome (solution) to a real number that indicates its quality: $f: P \to \mathbb{R}$.
    3. Selection:
        - During selection, a subset of the population is chosen as parents for the next generation.
        - The probability of selecting a chromosome x as a parent is often based on its fitness:
        - $P(\text{select } x \text{ as parent}) \propto f(x)$
        - Various selection methods can be used, such as roulette wheel selection or tournament selection.
    4. Crossover (Recombination):
        - Crossover combines genetic material from two parent chromosomes to produce offspring.

- The crossover operator can be defined as a function crossover(x, y) that takes two parent chromosomes x and y and produces one or more offspring chromosomes.

5. Mutation:
   - Mutation introduces random changes to individual genes in a chromosome.
   - The mutation operator can be defined as a function mutation(x) that modifies a chromosome x by altering one or more of its genes.
6. New Population:
   - The new population is created by combining offspring chromosomes and potentially retaining some of the best-performing parent chromosomes.
   - The size of the population remains relatively constant throughout the evolution.
7. Termination:
   - The GA iterates through multiple generations until a termination condition is met, such as a maximum number of generations, a satisfactory solution is found, or a resource limit is reached.
8. Best Solution:
   - The best solution found throughout the evolution is often the output of the GA and represents the solution to the optimization problem.

While GAs themselves are not defined by explicit mathematical equations, the details of the genetic operators (crossover and mutation) and the fitness function are typically problem-specific and may involve mathematical modeling and equations tailored to the particular optimization problem being addressed. The effectiveness of GAs lies in their ability to explore and evolve solutions based on the mathematical fitness landscape defined by the fitness function.

b. Particle Swarm Optimization (PSO): PSO is a population-based optimization algorithm inspired by the social behavior of birds flocking or fish schooling. In the context of resource allocation for industrial robotics, PSO can be used to optimize the allocation of robotic assets and energy consumption. Particles in the algorithm represent potential solutions and adjust their positions based on their individual and group experiences. Here is the basic mathematical model for PSO:

1. Initialization:
   - Initialize a population of particles, each representing a potential solution. For each particle i:
   - Initialize the position X_i in the search space.
   - Initialize the velocity V_i to zero.
2. Objective Function:
   - Define the objective function to be optimized: f(X), where X is the particle's position.
3. Particle Update Equations:
   - The position and velocity of each particle are updated based on its previous position, velocity, and its best-known position (individual best) and the best-known position among the entire swarm (global best). The update equations are as follows:
   - Update the velocity of particle i at each time step:
   - V_i(t+1) = w * V_i(t) + c1 * r1 * (Pbest_i - X_i) + c2 * r2 * (Gbest - X_i)
   - Update the position of particle i at each time step:

- X_i(t+1) = X_i(t) + V_i(t+1)

Where:

- V_i(t+1) is the new velocity of particle i.
- X_i(t+1) is the new position of particle i.
- w is the inertia weight, controlling the impact of the previous velocity.
- c1 and c2 are acceleration coefficients.
- r1 and r2 are random values between 0 and 1.
- Pbest_i is the best-known position of particle i (individual best).
- Gbest is the best-known position among the entire swarm (global best).

4. Termination:

- PSO iterates through multiple time steps (generations) or until a termination condition is met, such as a maximum number of iterations or a satisfactory solution is found.

5. Best Solution:

- The best-known position found by any particle in the swarm, often referred to as Gbest, and represents the solution to the optimization problem.

The effectiveness of PSO lies in its ability to explore the search space by adjusting particle positions and velocities based on both individual and swarm-wide knowledge. The balance between exploration and exploitation is controlled by the inertia weight w and the acceleration coefficients c1 and c2. PSO's strength is its ability to efficiently converge to optimal or near-optimal solutions in various optimization problems, making it a popular choice in optimization tasks and parameter tuning for machine learning models.

c. Linear Programming (LP): Linear programming is a mathematical technique used to allocate resources in a way that optimizes a linear objective function while satisfying linear constraints. LP can be applied to allocate resources like labor and materials efficiently within the constraints of the industrial operation. The general form of an LP problem can be expressed with mathematical equations as follows:

**Objective Function:** Maximize (Max) or Minimize (Min) a linear objective function:

$$Z=c1x1+c2x2+\ldots+c_nx_n$$

where:

- Z is the objective function value to be optimized.
- c_1, c_2, ..., c_n are coefficients representing the contribution of each decision variable to the objective function.
- x_1, x_2, ..., x_n are decision variables.

Constraints: Subject to a set of linear equality and inequality constraints:
Equality Constraints:

$$a_{11}x_1+a_{12}x_2+\ldots+a_{1n}x_n=b1 \quad (1)$$

$$a_{21}x1+a_{22}x_2+\ldots+a_{2n}x_n=b2 \quad (2)$$

$$a_{m1}x_1+a_{m2}x_2+\ldots+a_{mn}x_n=b_m \text{ (N)}$$

where:
Inequality Constraints:

$$g_{11}x_1+g_{12}x_2+\ldots+g_{1n}x_n\leq d_1 \quad (3)$$

$$g_{21}x_1+g_{22}x_2+\ldots+g_{2n}x_n\leq d_2 \quad (4)$$

$$g_{p1}x_1+g_{p2}x_2+\ldots+g_{pn}x_n\leq d_p \text{ (N)}$$

where:

- g_{ij} are coefficients representing the contribution of each decision variable to the left-hand side of the constraints.
- d_i are constants representing the right-hand side values of the constraints.
- p is the number of inequality constraints.

Bounds: Additionally, decision variables often have lower and upper bounds:

$x_i\geq 0$, for $i=1,2,\ldots,n$

These mathematical equations collectively define an LP problem, where the goal is to find values for the decision variables x_i that optimize the objective function Z while satisfying the constraints.

The solution to an LP problem provides the values of the decision variables that maximize or minimize the objective function while adhering to the given constraints. Linear Programming is a fundamental tool in operations research, economics, engineering, and various other fields for solving optimization problems of this nature.

d. Integer Programming (IP): Integer programming extends linear programming to situations where decisions must be made in whole numbers. In the context of industrial robotics, IP can be used to optimize the allocation of discrete resources, such as the allocation of robotic assets for specific tasks.
e. Mixed-Integer Linear Programming (MILP): MILP combines linear programming and integer programming, making it suitable for complex problems that involve both continuous and discrete

decision variables. It can be used to optimize resource allocation in situations where some resources can be allocated in fractions, while others must be allocated as a whole.

f. Dynamic Programming (DP): Dynamic programming is a method for solving complex problems by breaking them down into simpler, overlapping sub problems. In resource allocation, DP can be used to optimize the allocation of resources over time by considering the interdependencies and constraints involved in industrial processes.

These optimization algorithms can be integrated into the intelligent resource allocation system, where AI-driven decision-making processes leverage them to make informed and dynamic resource allocation decisions. The choice of the most appropriate algorithm depends on the specific characteristics and requirements of the industrial robotics application, such as the nature of the resources to be allocated, the constraints, and the objectives of the allocation process. The goal is to create a resource allocation system that maximizes efficiency, minimizes costs, and responds in real-time to changing conditions, ultimately enhancing industrial operations and supply chain management.

Real-time Monitoring: Resource allocation methods can be enhanced by real-time monitoring of resource usage and task progress. Sensors and IoT devices can provide data that AI algorithms use to make on-the-fly resource allocation decisions. For example, if a robotic system encounters a malfunction, AI can quickly allocate maintenance resources to address the issue.

In the context of industrial robotics using AI and blockchain, the ideal resource allocation method is a combination of AI-driven decision-making, dynamic scheduling, smart contracts, and blockchain-enhanced security and transparency. This approach ensures that resource allocation is data-driven, efficient, secure, and adaptable to changing conditions, contributing to the overall optimization and cost-effectiveness of industrial operations. It aligns with the goals of Industry 4.0 and the evolution of smart manufacturing.

## 7. OPTIMIZATION METHODS FOR INDUSTRIAL ROBOTICS

Optimization methods for industrial robotics are essential for improving the efficiency, productivity, and cost-effectiveness of robotic operations in manufacturing and other industrial applications. These methods encompass a range of techniques and approaches tailored to address specific optimization goals. Here are some common optimization methods for industrial robotics:

a. Path Planning and Trajectory Optimization: Path planning algorithms aim to find the most efficient and collision-free path for a robot to follow while performing a task. Trajectory optimization methods determine the robot's motion profile, considering factors like speed, acceleration, and energy consumption. These methods help robots move smoothly and efficiently.

b. Task Scheduling: Task scheduling optimization ensures that robotic tasks are allocated to robots in a way that maximizes utilization and minimizes idle time. It considers factors like task priorities, deadlines, and the availability of robots, optimizing the overall production process.

c. Resource Allocation: Resource allocation optimization involves determining how to assign available robotic assets, such as robots and tools, to different tasks to maximize throughput and minimize resource idle time. It ensures that resources are efficiently utilized, reducing operational costs.

d. Energy Efficiency: Optimizing energy consumption is critical for industrial robotics. Energy-efficient optimization methods aim to reduce the power consumption of robots during operation by adjusting motion profiles and optimizing the use of energy-saving features.
e. Load Balancing: Load balancing optimization ensures that robotic tasks are evenly distributed among robots to prevent overloading and underutilization. This method helps maintain a balanced workflow and maximizes the utilization of available resources.
f. Collision Avoidance: Collision avoidance optimization prevents collisions between robots, obstacles, and other objects in the workspace. It uses sensors and algorithms to plan paths and trajectories that avoid collisions, ensuring safety and efficiency.
g. End-Effector Optimization: Optimizing the design and control of robot end-effectors (grippers, tools, etc.) can significantly improve the performance of industrial robotics. This includes selecting the right end-effector for a specific task and optimizing its design and control algorithms.
h. Multi-Robot Coordination: In scenarios with multiple robots working together, multi-robot coordination optimization methods ensure efficient collaboration and task allocation. These methods consider factors like communication, resource sharing, and task allocation.
i. Quality Control and Inspection: Optimization in quality control and inspection tasks involves designing algorithms that can efficiently inspect and ensure the quality of products produced by robots. This includes visual inspection, defect detection, and quality assurance processes.
j. Maintenance Scheduling: Maintenance optimization aims to minimize downtime by scheduling robot maintenance activities based on predictive maintenance models. These models use data to predict when maintenance is required, reducing unplanned downtime.
k. Real-time Adaptation: Real-time optimization methods involve continuously adjusting robot parameters and decisions based on real-time data. This allows robots to adapt to changing conditions and unexpected events, improving flexibility and efficiency.
l. Machine Learning and AI-Based Optimization: Machine learning and artificial intelligence (AI) techniques are increasingly used for optimization in industrial robotics. They can adapt to complex and dynamic environments, learn from data, and make informed decisions for resource allocation, path planning, and other optimization tasks.

The choice of optimization method depends on the specific objectives and challenges of the industrial robotic application. Combining multiple optimization methods and leveraging advanced technologies like AI and machine learning can lead to more robust and efficient industrial robotic systems, contributing to improved productivity and cost-effectiveness in manufacturing and other industrial processes.

## 8. AI-DRIVEN RESOURCE ALLOCATION

### 8.1 Real-Time Data Analysis and Decision-Making

a) Real-Time Data Streams: Explain the concept of real-time data streams in industrial robotics. Discuss how data from sensors, robots, and other sources are continuously generated and processed.
b) Data Processing Frameworks: Introduce data processing frameworks such as Apache Kafka, Apache Flink, and Apache Spark for handling real-time data streams. Explain their significance in managing and analyzing large volumes of data in real time.

*Figure 4. Real-time data analysis and decision-making*

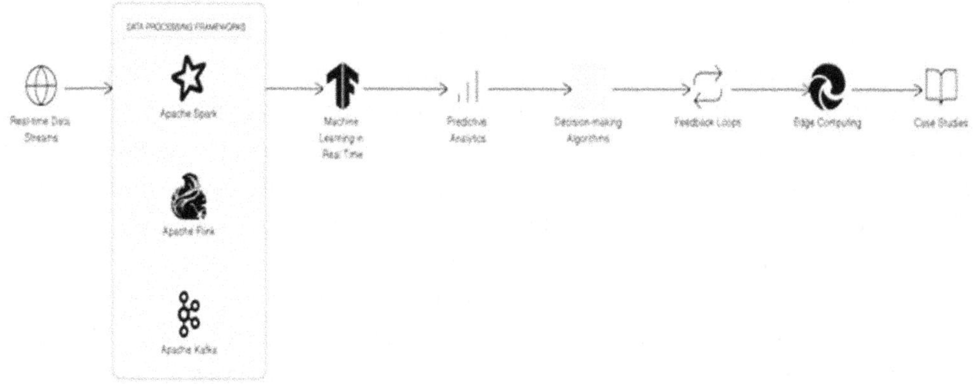

c) Machine Learning in Real Time: Discuss the challenges and opportunities of applying machine learning algorithms to analyze real-time data. Emphasize the need for low-latency, high-throughput models that can make predictions and decisions on the fly.
d) Predictive Analytics: Explore how predictive analytics can be integrated into real-time data analysis. Describe how machine learning models are trained to make resource allocation predictions in real time.
e) Decision-Making Algorithms: Discuss decision-making algorithms used in real-time resource allocation. Explain their role in selecting the most suitable allocation strategy based on incoming data.
f) Feedback Loops: Highlight the importance of feedback loops in the context of real-time decision-making. Explain how systems can learn and adapt their resource allocation strategies based on the outcomes of previous decisions.
g) Edge Computing: Introduce the concept of edge computing in industrial robotics. Explain how edge devices process data locally, reducing latency and enabling faster decision-making.
h) Case Studies: Provide examples of real-time data analysis and decision-making in industrial robotics. Showcase instances where immediate adjustments in resource allocation led to improved operational efficiency.

## 8.2 AI-Driven Resource Allocation in Robotics

a) Manufacturing Industry: Provide a case study from the manufacturing sector where AI-driven resource allocation improved production efficiency. Discuss how AI algorithms analyzed real-time data from robotic arms, conveyor belts, and other equipment to allocate resources optimally.
b) Logistics and Warehousing: Present a case study involving robotics in logistics and warehousing. Describe how AI-powered robots efficiently managed inventory, allocated tasks, and optimized their routes to fulfill orders in the least amount of time.
c) Healthcare Robotics: Explore a case study in healthcare where AI-driven resource allocation enhanced patient care. Explain how robots in hospitals optimized their schedules, ensuring timely delivery of medications and supplies to different wards.

d) Agriculture: Discuss a case study in the agricultural sector, showcasing how AI and robotics were used for precision farming. Explain how autonomous drones and robotic harvesters allocated resources, such as water and fertilizers, based on data from soil sensors and crop health analysis.
e) Energy Sector: Present a case study from the energy industry where AI-driven robots were employed for maintenance and inspections. Describe how these robots used predictive maintenance algorithms to allocate resources like inspection schedules and spare parts, reducing downtime.
f) Environmental Monitoring: Explore a case study related to environmental monitoring and conservation. Discuss how AI-equipped robots allocated resources to collect data on wildlife movements, pollution levels, or climate conditions in remote and challenging environments.
g) E-commerce and Delivery: Provide a case study from the e-commerce sector, highlighting how AI-powered robots optimized last-mile delivery. Explain how these robots determined the best routes, managed inventory in delivery centers, and scheduled deliveries efficiently.
h) Performance Metrics: Include key performance metrics and outcomes for each case study. Emphasize improvements in efficiency, cost reduction, resource utilization, and any other relevant factors achieved through AI-driven resource allocation.

## 9. BLOCKCHAIN-ENHANCED RESOURCE MANAGEMENT

### 9.1 Blockchain for Secure Resource Tracking

a) Immutable Record Keeping: Describe how blockchain's immutable ledger ensures that every transaction related to resource tracking is recorded securely. Once recorded, these transactions cannot be altered, providing a tamper-resistant history of resource movements.
b) Traceability: Emphasize the ability of blockchain to enable end-to-end traceability of resources. Explain how stakeholders can track the origin, movement, and status of resources in real-time, enhancing transparency and accountability.
c) Authentication and Authorization: Discuss the role of cryptographic techniques in blockchain that enable secure authentication and authorization for accessing and tracking resources. Explain how this prevents unauthorized access and maintains data integrity.
d) Real-time Updates: Highlight how blockchain allows for real-time updates on the status and availability of resources. This ensures that all relevant parties have access to the most current information, reducing delays and errors.
e) Decentralization: Explain how blockchain's decentralized nature eliminates the need for a central authority, making it resistant to single points of failure or manipulation. Describe how this decentralization contributes to increased security.
f) Smart Contracts Integration: Mention how smart contracts, which are self-executing agreements triggered by predefined conditions, can be integrated into the blockchain. Explain how they can automate resource tracking and allocation based on predefined rules.
g) Auditing and Compliance: Discuss how blockchain's transparent and auditable nature simplifies auditing processes and ensures compliance with regulations. Explain how organizations can easily demonstrate adherence to resource management protocols.

*Figure 5. Blockchain for secure resource tracking*

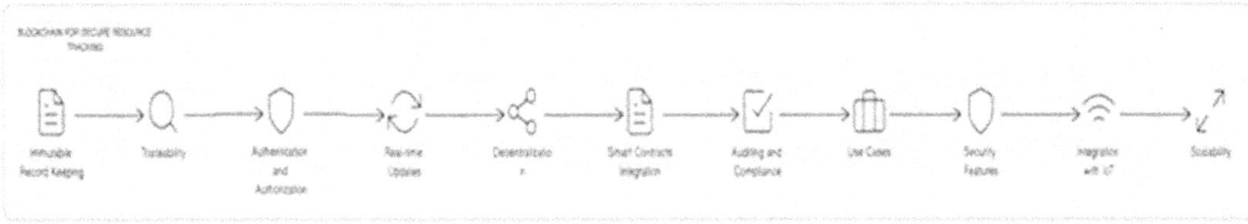

h) Use Cases: Provide examples of real-world use cases where blockchain has been used for secure resource tracking in industrial robotics. These could include scenarios in manufacturing, logistics, or supply chain management.
i) Security Features: Mention blockchain's security features such as consensus mechanisms, encryption, and public-private key pairs that safeguard resource tracking data from unauthorized access and tampering.
j) Integration With IoT: Explain how the integration of blockchain with the Internet of Things (IoT) can enhance resource tracking by enabling direct communication between IoT devices and the blockchain, ensuring real-time data updates.
k) Scalability: Address how blockchain networks are scalable, allowing them to handle large volumes of resource tracking data generated by industrial robotics operations.

## 10. DECENTRALIZED LEDGER FOR TRANSPARENCY

a) Introduction to Blockchain Ledger: Begin by explaining the concept of a blockchain ledger as a distributed, immutable, and secure record-keeping system.
b) Transparency and Trust: Emphasize how blockchain's decentralized ledger ensures transparency and trust in resource management. Transactions and resource allocation are recorded in a tamper-resistant manner visible to authorized participants.
c) Immutable Record-Keeping: Highlight the immutability of blockchain records. Once a transaction or resource allocation is recorded, it cannot be altered or deleted. This feature ensures the integrity of historical data.
d) Decentralization: Explain how blockchain operates on a network of nodes, with each node having a copy of the ledger. This decentralized nature eliminates the need for a central authority and reduces the risk of single points of failure.
e) Permissioned vs. Permissionless Blockchains: Differentiate between permissioned and permissionless blockchains. In industrial settings, permissioned blockchains are often preferred due to their controlled access, making them suitable for private and consortium environments.
f) Resource Allocation Transparency: Describe how blockchain provides transparency in resource allocation. Participants can view the allocation of robots, machinery, or other resources in real time, fostering trust among stakeholders.

*Figure 6. Decentralized ledger*

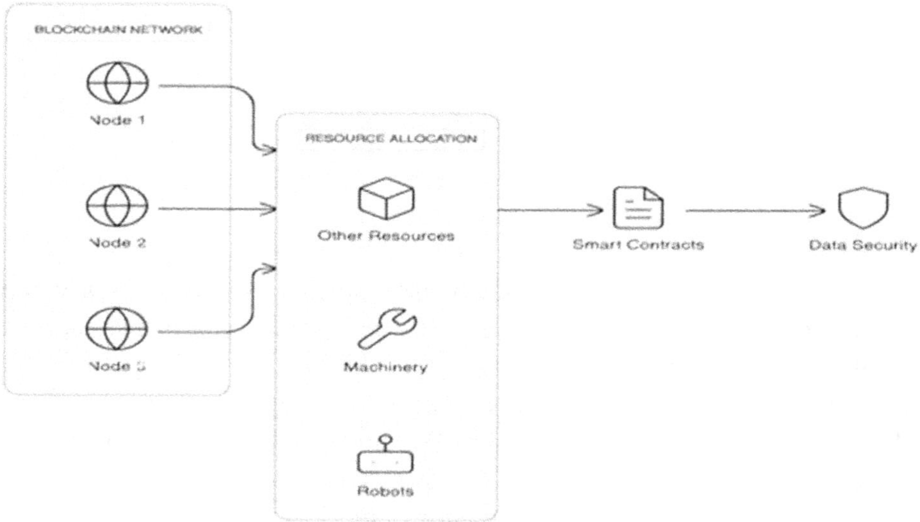


g) Auditing and Compliance: Discuss how blockchain's transparent ledger simplifies auditing and compliance efforts. Regulatory bodies can easily verify resource allocation and adherence to industry standards.

h) Provenance Tracking: Explain how blockchain can track the provenance of resources, including their maintenance history, usage patterns, and ownership changes. This is particularly valuable in industries where traceability is critical, such as aerospace or healthcare.

i) Smart Contracts Integration: Mention the integration of smart contracts with the blockchain ledger. Smart contracts execute resource allocation based on predefined rules, and their actions are transparently recorded on the blockchain.

j) Data Security: Highlight blockchain's cryptographic security mechanisms, which protect data from unauthorized access or tampering. This ensures that resource management information remains confidential and secure.

## 11. INTEGRATION OF AI AND BLOCKCHAIN IN INDUSTRIAL ROBOTICS

### 11.1 Synergies Between AI and Blockchain

In the context of industrial robotics, the integration of AI (Artificial Intelligence) and blockchain technologies offers several synergies that can significantly enhance the capabilities and functionalities of robotic systems.

These synergies primarily revolve around data management, security, and automation. Here's an overview:

*Figure 7. Synergies between AI and blockchain*

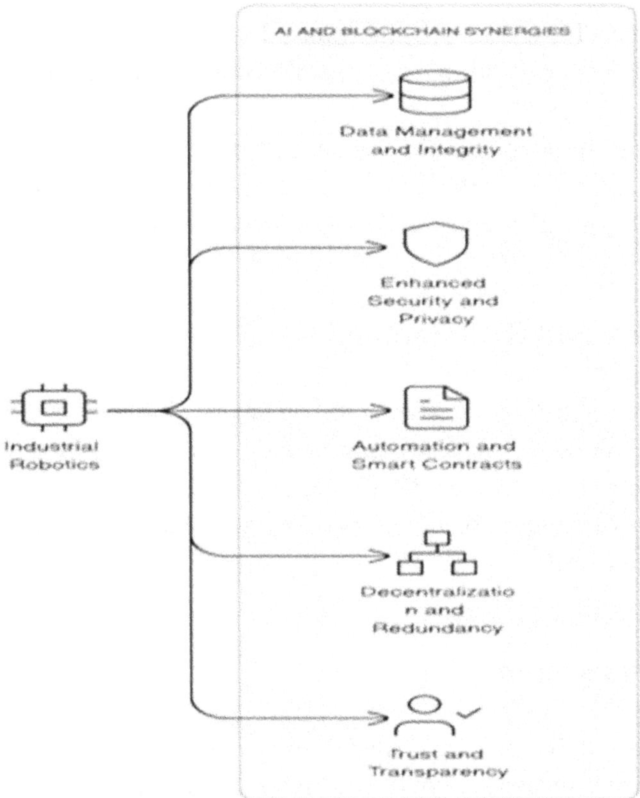


## 11.2 Data Management and Integrity

- AI's Data Processing: AI excels at processing and analyzing vast amounts of data generated by robotic sensors. It can detect patterns, predict failures, and optimize resource utilization.
- Blockchain's Data Immutability: Blockchain ensures the immutability and tamper-proof nature of data. Once data is recorded on the blockchain, it cannot be altered or deleted, providing a reliable and auditable data history.

## 11.3 Enhanced Security and Privacy

- AI-Driven Cybersecurity: AI systems can continuously monitor network traffic and system behavior to detect anomalies and potential security threats.
- Blockchain's Transparency: Blockchain provides a transparent and distributed ledger where all transactions and data access are recorded. This transparency enhances security by reducing the risk of unauthorized access and data breaches.

## 11.4 Automation and Smart Contracts

- AI-Driven Resource Allocation: AI algorithms can autonomously allocate resources such as robotic tasks, energy usage, and maintenance schedules based on real-time data and predictive analytics.
- Smart Contracts for Automation: Blockchain's smart contracts enable automated and self-executing agreements. In robotics, this means that certain tasks, maintenance routines, or resource allocations can be triggered automatically when predefined conditions are met.

## 11.5 Decentralization and Redundancy

- AI's Distributed Decision-Making: AI can distribute decision-making processes across a network of robotic systems, allowing for decentralized control and coordination.
- Blockchain's Decentralized Ledger: Blockchain operates on a decentralized network of nodes, ensuring redundancy and fault tolerance. This can be valuable in scenarios where uninterrupted robotic operations are critical.

## 11.6 Trust and Transparency

- AI-Driven Predictive Maintenance: AI models can predict when robotic equipment needs maintenance, reducing downtime and improving operational efficiency.
- Blockchain for Maintenance Records: Maintenance records stored on a blockchain provide a transparent and unforgeable history of equipment maintenance, enhancing trust between stakeholders.

These synergies highlight the potential for AI and blockchain to transform industrial robotics by improving efficiency, security, transparency, and automation. However, it's essential to address challenges such as integration complexity, data privacy, scalability, and regulatory compliance when implementing these technologies in industrial settings.

# 12. COLLABORATIVE RESOURCE ALLOCATION AND DECISION-MAKING

The collaborative integration of AI (Artificial Intelligence) and blockchain in industrial robotics opens up new possibilities for resource allocation and decision-making. This synergy enables more efficient and decentralized management of resources within robotic systems. Here's an overview of how collaborative resource allocation and decision-making work in this context:

## 12.1 Distributed Decision-Making

- AI's Role: AI algorithms can be distributed across robotic nodes or devices within a network. These AI agents are responsible for local decision-making based on real-time data from their respective sensors and tasks.

- Blockchain's Role: Blockchain acts as a decentralized ledger that records decisions and resource allocations made by AI agents. These decisions are transparently and immutably stored on the blockchain.

## 12.2 Resource Allocation Optimization

- AI-Driven Allocation: Each AI agent is responsible for optimizing the allocation of resources, such as task assignments, energy usage, maintenance schedules, and computing power.
- Collaboration via Blockchain: When an AI agent determines an optimal resource allocation, this decision is recorded on the blockchain. Other AI agents in the network can access this information, which may influence their own resource allocation decisions. This collaborative approach ensures that resources are allocated efficiently across the entire robotic system.

## 12.3 Real-Time Data Sharing

- AI's Real-Time Data Analysis: AI agents continuously analyze real-time data from sensors, monitoring the state and performance of robotic systems.
- Blockchain for Data Sharing: Relevant data, such as sensor readings, task progress, or maintenance needs, can be securely shared among AI agents through the blockchain. This real-time data sharing enhances situational awareness and helps AI agents make informed decisions.

## 12.4 Transparent Decision History

- AI's Decision Records: Each AI agent maintains a record of its resource allocation decisions and the rationale behind them.
- Blockchain's Transparency: These decision records are transparently stored on the blockchain. This history of decisions becomes a valuable resource for auditing, accountability, and learning within the robotic network.

## 12.5 Consensus Mechanisms

- AI's Local Consensus: AI agents may reach consensus on resource allocation based on their local observations and objectives.
- Blockchain's Consensus: Blockchain employs its consensus mechanism (e.g., Proof of Work or Proof of Stake) to validate and secure the decisions recorded on the ledger. This ensures that decisions are tamper-proof and trustworthy.

*Figure 8. Collaborative integration of AI (artificial intelligence) and blockchain in industrial robotics*

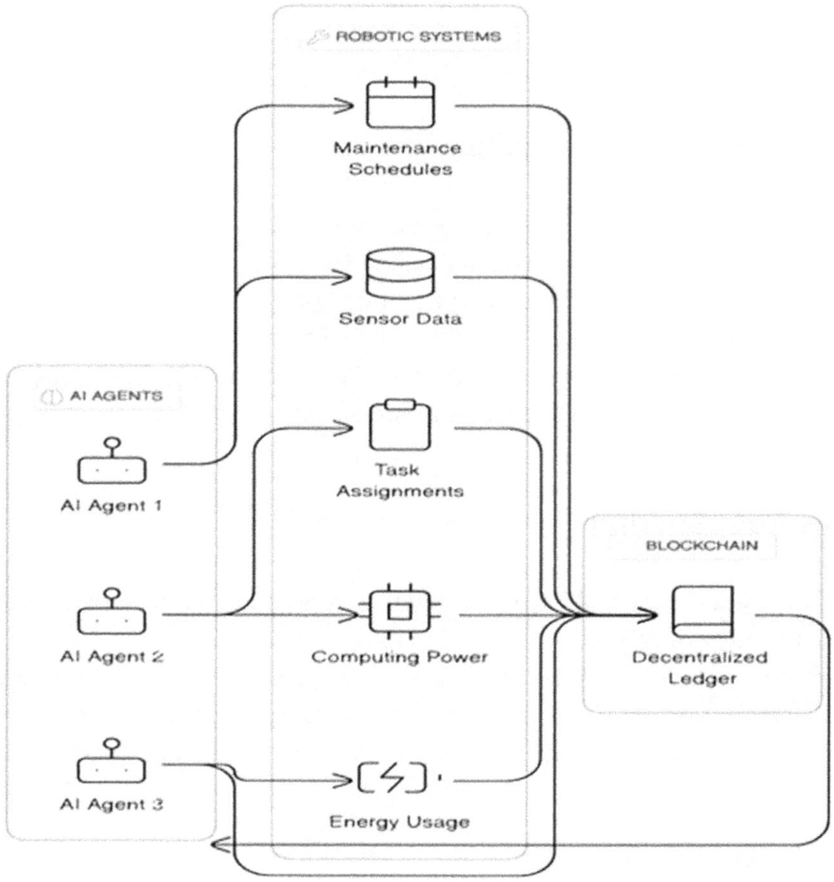


## 13. CASE STUDIES AND PRACTICAL IMPLEMENTATIONS

Case studies and practical implementations of AI and blockchain in robotics provide valuable insights into the real-world impact and challenges of this integration. Here are examples and considerations for each aspect:

### 13.1 Successful Implementations of AI and Blockchain in Robotics

a) Automated Manufacturing with Quality Assurance: A manufacturing company integrated AI-enabled robots for quality control and blockchain for recording inspection results. This reduced defects by 30%, and blockchain ensured the accuracy and integrity of quality data. Lessons: Integrating AI for real-time quality checks with blockchain can improve product quality and compliance.

b) Supply Chain Optimization: An automotive manufacturer utilized AI algorithms in its logistics robots to optimize routes and cargo allocation. Blockchain tracked the movement of components and vehicles across suppliers, warehouses, and assembly lines. This streamlined operations and reduced transportation costs. Lessons: AI-driven logistics enhanced by blockchain can lead to significant cost savings.

c) Pharmaceutical Traceability: A pharmaceutical company implemented AI-equipped robots for drug manufacturing and used blockchain to record the entire production process. This provided end-to-end traceability, improved regulatory compliance, and built trust with consumers. Lessons: Blockchain can be a game-changer in industries with stringent traceability requirements.

d) Robotics as a Service (RaaS): A robotics service provider adopted AI for predictive maintenance and used blockchain for transparent billing and maintenance history. This increased customer trust and reduced downtime. Lessons: Combining AI analytics with blockchain for service transparency can enhance customer satisfaction.

## 13.2 Impact on Industrial Operations and Efficiency

a) Enhanced Efficiency: Across these case studies, the integration of AI and blockchain consistently improved efficiency by automating tasks, optimizing processes, and reducing errors. AI-driven robots performed tasks more accurately and made data-driven decisions in real-time, while blockchain ensured data reliability.

b) Cost Reduction: Efficiency gains, reduced defects, and streamlined logistics led to cost reductions. Businesses saved on labor, maintenance, transportation, and compliance-related expenses.

c) Transparency and Trust: Blockchain's immutable ledger and transparent record-keeping fostered trust among stakeholders. Customers and regulatory bodies gained confidence in the integrity of data, leading to increased market competitiveness.

## 13.3 Lessons Learned and Best Practices

a) Data Quality: Accurate and reliable data are essential for AI and blockchain integration. Ensure that data collected by robots is clean and relevant to prevent inaccurate conclusions or blockchain entries.

b) Scalability: Consider the scalability of both AI and blockchain solutions. As operations grow, AI models and blockchain networks may need to handle increasing data and transactions.

c) Privacy and Security: Maintain the privacy and security of sensitive data. Implement encryption, access controls, and privacy-preserving AI techniques to protect information from unauthorized access.

d) Interoperability: Ensure that AI and blockchain systems can interoperate with existing technologies and platforms. This prevents silos and promotes data exchange.

e) Regulatory Compliance: Understand industry-specific regulations and compliance requirements. Design AI and blockchain solutions to facilitate compliance and reporting.

These practical implementations demonstrate that the synergy between AI and blockchain can drive efficiency, transparency, and trust in industrial robotics. By learning from successful cases and addressing associated challenges, organizations can harness the full potential of this integration for operational excellence.

## 14. FUTURE TRENDS AND RESEARCH DIRECTIONS

Future trends and research directions in the integration of AI and blockchain in industrial robotics are crucial for staying at the forefront of this rapidly evolving field. Here are some considerations:

### 14.1 Emerging Technologies in Industrial Robotics

a) Edge Computing: With the increasing complexity of AI algorithms, edge computing enables robots to process data locally, reducing latency and dependence on cloud services.
b) 5G Connectivity: High-speed, low-latency 5G networks will enable robots to communicate and share data faster and more reliably, enhancing collaborative tasks.
c) Quantum Computing: Quantum computing holds the potential to revolutionize AI algorithms, allowing for even more complex and efficient resource allocation and optimization.
d) Human-Robot Collaboration: Advancements in human-robot interaction, such as natural language processing and gesture recognition, will make robots more intuitive to work with in various industries.
e) Biologically Inspired Robotics: Bio mimicry and bio hybrid robotics are emerging fields that draw inspiration from nature to create more adaptable and efficient robots.

### 14.2 Potential Impact on Resource Allocation and Optimization

a) Real-Time Resource Allocation: Future research may focus on developing AI algorithms capable of dynamic and real-time resource allocation, ensuring robots adapt to changing environments and demands.
b) Energy Efficiency: As sustainability becomes a primary concern, research may aim to optimize resource allocation with a focus on energy-efficient operations.
c) Multi-Robot Systems: Investigating how AI and blockchain can enhance resource allocation in multi-robot systems, enabling collaborative decision-making among robots.
d) Human-Robot Collaboration: Understanding how AI-driven resource allocation can improve collaboration between humans and robots in shared workspaces.
e) Scalability: With the growth of robotics applications, scalability will be a critical consideration. Research may explore how AI and blockchain can efficiently scale resource allocation for large fleets of robots.

### 14.3 Recommendations for Researchers and Practitioners

a) Interdisciplinary Collaboration: Encourage collaboration between robotics experts, AI researchers, and blockchain specialists to tackle the complexities of integration effectively.
b) Data Privacy and Security: Focus on developing robust privacy-preserving techniques, as sensitive data is involved in both AI and blockchain systems.
c) Standards and Interoperability: Promote the development of industry standards to ensure interoperability between different robotic platforms, AI models, and blockchain networks.

*Figure 9. Impact on resource allocation and optimization*

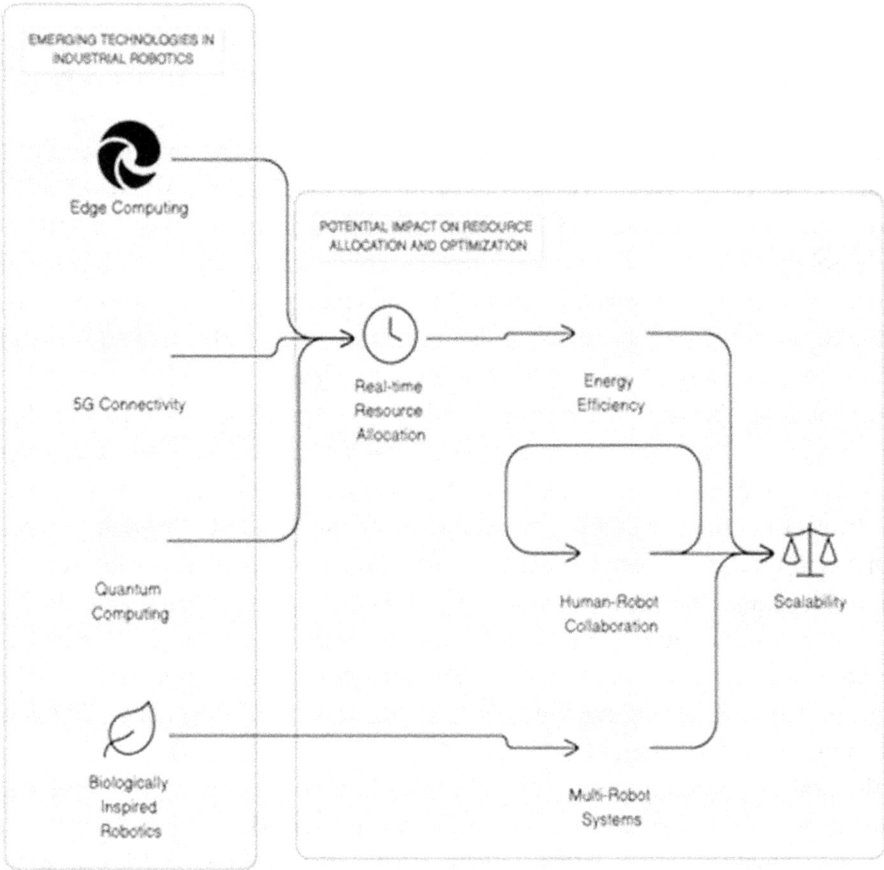


d) Regulatory Compliance: Stay updated on evolving regulations related to robotics, AI, and blockchain, especially in industries with strict compliance requirements like healthcare and finance.
e) Ethical Considerations: Research the ethical implications of AI-driven resource allocation and blockchain-enhanced transparency, addressing issues such as bias and fairness.
f) Education and Training: Invest in educating the workforce in AI, blockchain, and robotics to bridge the skills gap and fully leverage the potential of these technologies.
g) Prototyping and Pilots: Actively engage in pilot projects and prototyping to test and refine AI and blockchain applications in real-world industrial settings.

By embracing emerging technologies, considering potential impacts, and following best practices, researchers and practitioners can contribute to the continued evolution of AI and blockchain integration in industrial robotics, opening up new horizons for efficiency, transparency, and innovation in various industries.

## 15. CONCLUSION

In conclusion, the integration of Artificial Intelligence (AI) and blockchain technology in the context of resource allocation and optimization for industrial robotics offers a powerful framework for enhancing efficiency, transparency, and adaptability in manufacturing and industrial processes. This synergy of advanced technologies revolutionizes the way industrial resources are managed and allocated, resulting in a more intelligent and responsive industrial ecosystem. The incorporation of AI within this framework enables real-time data analysis and decision-making, optimizing resource allocation strategies, and ensuring that industrial robots operate at peak performance. Machine learning algorithms can forecast demand, track production, and identify patterns that would be challenging for human operators to discern. This leads to dynamic resource allocation, ensuring that robots are deployed where and when they are most needed, ultimately improving productivity and cost-effectiveness.

Blockchain technology plays a pivotal role in providing a transparent and immutable ledger for resource allocation and utilization. Through the integration of smart contracts, it ensures that agreements and resource assignments are executed automatically when predetermined conditions are met. Additionally, blockchain offers a secure and tamper-resistant system for recording and sharing data among stakeholders. This transparency and trust in the data facilitate collaboration and resource sharing across the industrial network, further optimizing resource utilization. The partnership between AI and blockchain empowers industries to adapt to changing circumstances and unforeseen disruptions effectively. For instance, during unexpected downtime or maintenance, AI can quickly reassign robots to alternative tasks based on real-time data, while blockchain maintains an auditable record of the changes. This adaptability is crucial in today's fast-paced industrial landscape.

Moreover, the ability to monitor and evaluate resource allocation strategies continually allows for further refinement and optimization. The data collected by AI-driven systems and recorded on the blockchain provides valuable insights into resource utilization, helping industries make informed decisions and enhancements.

However, it is essential to consider the security and privacy challenges that may arise in this intelligent resource allocation framework. Protecting sensitive data and ensuring the integrity of blockchain transactions are critical aspects. As the system evolves, it must adhere to robust security protocols and compliance with data privacy regulations.

In conclusion, the integration of AI and blockchain for intelligent resource allocation in industrial robotics represents a significant advancement in optimizing industrial processes. It enhances productivity, adaptability, and transparency, and it allows industries to thrive in a rapidly changing environment. By continuously refining strategies based on real-time data, this innovative framework is poised to shape the future of industrial resource allocation and drive increased efficiency and competitiveness. As technology further matures, we can expect even more sophisticated and streamlined solutions for resource allocation and optimization, providing industries with a competitive edge in the global market.

## REFERENCES

AlQerm, I., & Pan, J. (2022). I-HARF: Intelligent and Hierarchical Framework for Adaptive Resource Facilitation in Edge-IoT Systems. *IEEE Internet of Things Journal, 10*(5), 3954–3967. doi:10.1109/JIOT.2022.3151667

Du, J., Zhao, L., Feng, J., & Chu, X. (2017). Computation offloading and resource allocation in mixed fog/cloud computing systems with min-max fairness guarantee. *IEEE Transactions on Communications, 66*(4), 1594–1608. doi:10.1109/TCOMM.2017.2787700

Duc, T. L., Leiva, R. G., Casari, P., & Östberg, P. O. (2019). Machine learning methods for reliable resource provisioning in edge-cloud computing: A survey. *ACM Computing Surveys, 52*(5), 1–39. doi:10.1145/3341145

He, Y., Wang, Y., Qiu, C., Lin, Q., Li, J., & Ming, Z. (2020). Blockchain-based edge computing resource allocation in IoT: A deep reinforcement learning approach. *IEEE Internet of Things Journal, 8*(4), 2226–2237. doi:10.1109/JIOT.2020.3035437

Huang, Y., Zhang, J., Duan, J., Xiao, B., Ye, F., & Yang, Y. (2021). Resource allocation and consensus of blockchains in pervasive edge computing environments. *IEEE Transactions on Mobile Computing, 21*(9), 3298–3311. doi:10.1109/TMC.2021.3053230

Sharma, A., Rani, R. P., Abdulrahman, I. S., Sreeram, A., & Shukla, S. K. (2023, May). Resource Allocation Technology Mobile Edge Computing System Using Blockchain Technology. In *2023 3rd International Conference on Advance Computing and Innovative Technologies in Engineering (ICACITE)* (pp. 632-636). IEEE. 10.1109/ICACITE57410.2023.10183102

Sharma, V., Tan, T. G., Singh, S., & Sharma, P. K. (2021). Optimal and Privacy-Aware Resource Management in Artificial Intelligence of Things Using Osmotic Computing. *IEEE Transactions on Industrial Informatics, 18*(5), 3377–3386. doi:10.1109/TII.2021.3102471

Singh, S. K., Rathore, S., & Park, J. H. (2020). Blockiotintelligence: A blockchain-enabled intelligent IoT architecture with artificial intelligence. *Future Generation Computer Systems, 110*, 721–743. doi:10.1016/j.future.2019.09.002

Sun, Y., Peng, M., & Mao, S. (2018). Deep reinforcement learning-based mode selection and resource management for green fog radio access networks. *IEEE Internet of Things Journal, 6*(2), 1960–1971. doi:10.1109/JIOT.2018.2871020

Wang, C., Jiang, C., Wang, J., Shen, S., Guo, S., & Zhang, P. (2022). Blockchain-Aided Network Resource Orchestration in Intelligent Internet of Things. *IEEE Internet of Things Journal, 10*(7), 6151–6163. doi:10.1109/JIOT.2022.3222911

Xu, H., Klaine, P. V., Onireti, O., Cao, B., Imran, M., & Zhang, L. (2020). Blockchain-enabled resource management and sharing for 6G communications. *Digital Communications and Networks, 6*(3), 261–269. doi:10.1016/j.dcan.2020.06.002

Zhang, H., Xiao, Y., Bu, S., Niyato, D., Yu, F. R., & Han, Z. (2017). Computing resource allocation in three-tier IoT fog networks: A joint optimization approach combining Stackelberg game and matching. *IEEE Internet of Things Journal*, *4*(5), 1204–1215. doi:10.1109/JIOT.2017.2688925

## KEY TERMS AND DEFINITIONS

**Artificial Intelligence (AI):** Artificial Intelligence (AI) is a technology that enables computers and machines to think and act like humans. It involves creating smart systems that can learn from data, solve problems, and make decisions. AI helps computers recognize images, understand speech, translate languages, and even play games. It is used in many areas, such as self-driving cars, voice assistants like Siri or Alexa, and personalized recommendations on websites. AI is constantly improving and has the potential to revolutionize how we live and work.

**Deep Learning:** Deep learning is a subfield of machine learning that focuses on teaching computers to learn and make decisions in a way inspired by the human brain. It uses artificial neural networks, which are computational models composed of interconnected nodes called "neurons." These neural networks are structured in multiple layers, hence the term "deep" learning.

**Internet of Things (IoT):** The Internet of Things (IoT) is a concept that refers to the connection of everyday objects to the internet allowing them to send and receive data. These objects can include devices like smartphones, thermostats, wearables, home appliances, and even vehicles. The idea behind IoT is to create a network where these objects can communicate with each other, collect and share data, and perform tasks more efficiently.

**Natural Language Processing (NLP):** Natural Language Processing (NLP) is a branch of artificial intelligence that focuses on enabling computers to understand, interpret, and interact with human language in a natural and meaningful way. NLP involves the development of algorithms and models that allow computers to process, analyze, and generate human language.

# Chapter 5
# Multi-Layered Blockchain-Based Security Model for Electronic Health Record Systems

**Mrutyunjaya S. Hiremath**
https://orcid.org/0000-0002-1338-9946
*eMath Technology, Bangalore, India*

**Rajashekhar C. Biradar**
*Reva University, India*

## ABSTRACT

*The multi-layered blockchain-based electronic health records (ML-BbEHR) framework aims to address persistent healthcare issues such as data privacy and access control. Utilizing a multi-layered blockchain architecture, ML-BbEHR segments EHR data into categories like medical histories and clinical datasets, each managed within its distinct blockchain layer. This enhances security and optimizes data management. Unlike single-layer systems, this architecture mitigates latency issues. Additionally, the framework incorporates fingerprint authentication as an alternative to traditional cryptographic keys, bolstering patient identity verification. Smart contracts are also deployed across layers to delineate rules and permissions for healthcare data access, enhancing data integrity and security. This architecture is built on a general, non-specific blockchain platform chosen for its robust security features.*

## 1. INTRODUCTION

The transition to Electronic Health Record (EHR) systems has been accelerating in the healthcare industry, driven by the anticipated benefits for patients and healthcare providers. The U.S. Department of Health and Human Services, through its Office of the National Coordinator (ONC) for Health Information Technology, defines an EHR as a digital counterpart to a patient's traditional paper chart. Unlike static paper records, EHRs are dynamic, patient-centric digital repositories that enable real-time and secure access to medical information.

DOI: 10.4018/979-8-3693-0659-8.ch005

EHR systems are designed to go beyond merely storing standard clinical data generated during provider visits; they offer a comprehensive view of a patient's healthcare journey. These systems facilitate seamless information sharing with other entities in the healthcare ecosystem, including laboratories and specialized medical practitioners. As a result, an EHR encapsulates data from multiple healthcare providers involved in a patient's care, promoting a more integrated and holistic approach to healthcare delivery. The advantages of Electronic Health Record (EHR) systems as outlined by the Office of the National Coordinator (ONC) for Health Information Technology:

- EHR systems serve as comprehensive repositories that continuously update and synchronize key aspects of patients' medical profiles, including but not limited to medical history, diagnostic outcomes, medication regimens, treatment protocols, immunization timelines, allergenic susceptibilities, radiological imagery, and results of laboratory analyses.
- The systems are engineered to facilitate real-time access to evidence-based decision-making tools, thereby empowering healthcare providers to render more informed and effective patient care.
- Through automation and computational intelligence, EHR systems are designed to optimize the workflow of healthcare providers, enhancing operational efficiency and reducing the scope for manual errors.

Concurrent with advancements in Electronic Health Record (EHR) systems, the inception of blockchain technology (Nakamoto et al., 2019) heralded a significant technological shift. This innovation has attracted extensive scrutiny from both academic and commercial research sectors. Characterized by its decentralized structure and cryptographic security measures, blockchain technology has substantial potential to serve as the underlying framework for many distributed systems. While initially restricted to financial use cases, most notably in the form of cryptocurrencies such as Bitcoin, the scope of blockchain technology has substantially broadened over time.

The inclusion of smart contract functionalities in blockchain platforms (Rathee et al., 2022) has introduced programmability features, amplifying the technology's versatility. This enhancement has extended the applicability of blockchain technology beyond the financial sector, offering new avenues in diverse domains, including healthcare, education, governance, and manufacturing. Specifically, blockchain technology in healthcare and EHR systems (Rathee et al., 2021) provides a range of features capable of meeting the complex requirements associated with EHR systems (McGhin et al., 2019). These features and capabilities are expounded upon in the following sections.

## 1.2 EHR System Requirements

***Data Integrity and Security:*** The imperative for data integrity and robust security measures cannot be overstated in the healthcare sector, especially in EHR systems. Blockchain technology is viable for ensuring data integrity through cryptographic algorithms (McGhin et al., 2019).

***Interoperability:*** The capacity for diverse healthcare systems to communicate and exchange information seamlessly, known as interoperability, represents a critical requirement for EHR systems. Blockchain technology fosters interoperability by providing a standardized protocol for data exchange across disparate systems.

***Access Control and Authorization:*** Effective access control and authorization mechanisms are paramount in safeguarding sensitive medical data. With its intrinsic decentralized architecture, Blockchain technology enables granular control over data access rights, fulfilling this essential EHR requirement.

***Scalability:*** As healthcare systems evolve, so does the need for EHR systems to scale in response to increasing data volumes and user access requirements. Blockchain technology offers scalable solutions that can be adapted to meet the dynamic needs of healthcare institutions (McGhin et al., 2019).

***Real-time Access:*** Immediate and real-time access to healthcare records is indispensable for timely and effective medical interventions. The decentralized and distributed nature of blockchain technology ensures that EHR data is readily accessible when required, thereby fulfilling this crucial requirement.

***Auditability:*** The capacity to track changes and access to healthcare records is critical for ensuring compliance with regulatory standards. Blockchain technology inherently supports auditability by maintaining an immutable record of all transactions and modifications.

## 1.3 Blockchain Opportunities

***Enhanced Security Measures:*** Utilizing cryptographic algorithms intrinsic to blockchain technology offers an elevated level of data security, effectively mitigating unauthorized access to Electronic Health Record (EHR) systems.

***Immutable Data Provenance:*** The indelible nature of blockchain ledgers facilitates traceable, non-alterable data transaction histories, meeting stringent auditing and compliance requirements.

***Decentralization Paradigm:*** The decentralized architecture inherent in blockchain technology nullifies single points of failure, thereby augmenting the resilience and reliability of EHR systems.

***Automated Workflows via Smart Contracts:*** Deployment of smart contracts in blockchain frameworks automates multiple healthcare workflows, including but not limited to insurance claims and permissioned data sharing, thereby diminishing administrative overhead.

***Standardization and Interoperability:*** Blockchain technology is a unified layer, enabling disparate healthcare information systems to collaborate and exchange data, resolving prevalent data siloing issues.

***Patient-Centric Data Control:*** Implementation of blockchain technology empowers individual patients with enhanced control over personal healthcare data, fostering a more patient-centric healthcare ecosystem.

***Cost-Efficiency Metrics:*** Automation of diverse administrative and data-sharing functionalities via blockchain technology holds the potential for significant reductions in healthcare operational costs.

***Real-Time Data Availability:*** Decentralized architecture engenders real-time data-sharing capabilities among authorized entities, a feature of critical importance for time-sensitive medical decisions.

***Integrity of Data Records:*** The immutable, timestamped records within blockchain ledgers prevent data tampering, instilling higher levels of data trust and integrity.

***Global Research Collaboration:*** The borderless characteristics of blockchain technology facilitate international collaborative efforts in medical research, enabling secure and seamless data sharing across geopolitical boundaries.

## 1.4 Convergence of EHR Requirements and Blockchain Capabilities

The alignment between the multifaceted requirements of EHR systems and blockchain technology's inherent capabilities is a pivotal factor. This confluence has catalyzed numerous proposals advocating integrating blockchain-based applications into EHR systems. Such applications span a variety of functionalities, including but not limited to EHR monitoring, auditing protocols, mobile applications for healthcare, and mechanisms for secure information exchange (Barka et al., 2021).

### 1.5 Identified Limitations in Blockchain-EHR Integration

While the synergy between blockchain technology and EHR systems offers promising avenues for innovation and improved healthcare delivery, certain limitations have been delineated. These constraints pertain specifically to integrating blockchain technology within existing EHR systems and require critical examination for effective implementation.

This dual-edged perspective—highlighting both the potential and the limitations—provides a comprehensive overview of the current state of blockchain technology's applicability in the realm of EHR systems.

The subsequent sections of this manuscript are arranged in a structured manner to facilitate comprehensive understanding. Section 2 delves into preliminary concepts related to blockchain technology and existing Electronic Health Record (EHR) systems implementations. Section 3 elucidates the architecture of the proposed ML-BbEHR system, followed by a detailed presentation of its implementation. Section 4 encompasses functional and security assessments, augmented by a performance evaluation and an analytical discussion of the results. The manuscript culminates with concluding remarks in Section 5.

## 2. BACKGROUND AND RELATED WORK

This investigation elucidates the potential utilities, intricacies, and indispensable features of multi-layered blockchain technology within Electronic Health Records (EHR). The primary avenues of inquiry for this scholarly review encompass:

- Challenges associated with migrating from traditional, centralized EHR architectures to decentralized, multi-layered blockchain infrastructures.
- Requisite characteristics for multi-layered blockchain-enabled EHR systems to achieve broad-based adoption.
- Groundbreaking conceptualizations for EHR systems predicated on multi-layered blockchain technology, deliberately omitting specific architectures like Ethereum to concentrate on general blockchain frameworks.

This scholarly work aims to substantially contribute to the EHR field by scrutinizing how multi-layered blockchain technology can fortify security protocols. Through an exhaustive review of extant literature and methodological frameworks, this investigation identifies existing voids in EHR solutions. These voids include data integrity, system interoperability, and patient-centric data control. Instead of relying on Ethereum-based solutions, this study advocates a multi-layered blockchain architecture as a viable strategy to address these identified shortcomings (Ho et al., 2016).

With the escalating adoption of Electronic Health Records (EHR), there has been a concomitant rise in the digital storage of sensitive health-related data. Although EHR systems offer myriad benefits, such as streamlined access to patient information and optimized healthcare workflows, they also introduce significant challenges. These challenges include, but are not limited to, issues with data integrity, system interoperability, and user governance over personal data. Despite the availability of centralized solutions and cryptographic methodologies, existing EHR systems manifest inherent security vulnerabilities.

## 2.1 Literature Review

Research (Shi et al., 2020) presents an in-depth analysis of blockchain strategies designed for EHR systems, emphasizing privacy and security. This study identifies several research opportunities and challenges that need addressing. (Shahnaz et al., 2019) Propose a decentralized storage platform for patient records, addressing the issue of scalability through off-chain scaling solutions.

A study (Tanwar et al., 2020) surveys various blockchain solutions to overcome healthcare system limitations, specifically EHRs. The paper introduces algorithms for access control to enhance data accessibility among healthcare professionals.

A comprehensive study by (Fatima et al., 2022) discusses state-of-the-art blockchain-based healthcare systems. The paper acknowledges the challenges associated with the acceptance and adaptability of this novel technology, emphasizing the need for further research to ensure data security and privacy.

Various research endeavors, such as those described in (Khan et al., 2018), have proposed methods to secure EHR systems. Nonetheless, these methods predominantly focus on access control mechanisms, often employing public-private key cryptography for user authentication and data encryption (Guerbouj et al., 2019). However, these mechanisms are susceptible to certain security breaches (Ferrag et al., 2019). Furthermore, existing systems largely fail to comply with healthcare-specific requirements, such as HIPAA regulations, and do not offer Turing-complete platforms to accommodate complex healthcare workflows (Mistry et al., 2020).

In recent years, blockchain technology has emerged as a viable alternative for securing digital transactions and maintaining data integrity. A few notable proposals include MedRec, BBDS, MedShare, OmniPHR, and MedBlock, each with unique attributes and limitations (Moin et al., 2019). Notably, most of these solutions do not discuss the application of multi-layered architectures, a focus of this research.

## 2.2 Challenges in Centralized EHR Systems

(Magyar et al., 2017) Outline four key challenges of integrating centralized EHR systems into blockchain-based solutions. These challenges include disparate data formats, complex integration protocols, potential security vulnerabilities, and unclear regulatory landscapes.

## 2.3 Essential Features for Blockchain-based EHR Systems

(Zhang et al., 2017) identify seven critical metrics required to accept blockchain-based EHR systems: HIPAA compliance, Turing completeness, user identification, and cost-effectiveness. The researchers offer insights into the limitations of current encryption methods in blockchain, suggesting alternative approaches for secure data storage.

## 2.4 Existing Blockchain-Based EHR Systems

**MedRec: A Case Study -** Although an in-depth analysis is provided by (Azaria et al., 2016) concerning the use of blockchain for integrating centralized EHRs among healthcare providers, the focus remains on general blockchain solutions that could substitute or improve MedRec's functionalities.

**BBDS: A Permissioned Approach –** (Xia et al., 2017) present BBDS, a permissioned blockchain system for EHR access control. Despite its advantages, BBDS faces challenges regarding decentralization and scalability.

**MedShare: A Hybrid Cloud Infrastructure -** The study by (Yang et al., 2018) introduces MedShare, a system that utilizes a hybrid cloud infrastructure for EHR exchange. Although practical, MedShare lacks certain features like immutable access control and privacy isolation that could be facilitated by blockchain technology.

**OmniPHR: A Non-Turing-Complete Blockchain Solution -** (Roehrs et al., 2017) propose OmniPHR, designed to offer a unified view of healthcare records. Despite its novel approach, the system's scalability and data storage capabilities are limited.

**MedBlock: A Role-Based Model –** (Fan et al., 2018) describe MedBlock as a system designed for efficient medical data sharing. While MedBlock does address some challenges, it lacks features like programmability and efficient access control mechanisms.

## 2.5 Proposed Multi-Layered Blockchain Architecture for HER

In light of the limitations in existing solutions, this research proposes a novel multi-layered blockchain architecture specifically tailored for EHR systems. This architecture aims to improve data integrity, enable better interoperability between healthcare providers, and give users more control over their data. The proposed architecture is not based on Ethereum but leverages general blockchain technology for enhanced scalability and efficiency.

The multi-layered approach allows for the customization of security protocols and access permissions at various levels of data interaction. For instance, one layer could be exclusively for healthcare providers insurance companies, and yet another for patients, each with its corresponding blockchain maintaining the integrity of interactions within that layer.

Table 1 presents a detailed comparative assessment of extant blockchain-based EHR implementations. Current solutions predominantly rely on public/private key methodologies and Ethereum-centric frameworks. Contrarily, the scope of this scholarly review concentrates on the applicability of multi-layered blockchain technology to meet the generic requirements of EHR, including defenses against authorized internal intrusions.

This scholarly examination underscores the imperative for further research in multi-layered blockchain technology, specifically engineered for EHR applications. Such a focus establishes a foundational base for forthcoming investigations in this rapidly expanding field.

This research aims to redress the prevailing deficiencies in EHR security paradigms by proposing a multi-layered blockchain architecture. These deficiencies encompass data integrity, system interoperability, and end-user governance of personal information. Through methodical experimentation and comparative evaluations, this study endeavors to furnish empirical substantiation that corroborates the effectiveness of a multi-layered blockchain architecture in augmenting EHR security mechanisms.

*Table 1. Existing blockchain implementations for EHRs*

| Reference | Blockchain Type | Main Feature | Targeted Security Service |
|---|---|---|---|
| Shahnaz et al. [8] | Ethereum | Off-chain data storage | Access control |
| Tanwar et al. [9] | Hyperledger Fabric | Fast data accessibility | Access control |
| Xia et al. [19] | Any permission blockchain | Secure data sharing on the blockchain | Access control and availability |
| Yang et al. [20] | Hybrid cloud infrastructure | Real-time testbed in three hospitals | Privacy and access control |
| Roehrs et al. [21] | Not specified | Efficient data block distribution | Authentication, privacy, and digital signature management |
| Fan et al. [22] | Not specified | Scalable and secure interactions using double-signing | Confidentiality and access control |
| Dagher et al. [23] | Ethereum | Easy system integration with collaborative mining | Access control and availability |
| Proposed solution | Ethereum | Independent from private/public key standard approach | Privacy, access control, and confidentiality |

## 3. SYSTEM ARCHITECTURE

In the following discussion, an overview of Multi-Layered Blockchain-Based Electronic Health Records (ML-BbEHR) architecture is delineated in line with the design principles outlined in the referenced work. The ML-BbEHR system adopts a layered architectural paradigm, incorporating modular and distributed functionalities. Specifically, the system's architecture encompasses four layers: User Interface (UI), Middleware, Multi-layered Blockchain, and Cloud Store, as represented in Figure 1 and their associated modules.

The UI layer serves as the system's front end, facilitating interactions between ML-BbEHR and its diverse user base, including but not limited to medical practitioners, pharmacists, administrative staff, researchers, and insurance entities. This layer is localized to each healthcare provider participating in ML-BbEHR and comprises obligatory and customizable components tailored to the specific requirements of each provider.

Occupying a pivotal role in the ML-BbEHR architecture, the Middleware layer acts as the system's backbone. Its primary function includes the interconnection of all other layers within the ML-BbEHR framework while also delivering key services such as directory and database functionalities.

Distinct from the Middleware, the Multi-layered Blockchain layer is a shared resource across all participating healthcare providers. This layer harbors two core functionalities:

1. Immutable smart contracts for the execution of predefined logical operations.
2. Immutable tables manifested as chained blocks designed to securely store diversified data types resistant to unauthorized alterations.

Simultaneously, the Cloud Store layer, another shared resource among healthcare providers engaged in ML-BbEHR, is responsible for housing patients' unprocessed Electronic Health Records (EHRs). Indexing keys for these raw EHRs are securely stored in the Multi-layered Blockchain's immutable tables.

*Figure 1. Overview of the ML- BbEHR architecture*

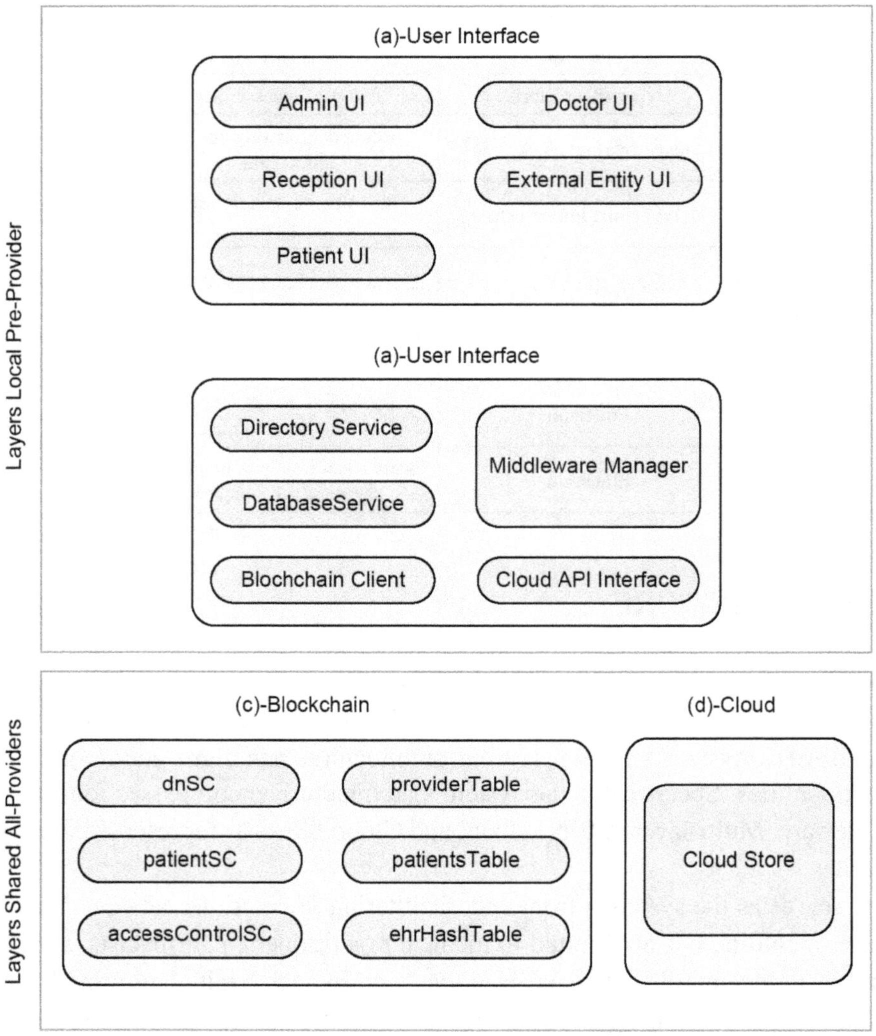


The reader is referred to for an in-depth understanding of the constituent elements and their intricate interconnections, including functional use cases (Baqari et al., 2020).

## 3.1 Functional Requirements

In designing the ML-BbEHR prototype, component selection adhered to multiple criteria: ease of setup and configuration, interoperability with other technological solutions, feature robustness—including built-in security mechanisms, and consistent reliability and operational stability. In alignment with these criteria, the following technologies were integrated:

**Django Web Development Platform:** Employed in constructing user interfaces for healthcare practitioners, administrators, and reception staff. Django, a Python-based framework, offers inherent directory services, facilitating its integration with custom Python modules for extended functionalities like API-Ethereum interaction.

**Python Interpreter:** Situated at the heart of the ML-BbEHR architecture, Python serves multiple roles, including but not limited to programming language for Django-based web development, facilitator for interaction with built-in directory services, integrator for blockchain and cloud storage APIs, and implementer of SQL interfaces for healthcare databases. Python also handles the integrity validation and hashing processes for Electronic Health Records (EHRs).

**SQLite:** Deployed for the local healthcare provider's database service module, SQLite principally stores data types such as department IDs, appointment schedules, and blockchain-specific information. Its native connectivity with Django is mediated via Python.

**Ethereum Blockchain:** Chosen for its public nature, Ethereum contributes immutable smart contracts and databases to the ML-BbEHR ecosystem. It is particularly useful for enabling wide-ranging EHR accessibility, even accommodating scenarios where patients may change geographical locations, thus requiring new healthcare providers to retrieve historical EHRs.

**Microsoft Azure Files:** Selected for cloud-based EHR storage, this service offers advantages such as ease of use, format-agnostic data storage, and secure communication channels via the SMB protocol (Microsoft, 2020).

## 3.2 Django Architecture

In the architecture of Django, a bifurcation into front-end and back-end layers is observed. The front-end layer is constructed using HTML templates that serve as the interface for client interactions. Conversely, the back-end layer receives inputs from these templates and executes the corresponding programmed logic. A block diagram delineating this architecture is presented in Figure 2.

**Forms in Django:** These Python-based functions, contained within the `forms.py` file, facilitate HTML coding by defining input fields, their labels, types, and constraints, such as maximum lengths. It is noteworthy that Forms are devoid of HTML styling and JavaScript functionalities.

**HTML Templates:** Client-facing HTML pages are generated by amalgamating stylesheets, JavaScript, bootstraps, and other HTML-defined elements, integrated with inputs sourced from `forms.py.`

**Models in Django:** Defined within the `models.py` file, Models serve as abstraction layers that translate Python instructions into database queries based on the specific type of database integrated. These Models can receive input values from HTML templates via Views (`views.py`) and pre-filling Forms with database-polled data for client presentation (e.g., dropdown selections).

**Tables in Django:** Located in the `tables.py` file, these structures facilitate the client-side presentation of database-derived information in tabular format, simplifying HTML table creation. Styling remains external to Tables and is managed via HTML templates.

**Views in Django:** Acting as the orchestrators, Views interconnect Forms, Tables, and Models with Templates. They regulate the application's web flow logic and manage client-web application request-response mechanisms. Within Views, the ML-BbEHR system's essential functionalities are instantiated. Furthermore, Views enables the integration of custom Python models to imbue the system with functionalities beyond the native capabilities of Django.

*Figure 2. Django web development architecture*

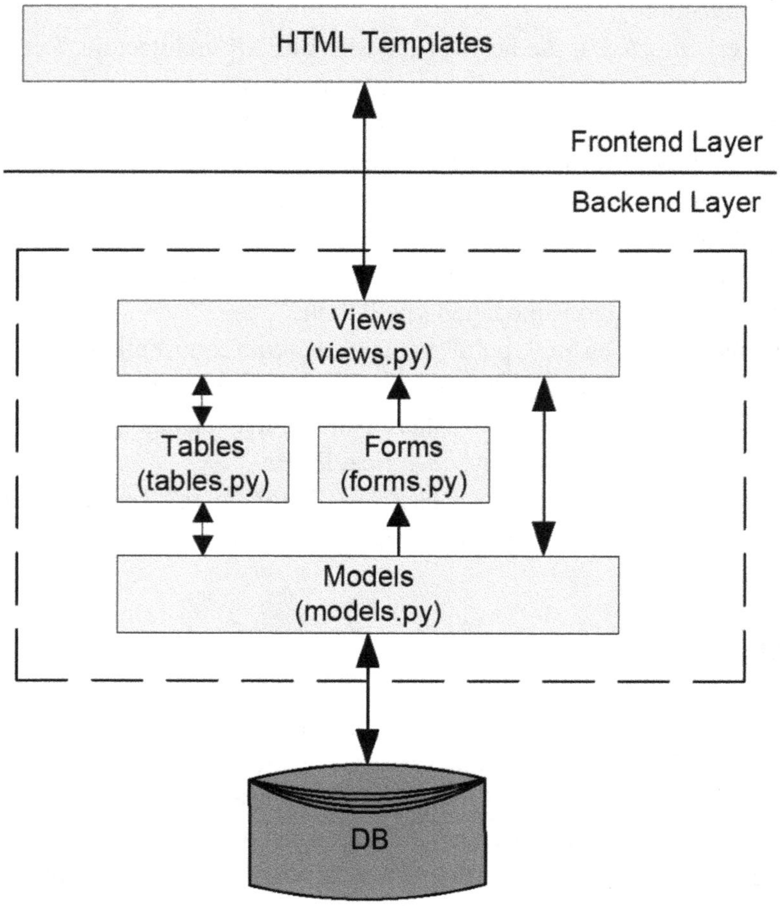


## 3.3 Blockchain Architecture

### 3.3.1 Development Environment

Simulation models were deployed in two distinct environments to validate the feasibility and practicality of the proposed multi-layered blockchain framework focused on Electronic Health Records (EHR). The initial model employed a Hyperledger Fabric (HLF) blockchain at the second layer, encapsulating computational healthcare nodes (peers), Application Programming Interfaces (APIs), and organizational structures. A parallel simulation at the third layer utilized a Base Server (BS) workstation to execute blockchain applications for a comparative analysis of Ethereum and HLF implementations.

IBM Cloud served as the development environment, hosting tools, and technologies crucial for the implementation. The HLF network, in this experimental setup, comprised four peers and an orderer node, executed as Docker images within Docker containers (Docker, 2017). Utilizing Ubuntu Linux 18.4 LTS, hosted on an Intel Core i7-3770 @ 3.4 GHz processor with 16 GB memory, the Docker engine (version 19.03.8) executed the environment, with Docker Compose (version 1.17.0) managing the configuration of images and containers.

Smart contracts were instantiated on peer nodes, and data storage resources were allocated for blockchain transaction logging. Composer Command Line Interface (CLI) facilitated the deployment, implementation, and execution of smart contracts and associated definitions. Peer nodes utilized CouchDB to manage state data capable of handling complex queries against transaction logs. ChainCode (CC), modeled as JavaScript Object Notation (JSON), could be invoked by client applications to access the state database and perform various operations. RESTful APIs were deployed through a REST server, facilitating transaction submissions through HTTP GET or POST requests. The orderer node maintained ledger consistency by implementing the Practical Byzantine Fault Tolerance (PBFT) algorithm. The HLF framework supported the Execute-Order-Validate and Commit transaction model ((Du et al., 2016)).

For the third-layer simulation, a workstation ran the blockchain applications as the Base Server (BS). This setup enabled the measurement of throughput and latency parameters of both Ethereum and HLF private networks under similar conditions. Workstations equipped with an Intel Core i7-3770 @ 3.4 GHz processor and 16 GB memory functioned as the BS. Only one mining node was deployed in the Ethereum network to maintain simplicity.

### 3.3.2 Multi-Layer Security Framework

The primary objective of the proposed multi-layered blockchain architecture is to establish a secure, reliable, and efficient mechanism tailored for EHR systems. Utilizing advanced machine learning and intelligent clustering techniques based on Swarm Intelligence (SI) and Evolutionary Computation (EC) algorithms (Nykyri et al., 2019), the architecture is designed to instantiate a multi-layered structure. This hierarchical framework segregates the network into distinct tiers, each serving a specialized function. Layer 1 is composed of various clusters encompassing diverse EHR data sets. Layer 2 consists of control nodes responsible for administrative and governance roles, while Layer 3 serves as the backbone, providing substantial computational infrastructure for the network.

Although integrating blockchain technology inherently introduces potential scalability and overhead concerns (Khan et al., 2018), this multi-layered approach mitigates such limitations. As illustrated in Figure 3, the architecture aims to minimize system overhead, reduce latency, facilitate data-specific channeling, secure communications, and address scalability imperatives. Within this schema, Layer-1 nodes utilize localized authentication and authorization services managed by administrative nodes. Layer 2 operates within a localized, permissioned blockchain environment, enabling secure and efficient transactions. Layer 3, with its high computational capabilities, is structured to deploy robust asymmetric cryptographic mechanisms, thereby reinforcing system-wide security and privacy through utilizing a global multi-layered blockchain and sophisticated security protocols.

#### *3.3.2.1 Layer-1*

The proposed architecture features an intricate multi-layered blockchain system designed to cater specifically to the complexities of EHR systems. This hierarchical model includes communication network elements, procedures, and protocols orchestrated by unsupervised hybrid clustering algorithms. These algorithms transform the overall network into multiple clusters, each associated with an administrative control node, commonly called Cluster Head (CH). The privacy and security within each cluster are managed through local authorization and authentication services, thereby mitigating risks associated with globally centralized models.

*Figure 3. Multi-layer network model with local authorization and integrated blockchain layers*

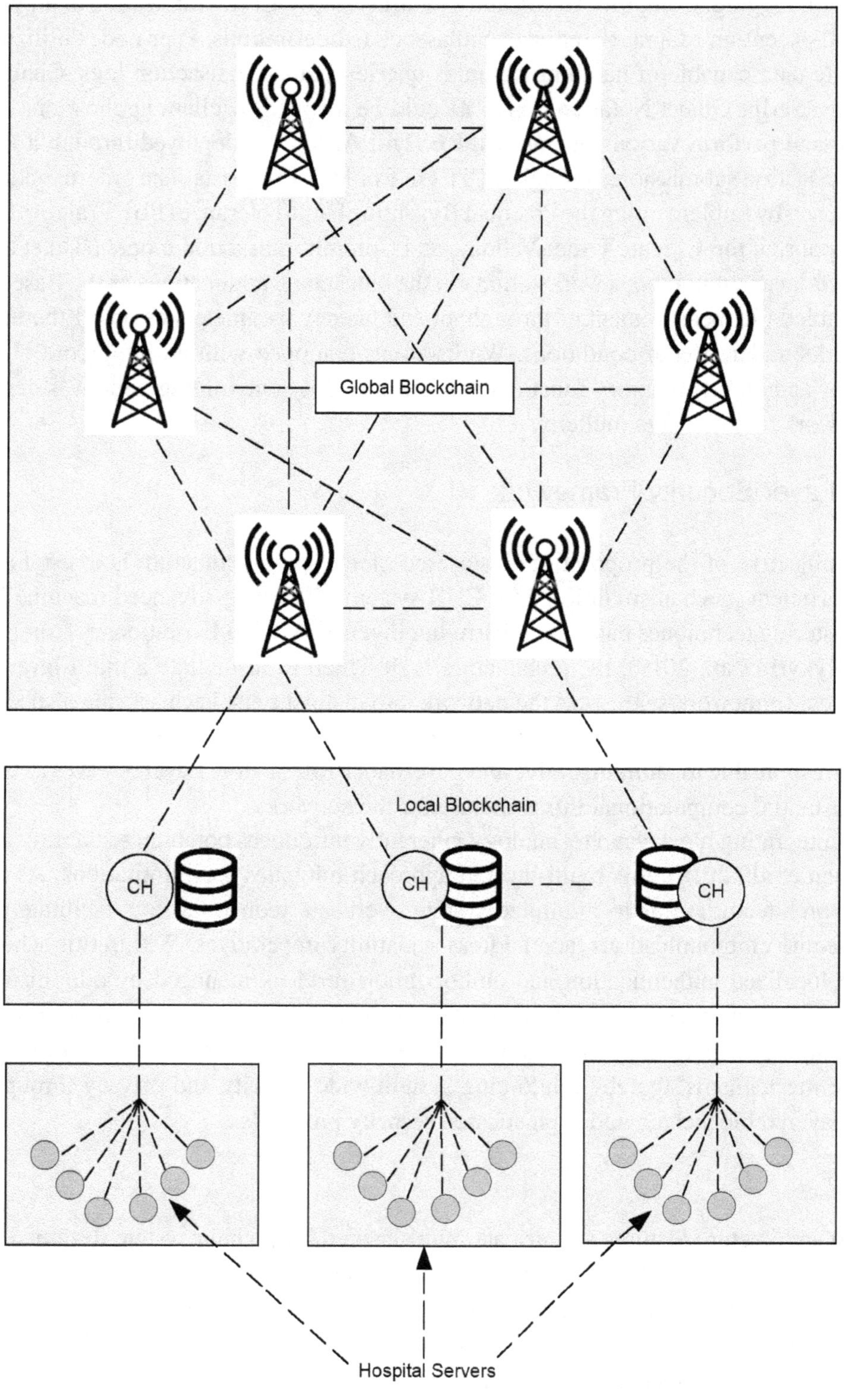

*Figure 4. Local authorization service among IoT entity and Cluster Head (CH) nodes for secure communication*

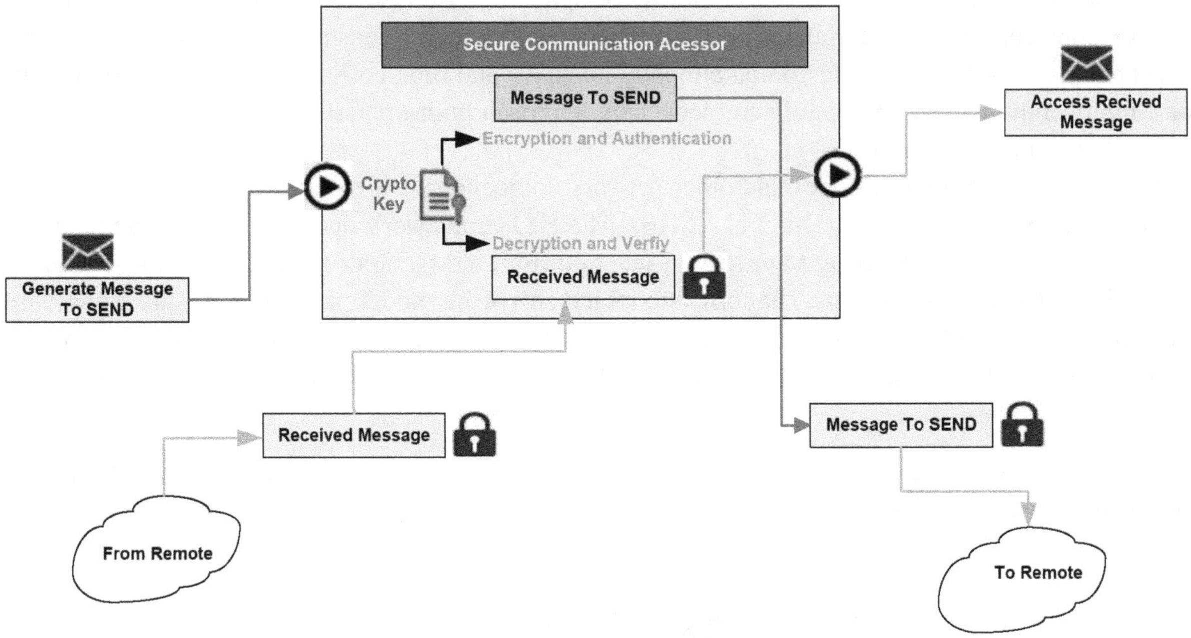


Intra-cluster security is enhanced by CH nodes that also serve as edge processing units, as depicted in Figure 3. This design elevates implementation efficiency and addresses the challenges related to centralized cloud computing solutions. This framework assigns lightweight session keys to establish secure communication channels during authentication. These session keys undergo periodic validation by CH nodes, which also manage local registration and authentication services to improve scalability and manage device heterogeneity.

The architecture employs symmetric keys and lightweight cryptography to tackle scalability challenges. CH nodes are responsible for critical tasks such as new node registration, session key distribution, communication initiation, and secure communication establishment. Encryption protocols, including symmetric key-wrapping, secure these lightweight session keys, termed distribution keys. Every communication transaction is safeguarded through these session keys, each with a unique ID and a defined validity period. Overall, the cryptographic operations are managed by selected CH nodes for secure communication, as outlined in Figure 4.

### 3.3.2.2 Layer-2

The architecture advances to its second layer, wherein selected Cluster Heads (CH) nodes operate under the jurisdiction of Base Station (BS) units. These CH nodes are responsible for data aggregation and forwarding to the higher computational layer. Within this tier, each node executes a localized version of a private Ledger-Based Blockchain (LBC), achieving consensus through a predefined algorithm (Khan et al., 2018). BS and CH units are tasked with block generation and verification, intra-layer communications, and block broadcasting across respective clusters. Trusted nodes within this layer serve as interfaces to subordinate and superior architectural layers, utilizing a blockchain protocol. The Hyperledger Fabric (HLF) blockchain platform is recommended for integration at this level.

The proposed network model is meticulously designed to ensure secure CH communications, particularly considering the decentralized node distribution and resource constraints. Given that traditional blockchain mining processes are computationally intensive and time-consuming, the architecture opts for a lightweight, private, decentralized blockchain approach built upon distributed consensus for efficient data communications.

Interactions between CH nodes and other network components are facilitated through a localized, permissioned HLF blockchain (Du et al., 2016). The HLF framework operates on an execute-order-validate architecture, segregating transaction execution from transaction ordering to achieve enhanced scalability and modular consensus. The cornerstone elements of this model encompass organizations (base stations), network nodes, ordering clusters, and a variety of peers, including endorsers and committers, Membership Service Providers (MSPs), and channels (De Angelis et al., 2018), as illustrated in Figure 5.

*Figure 5. The network architecture of the IoT-enabled cellular system combined with blockchain technology and smart contracts*

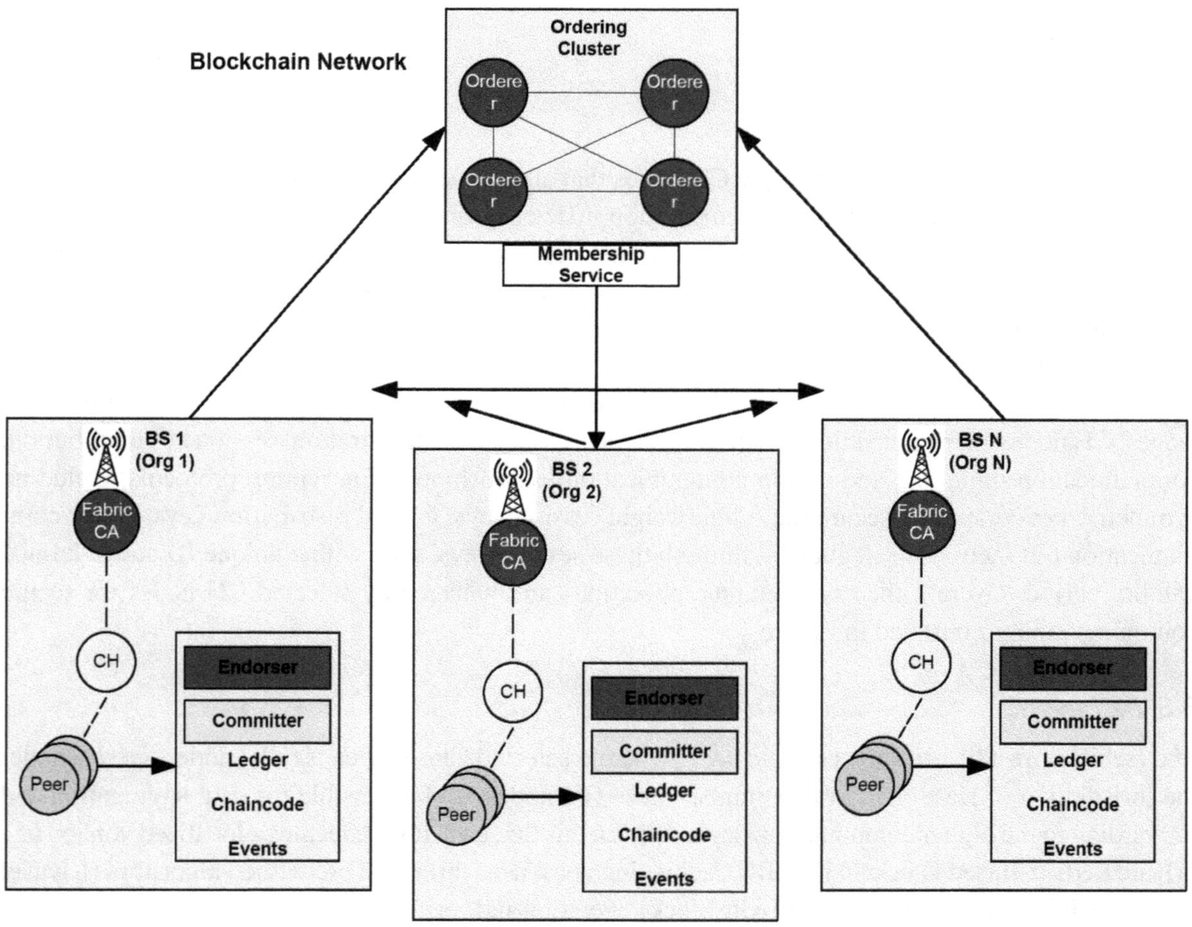

In this advanced multi-layered blockchain architecture tailored for EHR systems, peer nodes play pivotal roles in maintaining the distributed ledger and executing transactions. These peers can function as endorsers, committers, or both. Orderers in the network are tasked with transaction sequencing and new block proposals as part of the aggregate ordering service. By default, all peers act as committers, validating transactions in newly received blocks, committing the changes locally, and updating the blockchain by appending new blocks. Some peers also serve as endorsers, executing smart contracts—called ChainCode in the Hyperledger Fabric (HLF) framework—and appending the results with cryptographic signatures or endorsements.

Membership Service Providers (MSP) administer authentication services within the Hyperledger network, verifying the identities of network nodes. Organizations within this framework represent logical entities that manage network members via MSP. Channels within the network facilitate discrete connections between various network elements using private or dedicated pathways.

Transactions within the Hyperledger Blockchain are orchestrated through smart contracts (Cachin et al., 2016). Base stations in the upper layers serve as organizational hubs for blockchain implementation, interfacing with Cluster Heads (CH) and other network nodes. Ordering clusters manage transactions and queue orders, facilitating peer-to-peer communications through a shared channel. The ordering service additionally handles message broadcasting, transaction block creation, and distribution to CH nodes.

In this architecture, the role of a node—whether as an endorser or a committer—is determined based on predefined roles and various metrics such as network configuration. Aside from their validation responsibilities, committer nodes are specifically tasked with appending new blocks to the blockchain ledger. Endorser roles are established by submitting endorsement requests approved after a consistency check via smart contract execution. Transaction blocks are generated by ordering clusters and disseminated to all CH nodes. This ensures a synchronized update of the blockchain ledger across the network, including transaction details and specific EHR data attributes. After validation, a copy of the blockchain ledger is disseminated to all network peers, maintaining a unified and secure data management system for healthcare records.

#### 3.3.2.3 Layer-3

Base Station (BS) nodes operate as organization owners within a distributed network framework in the upper echelons of this multi-layered blockchain architecture designed for EHR systems. These BS nodes possess substantial computational resources, enabling robust asymmetric cryptographic mechanisms to be deployed. This capability is augmented by utilizing a global blockchain, such as the Ethereum blockchain framework, and advanced cryptographic techniques like Elliptic Curve Cryptography (ECC) (Joglekar et al., 2019). The blockchain implementation in this layer bolsters privacy and security while ensuring data integrity.

Notably, this high-level layer lacks any central node, and the devices operate independently regarding data. The blockchain ledger records transactions among nodes at this level, facilitating a globally distributed trust mechanism. Cluster heads, base stations, and computational edge nodes within the network initialize this.

The inherent peer-to-peer structure of blockchain technology provides an optimal framework for establishing a globally distributed, secure communication network among diverse entities, including BS nodes. Communication among Cluster Head (CH) and computational edge nodes is mediated through blockchain-based protocols, employing certificates as a trust mechanism. Smart contracts oversee the distribution of these certificates, thus enabling trusted communication within the blockchain network. CH nodes are mandated to sign these certificates to validate their authenticity digitally.

This architecture enhances distributed trust between CH nodes and corresponding BS nodes, especially during collaborative authorization processes. Additionally, it amplifies trust levels when an entity initiates communication with nodes located in different clusters under various BS nodes.

The system eliminates the need for fixed addresses and domain names for cluster heads during communications with edge devices and for the execution of smart contracts. Smart contracts within the blockchain network obviate the requirement for domain names and fixed addresses, thereby simplifying the communication protocols among nodes. This approach yields a more streamlined, secure, and efficient mechanism for managing electronic health records in a decentralized fashion.

## 3.4 Implementation Steps

The development process of the ML-BbEHR prototype, intended for streamlining Electronic Health Records (EHRs) through a multi-layered blockchain architecture, encompasses several pivotal steps outlined below:

**Runtime Environment Construction:** The initial phase involves setting up the runtime environment essential for the ML-BbEHR application, laying the foundation for further development activities.

**Initialization of Web and Database Components:** Web and database components crucial to the ML-BbEHR system are initialized. This core structure supports the application logic and data management functionalities.

**Development of ML-BbEHR-specific Django Code:** Customized code, adapted to ML-BbEHR functionalities, is developed within the Django framework. This code forms the crux of logic responsible for EHR management, from user interface elements to back-end operations.

**Blockchain Smart Contracts Creation:** Smart contracts on a generic blockchain platform are developed to provide immutable logic functions and secure data storage for EHRs. These contracts serve to automate various operations and validations within the ML-BbEHR system.

**Integration of Django with Blockchain:** The concluding step is integrating the Django framework with blockchain capabilities through API connections and tailored Python modules. This enables the ML-BbEHR system to exploit the blockchain's robust, immutable data structures for EHR management.

### 3.4.1 Building the Runtime Environment

**Runtime Environment Configuration:** MacOS Catalina is the selected operating system for hosting the ML-BbEHR system's prototype components, as delineated in Section A. This OS choice facilitates the execution of key technologies, including Django, Python, and SQLite, and ensures compatibility for communication with a generic blockchain and MS Azure.

**Python and IDE Installation:** Python 3.8 and PyCharm 2019.3.2 IDE were downloaded and installed, laying the foundational software stack for the subsequent development phases.

**Virtual Environment and Project Initialization:** Within the PyCharm IDE, a new virtual environment was instantiated, designating Python 3.8 as the interpreter. This environment serves as the runtime for both Django and custom Python modules. Subsequently, a new project, ML-BbEHR, was initiated within this virtual environment.

*Table 2. The libraries used in the implementation of ML-EBbEHR*

| Library Name | Purpose |
|---|---|
| Django | Web development framework: This installation includes SQLite 3 |
| django-Tables2 | For formatting and styling tables in Django |
| crispy | For formatting and styling HTML templates in Django |
| web3 | For API communication with Ethereum |
| azure-storage-file | For API communication with MS Azure |

**Library Installation:** Upon establishing the ML-BbEHR project, a compilation of requisite Python libraries was installed. These libraries, summarized in Table 2, were fetched and installed using the PIP3 utility.

### 3.4.2 Django Web Framework and Database

**Django Web Framework and Database Initialization:** After configuring the runtime environment, steps were taken to initialize the Django web framework and its corresponding database. The aim was to build a multi-layered blockchain-based Electronic Health Records (EHR) management system. The initialization sequence was as follows:

**SQLite3 Database Initialization:** Commands from the PyCharm terminal within the ML-BbEHR project were executed to initialize the SQLite3 database. This database would hold information about the healthcare provider's Ethereum details and clinic information. The commands for initialization were: `Python manage to make migrations and `Python manage to migrate.`

**Django Admin User Creation:** An admin user was created within the Django framework to handle tasks such as account creation for various healthcare personnel, including doctors, nurses, and receptionists. The admin user also populated the SQLite3 database with pertinent information. The command for this step was: `Python manages to create a superuser.`

**User Account and Group Management:** The admin interface, accessed at `http://localhost:8000/admin` on 31 October 2022, created ML-BbEHR user accounts and groups. Various permissions were assigned to these groups, and users were associated with their respective groups.

**Local Database Tables for Healthcare Providers:** A structure was defined for the SQLite3 database tables to store healthcare provider information. These tables included:

- ***Appointments Table:*** Included columns for patient name, National ID, appointment date and time, and the clinic to be visited.
- ***Departments Table:*** Incorporated clinic code and clinic name.
- ***Provider Table:*** Consisted of Ethereum public and private keys.
- The Python code for table creation was located within the `models.py` file.

**Data Population:** The Django admin console populated the Departments and Provider tables with the healthcare providers' clinic details and Ethereum information. The appointment database, however, was populated by receptionists at a later stage.

### 3.4.3 Django Python Code Implementation

Following the successful initialization of the Django framework, the development process shifted to writing the Python code required for the ML-BbEHR prototype, focusing specifically on enhancing Electronic Health Records (EHR) security through multi-layered blockchain technology.

**Front-End HTML Templates:** The initial coding phase involved crafting HTML templates for the front end. A master template (`base.html`) was created to contain elements common to all pages, such as headers and footers. Child templates were subsequently developed to contain page-specific elements integrated with the master template. The complete HTML coding approach is summarized in Figure 6.

**Forms and Views:** After HTML template creation, functions were built in `forms.py` to define the input fields associated with each template. These forms were then linked to their corresponding HTML templates through functions within `views.py.` Figures 7 - 9 illustrate the structure and relationships between `forms.py,` `views.py,` and the associated User Interfaces.

**Role-Based Access Control (RBAC):** Access to different User Interfaces was managed using Python decorators to implement RBAC. The decorators captured the session username, verified its group membership (configured during Django initialization), and granted access based on the user's group.

**EHR Access Control:** Further layers of access control were applied to EHRs. Permissions were granted to clinics based on the session user's membership in the ML-BbEHR Reception group and

*Figure 6. Summary of the HTML ML-BbEHR template structure*

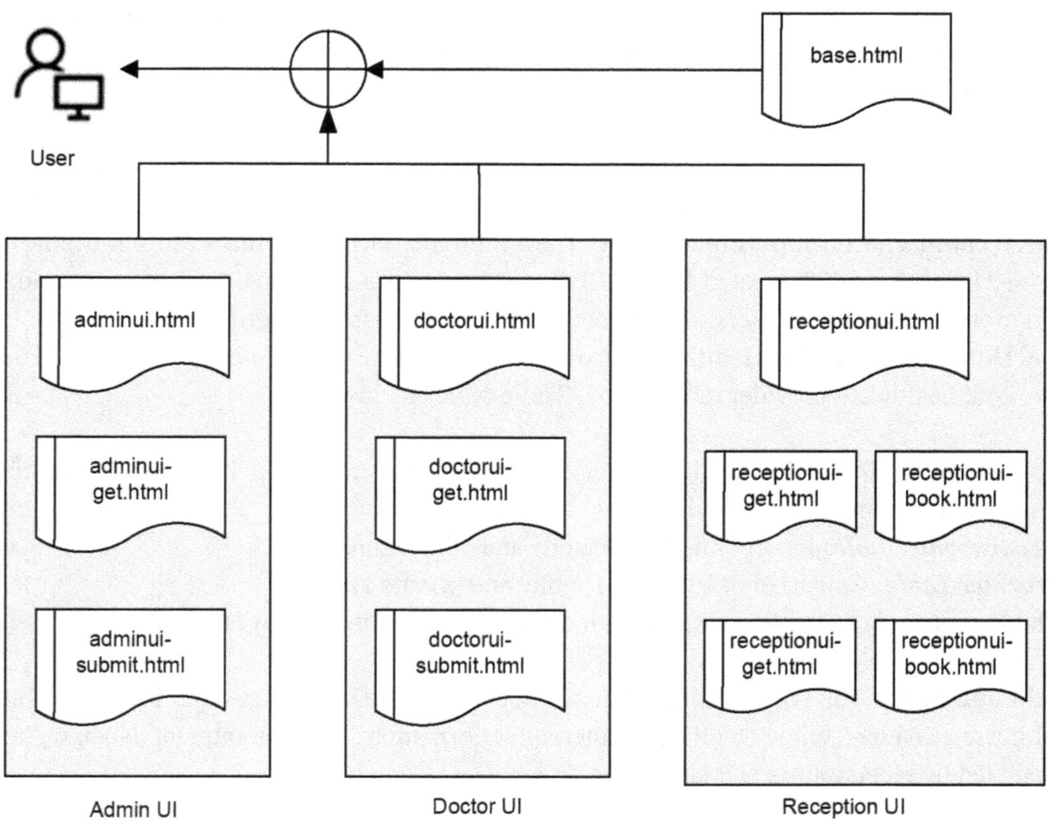

*Figure 7. Admin UI with forms and views*

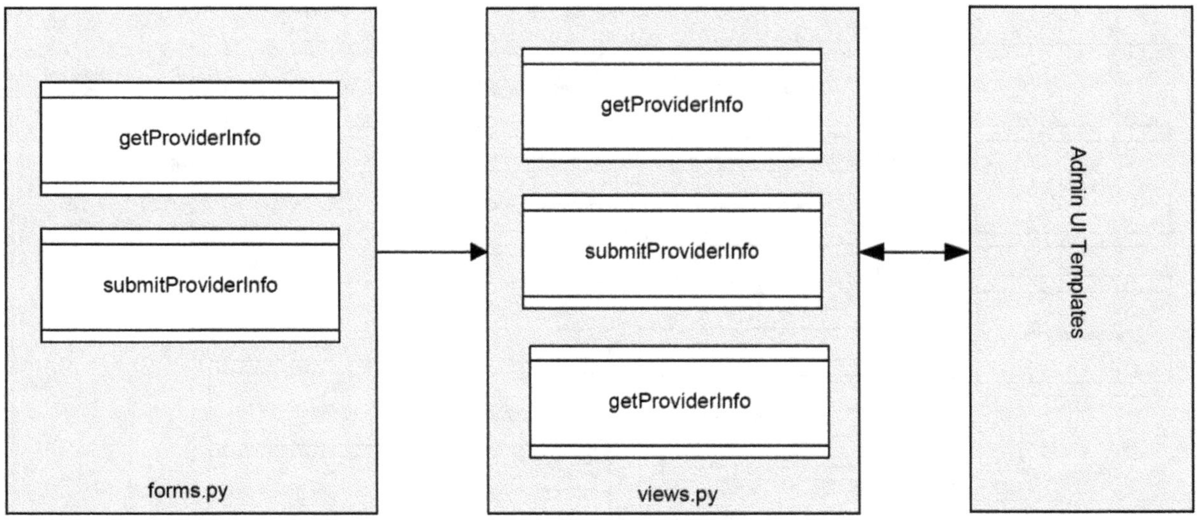


*Figure 8. Doctor UI with forms and views*

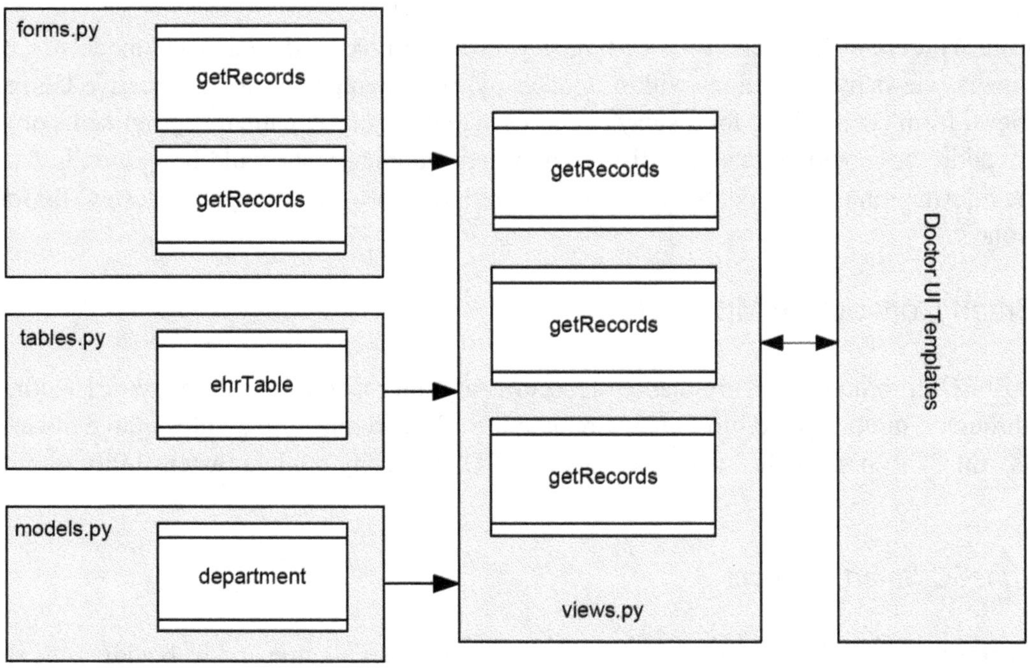

*Figure 9. Reception UI with forms and views*

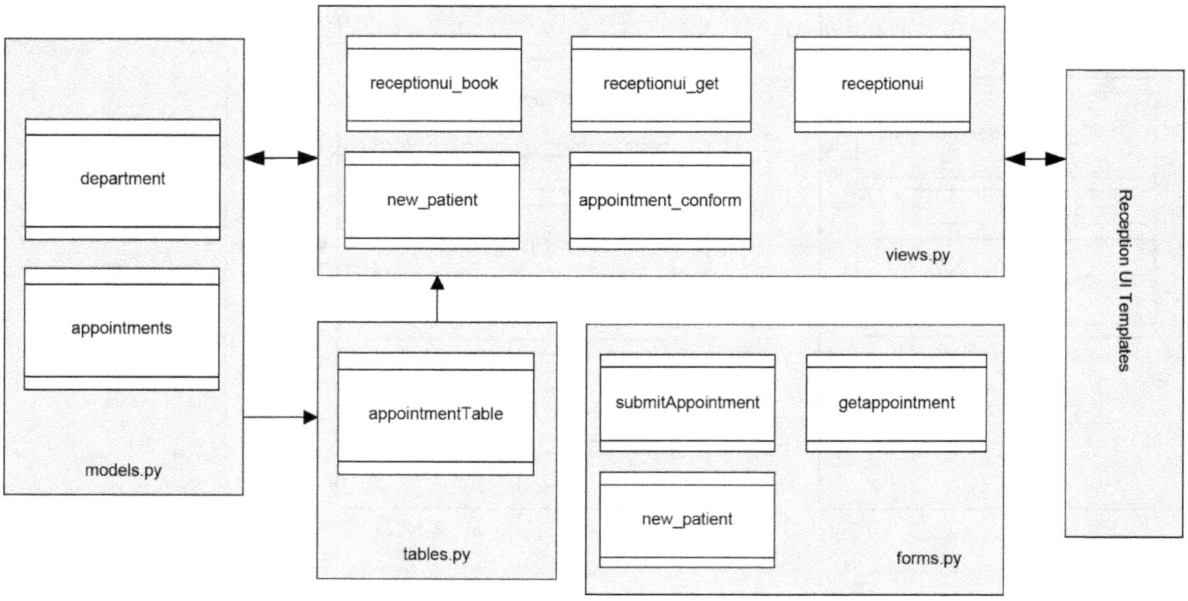


patient fingerprint verification. Another layer of access control specific to doctors is discussed in a subsequent section.

**Linking Models to UIs:** The final coding step involved linking the database models to the User Interfaces via `views.py.` Functions within `models.py` were mapped to their respective UIs, enabling data retrieval from the database for display to end-users. For instance, patient appointments or provider Ethereum addresses could be read and displayed, or new appointments could be written into the database. The information from `models.py` was transferred to `views.py` through `tables.py` for formatted table displays.

### 3.4.4 Smart Contracts in ML-BbEHR

The ML-BbEHR prototype aims to enhance the security and interoperability of Electronic Health Records (EHR) through a multi-layered blockchain architecture. This architecture employs three distinct smart contracts: `dnsSC,` `patientsSC,` and `accessControlSC,` fostering modular extensibility for future use cases.

#### *3.4.4.1 dnsSC Smart Contract*

- **`createProvider` Function:** Accepts a healthcare provider's name and web address as string inputs. These entries are stored in an immutable blockchain table called `providersTable,` indexed by the healthcare provider's blockchain address (OHP). In case of duplicate providers, an exception error is triggered.
- **`getProvider` Function:** Takes a blockchain address (OHP) as input and retrieves the healthcare provider's name and web address from the `providersTable.`

#### *3.4.4.2 accessControlSC Smart Contract*

- **`addClinic` Function:** Utilizes the patient's fingerprint hash and clinic ID as string inputs. Both parameters are stored in an array, indexed by the fingerprint hash, within the `patientsTable.` This array signifies that clinics have authorized access to a specific patient's EHR.
- **`grantClinicAccess` Function:** Takes in the patient's fingerprint hash and the clinic ID. It searches for the clinic ID within the `patientsTable` using the patient's fingerprint hash. A 'True' or 'False' value is returned, depending on whether the clinic can access the patient's EHR.

#### *3.4.4.3 patientsSC Smart Contract*

- **`createPatient` Function:** Incorporates several string inputs, including the patient's fingerprint hash, name, date of birth, healthcare provider's blockchain address, and reception ID. This information is stored in the `patientsTable,` indexed by the fingerprint hash.
- **`getPatient` Function:** Requires the patient's fingerprint hash and fetches the associated information from the `patientsTable`.
- **`createEHR` Function:** Accepts multiple parameters such as the fingerprint hash, EHR name, date, status, and Merkle root hash. These entries are stored in a separate blockchain table named `ehrHashTable,` indexed by the fingerprint hash.
- **`get EHR` Function:** Takes the fingerprint hash as input and retrieves the list of EHRs for the patient. The data returned includes the EHR name, date, Merkle root hash, and status.

The smart contracts are deployed on a blockchain platform, each having a unique blockchain address to facilitate secure and effective communication. The healthcare provider is responsible for the initial deployment of these smart contracts, which are pivotal to the secure and efficient management of EHRs within the ML-BbEHR framework.

### 3.4.5 Blockchain with Django Web Application

The ML-BbEHR prototype employs a multi-layered blockchain architecture to optimize the handling of Electronic Health Records (EHR). The Django web application integrates with this blockchain platform via customized Python modules, utilizing a dedicated mining node service to ensure patients' privacy. Figure 10 provides an overview of this integration.

#### *3.4.5.1 Dedicated Mining Node for Privacy*

- A dedicated mining node runs proprietary blockchain mining software, ensuring patients' EHRs remain private. A third-party mining pool service was used for the prototype, offering up to 100,000 blockchain transactions within 24 hours.

*Figure 10. Integrating Django with Ethereum*

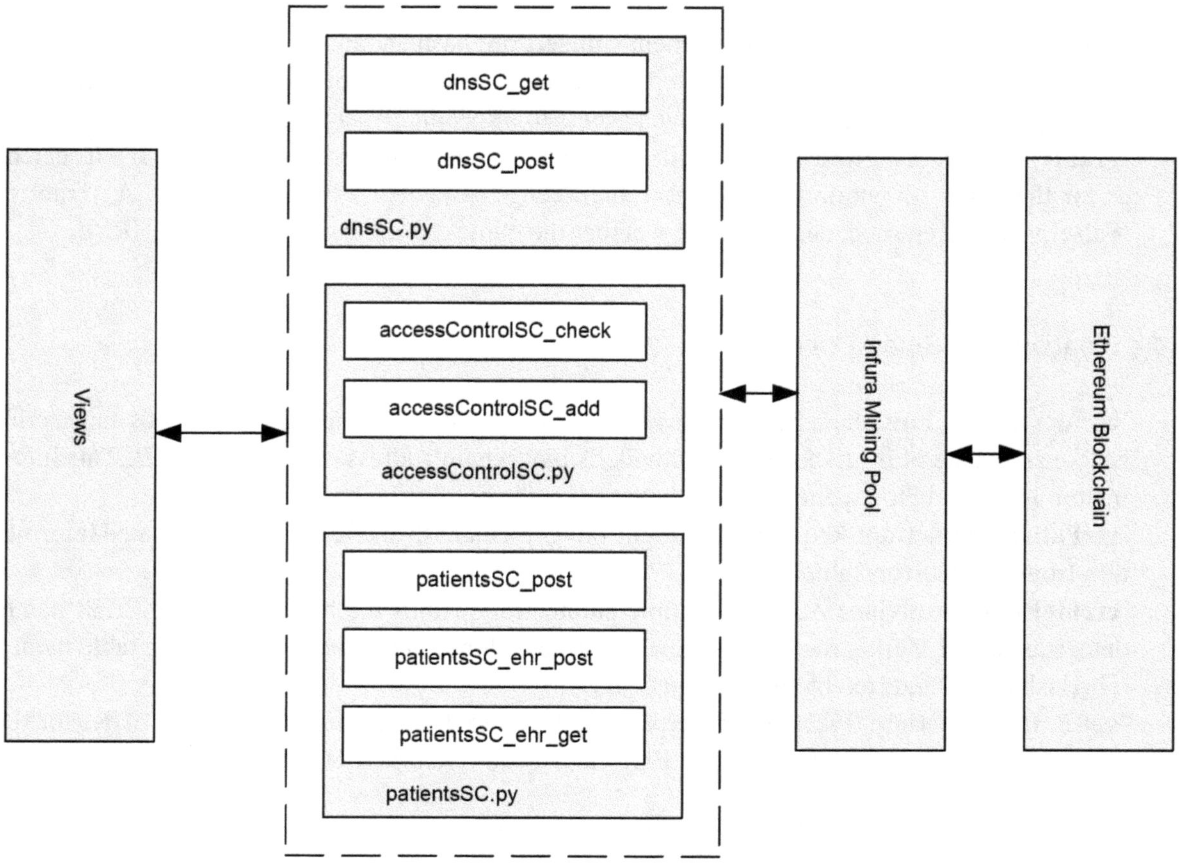


### *3.4.5.2 Unique URL for Blockchain Communication*

- A unique URL is generated upon setting up an account with the third-party mining pool. This URL is the communication channel for posting and reading blocks to and from the blockchain.

### *3.4.5.3 Custom Python Modules for Blockchain Interaction*

- Django views interact with the blockchain through custom Python modules. These modules utilize Web3-like APIs to enable communication with the smart contracts on the blockchain.
- Each Python module contains functions corresponding to the respective smart contract functions deployed on the blockchain. Figure 11 outlines the workflow of these Python modules.

Through this integration, the Django web application can securely and efficiently manage EHRs, fulfilling the requirements of the ML-BbEHR prototype in a healthcare setting. This approach ensures that all EHR transactions are immutable, transparent, and compliant with privacy regulations.

*Figure 11. Custom Python module operation*

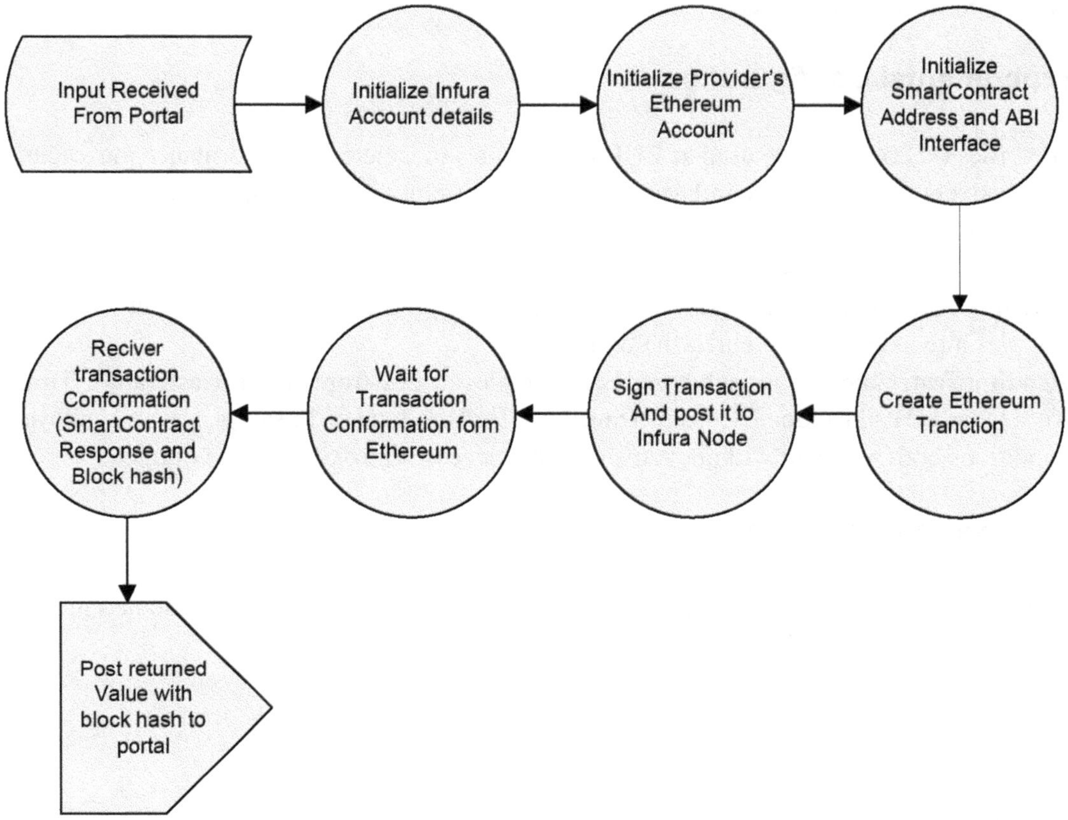


## 4. PERFORMANCE EVALUATION

A series of simulations and real-world tests will be conducted to validate the efficacy of the proposed multi-layered blockchain architecture. These tests measure various metrics such as latency, throughput, and HIPAA compliance. This research phase will include a comparative analysis of existing EHR solutions.

Integrating blockchain technology into Electronic Health Records (EHR) management remains a subject of considerable research interest. This scholarly evaluation focuses on the ML-BbEHR architectural model, specifically addressing performance bottlenecks arising from the Proof of Work (PoW) consensus algorithm utilized in the foundational blockchain layer. Smart contracts, designated as 'accessControlSC' and 'patientsSC,' regulate this layer.

### 4.1 Performance Assessment

Endorsed by the Hyperledger Performance and Scale Working Group (Caliper, 2019), two quintessential metrics have been employed to assess the blockchain performance in EHR management:

1. **Transaction Throughput:** Defined as the aggregate number of transactions successfully committed by the System Under Test (SUT) within a designated time frame, usually quantified in seconds.

2. **Transaction Latency:** This represents the period needed for a transaction to be permanently recorded on the blockchain ledger.

## 4.2 Experimental Configuration

Small text files (< 20 KB) were used as EHR test samples to isolate the performance intricacies attributable exclusively to the blockchain layer, independent of cloud storage speed variables. Two distinct tests were conducted:

1. **Writing Test:** Involving the creation of a test patient profile and the subsequent writing of EHR samples into a designated `ehrHashTable.`
2. **Reading Test:** Comprising each stored EHR sample retrieval from the `ehrHashTable.` Time delays were recorded using Google Chrome Developer Tools. A total of 15 samples were employed under conditions both with and without activating the 'accessControlSC' smart contract.

## 4.3 Observational Findings

Performance metrics were evaluated under consistent internet connectivity, as presented in Figures 12 and 13. Key observations include:

*Figure 12. Reading performance testing*

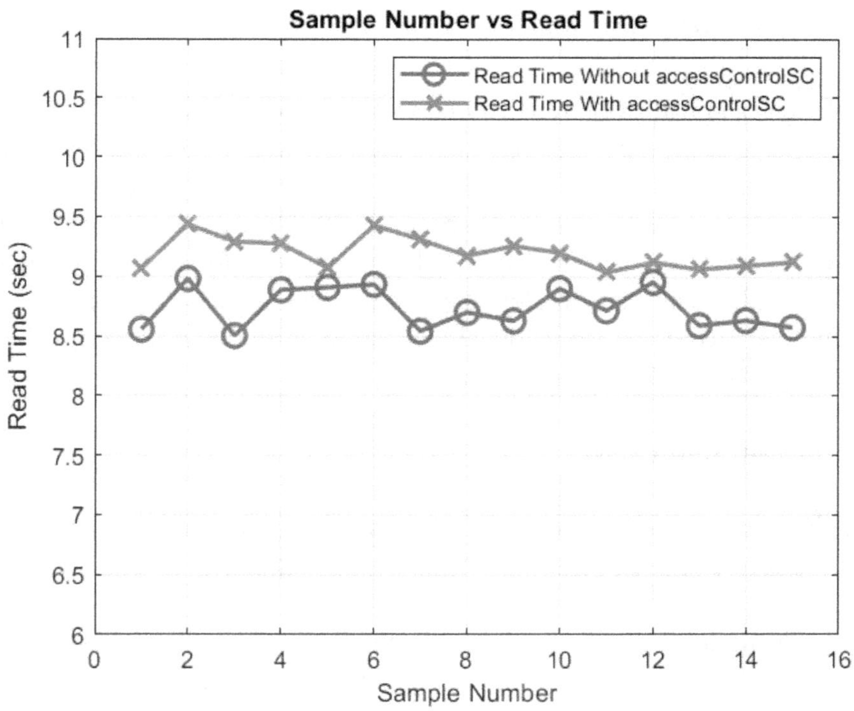

*Figure 13. Writing performance testing*

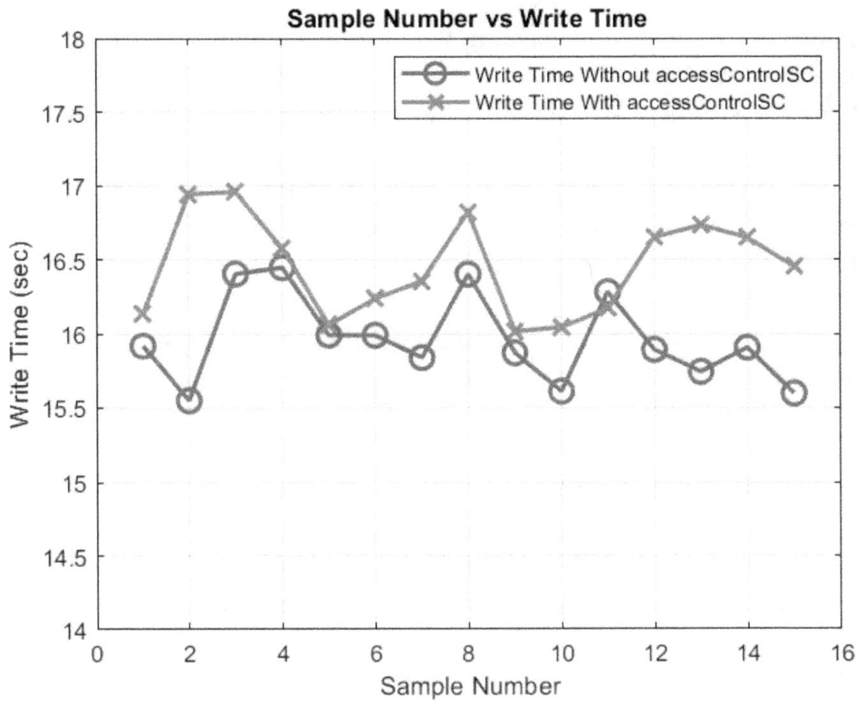


*Figure 14. Reading time analysis with a larger ehrHashTable*

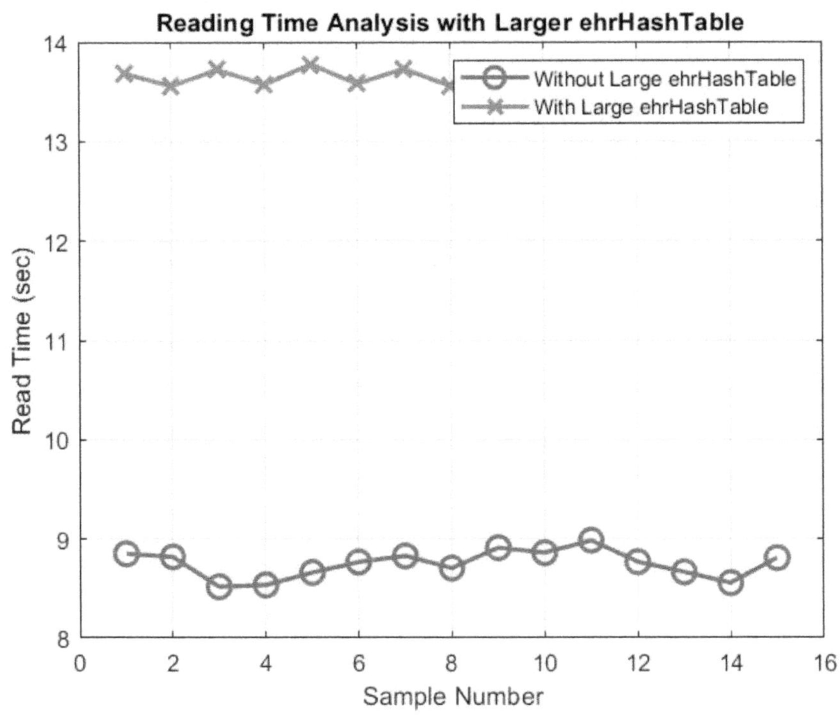

1. **Reading vs. Writing Latency:** Average reading time recorded was 8.81 seconds, while writing time was nearly twice as long, at 16 seconds.
2. **Impact of Access Control:** Implementing the 'accessControlSC' smart contract introduced no discernible latency overhead.
3. **Writing Time Consistency:** Writing time remained stable, irrespective of the volume of records in `ehrHashTable.`
4. **Reading Time Variability:** A noticeable increase in reading time, from 8.81 seconds to 13.67 seconds, was observed upon expanding the `ehrHashTable` by adding five more EHR samples, as shown in Figure 14.

While not aligning with modern internet speed expectations, these delays are generally acceptable in healthcare settings. Further refinement through code and database optimization could offer performance improvements.

## 4.4 Resource Allocation Strategy

An in-depth examination of the multi-layered blockchain architecture was conducted, employing Hyperledger (Caliper, 2019) to facilitate blockchain configuration. The performance benchmarks were grounded in academic literature to validate the system's efficacy. Resource allocation was meticulously designed to meet latency and throughput criteria, as summarized in Figures 15 and 16. Experimental outcomes indicate that the SUT can sustain nearly 100% throughput and an average latency of 500 milliseconds up to a sending rate of 110 TPS. Beyond this, a decline in throughput and increased maximum latency to nearly 15 seconds were observed, attributed to container resource constraints on peer nodes. Notably, all transactions in all test cases were completed, reinforcing the robustness of this architecture.

*Figure 15. Latency vs. transaction sending rate*

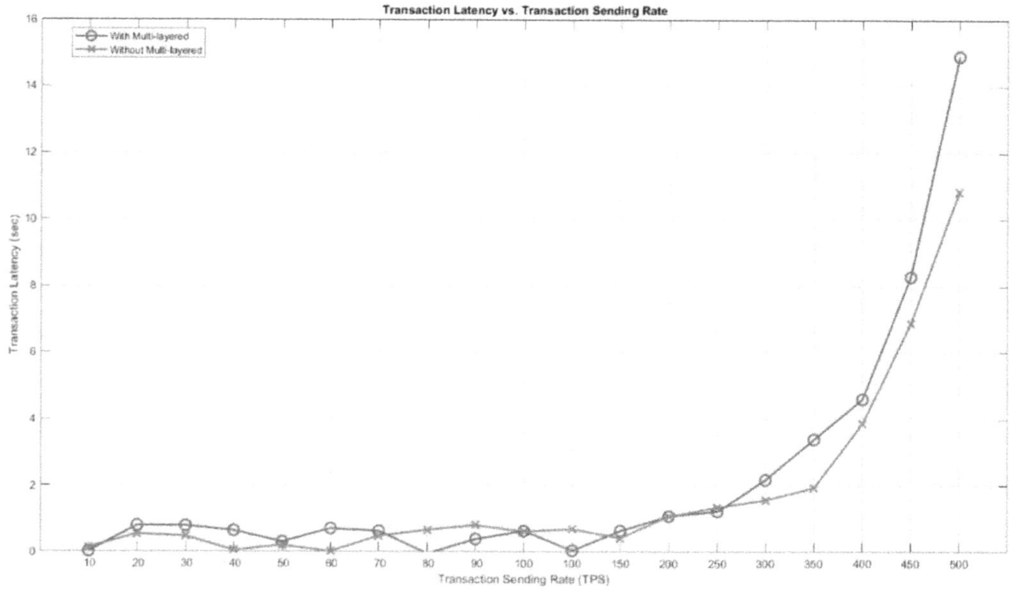

*Figure 16. Throughput vs. transaction sending rate*

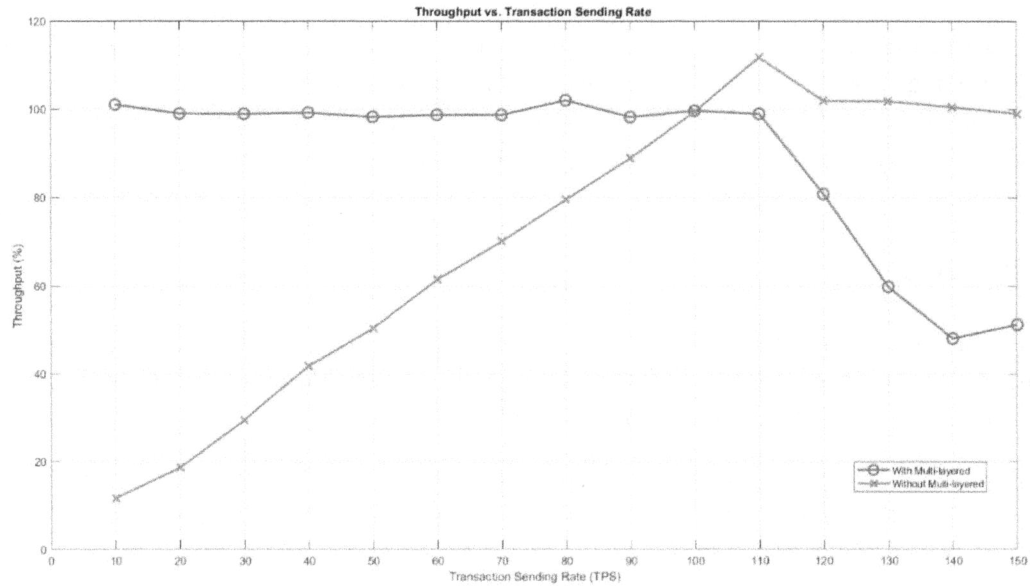


This examination substantiates the potential of the multi-layered blockchain architecture in the efficient and secure management of Electronic Health Records (EHR). Despite the inherent limitations of the PoW consensus algorithm and other configuration parameters, the architecture demonstrates resilience and robustness, affirming its viability for scalable EHR management. Future research should focus on optimizing code and databases to enhance performance metrics.

## 5. CONCLUSION

The research explores the development of a multi-layer blockchain model aimed at enhancing access control and recovery mechanisms for Electronic Health Records (EHRs) distributed among various healthcare providers. The study began with a comprehensive review of blockchain's applications in healthcare, subsequently narrowing its focus to address challenges and requirements specific to EHR systems. This led to the proposal of a novel multi-layer system that delineates roles among healthcare providers, the blockchain, and cloud storage, enabling secure and flexible access recovery for EHRs.

One of the unique features of this model is its independence from traditional public/private key strategies, thereby enhancing its resilience against internal and external security threats. Designed to improve security and interoperability, the model provides external User Interface (UI) integration options to help healthcare providers transition smoothly to blockchain technologies. A prototype was developed to validate the proposed system using Django, Python, Ethereum, and MS Azure technologies. While the prototype met most functional and security requirements, it did experience some latency, primarily due to the Ethereum blockchain. However, this latency was considered acceptable within healthcare settings.

## 6. FUTURE SCOPE

Future work aims to evaluate the system against hybrid blockchains that utilize faster consensus algorithms, thereby improving its performance. The team also plans to introduce advanced biometric security features and functionalities in smart contracts for more nuanced access control. The prospect of using mobile-based biometric authentication to give patients greater control over their EHRs is also being considered.

## REFERENCES

Azaria, A., Ekblaw, A., Vieira, T., & Lippman, A. (2016). Medrec: Using blockchain for medical data access and permission management. In *Proceedings of the 2016 2nd International Conference on Open and Big Data* (pp. 25–30). IEEE. 10.1109/OBD.2016.11

Baqari, M., & Barka, E. (2020). Biometric-Based Blockchain EHR System (BBEHR). In *Proceedings of the 2020 International Wireless Communications and Mobile Computing* (pp. 2228–2234). IEEE. 10.1109/IWCMC48107.2020.9148357

Barka, E., Dahmane, S., Kerrache, C. A., Khayat, M., & Sallabi, F. (2021). STHM: A Secured and Trusted Healthcare Monitoring Architecture Using SDN and Blockchain. *Electronics, 10*, 1787.

Cachin, C. (2016). *The architecture of the hyper ledger blockchain fabric*. In *Workshop on Distributed Cryptocurrencies and Consensus Ledgers*. Chicago, IL, USA.

Caliper, H. (2019). Hyperledger Caliper Architecture. *Hyperledger*. (https://hyperledger.github.io/caliper/docs/2_Architecture) (accessed on 15 September 2020).

Dagher, G. G., Mohler, J., Milojkovic, M., & Marella, P. B. (2018). Ancile: Privacy-preserving framework for access control and interoperability of electronic health records using blockchain technology. *Sustainable Cities and Society, 39*, 283–297.

De Angelis, S., Aniello, L., Baldoni, R., Lombardi, F., Margheri, A., & Sassone, V. (2018). PBFT vs. proof-of-authority: Applying the CAP theorem to permissioned blockchain. In *Italian Conference on Cyber Security*. Milan, Italy.

Docker, Inc. (2017). What is Docker? Docker, Inc. https://www.docker.com/what-docker

Du, K. L., & Swamy, M. (2016). Search and optimization by metaheuristics. In *Techniques and Algorithms Inspired by Nature*. Birkhauser. doi:10.1007/978-3-319-41192-7

Fan, K., Wang, S., Ren, Y., Li, H., & Yang, Y. (2018). Medblock: Efficient and secure medical data sharing via blockchain. *Journal of Medical Systems, 42*, 136.

Fatima, N., Agarwal, P., & Sohail, S. S. (2022). Security and Privacy Issues of Blockchain Technology in Health Care—A Review. In ICT. *Analysis and Applications*, 193–201.

Ferrag, M. A., Derdour, M., Mukherjee, M., Derhab, A., Maglaras, L., & Janicke, H. (2019). Blockchain Technologies for the Internet of Things: Research Issues and Challenges. *IEEE Internet of Things Journal, 6,* 2188–2204.

Guerbouj, S. S. E., Gharsellaoui, H., & Bouamama, S. (2019). A Comprehensive Survey on Privacy and Security Issues in Cloud Computing, Internet of Things and Cloud of Things. *International Journal of Service Science, Management, Engineering, and Technology, 10,* 32–44.

Ho, G., Leung, D., Mishra, P., Hosseini, A., Song, D., & Wagner, D. (2016). Smart locks: Lessons for securing commodity Internet of things devices. In *Proceedings of the 11th ACM on Asia Conference on Computer and Communications Security* (pp. 461–472). ACM. 10.1145/2897845.2897886

Hyperledger Blockchain. (2018). *Performance Metrics.* (White paper, Vol. 1). https://www.hyperledger.org/wpcontent/uploads/2018/10/HL_Whitepaper_Metrics_PDF_V1

Joglekar, J., Bhutani, S., Patel, N., & Soman, P. (2019). Lightweight Elliptical Curve Cryptography (ECC) for Data Integrity and User Authentication in Smart Transportation IoT Systems. In Sustainable Communication Networks and Application (pp. 270–278).

Khan, M. A., & Salah, K. (2018). IoT security: Review, blockchain solutions, and open challenges. *Future Generation Computer Systems, 82,* 395–411.

Magyar, G. (2017). Blockchain: Solving the privacy and research availability tradeoff for EHR data: A new disruptive technology in health data management. In *Proceedings of the 2017 IEEE 30th Neumann Colloquium* (pp. 135–140). IEEE. 10.1109/NC.2017.8263269

McGhin, T., Choo, K. K. R., Liu, C. Z., & He, D. (2019). Blockchain in healthcare applications: Research challenges and opportunities. *Journal of Network and Computer Applications, 135*, 62–75.

Microsoft Azure. (2020). *What is Azure Files?* Microsoft Azure. https://azure.microsoft.com/en-in/products/storage/files/

Mistry, I., Tanwar, S., Tyagi, S., & Kumar, N. (2020). Blockchain for 5G-enabled IoT for industrial automation: A systematic review, solutions, and challenges. *Mechanical Systems and Signal Processing, 135,* 106382.

Moin, S., Karim, A., Safdar, Z., Safdar, K., Ahmed, E., & Imran, M. (2019). Securing IoTs in the distributed blockchain: Analysis, requirements, and open issues. *Future Generation Computer Systems, 100*, 325–343.

Nakamoto, S. (2019). Bitcoin: A Peer-to-Peer Electronic Cash System. Decentralized Business Review. https://manubot.org/

Nykyri, M., Kuisma, M., Kärkkäinen, T. J., Hallikas, J., Jäppinen, J., Korpinen, K., & Silventoinen, P. (2019). IoT Demonstration Platform for Education and Research. *In Proceedings of the 2019 IEEE 17th International Conference on Industrial Informatics (INDIN)* (Vol. 1, pp. 1155–1162). IEEE. 10.1109/INDIN41052.2019.8972280

Rathee, G., Ahmad, F., Sandhu, R., Kerrache, C. A., & Azad, M. A. (2021). On designing and implementing a secure blockchain-based hybrid framework for Industrial Internet-of-Things. *Information Processing & Management, 58*, 102526.

Rathee, G., Kerrache, C. A., & Ferrag, M. A. (2022). A Blockchain-Based Intrusion Detection System Using Viterbi Algorithm and Indirect Trust for IIoT Systems. *Journal of Sensor and Actuator Networks, 11*(71).

Roehrs, A., da Costa, C. A., & da Rosa Righi, R. (2017). OmniPHR: A distributed architecture model to integrate personal health records. *Journal of Biomedical Informatics, 71*, 70–81.

Shahnaz, A., Qamar, U., & Khalid, A. (2019). Using blockchain for electronic health records. *IEEE Access, 7,* 147782–147795.

Shi, S., He, D., Li, L., Kumar, N., Khan, M. K., & Choo, K. K. R. (2020). Applications of blockchain in ensuring the security and privacy of electronic health record systems: A survey. *Computer Security, 97*, 101966.

Tanwar, S., Parekh, K., & Evans, R. (2020). Blockchain-based electronic healthcare record system for healthcare 4.0 applications. *Journal of Information Security and Applications, 50*, 102407.

Xia, Q., Sifah, E. B., Smahi, A., Amofa, S., & Zhang, X. (2017). BBDS: Blockchain-based data sharing for electronic medical records in cloud environments. *Information, 8,* 44.

Yang, Y., Li, X., Qamar, N., Liu, P., Ke, W., Shen, B., & Liu, Z. (2018). Mindshare: A novel hybrid cloud for medical resource sharing among autonomous healthcare providers. *IEEE Access, 6,* 46949–46961.

Zhang, P., Walker, M. A., White, J., Schmidt, D. C., & Lenz, G. (2017). Metrics for assessing blockchain-based healthcare decentralized apps. In *Proceedings of the 2017 IEEE 19th International Conference on e-Health Networking, Applications and Services (Healthcom)* (pp. 1–4). IEEE. 10.1109/HealthCom.2017.8210842

# Chapter 6
# IoT Sensors for Smart Automation:
## A Systematic Review

**C. L. Chayalakshmi**
*Basaveshwar Engineering College, Bagalkote, India*

**Mahabaleshwar S. Kakkasageri**
*Basaveshwar Engineering College, Bagalkote, India*

**Rajani S. Pujar**
*Basaveshwar Engineering College, Bagalkote, India*

**Nayana Hegde**
https://orcid.org/0000-0001-7467-2553
*Reva University, India*

## ABSTRACT

*The internet of things (IoT) has changed the comforts of every individual. Sensors and communication models play an important role in the effective implementation of IoT systems for any applications. Sensors can be classified based on the parameters to be measured and on the working principles. The second significant component in IoT-based systems is communication models, which help to exchange the data between the two IoT devices. Basically, there are four types of communication models that are commonly used: device to device, device to cloud, device to the gateway, and backend data sharing communication models. This chapter explains various sensors that can be connected to the internet for transferring data between IoT devices, application areas and communication models. The classification of IoT application areas is based on the scope of functionality, use, and adaptation. Sensors are classified on the basis of types of measurement such as vision, position, physical, and others. Basic communication models and open research challenges are explained at the end.*

DOI: 10.4018/979-8-3693-0659-8.ch006

## 1. INTERNET OF THINGS

Internet of Things (IoT) plays a significant role in our day to day life. It includes various application areas such as health care, vehicles, entertainment, home and industrial appliances, and sports (Ammar et al., 2018). IoT is a vast and a fast growing technology which connects the objects/things anywhere in the world in a network through internet. IoT is an amalgamation of sensor technology, computational intelligence, communication, artificial intelligence and microelectronics (Poornima, 2021). Things in IoT with the processing unit are able to aggregate, process, take decisions, and transmit/receive the generated data, without intervention of human beings (Alanzi et al., 2019). It is estimated that 15 billions of IoT devices in 2023 and may increase to 21.5 billion in 2025 (Tariq, 2018). The main motive of IoT is to transform real devices to smart and intelligent virtual devices and to connect all devices under a common network. The basic components on every IoT are devices, data processing, connectivity and user interface. Collect, communicate, analyse and act are the four distinct steps of IoT life cycle. IoT devices are collecting the data from various areas of application. Collected data will be transmitted to other devices through the network. Received data will be converted to information and the action is based on the information and data. Consumer IoT (CIoT) and Industrial IoT (IIoT) are the basic types. Consumer IoT focuses on the communication between consumers and Internet of Things. Industrial IoT is concerned with employing IoT for designing industrial systems (Parthasarathy, 2022).

IoT architecture is a basic method to devise the components of IoT for delivering the services effectively over the network. Various types of IoT architecture exist to cater the needs of different applications. Three layer, four layer, five layer and Service-oriented Architecture (SoA) are the types as shown in Figure 1. The most fundamental IoT architecture is three layered architecture with perception layer, network layer and application layer (Al-Qaseemi et al., 2016; Wu et al., 2010). Three layered architecture was not sufficient for large scale IoT applications. So four layered architecture came into existence with perception layer, network layer, data processing layer and application layer. To provide services to various industries, institutes, companies, and government sectors and to enhance the security features, five layered IoT architecture was designed by researchers. Along with all layers of three layered architecture, business layer and middle ware layer are added in five layered IoT architecture (Zhong et al., 2015; Kakkar et al., 2021). SoA architecture is a top layered architecture. System functions are provided to the end users (Miorandi et al., 2012; Li et al., 2015).

### A. IoT Applications

Internet of Things has expanded its status in modern days for connecting all the things such as devices, sensors, equipments, software, and information services through internet (Samani & Wahaishi, 2015; Atanasovski et al., 2015). Applications of IoT are many and endless. Some of prominent IoT applications are as follows.

- **Smart Mobility:** Smart mobility is the intelligent transport and mobility network. The basic objective of the smart mobility is to provide reduction in noise generated due to automobiles, more flexible, multi modal transport system, preventing or reducing road accidents thereby enhancing the road safety, and to develop an advanced traffic flow (Zorzi et al., 2010).
- **Smart Grid:** Smart grid is an electrical network with digital and wireless communication technology to detect and react to the variations in usage of energy. The basic objective of smart grid is

*Figure 1. IoT architectures*

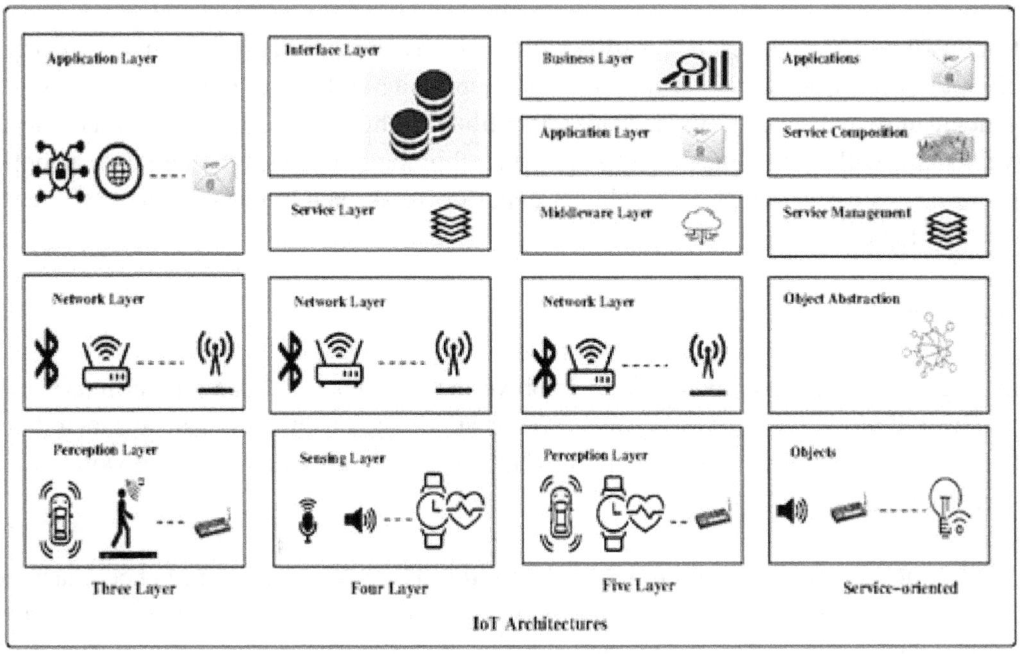


to match the supply and demand of energy more efficiently and at reduced cost. It has to monitor and control the possible faults.

- **Smart Homes/Buildings:** In smart homes, appliances and other devices are connected to each other and can be accessed through a single point like mobile phone or laptop or any other digital gadgets. The objective is to ease daily schedule, provide comfort, and easy remote access of all the devices to the user. The user has the flexibility of controlling the devices at his choice.
- **Public Safety and Environmental Monitoring:** It is the process of monitoring various parameters that are connected directly or indirectly with environment such as weather conditions, protection of endangered species, water and soil quality monitoring. Objective is to provide safety to the people by helping officials to take appropriate decisions based on the adverse situation.
- **Medical and Healthcare:** IoT plays an important and major role in medical and health care sector. IT sector is amalgamated with medical and health care field to provide better services to the patients, monitor the patient's vital parameter for better treatment, machine to machine communication, reduction in cost, and better allocation of resources in the hospital for better utilization.
- **Smart Agriculture and Breeding:** Smart agriculture utilizes the modern methods and digital technology in order to manipulate and manage the agricultural yield. Environmental parameters will be measured and monitored using sensors and associated electronic circuits. Based on the climatic conditions and the soil fertility the smart agriculture system helps farmers to take appropriate decisions to increase the yield.
- **Independent Living:** It aims at providing support to the elderly people in their daily routine in the best possible way. This helps elderly people to perform their works independently without anybody's help safely.

## 2. SMART AUTOMATION

Automation is usually applied for the repetitive tasks which can be easily programmed. Smart automation involves the use of advanced technologies for improving and streamlining the processes so that the efficiency can be enhanced without the intervention of a human being. To build the adaptive and responsive systems with intelligence, these advanced technologies need to be integrated. Robotics plays a significant role in smart automation as robots can perform tasks and also can be integrated easily with advanced technologies.

A. Smart Automation Techniques

The main goal of smart automation is to improve the efficiency, reduce human intervention, and enhance the overall productivity of the systems. Smart automation techniques can be adapted in various applications such as smart cities, smart buildings/homes, and smart industries. Following are some of the smart automation techniques:

- **Artificial intelligence (AI) and machine learning (ML):** Artificial intelligence, machine learning, natural language processing, and computer vision help to learn from the historical data, automate the process of decision making and enhance the performance.
- **Block chain mechanism:** Block chain technology can play a vital role in smart automation by improving security, transparency and trust in automated processes and data exchange.
- **Bio-inspired computing:** Bio inspired computing is also known as biologically inspired computing or nature inspired computing is a field of study that draws the inspiration from biological systems, and processes for the development of computational methods and algorithms. These help in solving complex problems and improve various aspects of smart automation.
- **Internet of things (IoT):** IoT connects various physical devices to the internet, which allows them to connect and exchange their data. IoT is preferred in smart automation for real time data acquisition and control.
- **Robotics and process automation:** Autonomous or semi autonomous robots are used in manufacturing industries, logistics and other industries to perform the tasks with precision and enhanced efficiency. Robotic process automation and work fow automation tools help in automation of various routines and rule based tasks.
- **Data Analytics and cloud computing:** Big data analytics plays a vital role in smart automation by helping the people to optimize the process. Cloud services provide cost effective storage and data processing resources for smart automation solutions.

## 3. IoT SENSORS

Sensors are the devices that convert one form of energy into another form. These sensors notices and react to the input from environment. Sensors for IoT are devices which detect the changes in the surrounding environment and also collect the data. Sensors are one of the fundamental and most crucial components of IoT. Basically sensors are used for the measurement of various parameters (Pflanzner & Kertész,

*Figure 2. Sensors for smart mobility*

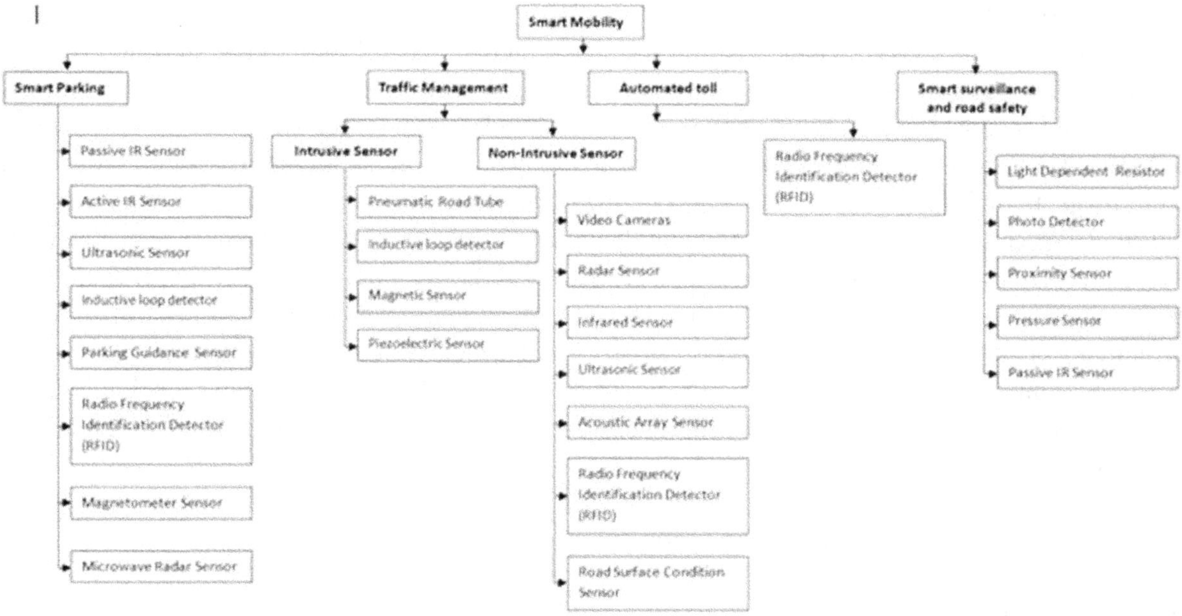


2018). Sensors are capable of absorbing the environmental situation and capable of measuring them. This section explains the sensors that are based on the various application areas (Khanna & Kaur, 2020).

## 3.1 IoT Sensors for Smart Mobility

Smart mobility is one of the best application areas of Internet of Things that allows flawless, effective, efficient and flexible travel across various modes of transportation. Vehicular Ad-hoc Networks (VANETs) are popular due to the huge demand of the society that tends to improve road safety either by reducing the accidents or preventing them. Following are the sub components of smart mobility (Paidi et al., 2018). Figure 2 depicts the sensors used for smart mobility. Smart mobility can be broadly classified as smart parking, traffic management, automated toll and smart surveillance and road safety.

### 3.1.1 IoT Sensors for Smart Parking

Time consumption in parking a vehicle in urban and metropolitan cities is increasing day by day due to the increase in the population and income. Enhancement in sensor technologies and advancements in the internet usage, helps human beings to add intelligence to parking systems. The commonly used sensors for smart parking are listed in Figure 2. Passive infrared sensors measure infrared radiation from objects, identifying empty or occupied spaces. They are suitable for indoor parking lots but not outdoor ones (Shaheen, 2005). Active infrared sensors measure the amount of infra-red radiation reflected from objects or vehicles, detecting empty or occupied spaces (Mouskos et al., 2007). Ultrasonic sensors emit sound waves ranging from 25 kHz to 50 kHz, detecting objects or vehicles based on reflected energy. They are cost-effective but expensive to install, maintain, and connect to networks. Wireless ultrasonic sensors are available and to be connected to wireless sensor networks such as ZigBee and others (Idris et al., 2009; José, 2018). Inductive loop detectors work on electromagnetic induction and are installed below the ground to identify vehicle entries and exits. When these detectors are used with computational techniques, we can identify various types of vehicles (Fahim et al., 2021). Parking guidance systems display the number of vehicles in the parking lot, but drivers' needs to navigate around the lot to determine free and occupied spaces. But with the adaptation of optimization algorithms, the optimal path and shortest time to reach the parking space can be found easily (Liu et al., 2020). RFID technology uses an electromagnetic field to identify objects, but does not provide individual occupancy status or help drivers find empty spaces. It does not provide the individual parking occupancy status and does not help driver to find the empty parking space (Zhu et al., 2012; Atta et al., 2018). Magnetometers detect changes in electromagnetic fields and can be used indoor/closed and outdoor/open. As they are insensitive to the surrounding environment, can be used both indoor/ closed and outdoor/ open parking lots (Floris et al., 2020). Dual microwave Doppler radar is suggested for detecting both stationary and moving objects/ vehicles, installed in parking spaces below vehicle surfaces, but expensive to install and maintain in large scale (Fadi, Al-Turjman, 2019).

### 3.1.2 IoT Sensors for Traffic Management

Traffic management can be avoided using data analytics effectively. Real time notifications can be forwarded to the related vehicles about the parking status, shortest routes and so on. Sensors for traffic management is classified based on the position of installation as intrusive and non-intrusive (Guerrero-Ibáñez et al., 2018).

**A. Intrusive sensors:** Intrusive sensors are placed on the footpath surfaces to control traffic. These sensors are further classified as in figure 2. Pneumatic road tubes, inductive loop detectors, magnetic sensors, piezoelectric sensors, radar sensors, infrared sensors, ultrasonic sensors, and acoustic array sensors are used to detect vehicles and their speed. Intrusive sensors adapt different techniques for measurement. These sensors are insensitive to weather conditions but can be affected by heavy vehicles.

Pneumatic road tubes are used to count the number of vehicles passing over roads by detecting changes in pressure inside the tube with great precision (Divatankar & Hivarkar, 2019). Inductive loop detectors use coil-emitting wires to measure the change in electromagnetic field, detecting individual vehicles, speed, occupancy, and gaps between vehicles (Chiplunkar et al., 2021). Magnetic sensors use a square-shaped magnetic field to identify vehicles and count the number of vehicles on the road. These sensors are highly employed in traffic monitoring due to their high sensitivity and low cost (Humayun et al., 2022; Wang et al., 2017). Piezoelectric sensors collect information by changing mechanical signals to electric signals, creating vibrations and voltages when vehicles move on them. These sensors are insensitive to weather conditions but can be affected by heavy vehicles (Krishnendhu & Mohandas, 2023).

**B. Non-intrusive sensors:** Non-intrusive sensors are placed at various locations for detection of the vehicle, its speed, lane coverage. Video cameras are used for monitoring traffic density and vehicle types, while radar sensors work on the principle of Doppler's effect. Infrared sensors measure speed, occupancy, density, and classification of vehicles by collecting reflected radiations. Ultrasonic sensors approximate vehicle distance based on time elapsed between transmitted and reflected sound. Acoustic array sensors consist of microphones and other sensors (Ray et al., 2019; Tasgaonkar et al., 2020).

### 3.1.3 IoT Sensors for Automated Toll Collection

Toll taxes are collected on highways to maintain road conditions and avoid long queues. Automated processes use RFID technology, providing tags to vehicles linked to individual bank accounts. Authorities collect automatic toll taxes when vehicles cross toll plazas. RFID also helps identify vehicles in collisions, with overloaded vehicles receiving extra tax. Vehicle weighing facilities at toll collection units can also help identify stolen vehicles (Bhavke & Pai, 2017).

### 3.1.4 IoT Sensors for Smart Surveillance and Road Safety

Smart city traffic management involves high-precision cameras for monitoring road traffic and data analysis. This helps in identifying congestion and makes decisions. Categories include smart highway lighting, emergency management, road safety, renewable energy resources, and smart information displays. Sensors like LDR, photo detectors, proximity, pressure, and PIR are used for energy conservation (Lee & Chiu, 2020). Automatic street light systems use an IR sensor, LDR, and other electronic components to measure environmental light intensity and control lights. These systems can be controlled automatically based on sunlight illumination, reducing energy requirements (Mukta, Rahman, Asyhari, & Md, 2020; Abinaya et al., 2017). Photo-detector collects and measures light intensity, while proximity sensors detect vehicle presence and absence on roads (Rahman et al., 2021; Mukta, Rahman, Asyhari, & Md, 2020). Pressure sensors identify vehicles on highways by applying pressure and developing transient voltage. Microcontrollers can identify voltages and turn lights on or off automatically based on threshold values (Kaur et al., 2016). Passive infrared sensors measure light radiated from nearby objects, used for motion detection of people, animals, and cars. These systems help reduce energy requirements and improve road safety (Ma et al., 2020; Yang et al., 2015).

## 3.2 IoT Sensors for Smart Grid

With the help of a digital technology and other advanced technologies, the electricity network can be called as a smart grid. Smart grids are used to monitor and manage the electricity generated from all the sources and satisfy the demand of the end users. Basically smart grid has transmission and distribution system. Sensors are the primary elements in both transmission and distribution. The sensors that are used for smart grid can be classified based on the type of measurement carried out such as energy flow sensors, environmental sensors, and working condition sensors (Dileep, 2020). Figure 3 depicts the sensors used for smart grid.

### 3.2.1 Energy Flow Sensors

In smart grid development, real-time and precise data about the major grid parameters such as voltage and current in each node are utmost important. Sensors that are used for measuring energy flow in a grid such as voltage, current, energy, power factor; magnetic/electric fields are grouped under this category. Smart grid development relies on real-time data on grid parameters like voltage and current. Sensors are used to measure energy flow in a grid, including voltage, current, energy, power factor, and magnetic/electric fields. Voltage sensors, such as transformers and capacitive dividers, are used for medium and large voltage measurements, but are limited by frequency band (Wetula et al., 2021). Current sensors,

*Figure 3. Sensors for smart grid*

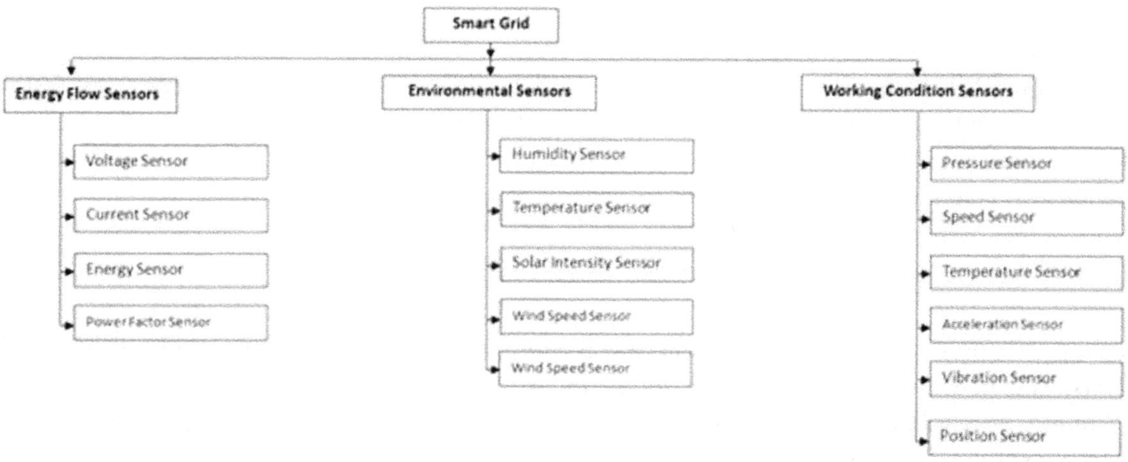


like Hall current sensors, are essential for current measurement in the smart grid, as they monitor high voltage lines and produce an Alternating Current (AC) in secondary windings (Ripka, 2010). Anisotropy magneto-resistance (AMR) is another type of sensor used for current measurement in smart grids.

Current shunts generate a small voltage at both ends of the shunt, providing the DC current value in the circuit (Bernieri et al., 2017; Rind et al., 1974). The Rogowski coil is a current sensor used in smart grids for detecting arc faults. Magneto-optical current sensors, which use Faraday's effect, are contactless and suitable for measuring current (Naseri et al., 2019; Chen et al., 2023). Energy meters, installed at consumer end, can monitor energy consumption without manual intervention. Smart meters can transmit mistake-free consumption data to suppliers. Power factor sensors measure energy efficiency in a power grid, providing the cosine of phase angle between supply voltage and load current. A high power factor indicates that only 80% of incoming current works, with the remaining 20% wasted through heating the conductor. Power quality analyzers are used for measuring power factors.

### 3.2.2 Environmental Sensors

Sensors to measure the environmental impact on the grid like humidity, temperature, luminance movement and occupancy, solar intensity, wind speed, smoke and gas are grouped. Humidity sensors measure the water content in air or other gases, with three types: absolute humidity, dew point, and relative humidity. Thermal conductivity humidity sensors measure the difference in thermal conduction between humid and dry air, while resistive humidity sensors measure resistance changes inversely proportional to humidity (Rivas, 2020; Wilson & Fontes, 2005). Temperature sensors are used to measure environmental temperature, with thermistor, thermocouple, and Resistance Temperature Detector (RTD) being recommended. Thermocouples provide output in terms of voltage based on changes in environmental temperature, while RTDs measure resistance changes with temperature. Thermistors are similar to RTDs but are less accurate (Samimi & Ilkhechit, 2020). Solar intensity sensors measure the power per unit area received from the sun, with Pyranometers and Pyrheliometers commonly used (Pazikadin et al., 2020). Wind speed sensors measure the time difference between the transmission and reception of ultrasonic waves, with anemometers providing 1-5% accuracy (Azorin-Molina et al., 2018). Smoke and gas sen-

*Figure 4. Sensors for smart home/building*

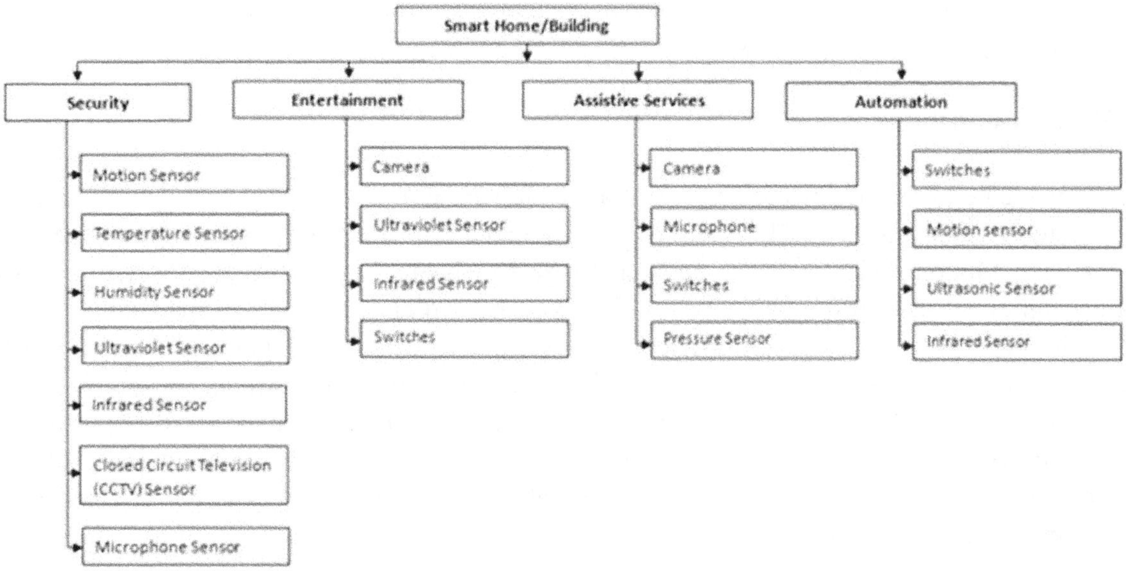


sors measure the electrical resistance of semi-conducting materials, such as metal oxide, used for fire alarms. These sensors measure physical, visual, and dynamic properties of smoke (Peng & Wang, 2019).

### 3.2.3 Working Environment Based Sensors

Monitoring the health of the transformer, monitoring poles and high tension towers and ability to handle uncertainty conditions are some of the areas of the smart grid. To monitor the working conditions of the grid, frequently used sensors are pressure, speed, temperature, acceleration, vibration and position.

## 3.3 IoT Sensors for Smart Home/Building

Smart home systems help human beings to reduce the cost of living by providing information about various factors of a home. The sensors also help us to take appropriate decisions in our daily life to manage and control various aspects of our home. Figure 4 depicts the sensors used for smart home/building.

### 3.3.1 Security

Personal and private asset security is becoming difficult day by day. Safety is utmost important to avoid various kinds of crimes. Sensors can detect the intrusions, environmental hazards and security threats. Major sensors available in literature to provide security are listed as in Figure 4 (Fadhil et al., 2020; Bt Aripin & Othman, 2014). Door and window sensors, motion sensors, glass break sensors, security cameras, flood and water leak sensors, smoke and carbon monoxide sensors, temperature sensors, and panic buttons are essential security measures in homes. These sensors detect unauthorized entry, detect water leaks, and detect fires. Motion sensors measure distance and can be used in home automation. Glass break sensors capture sound and vibration frequencies within 20 feet, triggering alarms if they are within 4-5 kHz. Security cameras with night vision and built-in microphones and speakers provide two-way communication. Temperature sensors detect temperature increases during fires, and panic buttons provide security during emergencies.

### 3.3.2 Entertainment

Smart home entertainment aims to create a relaxing atmosphere for returning home after a long day. Sensors such as cameras, UV or IR motion detectors, and eye tracking sensors are used to control the environment. Motion detectors detect entry and adjust ambient lights, while light intensity sensors measure light intensity to control brightness. Temperature and humidity sensors maintain a comfortable environment, while sound sensors detect noise levels and adjust audio volume. PIR sensors detect room occupancy and conserve power by generating pulses. Gesture and proximity sensors allow users to adjust volume, play back, or navigate menus using gestures. RFID tags can personalize playlists and settings, allowing family members to create their own playlists and audio video settings. Overall, smart home entertainment system helps to create a comfortable and inviting atmosphere for all users (Zaman & Rafiunnisa, 2018; Huang & Huang, 2014).

### 3.3.3 Assistive Services

The nuclear family concept has made it difficult to care for elderly people at home, leading to the need for assistive services through electronic gadgets. These gadgets include fall detection sensors, environmental sensors, wearable health sensors, voice and speech recognition sensors, eye tracking sensors, pressure and weight sensors, gyroscope and accelerometers, and wheel chair controls. Fall detection sensors detect the speed of falling and trigger alarms in case of emergencies. Wearable health sensors monitor vital health parameters like heart rate and blood pressure. Voice and speech sensors assist mobility by controlling appliances, communicating with others, and accessing relevant information. Eye tracking sensors convert pupil position into binary data, while pressure and weight sensors monitor posture and assist users. Gyroscope and accelerometers are used in wheel chairs for stability and balance (Bajpai & Radha, 2019; Asham et al., 2018).

### 3.3.4 Automation

Home automation is a system that measures and controls physical parameters of home appliances, transmitting data to the user for control. It can be wired or wireless, with each having its advantages and disadvantages. Sensors used in home automation include motion sensors for energy conservation and security, contact sensors for object detection, light sensors for ambient light intensity, gas sensors for detecting harmful gases, occupancy sensors for lighting and security, proximity sensors for touch less faucets and automatic doors, smart thermostats for heating and cooling optimization, and soil moisture sensors for optimizing watering schedules. Each type of sensor has its own advantages and disadvantages (Stolojescu-Crisan et al., 2021).

*Figure 5. Sensors for public safety and environmental monitoring*

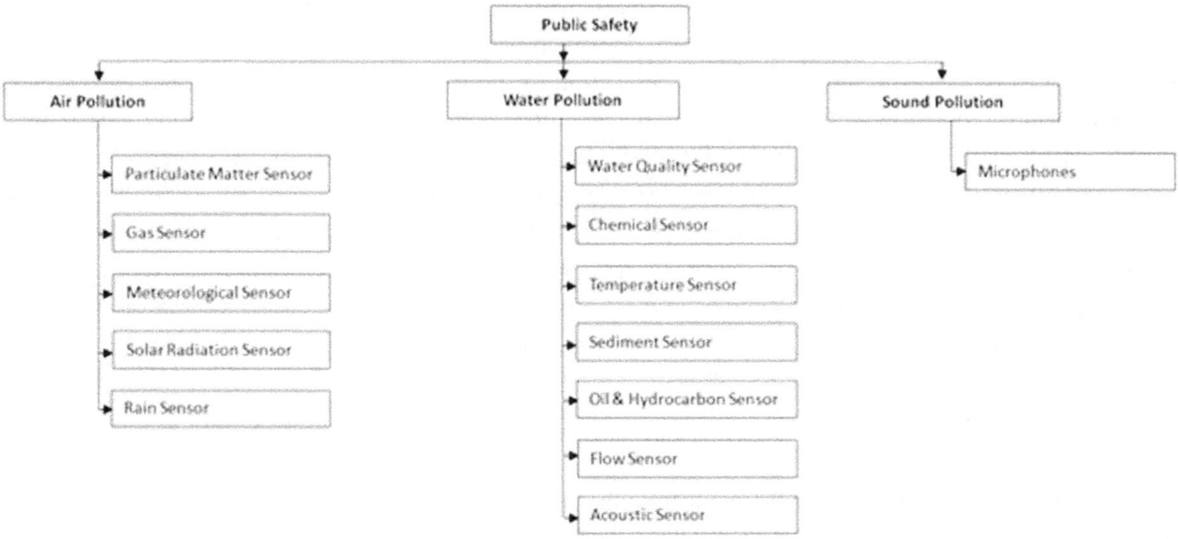


## 3.4 IoT Sensors for Public Safety and Environment Monitoring

Environmental pollution is most important concern for public health and safety. Air, water, and sound pollution are the basic types of pollution which degrades the quality of life. To avoid health risks, level of these pollutions need to be measured and monitored.

### 3.4.1 Air Pollution

Air is composed of oxygen and nitrogen, with 21% being oxygen and 78% nitrogen. Indoor air is 100 times more contaminated than outside air, with most humans spending 80-90% of their daily routine in indoor activities. Major indoor air pollutants include NOx, VOCs, SO2, O3, CO, PM, radon, toxic metals, and microorganisms. IoT sensors help monitor air quality and public safety (Leung, 2015; Pantelic et al., 2022). Figure 5 depicts various IoT sensors used for environmental monitoring and helps in providing the public safety. Particulate matter sensors measure fine particulate matter, which can cause adverse health effects. Gas sensors monitor gases like NO2, SO2, CO, ozone, and VOC, which contribute to air pollution. Meteorological sensors study climatic conditions, such as temperature, humidity, wind speed, and pressure. Solar radiation sensors measure the amount of solar radiation reaching the Earth's surface, which can manipulate atmospheric chemical reactions and contribute to air pollution. Rainfall sensors detect precipitation levels, helping to understand the impact of rainfall on air quality by removing pollutants.

### 3.4.2 Water Pollution

Water pollution monitoring in water bodies such as lakes, rivers, oceans and ground water, plays a crucial role in accessing and maintaining the quality of water. The pathogenic microorganisms and other contaminants of water cause serious health issues such as cholera, diarrhoea, typhoid, amebiasis, hepatitis, and worm infections. Many sensors are available to measure various water pollutants (AlZubi, 2022; Lakshmikantha et al., 2021). Water quality sensors measure various parameters such as pH, dissolved oxygen (DO), turbidity, conductivity, and total dissolved solids (TDS). pH sensors detect the alkalinity of water, while DO sensors measure the concentration of dissolved oxygen. Turbidity is the scattering of light in water, while conductivity measures the ability of water to conduct electric current. Chemical sensors detect heavy metals from industrial wastes and chemicals, while temperature sensors measure water's chemistry, biological processes, and aquatic ecosystem. Sediment sensors measure the amount of light scattered by particles at the bottom of water. Oil and hydrocarbon sensors detect oil or hydrocarbons from spills, runoff, or industrial discharges. Flow sensors measure the flow of water in rivers and streams, assessing pollution transportation. Acoustic sensors monitor underwater vibrations and sounds, gathering information about water quality and aquatic animals.

### 3.4.3 Sound Pollution

Sound pollution is one of the causes for environmental pollutions. Excessive and unbearable sound affects the health of a human being. These excessive sounds are usually produced by vehicle horns, aircraft, crackers, industrial machines and loud speakers in urban areas and industrial areas. Repeated exposure to excessive noise may lead to permanent hearing loss. Repetitive loud noise can create physical and psychological stress, reduction in productivity, problem in concentration span, and may lead to accidents and injuries. Hence these excessive and unwanted noise level need to be measured and monitored in order to identify the sources of noise pollution (Cornelius et al., 2020). Microphones are used to measure various types of sounds, including noise intensity, by vibrating the diaphragm of the microphone. There are two types: omni-directional and directional. Omni-directional microphones capture sound from all directions, suitable for outdoor noise, while directional microphones capture sound waves from a specific direction. Noise level meters, microphones, and ambient noise sensors are used to measure noise intensity in decibels. Traffic noise sensors measure noise on roads, industrial noise sensors identify sound intensity in industries, and aircraft noise sensors measure noise impact on society.

## 3.5 IoT Sensors for Medical and Health Care

Internet of things is having varieties of applications for medical and health care. These sensors help patient, elderly people and people who want to be fit. These sensors can be classified based on the operations performed as patient monitoring, managing health facilities, tracking medical equipment and improving overall healthcare outcomes (Javaid & Khan, 2021; Farahani et al., 2020). Various sensors for medical and health care monitoring are listed in Figure 6.

### 3.5.1 Patient Monitoring Sensors

Remote patient monitoring is one of the trending applications of IoT in which various health parameters are provided to the cloud for data storage and further analysis. Continuous monitoring of patients is necessary in case of chronic conditions, people recovering from injuries, patients in intensive care unit.

IoT sensors are essential for monitoring various health parameters, including heart beat, blood pressure, pulse oximeter, respiratory rate, temperature, blood glucose, accelerometer, medication adherence, infant monitoring, fertility tracking, and implantable sensors. Heart beat sensors detect cardiovascular

*Figure 6. Sensors for medical and healthcare monitoring*

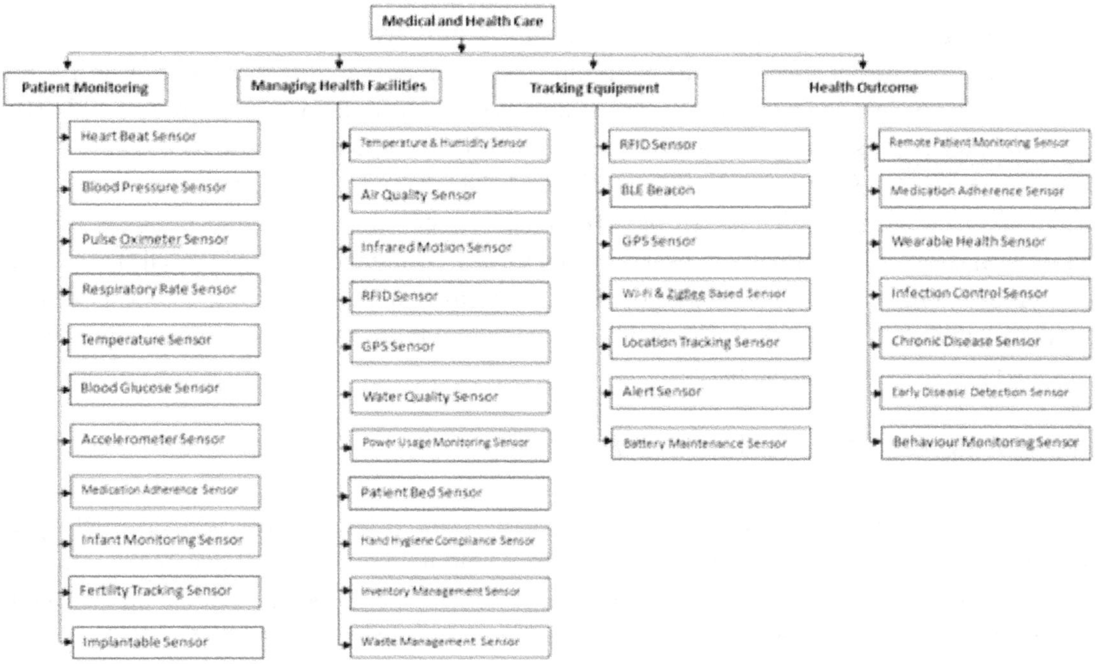


health by measuring blood vessel volume, while blood pressure sensors help in identifying blood pressure abnormalities. Pulse oximeters measure oxygen saturation levels in blood, while respiratory rate sensors assess lung function and detect respiratory distress. Temperature sensors track body temperature changes during infection, while blood glucose sensors help monitor diabetes patients' blood glucose levels. Accelerometers help diagnose sleep disorders and track medication adherence. IoT sensors can also monitor infants, fertility, and neurological disorders like epilepsy or Parkinson's disease. These sensors can be used in various applications, such as NICU monitoring, fertility tracking, and implantable sensors.

### 3.5.2 Managing Health Facilities

Health facility management covers a vast range of activities by increasing the efficiency, paving way to patient safety, and enhancing overall facility management. Sensors play a crucial role in monitoring environmental conditions in hospitals, ensuring comfort and safety. They monitor temperature and humidity, air quality, PIR, RFID, GPS, water quality, power usage, patient bed occupancy, hand hygiene compliance, inventory management, and waste management. DHT sensors are commonly used to maintain optimal humidity levels, while PIR sensors detect movement and optimize HVAC systems and lighting. RFID and GPS sensors track medical equipment locations and status, while water quality sensors monitor Legionella outbreaks. Power usage monitors identify power wastage and control it. Patient bed sensors provide information on patient occupancy, while hand hygiene compliance sensors monitor hygiene compliance among healthcare workers. Inventory management sensors track medical supplies and pharmaceuticals, and waste management sensors monitor dust/waste bins, sending alarms to maintain cleanliness and prevent overflow.

### 3.5.3 Tracking Medical Equipment

Medical equipment is expensive and crucial for efficient healthcare operations. IoT sensors provide real-time visibility of equipment location, status, and usage, optimizing inventory and ensuring critical devices are available. Implementing these systems requires careful planning, integration with existing infrastructure, and cyber security for patient data. Common sensors used include RFID tags, Bluetooth low energy beacons, GPS sensors, Wi-Fi and Zigbee-based tracking, location tracking sensors, alert sensors, and battery and maintenance sensors. RFID tags detect movement and locations of high-value and mobile equipment, while Bluetooth beacons send unique ID numbers to devices. GPS sensors track equipment movement between facilities or outdoor locations, while Wi-Fi and Zigbee-based tracking locates devices in indoor facilities. IoT sensors can also provide alerts about restricted zone entry and monitor battery life or consumable usage.

### 3.5.4 Improving Overall Healthcare Outcomes

IoT sensors can significantly improve healthcare outcomes by providing real-time data, predicting future emergencies, and enhancing patient care. These sensors help monitor patients, manage chronic conditions, prevent diseases, and optimize healthcare processes. Examples include remote patient monitoring sensors, medication adherence sensors, wearable health sensors, infection control sensors, chronic disease sensors, early disease detection sensors, and health behavior monitoring sensors. These devices help detect abnormalities, manage chronic conditions, and optimize healthcare processes, leading to better patient outcomes, reduced healthcare costs, and improved quality of care. By implementing these sensors, healthcare providers can detect and intervene early, prevent diseases, and improve patient care.

## 3.6 IoT Sensors for Agriculture and Breeding

IoT sensors play a significant role in transforming convention agriculture to precision agriculture. This is due to their ability to collect the real time data, automation capabilities to provide information to cultivate the precision methods for agriculture based on the soil quality, climatic conditions and the quality of seeds. IoT sensors for agriculture can be classified based on different categories based on their functions, applications and type of data collection (Khujamatov et al., 2021; Xu et al., 2022). Figure 7 shows the possible sensors for agriculture and breeding.

### 3.6.1 Environmental Sensors

IoT sensors are crucial for farmers to monitor climatic conditions, predicting weather events and responding promptly. Soil moisture sensors measure the moisture content of soil, which is essential for plant growth. Temperature sensors, like LM 35, help maintain optimal temperature for seed germination. Soil pH sensors measure the acidity or alkalinity of soil, which determines crop growth. Wind speed and direction sensors measure wind speed and direction, enhancing photosynthesis and plant productivity. Atmospheric gas sensors measure air quality, such as oxygen, carbon dioxide, or ammonia, to assess air quality. Cameras help identify crop health, detect diseases, pests, and nutrient deficiencies. Stunted growth, black spots on leaves, decay, or rot are indicators of plant diseases, which can be processed

*Figure 7. Sensors for agriculture and breeding*

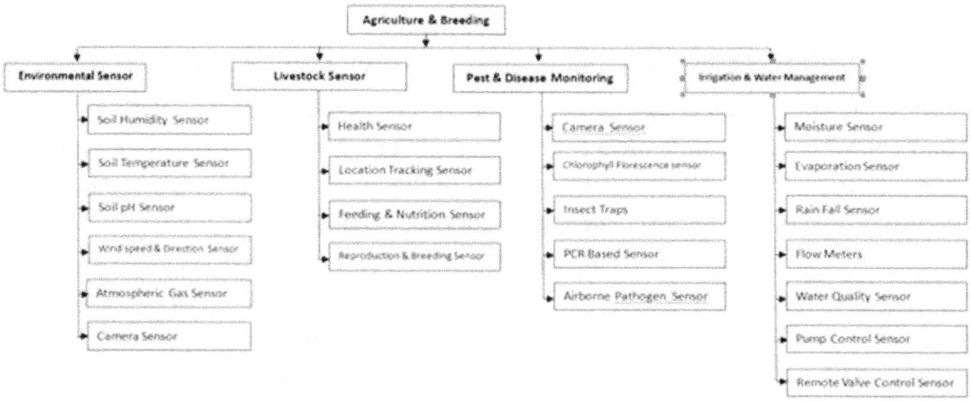


after processing images captured by cameras. These sensors help farmers take preventive measures to protect crops and minimize losses.

### 3.6.2 Livestock Sensors

Agriculture and animal husbandry are two faces of a coin. Both are practiced by a farmer in conventional agriculture for optimizing the overall productivity. Many animals are raised for meat, milk, fibre and other products. It requires day to day care, selective breeding and the raising of livestock. Livestock monitoring includes use of technology to track and control the health of farm animals. Wearable sensors on animals monitor temperature, heart rate, respiration, and chewing patterns, providing insights into their digestive system. Location tracking sensors use GPS or RFID tags to identify animals and their movements. Feed intake monitoring systems help farmers schedule feedings using proximity sensors, scales, RFID antennas, and weighing balances. Heat detection sensors detect unusual ovulation patterns, and ultrasound devices and hormone monitors confirm pregnancy in livestock. These technologies help in managing livestock health, nutrition, and reproduction.

### 3.6.3 Pest and Disease Monitoring Sensors

Pest and disease detection in both plants and animals play a crucial role in helping farmers to identify and manage threats. These sensors help in early detection of disease and help to minimise the spread of pathogens. Multispectral and hyperspectral cameras capture images of crops to detect stress, disease, or nutrition deficiencies. Chlorophyll fluorescence sensors measure florescence in leaves, indicating photosynthesis activity. Insect traps detect insect types and minimize damage. PCR-based sensors detect specific pathogen DNA, identifying species and monitoring diseases. Airborne pathogen sensors detect and monitor microorganisms like bacteria, viruses, and fungi. Biosensors generate florescence or electrical current, optical sensors use light scattering, and particle counters detect airborne pathogens. These devices help farmers monitor crop health and reduce damage caused by insects.

### 3.6.4 Irrigation and Water Management Sensors

IoT sensors are precious tools for irrigation and water management in agriculture. They help farmers to optimize water usage, reduce wastage and ensure that the crops receive the optimal amount of moisture. By utilizing these sensors, farmers can reduce water consumption, lower the operational cost, improve crop yield, and contribute to sustainable agricultural practices. Moisture sensors measure soil moisture levels to determine irrigation levels, preventing over- or under-irrigation. Evaporation sensors measure water loss from soil due to evaporation and transpiration by plants. Rainfall sensors detect rainfall and trigger irrigation systems to turn off during raining periods. Flow meters monitor water consumption to optimize water consumption. Water quality sensors quantify the chemical composition of irrigation water, ensuring it is suitable for crops. Pump control sensors monitor pump performance and prevent crop stress due to variations in water supply. Remote valve control sensors allow farmers to remotely control irrigation valves, optimizing scheduling for precise farming and enhanced yield. These sensors help farmers make decisions about irrigation scheduling based on sensor outputs, ensuring optimal soil moisture and plant growth.

## 4. IoT COMMUNICATION MODELS

IoT devices are basically designed to collect, process, communicate, and exchange the data between each other. They are also capable of communicating with either gateway or the internet. Based on this the communication models can be largely classified as device to device, device to cloud, device to gateway and back-end data sharing communication model (Sharma et al., 2019; Nour et al., 2019). These models help people to access the data anywhere and anytime.

### 4.1 Device to Device Communication Model

In this model, one device will communicate with the other without an application server (Nayak & Vijayalakshmi, 2016; Bello et al., 2017). The model follows publish/subscribe architecture for machine to machine communication (El Ouadghiri et al., 2020). Smart light communication with a switch in home automation and smart watch communication with mobile phone are few examples. Various wireless protocols used for communication include Bluetooth, ZigBee, and Z-wave (Bluetooth, n.d.). Communication happens through IP network or Internet. Devices can transmit/receive only small data packets. Different manufacturer uses different protocols and the devices can only communicate with other devices that are designed with the same protocol.

### 4.2 Device to Cloud Communication Model

Wireless IoT devices exchange the data and to receive message traffic control with other devices through the internet cloud service (application server). Users can access the sensor data remotely using smart phones through cloud service. The protocols such as HTTP, TCP/IP, TLS are used by the devices to connect to the cloud services. Smart home appliances such as smart television, smart thermostat device and smart speakers are few examples (Meet the Nest Thermostat & Nest Labs, 2015). IoT devices in this model, collects the data from the sensors and transmits to the cloud data base where it is accumulated after processing and analyzing the data for future use. IoT devices can remotely access the data accumulated in the cloud through web interface or by a smart phone. Communication system to function properly, IoT devices and the cloud should be from the same vendor. The system suffers from interoperability problem if the device and cloud are from different vendors.

## 4.3 Device to Gateway Communication Model

This model is called also as device to application layer model. IoT devices are connected to the internet cloud services for data exchange via the gateway. Gateway is loaded with application software and is an intermediate between the IoT devices and the cloud server. Gateway is as essential component of this model as it provides security and data or protocol translation. This type of communication models are used in smart homes, smart cities. In many applications, the local gateway is the mobile/smart phone with an App for communicating between IoT devices and the cloud services. The main disadvantage of this model is more cost and complexity in designing the system due to addition of application layer software.

## 4.4 Back-End Data Sharing Communication Model

This model allows users to share cloud-based sensor data with third parties, enabling data aggregation and analysis from multiple IoT devices (Reddy, 2021). It overcomes device-to-cloud constraints and ensures interoperability with cloud services or APIs. The choice of communication model depends on the application, enhancing user expectations. Communication models are critical for IoT, providing data collection, analysis, visualization, and prediction, with each model offering advantages and tradeoffs.

# 5. OPEN RESEARCH CHALLENGES

The practical IoT system is more complex as it involves heterogeneous devices interacting with each other in large scale (Hanes et al., 2017). Following are the general challenges for effective implementation of any IoT systems:

- **Security and trust:** It is very difficult achieve the essential level of security in IoT systems. The security features required for IoT are data confidentiality, authentication of the user or the device accessing the IoT system, trust, and data integrity (Chanal & Kakkasageri, 2021). An IoT system uses trillions of smart devices which can access the sensors data and analyse it, take appropriate action and communicate the data to the other devices in the network. So maintaining the data confidentially is an extremely huge task. There are various encryption and decryption algorithms such as light weight encryption method, elliptical curve cryptography, which can help to provide confidentiality to the massive data of the IoT systems (Oriol et al., 2015; Pedro et al., 2018). Data authenticity deals with allowing a right user or the device to access the data in the network. Rivest Shamir Adleman (RSA), Hybrid Advanced Encryption Standard (AES) and Hybrid RSA and AES algorithms help in providing authenticity to the data (Samiksha & Vinay, 2016; Trivedi & Raval, 2018). Data integrity is one of the major concerns of IoT. Data integrity is to ensure the correctness, completeness, consistency, and validity of the data of an application. AES, hybrid RSA and hybrid AES algorithms are used by the researchers to provide data integrity to the IoT systems (Kasgar et al., n.d.; Kavitha & Alphonse, 2019). Hence, IoT system needs to solve the domain dependent data security issues.

- **ML and IoT:** Any system in today's life is updated with automation. Artificial intelligence (AI) and machine learning (ML) is playing an important role in providing decision making capabilities and respond to the unusual situation. Integration of AI and ML for IoT systems will provide an additional potential to the applications. Almost all systems are integrated with either AI or ML. There are few ML algorithms that are providing the solutions to the performance related challenges such as congestion control, fault detection, resource management and security. Development of ML model from the collected data set using supervised method and to validate the model requires humans. Researchers need to still exploit the advantage of other ML algorithms for improving or enhancing the performance of IoT systems (Alam et al., 2020; Camarinha-Matos et al., 2020).
- **Data privacy:** IoT-based systems generate vast amounts of sensitive data, which needs to be protected. To achieve data privacy, devices, storage, processing, and communication must be addressed. Addressing issues like location, user information, theft, and non-identifiability can be achieved using algorithms like multi-routing random walk, QR Code, and synchronous CPUs. Data storage should be minimized, personal data retained, and information displayed based on need. Data processing must ensure user and data owners' permission before exposing it to third parties. Communication can be protected through encryption and decryption (Alferidah, 2020; Abomhara & Koien, 2014).
- **Energy efficiency:** Energy is crucial for IoT systems' tasks like sensing, processing, and communication. Energy-efficient techniques include sleep off scheduling algorithms, selective sensor sensing, data buffer delivery, wireless communication, and advanced node distribution. Edge devices require constant power, requiring optimization. Edge devices in IoT systems require constant power. Powering the edge devices and manage the optimization of power consumption is still a challenge (Tahiliani & Dizalwar, 2018; Kaur & Sood, 2017).
- **Connectivity and communication:** IoT systems require reliable, secure connectivity, but current cellular networks struggle to accommodate increasing devices. 5G and 6G technologies are crucial, but bandwidth is a bottleneck. Integrating IoT devices with ML training is another challenge, necessitating further research.
- **Legacy systems:** Day by day the communication technologies are changing. Old operational technologies need to be considered with ever changing new information technology. Research is needed to build the reliable and efficient system where a new information technology is blended the old operational technology.
- **Regulations and certifications:** Internet of Things has changed the life of every person on this world. The way of interacting with the devices is changed drastically due to the evolution of IoT. Number of devices connected to the internet is increasing exponentially. They transfer the data between them, to enhance the functionality to the end users. Increase in the number of devices yield malicious activities. This is due to the lack of regulations in IoT industry, between various manufacturers. No policies for security exist and manufacturers may not opt for adapting to install security mechanisms. Regulations are difficult to impose but it can bring the awareness in users and administrators. To the possible extent these IoT component technologies must be certified.

## 6. CONCLUSION

IoT systems are helping people to gather the data from the sensors, process, analyse and transfer it to other devices connected through the communication models. In IoT based systems, sensors and the communication models play an important role in successful implementation. Various sensors are in market that supports IoT devices. After gathering data from various sensors, IoT device will process and analyse the data and convert it to valuable information. The essential information will be shared between various other IoT devices through the internet. Various sensors and communication models are discussed in this paper based on the existing research. Open research challenges are listed which requires further research.

## REFERENCES

Abinaya, B., Gurupriya, S., & Pooja, M. (2017). IoT based smart and adaptive lighting in street lights. In *Second International Conference on Computing and Communications Technologies, ICCCT17)* (pp. 195-198). IEEE. 10.1109/ICCCT2.2017.7972267

Abomhara, M., & Koien, G. M. (2014). Security and Privacy in the Internet of Things: Current Status and Open Issues. *International Conference on Privacy and Security in Mobile Systems (PRISMS)*, (pp. 1-8). IEEE. 10.1109/PRISMS.2014.6970594

Aditya, M. S., & Gupta, S. C. (2018). An Internet of Things Based Smart Surveillance and Monitoring System using Arduino. *International Conference on Advances in Computing and Communication Engineering(ICACCE)*, (pp. 428–433). IEEE. 10.1109/ICACCE.2018.8441725

Al-Qaseemi, S. A., Almulhim, H. A., Almulhim, M. F., & Chaudhry, S. R. (2016). IoT architecture challenges and issues: Lack of standardization. *2016 Future Technologies Conference (FTC)*, San Francisco, CA, USA. 10.1109/FTC.2016.7821686

Alanzi, H., Alshammri, A. S., Alzu'bi, R., & Hamed, E. (2019). Internet of Things Objectives, Benefits, and Applications. *Oral Science International, 31*(1), 95-100.

Alferidah, D. K. (2020). A Review on Security and Privacy Issues and Challenges in Internet of Things. *International Journal of Computer Science and Network Security IJCSNS*, *20*(4), 263–286.

AlZubi, A. A. (2022). IoT-Based Automated Water Pollution Treatment Using Machine Learning Classifiers. *Environmental Technology*, 1–9. doi:10.1080/09593330.2022.2034978 PMID:35083949

Ammar, M., Russello, G., & Crispo, B. (2018). Internet of Things: A survey on the security of IoT frameworks. *Journal of Information Security and Applications*, *38*, 8–27. doi:10.1016/j.jisa.2017.11.002

Andriansyah, M., Subali, M., Purwanto, I., Irianto, S. A., & Pramono, R. A. (2018). e-KTP as the basis of Home Security System using Arduino UNO. *4th International Conference on Computer Applications and Information Processing Technology (CAIPT)*. IEEE.

Antunes, L. (2022). IoT security certifications: Challenges and Potential Approaches. *Computers and Security, 116*. doi:10.1016/j.cose.2022.102669

Asham, A. D., Hanaa, M., Alyoubi, B., Badawood, A. M., & Alharbi, I. (2018). A Simple Integrated Smart Green Home Design. *Intelligent Systems Conference (IntelliSys),* (pp. 194–197).

Atta, A., Abbas, S., Khan, M., Ahmed, G., & Farooq, U. (2018). An Adaptive Approach: Smart Traffic Congestion Control System. *Journal of King Saud University- Computer and Information Science, 32*(9), 1012-1019.

Azorin-Molina, C., Asin, J., McVicar, T. R., Minola, L., Lopez-Moreno, J. I., Vicente-Serrano, S. M., & Chen, D. (2018). Evaluating Anemometer Drift: A Statistical Approach to Correct Biases in Wind Speed Measurement. *Atmospheric Research, 203*, 175-188. doi:10.1016/j.atmosres.2017.12.010

Babu, S. M., Lakshmi, A. J., & Rao, B. T. (2015). A Study on Cloud Based Internet of Things: Cloudiot, *Global Conference on Communication Technologies (GCCT),* Thuckalay, India. 10.1109/GCCT.2015.7342624

Bai, Y., & Lu, L. (2020). Acoustic-Based Sensing and Applications: A Survey. *Computer Networks, 181.* doi:10.1016/j.comnet.2020.107447

Bajpai, S., & Radha, D. (2019). Smart Phone as a Controlling Device for Smart Home using Speech Recognition. *IEEE International Conference on Communication and Signal Processing, ICCSP,* (pp. 701–705). IEEE. 10.1109/ICCSP.2019.8697923

Bala Anand Muthu, C. B. (2020). IOT Based Wearable Sensor for Diseases Prediction and Symptom Analysis in Healthcare Sector. *Peer-to-Peer Networking and Applications, 13*(6), 2123–2134. doi:10.100712083-019-00823-2

Bello, O., Zeadally, S., & Badra, M. (2017). Network layer Inter-operation of Device-to-Device Communication Technologies in Internet of Things (IoT). *Ad Hoc Networks, 57.* doi:10.1016/j.adhoc.2016.06.010

Bernieri, A., Ferrigno, L., Laracca, M., & Rasile, A. (2017). An AMR-Based Three-Phase Current Sensor for Smart Grid Applications. *IEEE Sensors Journal, 17*(23), 7704–7712. doi:10.1109/JSEN.2017.2725983

Bhavke, A., & Pai, S. (2017). Advance Automatic Toll Collection and Vehicle Detection During Collision Using RFID. *International Conference on Nascent Technologies in Engineering (ICNTE),* Vashi, India. 10.1109/ICNTE.2017.7947958

Bluetooth. (n.d.). *Home.* Bluetooth. https://www.bluetooth.com/learn-about-bluetooth/tech-overview/

Bt Aripin, N., & Othman, M. B. (2014). *Voice Control of Home Appliances using Android, Electrical Power, Electronics, Communications, Controls, and Informatics Seminar.* EECCIS.

Camarinha-Matos, L. M., & Katkoori, S. (2020). Challenges in IoT Applications and Research. In L. M. Camarinha-Matos, G. Heijenk, S. Katkoori, & L. Strous (Eds.), *IFIP Advances in Information and Communication Technology* (Vol. 641, pp. 3–10). Springer. doi:10.1007/978-3-030-96466-5_1

Chanal, P. M., & Kakkasageri, M. S. (2021). Preserving Data Confidentiality in Internet of Things. *SN Computer Science, 02*(53), 53. doi:10.100742979-020-00429-z

Chandu, Y., Rakesh Kumar, K. S., Khanolkar, N. V., Anish, A. N., & Rawal, S. (2017). Design and Implementation of Hybrid Encryption for Security of IoT Data. *International Conference on Smart Technologies for Smart Nation (Smarttechcon'17)*, Bangalore. 10.1109/SmartTechCon.2017.8358562

Chen, G., Jin, Z., & Chen, J. (2023). A Review: Magneto-Optical Sensor Based on Magnetostrictive Materials and Magneto-Optical Material. *Sensors and Actuators Reports, 5*. doi:10.1016/j.snr.2023.100152

Chen, S., Xu, H., Liu, D., Hu, B., & Wang, H. (2014, August). A Vision of IoT: Applications, Challenges, and Opportunities With China Perspective. *IEEE Internet of Things Journal, 1*(4), 349–359. doi:10.1109/JIOT.2014.2337336

Chowdhry, D., Paranjape, R., & Laforge, P. (2015). Smart Home Automation System for Intrusion Detection. *IEEE 14th Canadian Workshop on Information Theory (CWIT)*, (pp. 75–78). IEEE. 10.1109/CWIT.2015.7255156

Cognizant. (2016). *How the Internet of things will overcome a lack of standards*. Cognizant. https://www.cognizant.com/perspectives/how-the-internet-of-thingswill-overcome-a-lack-.......... mpaign=ThoughtLeadership.

Computing. (2016). *Internet of things undermined by a lack of standards, warns Pentaho VP EMEA Paul Scholey*. Computing.Co. http://www.computing.co.uk.

Cornelius, K., Kumar, N. K., Pradhan, S., Patel, P., & Vinay, N. (2020). An Efficient Tracking System for Air and Sound Pollution using IoT. *6th International Conference on Advanced Computing and Communication Systems (ICACCS)*, Coimbatore, India. 10.1109/ICACCS48705.2020.9074301

CSA. (n.d.). *Governing Documents and Support from the Alliance*. CSA. https://csa-iot.org/resources/security/

Dileep, G. J. R. E. (2020). A Survey on Smart Grid Technologies and Applications. *Renewable Energy, 146*, 2589–2625. doi:10.1016/j.renene.2019.08.092

Divatankar, S., & Hivarkar, U. N. (2019). Survey and Comparative Study of Various Approaches to Monitor the Road Traffic. *Journal of Emerging Technologies and Innovative Research, 6*(4), 248–253.

Duan, R., Chen, X., & Xing, T. (2011). *A QoS Architecture for IoT*. International Conference on Internet of Things and 4th International Conference on Cyber, Physical and Social Computing, Dalian, China. 10.1109/iThings/CPSCom.2011.125

El Ouadghiri, M., Aghoutane, B., & El Farissi, N. (2020). Communication model in the Internet of Things. *Procedia Computer Science, 177*, 72-77. doi:10.1016/j.procs.2020.10.013

Ericsson Working Group. (2016). *Cellular Networks for Massive IoT.* (Ericsson White Paper). Ericcson. https://www.ericsson.com/en/white-papers/cellular-networks-for-massive-iot& ndash;enabling-low-power-wide-area-applications

Fadhil, J. A., Omar, O. A., & Sarhan, Q. I. (2020). A Survey on the Applications of Smart Home Systems. *International Conference on Computer Science and Software Engineering (CSASE)*, Duhok, Iraq. 10.1109/CSASE48920.2020.9142103

Fadi, Al-Turjman. (2019). Smart Parking in IoT-Enabled Cities: A Survey. *Sustainable Cities and Society, 49*. doi:10.1016/j.scs.2019.101608

Fahim, A., Hasan, M., & Chowdhury, M. A. (2021). *Smart Parking Systems: Comprehensive Review Based on Various Aspects, Heliyon, 7*(5). doi:10.1016/j.heliyon.2021.e07050

Fakih, A., Naik, R., Rais, U., & Maniar, S. (2018). A Green Energy System for Smart abode. *International Conference on Smart City and Emerging Technology (ICSCET),* (pp. 1–5). IEEE. 10.1109/ICSCET.2018.8537365

Farahani, B., Firouzi, F., & Chakrabarty, K. (2020). *Healthcare IoT, Intelligent Internet of Things*. Springer. doi:10.1007/978-3-030-30367-9_11

Fareed, M. M. S., Pradhan, B., & Bhattacharyya, S. P. (2021). Kunal, IoT-Based Applications in Healthcare Devices. *Journal of Healthcare Engineering, 2021*, 1–18. doi:10.1155/2021/6632599

Feng, Y., Mao, G., & Cheng, B. (2020). MagMonitor: Vehicle Speed Estimation and Vehicle Classification Through a Magnetic Sensor. *IEEE Transactions on Intelligent Transportation Systems, 23*(2), 1311–1322. doi:10.1109/TITS.2020.3024652

Floris, A., Girau, R., Porcu, S., Pettorru, G., & Atzori, L. (2020). *Implementation of a Magnetometer based Vehicle Detection System for Smart Parking* applications, IEEE International Smart Cities Conference (ISC2), Piscataway, NJ, USA. 10.1109/ISC251055.2020.9239005

Fontes, J. (2005). CHAPTER 12 - Humidity Sensors. In J. S. Wilson (Ed.), *Sensor Technology Handbook* (pp. 271–284). Newnes. doi:10.1016/B978-075067729-5/50052-5

Fuzi, M. F. M., Ibrahim, A. F., Ismail, M. H., & Halim, N. S. A. (2014). HOME FADS: A Dedicated Fire Alert Detection System using Zigbee Wireless Network. *IEEE 5th Control and System Graduate Research Colloquium,* (pp. 53–58). IEEE. 10.1109/ICSGRC.2014.6908695

García, L., Parra, L., Jimenez, J. M., Lloret, J., & Lorenz, P. (2020). IoT-based Smart Irrigation Systems: An Overview on the Recent Trends on Sensors and IoT Systems for Irrigation in Precision Agriculture. *Sensors (Basel), 20*(4), 1042. doi:10.339020041042 PMID:32075172

Ghosh, P., & Dhar, P. K. (2019). GSM Based Low-cost Gas Leakage, Explosion and Fire Alert System with Advanced Security. *International Conference on Electrical, Computer and Communication Engineering (ECCE),* (pp. 1–5). IEEE. 10.1109/ECACE.2019.8679411

Gigli, M., & Koo, S. (2011). Internet of Things: Services and Applications Categorization. *Advances in Internet of Things, 1*(2), 27–31. doi:10.4236/ait.2011.12004

Gubbi, J., Buyya, R., Marusic, S., & Palaniswami, M. (2013). Internet of things (IoT): A Vision, Architectural Elements, and Future Directions. *Future Generation Computer Systems, 29*(7), 1645–1660. doi:10.1016/j.future.2013.01.010

Guerrero-Ibáñez, J., Zeadally, S., & Contreras-Castillo, J. (2018). Sensor Technologies for Intelligent Transportation Systems,. *Sensors (Basel), 18*(4), 1–24. doi:10.339018041212

Gunatilaka, D. (2021). *An IoT-Enabled Acoustic Sensing Platform for Noise Pollution Monitoring.* IEEE 12th Annual Ubiquitous Computing, Electronics & Mobile Communication Conference (UEMCON), New York, NY, USA. doi:10.1109/UEMCON53757.2021.9666534

Hanes, D., Salgueiro, G., Grossetete, P., Barton, R., & Henry, J. (2017). *IoT Fundamentals: Networking Technologies, Protocols, and Use Cases for the Internet-of-Things.* Cisco Press.

Hasan, M., Anik, M. H., & Islam, S. (2018). *Microcontroller Based Smart Home System with Enhanced Appliance Switching Capacity, 5th HCT Information Technology Trends.* ITT.

Huang, Y. C., & Huang, S. C. C. (2014). A Personalized Smart Living Room. Lecture Notes in Computer Science. Springer. doi:10.1007/978-3-319-07788-8_4

Humayun, M., Afsar, S., Almufareh, M. F., Jhanjhi, N. Z., & AlSuwailem, M. (2022). N. Z. Jhanjhi, Mashayel AlSuwailem, Smart Traffic Management System for Metropolitan Cities of Kingdom Using Cutting Edge Technologies, Journal of Advanced Transportation, Vol. 2022. *Journal of Advanced Transportation, 4687319*, 1–13. doi:10.1155/2022/4687319

IBM Think Academy. (2015). *How it works: Internet of things.* [Video]. Youtube. https://www.youtube.com/watch?v=QSIPNhOiMoE

Idris, M., Tamil, E. M., Noor, N. M., Razak, Z., & Fong, K. W. (2009). Parking Guidance System Utilizing Wireless Sensor Network and Ultrasonic Sensor. *Information Technology Journal, 8*(2), 138–146. doi:10.3923/itj.2009.138.146

Imario, A., Sudiharto, D. W., & Ariyanto, E. (2018). The Validated Voice Recognition Measurement Of Several Tribes In Indonesia Using Easy VR 3.0. Case Study: The Prototype Of Automated Doors. *International Seminar on Application for Technology of Information and Communication (iSemantic).* IEEE.

Ismail, W. H. W., Husny, H. R. M., & Abdullah, N. Y. (2016). Smoke Detection Alert System via Mobile Application. *10th International Conference on Ubiquitous Information Management and Communication (IMCOM)*, (pp. 1–6). ACM. 10.1145/2857546.2857581

Ismail, Y. (2019). I*nternet of Things (IoT) for Automated and Smart Applications.* IntechOpen, London. http://dx.doi.org/ doi:10.5772/intechopen/77404

Jain, N. K., Saini, R. K., & Mittal, P. (2019). A Review on Traffic Monitoring System Techniques. In K. Ray, T. Sharma, S. Rawat, R. Saini, & A. Bandyopadhyay (Eds.), *Soft Computing: Theories and Applications. Advances in Intelligent Systems and Computing* (Vol. 742, pp. 569–577). Springer., doi:10.1007/978-981-13-0589-4_53

Javaid, M., & Khan, I. H. (2021). Internet of Things (IoT) Enabled Healthcare Helps to Take the Challenges of COVID-19 Pandemic. *Journal of Oral Biology and Craniofacial Research, 11*(2), 209-214. doi:10.1016/j.jobcr.2021.01.015

José, M. (2018). Environmental Perception For Intelligent Vehicles. Intelligent Vehicles: Enabling Technologies and Future Developments (pp. 23–101). Elsevier.

Juhana, T., & Anggraini, V. G. (2016). Design and Implementation of Smart Home Surveillance System, *10th International Conference on Telecommunication Systems Services and Applications (TSSA)*, (pp. 1-5). IEEE. 10.1109/TSSA.2016.7871101

Kakkar, L., Gupta, D., Saxena, S., & Tanwar, S. (2021). IoT Architectures and Its Security: A Review. In: Goyal, D., Gupta, A.K., Piuri, V., Ganzha, M., Paprzycki, M. (eds.) *Proceedings of the Second International Conference on Information Management and Machine Intelligence.* Springer, Singapore. 10.1007/978-981-15-9689-6_10

Kalla, A. (2019). Pawani Prombage, Madhusanka Liyanage, Introduction to IoT. In M. Liyanage, A. Braeken, P. Kumar, & M. Ylianttila (Eds.), *IoT Security: Advances in Authentication* (pp. 1–25). Wiley. doi:10.1002/9781119527978.ch1

Kamelia, L., Effendi, M. R., & Pratama, D. F. (2018). Integrated Smart House Security System Using Sensors and RFID. *4th International Conference on Wireless and Telematics (ICWT)*, (Vol. 91, pp. 1–5). IEEE. 10.1109/ICWT.2018.8527803

Kamruzzaman, J., Wang, G., Karmakar, G., Ahmad, I., & Md, Z. A. B. (2018). Acoustic sensor networks in the Internet of Things applications. *Future Generation Computer Systems, 86,* 1167-1169. doi:10.1016/j.future.2018.05.019

Kasgar, A. K., Dhariwal, M. K., Tantubay, N., & Malviya, H. (2013). A Review Paper of Message Digest 5 (MD5). *International Journal of Modern Engineering and Management Research, 1*(4).

Kashani, M. H., Madanipour, M., Nikravan, M., Asghari, P., & Mahdipour, E. (2021). A Systematic Review of IoT In Healthcare: Applications, Techniques, and Trends. *Journal of Network and Computer Applications, 192*. doi:10.1016/j.jnca.2021.103164

Kaur, A., Saini, S. S., Singh, L., Sharma, A., & Sidhu, E. (2016). Efficient Arduino UNO Driven Smart Highway/Bridge/Tunnel Lighting System Employing Rochelle Piezoelectric Sensor. *International Conference on Control, Computing, Communication and Materials (ICCCCM)*, Allahbad, India. 10.1109/ICCCCM.2016.7918247

Kaur, N., & Sood, S. K. (2017). An Energy-Efficient Architecture for the Internet of Things (IoT). *IEEE Systems Journal, 11*(2), 796–805. doi:10.1109/JSYST.2015.2469676

Kavitha, S., & Alphonse, P. J. A. (2019). A Hybrid Cryptosystem to Enhance Security in IoT Health Care System. *International Journal of Wireless and Microwave Technologies, 9*(1), 1–10.

Kesavan, G., Sanjeevi, P., & Viswanathan, P. (2016). A 24 Hour IoT Framework for Monitoring and Managing Home Automation *International Conference on Inventive Computation Technologies (ICICT)*, (Vol. 1, pp. 1–5). IEEE. 10.1109/INVENTIVE.2016.7823205

Khan, M. A., & Khaled, S. (2018). IoT security: Review, Blockchain Solutions, And Open Challenges. *Future Generation Computer Systems*, *82*(May), 395–411. doi:10.1016/j.future.2017.11.022

Khanna, A., & Kaur, S. (2020). Internet of Things (IoT), Applications and Challenges: A Comprehensive Review. *Wireless Personal Communications*, *114*(2), 1–76. doi:10.100711277-020-07446-4

Khanna, A., & Kaur, S. (2020). Comprehensive Review. *Wireless Personal Communications, 114*(2), 1687–1762. doi:10.100711277-020-07446-4

Khattab, A., & Youssry, N. (2020). Machine Learning for IoT Systems. In M. Alam, K. Shakil, & S. Khan (Eds.), *Internet of Things*. Springer. doi:10.1007/978-3-030-37468-6_6

Khujamatov, K. E., Toshtemirov, T. K., & Lazarev, A. P. (2021). IoT and 5G Technology In Agriculture. *International Conference on Information Science and Communications Technologies (ICISCT)*, Tashkent, Uzbekistan. 10.1109/ICISCT52966.2021.9670037

Krishnendhu, S. P., & Mohandas, P. (2023). SAD: Sensor-based Anomaly Detection System for Smart Junctions. *IEEE Sensors Journal, 23*(17), 20368–20378. doi:10.1109/JSEN.2023.3297205

Kumar, J. & Patel, D. (2014). A Survey on Internet of Things: Security and Privacy Issues. *International Journal of Computer Applications, 90(*11).

Lakshmikantha, V., Hiriyannagowda, A., Manjunath, A., Patted, A., Basavaiah, J., & Anthony, A. A. (2021). IoT Based Smart Water Quality Monitoring System. *Global Transitions Proceedings*. Science Direct. doi:10.1016/j.gltp.2021.08.062

Lartigue, J. W., McKinney, C., Phelps, R., Rhodes, R., Rice, A. D., & Ryder, A. (2014). A Tablet-Controlled, Mesh-Network Security System. *ACM Southeast Regional Conference*. ACM. 10.1145/2638404.2638500

Lee, W.-H., & Chiu, C.-Y. (2020). Design and Implementation of a Smart Traffic Signal Control System for Smart City Applications. *Sensors (Basel), 20*(2), 508–526. doi:10.339020020508 PMID:31963229

Leung, D. Y. C. (2015). Outdoor-Indoor Air Pollution in Urban Environment: Challenges and Opportunity. *Frontiers in Environmental Science*, *2*(69), 1–7. doi:10.3389/fenvs.2014.00069

Li, S., Xu, L. D., & Zhao, S. (2015). The Internet of Things: A Survey. *Information Systems Frontiers, 17*(2), 243–259. doi:10.100710796-014-9492-7

Lian, K.-Y., Hsiao, S. J., & Sung, W. T. (2014). Smart Home Safety Handwriting Pattern Recognition with Innovative Technology. *Computers & Electrical Engineering*, *40*(4), 1123–1142. doi:10.1016/j.compeleceng.2014.02.010

Lin, H.-E., Zito, R., & Taylor, M. (2005). A review of travel-time prediction in transport and logistics. *Proceedings of the Eastern Asia Society for Transportation Studies*, *5*, 1433–1448.

Liu, J., Wu, J., & Sun, L. (2020). Control Method of Urban Intelligent Parking Guidance System Based on Internet of Things. *Computer Communications, 153*, 279-285. doi:10.1016/j.comcom.2020.01.063

Ma, R., Gong, J., Liu, G., & Hao, Q. (2020). Enabling Cognitive Pyroelectric Infrared Sensing: From Reconfigurable Signal Conditioning to Sensor Mask Design. *IEEE Transactions on Industrial Informatics*, *16*(7), 4436–4446. doi:10.1109/TII.2019.2944700

Mahmoud, R., Yousuf, T., Aloul, F., & Zualkernan, I. (2015). Internet of things (IoT) security: Current status, challenges and prospective measures. *10th International Conference for Internet Technology and Secured Transactions (ICITST)*, London, UK. 10.1109/ICITST.2015.7412116

Mathur, S., Jin, T., & Kasturirangan, N. (2010). Parknet: Drive-By Sensing of Road-Side Parking Statistics. *Proc. of the 8th Int. Conf. on Mobile Systems, Applications, and Services*, ACM. 10.1145/1814433.1814448

Miorandi, D., Sicari, S., De Pellegrini, F., & Chlamtac, I. (2012). Internet of Things: Vision, Applications and Research Challenges. *Ad Hoc Networks, 10*(7), 1497-1516. doi:10.1016/j.adhoc.2012.02.016

Mondal, M. A., & Rehena, Z. (2021). An IoT-Based Congestion Control Framework for Intelligent Traffic Management System. In N. N. Chiplunkar & T. Fukao (Eds.), *Advances in Artificial Intelligence and Data Engineering* (Vol. 1133, pp. 1287–1297). Springer. doi:10.1007/978-981-15-3514-7_96

Mouskos, K., Boile, M., & Parker, N. A. (2007). *Technical Solutions to Overcrowded Park and Ride Facilities*. New Jersey Department of Transportation.

Mukta, M. Y., Rahman, M. A., Asyhari, A. T., & Md, Z. A. B. (2020). IoT for energy efficient green highway lighting systems: Challenges and issues. *Journal of Network and Computer Applications, 158*. doi:10.1016/j.jnca.2020.102575

Naseri, F., Qaedi, R., Karimi, S., Ghanbari, T., & Farjah, E. (2019). An effective Arc Fault Detection Approach for Smart Grid Solar Farms Using Rogowski Coil Sensor. *Smart Grid Conference (SGC)*, Tehran, Iran. 10.1109/SGC49328.2019.9056622

Nath, P., & Pati, U. C. (2018). Low-Cost Android App Based Voice Operated Room Automation System. 3rd International Conference for Convergence in Technology (I2CT), (pp. 1–4). IEEE. 10.1109/I2CT.2018.8529157

Nayak, S., & Vijayalakshmi, M. N. (2016). Analysis on IoT Challenges, Opportunities, Applications and Communication Models. *International Journal of Advanced Engineering, Management and Science, 2*(4).

Nest Labs. (2015). Meet the Nest Thermostat. Nest Labs.

Nour, B., Sharif, K., Li, F., Biswas, S., Moungla, H., Guizani, M., & Wang, Y. (2019). A survey of Internet of Things communication using ICN: A use case perspective. *Computer Communications, 142–143*, 95-123. doi:10.1016/j.comcom.2019.05.010

Oriol, P., Piol, S. R., Joakim, E., & Thiemo, V. (2015). BSD Based Elliptic Curve Cryptography for the Open Internet of Things. *7th International Conference on New Technologies, Mobility and Security (NTMS'15)*, Paris.

Paidi, V., Fleyeh, H., Håkansson, J., & Nyberg, R. G. (2018). Smart Parking Sensors, Technologies and Applications for Open Parking Lots: A Review. *IET Intelligent Transport Systems, 12*(8), 735–741. doi:10.1049/iet-its.2017.0406

Pantelic, J., Nazarian, N., Miller, C., Meggers, F., Jason, K. W. L., & Licina, D. (2022). Transformational IoT Sensing for Air Pollution and Thermal Exposures. *Frontiers in Built Environment, 8*, 236. doi:10.3389/fbuil.2022.971523

Parthasarathy, A. R. (2022). Analysis of the Trust and Resilience of Consumer and Industrial Internet of Things (IoT) Systems in the Indian Context. Khanna, K., Estrela, V.V., Rodrigues, J.J.P.C., (eds.) Cyber Security and Digital Forensics, Lecture Notes on Data Engineering and Communications Technologies. Springer, Singapore. doi:10.1007/978-981-16-3961-6_21

Parul, G., Sahoo, A., & Sharma, T. (2020). Internet of Things: Applications, Security and Privacy: A survey. *Materials Today: Proceedings*, 1–8.

Pawar, A. B., & Ghumbre, S. (2016). A survey on IoT applications, security challenges and counter measures. *International Conference on Computing, Analytics and Security Trends (CAST)*, Pune, India. 10.1109/CAST.2016.7914983

Pazikadin, A. R., Rifai, D., Ali, K., Malik, M. Z., Abdalla, A. N., & Faraj, M. A. (2020). Solar Irradiance Measurement Instrumentation and Power Solar Generation Forecasting Based on Artificial Neural Networks (ANN): A Review of Five Years Research Trend. *Science of The Total Environment, 715*. doi:10.1016/j.scitotenv.2020.136848

Pedro, S. M., Nam, T., Brandon, C., Behnam, D., & Yuhong, L. (2018). Analyzing The Resource Utilization of AES Encryption on IoT Devices. *Asia Pacific Signal And Information Processing Association Annual Summit And Conference*, Hawaii, USA.

Peng, Y., & Wang, Y. (2019). Real-Time Forest Smoke Detection using Hand-Designed Features and Deep Learning. *Computers and Electronics in Agriculture*, *167*, 105029. doi:10.1016/j.compag.2019.105029

Perilla, F. S., Villanueva, G. R., Cacanindin, N. M., & Palaoag, T. D. (2018). Fire Safety and Alert System Using Arduino Sensors with IoT Integration, *7th International Conference on Software and Computer Applications (ICSCA)*. ACM. 10.1145/3185089.3185121

Pflanzner, T., & Kertész, A. (2018). A Taxonomy and Survey of IoT Cloud Applications. EAI Endorsed Transactions on Internet of Things, 3(12). doi:10.4108/eai.6-4-2018.154391

Poornima, M. (2021). *Security and privacy in the internet of things: computational intelligent techniques-based approaches, Recent Trends in Computational Intelligence Enabled Research*. Academic Press.

Pyingkodi, M., Thenmozhi, K., Nanthini, K., Karthikeyan, M., & Suresh Palarimath, V. (2022). Sensor Based Smart Agriculture with IoT Technologies: A Review. *International Conference on Computer Communication and Informatics (ICCCI)*, Coimbatore, India. 10.1109/ICCCI54379.2022.9741001

Qiu, T., Zhao, A., & Ma, R. (2018). A task-efficient sink node based on embedded multi-core SoC for Internet of Things. *Future Generation Computer Systems, 82*. doi:10.1016/j.future.2016.12.024

Rahman, M. A., Asyhari, A. T., Obaidat, M. S., Kurniawan, I. F., Mukta, M. Y., & Vijayakumar, P. (2021). IoT-Enabled Light Intensity-Controlled Seamless Highway Lighting System. *IEEE Systems Journal*, *15*(1), 46–55. doi:10.1109/JSYST.2020.2975592

Reddy, A. (2021). A Review on IoT Enabling Technologies and Back-End Data-Sharing Model. *Journal of Electronics, Computer Networking and Applied Mathematics(JECNAM), 01*(1), 41–47. doi:10.55529/jecnam11.41.47

Revathy, M., Ramya, S., Sathiyavathi, R., Bharathi, B., & Anu, V. M. (2018). Automation of Street Light for Smart City. *IEEE International Conference on Communication and Signal Processing*. IEEE. 10.1109/ICCSP.2017.8286503

Rind, Y. M., Raza, M. H., Zubair, M., Mehmood, M. Q., & Massoud, Y. (1974). Smart Energy Meters For Smart Grids, An Internet of Things Perspective. *Energies*, *16*(4), 2023. doi:10.3390/en16041974

Ripka, P. (2010). Electric Current Sensors: A Review. *Measurement Science & Technology, 21*(11), 1–32. doi:10.1088/0957-0233/21/11/112001

Rivas, A. (2020). Faults in Smart Grid Systems: Monitoring, Detection and Classification. *Electric Power Systems Research, 189*. 10.1016/j.epsr.2020.106602n

Samani, H. H. G. A., & Wahaishi, A. (2015). Privacy in Internet of Things: A Model and Protection Framework, *6th International Conference on Ambient Systems, Networks and Technologies*. Science Direct. 10.1016/j.procs.2015.05.046

Samiksha, S., & Vinay, C. (2016). Analysis of AES encryption with ECC. *International Conference on Engineering Science And Management*, Goa.

Samimi, M. H., & Ilkhechit, H. D. (2020). Survey of Different Sensors Employed for the Power Transformer Monitoring. *IET Science, Measurement & Technology, 14*(1), 1–8. doi:10.1049/iet-smt.2019.0103

Samsung Corp. (2015). *Privacy Policy—Smart TV Supplement.* Samsung Corp.

Sethi P. & Sarangi S. R. (2017). Internet of Things: Architectures, Protocols, and Applications. *Journal of Electrical and Computer Engineering*. . doi:10.1155/2017/9324035

Shaheen, S. (2005). *Smart Parking Management Field Test: A Bay Area Rapid Transit(Bart) District Parking Demonstration*. Institute of Transportation Studies.

Sharma, N., Shamkuwar, M., & Singh, I. (2019). The History, Present and Future With IoT. *Internet of Things and Big Data Analytics For Smart Generation, 154*, 27–51. doi:10.1007/978-3-030-04203-5_3

Stolojescu-Crisan, C., Crisan, C., & Butunoi, B. P. (2021). An IoT-Based Smart Home Automation System. *Sensors (Basel), 21*(11), 3784. doi:10.339021113784 PMID:34070717

Sukmana, H. T., Farisi, M. G., & Khairani, D. (2015). Prototype Utilization of PIR Motion Sensor for Real Time Surveillance System and Webenabled Lamp Automation. *IEEE Asia Pacific Conference on Wireless and Mobile (APWiMob)*. IEEE.

Suresh, S., Bhavya, J., Sakshi, S., Varun, K., & Debarshi, G. (2016). Home Monitoring and Security system. *International Conference on ICT in Business Industry & Government (ICTBIG),* (pp. 1–5). IEEE. 10.1109/ICTBIG.2016.7892665

Tahiliani, V., & Dizalwar, M. (2018). Green IoT Systems: An Energy Efficient Perspective. *Eleventh International Conference on Contemporary Computing*, Noida, India. 10.1109/IC3.2018.8530550

Tariq, A. (2018). Internet of Things: A Comprehensive Study of Security Issues and Defense Mechanisms. *IEEE Access : Practical Innovations, Open Solutions, 7*, 11020–11028.

Tasgaonkar, P. P., Garg, R. D., & Garg, P. K. (2020). Vehicle Detection and Traffic Estimation with Sensors Technologies for Intelligent Transportation Systems. *Sensing and Imaging, 21*(1), 29. doi:10.100711220-020-00295-2

Trivedi, D. M., & Raval, T. J. (2018). Improving Communication Security in IoT using Hybrid Encryption Decryption. *International Journal of Advance Research and Innovative Ideas in Education, 4*(2), 4175–4182.

Ullo, S. L., & Sinha, G. R. (2021). Advances in IoT and Smart Sensors For Remote Sensing And Agriculture Applications. *Remote Sensing (Basel), 13*(13), 2585. doi:10.3390/rs13132585

Wang, C. (2012). An IBE Based Security Scheme on Internet of Things. *2nd International Conference on Cloud Computing and Intelligence Systems*, China.

Wang, C., Daneshmand, M., Dohler, M., Mao, X., Hu, R. Q., & Wang, H. (2013). Guest editorial - special issue on Internet of things (IoT): Architecture, protocols and services. *IEEE Sensors Journal, 13*(10), 3505–3510. doi:10.1109/JSEN.2013.2274906

Wang, Q., Zheng, J., Xu, H., Xu, B., & Chen, R. (2017). Roadside Magnetic Sensor System for Vehicle Detection in Urban Environments. *IEEE Transactions on Intelligent Transportation Systems, 19*(5), 1365–1374. doi:10.1109/TITS.2017.2723908

Watson, I. (2015). Design and Implementation of an automatic street light Control System,. *International Journal of Emerging Technology and Advanced Engineering, 5*(3), 336–340.

Wetula A., Bień A., & Parekh M. (2021). New Sensor for Medium and High-Voltage Measurement. *Energies, 14*(15). doi:10.3390/en14154654

Wu, M., Lu, T.-J., Ling, F.-Y., Sun, J., & Du, H.-Y. (2010). Research on the architecture of Internet of Things. *3rd International Conference on Advanced Computer Theory and Engineering(ICACTE)*, Chengdu. 10.1109/ICACTE.2010.5579493

Xu, J., Gu, B., & Tian, G. (2022). Review of agricultural IoT technology. *Artificial Intelligence in Agriculture, 6.* doi:10.1016/j.aiia.2022.01.001

Yang, D., Rao, K., Xu, B., & Sheng, W. (2015). PIR Sensors Deployment with the Accessible Priority in Smart Home Using Genetic Algorithm. *International Journal of Distributed Sensor Networks, 11*(11), 1–10. doi:10.1155/2015/146270

Zaman, H. U., & Rafiunnisa, A. M. (2018). A User-Friendly Low-Cost Mobile App Based Home Appliance Control And Circuit Breaker, *2nd International Conference on Computing Methodologies and Communication (ICCMC)*, (pp. 203–208). IEEE.

Zhang, Q., & Fitzek, F. H. P. (2015). Mission Critical IoT Communication in 5G. In V. Atanasovski & A. Leon-Garcia (Eds.), *Future Access Enablers for Ubiquitous and Intelligent Infrastructures, FABULOUS 2015, Lecture Notes of the Institute for Computer Sciences, Social Informatics and Telecommunications Engineering* (Vol. 159). Springer. doi:10.1007/978-3-319-27072-2_5

Zhong, C. L., Zhu, Z., & Huang, R.-G. (2015). Study on the IoT Architecture and Gateway Technology, *14th International Symposium on Distributed Computing and Applications for Business Engineering and Science (DCABES)*, Guiyang, China. 10.1109/DCABES.2015.56

Zhu, X., Mukhopadhyay, S. K., & Kurata, H. (2012). A Review of RFID Technology and Its Managerial Applications in Different Industries. *Journal of Engineering and Technology Management, 29*(1), 152–167. doi:10.1016/j.jengtecman.2011.09.011

Zorzi, M., Gluhak, A., Lange, S., & Bassi, A. (2010). From Today's Intranet Of Things To A Future Internet of Things: A Wireless-And Mobility-Related View. *IEEE Wireless Communications*, *17*(6), 44–51. doi:10.1109/MWC.2010.5675777

zWave. (n.d.). *Z-Wave compatible products.* zWave. https://www.z-wave.com/shop-z-wave-smart-home-products

# Chapter 7

# Blockchain and Artificial Intelligence for Cyber Security in the Era of Internet of Things and Industrial Internet of Things Applications

**Amit Kumar Tyagi**

https://orcid.org/0000-0003-2657-8700

*National Institute of Fashion Technology, New Delhi, India*

## ABSTRACT

*As the internet of things (IoT) and industrial internet of things (IIoT) continue to expand, the need for robust cyber security solutions becomes increasingly critical. The convergence of blockchain technology and artificial intelligence (AI) offers promising opportunities to address the security challenges posed by IoT and IIoT applications. This chapter provides an overview of the potential synergies between blockchain and AI in the context of cyber security for IoT and IIoT. Blockchain technology, with its decentralized and immutable nature, can provide enhanced security and trust in IoT and IIoT networks. It offers features such as data integrity, transparency, and tamper resistance, making it well-suited for securing critical data and transactions. Additionally, blockchain can facilitate secure device identity management, access control, and secure communication among IoT and IIoT devices.*

DOI: 10.4018/979-8-3693-0659-8.ch007

# 1. INTRODUCTION

The rapid proliferation of IoT and IIoT devices has introduced new challenges in ensuring the security and privacy of connected systems. The interconnected nature of these devices, along with the vast amounts of data they generate, has created a complex cyber security landscape. In response to these challenges, the integration of Blockchain technology and AI has emerged as a potential solution (Gupta & Sehgal, 2019). Note that Blockchain, originally developed for cryptocurrency applications, is a decentralized and distributed ledger technology that ensures the transparency, integrity, and immutability of data. It provides a secure and tamper-resistant platform for recording and verifying transactions. The decentralized nature of Blockchain eliminates the need for a central authority, making it suitable for securing IoT and IIoT networks. Artificial Intelligence, on the other hand, leverages advanced algorithms and machine learning techniques to enable systems to analyze, learn, and make intelligent decisions (Xu et al., 2018). AI has the potential to enhance cyber security by detecting anomalies, identifying patterns, and predicting and mitigating cyber threats. It can adapt and improve its defense mechanisms based on continuous learning from data patterns and user behavior. The integration of Blockchain and AI in the context of cyber security for IoT and IIoT applications offers several advantages. Blockchain provides a secure and transparent infrastructure for recording and sharing security-related information, such as device identities and access permissions. It enhances trust and accountability in IoT and IIoT networks. AI, on the other hand, can leverage the data stored on the Blockchain to analyze and detect potential threats in real-time. By combining the strengths of both technologies, organizations can achieve a higher level of security and resilience in their connected systems.

However, there are challenges and issues to be addressed in the implementation of Blockchain and AI for cyber security. Scalability and interoperability issues need to be overcome to handle the increasing volume of IoT and IIoT data. The computational overhead of Blockchain and the complexity of AI models pose additional challenges. Moreover, the explainability, privacy, and ethical aspects of AI algorithms need careful attention. In summary, the integration of Blockchain and AI presents a promising approach to address the cyber security challenges in the era of IoT and IIoT applications. By leveraging the strengths of Blockchain's decentralized and immutable nature and AI's advanced analytics and decision-making capabilities, organizations can enhance the security, trust, and resilience of their connected systems. In last, this work has been discussed in 11 sections.

# 2. INTERNET OF THINGS (IoT) AND INDUSTRIAL INTERNET OF THINGS (IIoT)

## 2.1 Overview of IoT and IIoT

IoT and IIoT are two interconnected concepts that involve the integration of smart devices, sensors, and networks to enable communication and data exchange between physical objects and digital systems (Aljawarneh et al., 2020). The IoT refers to a network of interconnected devices, objects, and systems that are embedded with sensors, software, and connectivity capabilities. These devices can collect and exchange data with each other and with cloud-based platforms, enabling various applications and services. IoT devices can range from everyday consumer products like smart thermostats and wearable devices to industrial equipment and infrastructure. The IIoT, on the other hand, focuses specifically on the application of IoT technologies in industrial sectors such as manufacturing, energy, transportation, and agriculture. IIoT systems are designed to improve operational efficiency, optimize resource utilization, and enable data-driven decision-making in industrial processes. It involves the integration of sensors, automation, and data analytics in industrial equipment and infrastructure.

Note that both IoT and IIoT share common characteristics such as connectivity, data exchange, and the ability to remotely monitor and control devices. They enable real-time data collection, analysis, and interaction between physical objects and digital systems, creating new opportunities for efficiency, automation, and innovation. Hence, few key components of IoT and IIoT include:

- Devices and Sensors: These are physical objects equipped with sensors and actuators to collect and transmit data.
- Connectivity: IoT and IIoT devices rely on various communication technologies such as Wi-Fi, Bluetooth, cellular networks, or low-power wide-area networks (LPWAN) to connect and exchange data.
- Cloud Computing: Data collected from IoT and IIoT devices is often processed and stored in cloud-based platforms, providing scalability and accessibility for data analysis and application development.
- Data Analytics: IoT and IIoT generate large volumes of data, which can be analyzed using advanced analytics techniques such as machine learning and artificial intelligence to derive valuable insights and make informed decisions.
- Applications and Services: IoT and IIoT enable a wide range of applications and services across various industries, including smart home automation, predictive maintenance, remote monitoring, supply chain optimization, and energy management.

Hence, the growth of IoT and IIoT has the potential to revolutionize industries, improve efficiency, and create new business opportunities (Raza et al., 2017). However, it also poses challenges related to data privacy, security, interoperability, and scalability. Addressing these challenges requires a holistic approach that considers technical, regulatory, and ethical issues to ensure the successful implementation and adoption of IoT and IIoT technologies.

## 2.2 Applications and Use Cases of IoT and IIoT

The applications and use cases of the IoT and IIoT span across various industries and sectors. Here are some prominent examples:

- Smart Home Automation: IoT enables homeowners to control and automate various devices in their homes, such as lighting, temperature, security systems, and appliances. It enhances convenience, energy efficiency, and security.
- Smart Cities: IoT technologies can be used to optimize urban infrastructure and services, including smart parking, intelligent traffic management, waste management, air quality monitoring, and efficient energy distribution.
- Industrial Automation: IIoT facilitates the automation and optimization of industrial processes, such as manufacturing, supply chain management, and logistics. It enables real-time monitoring, predictive maintenance, and improved operational efficiency.

- Agriculture: IoT is utilized in precision farming to monitor and optimize crop conditions, irrigation, and livestock management. It helps farmers make data-driven decisions, reduce resource wastage, and enhance productivity.
- Energy Management: IoT and IIoT enable smart grid systems that monitor and manage energy distribution, consumption, and renewable energy sources. They promote energy efficiency, grid stability, and demand response programs.
- Healthcare: IoT devices are used for remote patient monitoring, wearable health trackers, and smart medical devices. They enable real-time health data collection, early detection of medical conditions, and personalized healthcare.
- Environmental Monitoring: IoT sensors and networks are deployed for monitoring and managing environmental conditions, such as air quality, water quality, and weather patterns. They aid in pollution control and natural resource management.
- Transportation and Logistics: IoT is applied to track and manage fleet vehicles, optimize routes and logistics operations, and improve supply chain visibility. It enhances efficiency, safety, and asset management.
- Smart Retail: IoT is utilized in retail environments for inventory management, personalized customer experiences, smart shelves, and automated checkout systems. It enhances customer satisfaction and operational efficiency.

Hence these are few examples, and the potential applications of IoT and IIoT continue to expand across industries. The ability to collect and analyze data from connected devices enables organizations to gain valuable insights, optimize processes, and create innovative solutions that improve efficiency, sustainability, and quality of life.

## 3. CYBERSECURITY IN THE IoT AND IIoT LANDSCAPE

### 3.1 Threat Landscape in IoT and IIoT

The threat landscape in IoT and IIoT is constantly evolving, with various types of threats targeting the connected devices and systems. Here are some key threats in the IoT and IIoT landscape:

- Unauthorized Access: Attackers may attempt to gain unauthorized access to IoT devices or networks to exploit vulnerabilities, steal sensitive data, or take control of devices for malicious purposes. Weak or default passwords, insecure communication protocols, and outdated software are common entry points for unauthorized access.
- Data Breaches: IoT devices collect and transmit vast amounts of sensitive data, including personal information and operational data. Data breaches can occur if security measures, such as encryption, access controls, and secure data storage, are not implemented properly. Stolen or compromised data can be used for identity theft, financial fraud, or other malicious activities.
- Botnets and DDoS Attacks: IoT devices can be compromised and recruited into botnets, which are networks of compromised devices controlled by a remote attacker. Botnets can be used to launch Distributed Denial of Service (DDoS) attacks, overwhelming targeted systems or networks with a flood of traffic and causing service disruptions.

- Device Tampering and Physical Attacks: Physical access to IoT devices can enable attackers to tamper with their functionality or extract sensitive information. This includes physically altering devices, manipulating sensors or actuators, or extracting cryptographic keys. Physical attacks can also target the infrastructure supporting IoT deployments, such as communication networks or data centers.
- Malware and Ransomware: IoT devices are susceptible to malware infections and ransomware attacks, where malicious software encrypts or locks down devices until a ransom is paid. Malware can spread through vulnerable devices or compromised networks, leading to system disruption, data loss, or unauthorized access.
- Supply Chain Attacks: IoT devices often rely on a complex supply chain involving multiple vendors and components. Attackers may exploit vulnerabilities in the supply chain to introduce malicious software or hardware, compromising the security and integrity of the devices.
- Privacy Violations: IoT devices constantly collect and transmit data, raising concerns about user privacy. Unauthorized access to personal information, unauthorized monitoring, or data leakage can result in privacy violations and compromise user trust.
- Insider Threats: Insider threats involve malicious actions or negligence from employees, contractors, or individuals with authorized access to IoT systems. Insider threats can result in data breaches, system manipulation, or theft of sensitive information.
- Lack of Security Updates and Patching: IoT devices often have long lifecycles and limited resources, making them vulnerable to security vulnerabilities that may go unpatched. Failure to regularly update and patch IoT devices can leave them susceptible to known security vulnerabilities.
- Lack of Standardization: The lack of standardized security protocols and practices across IoT devices and platforms can create inconsistencies and vulnerabilities. Lack of interoperability and standardization hinders the implementation of robust security measures.

Hence, addressing the IoT and IIoT threat landscape requires a comprehensive approach that includes implementing strong authentication mechanisms, encryption, access controls, regular security updates, and monitoring mechanisms. User awareness, industry collaboration, and regulatory frameworks are essential in mitigating the evolving threats in the IoT and IIoT ecosystem.

## 3.2 Security Challenges and Vulnerabilities in IoT and IIoT

The widespread adoption of IoT and IIoT brings about various security challenges and vulnerabilities (Biswas & Koo, 2020; Puthal et al., 2019). Here are some key security challenges and vulnerabilities in IoT and IIoT:

- Weak Authentication and Authorization: Many IoT devices and systems suffer from weak or default authentication mechanisms, making them susceptible to unauthorized access. Lack of strong authentication and authorization mechanisms can allow attackers to compromise devices, gain unauthorized control, or access sensitive data.
- Insecure Communication: IoT devices often rely on communication protocols that lack encryption or use weak encryption methods. This vulnerability can enable attackers to intercept and tamper with data during transmission, leading to data breaches or unauthorized control of devices.

- Lack of Device Management: IoT devices often have limited security features and lack robust device management capabilities. This makes it challenging to enforce security policies, apply security updates, and detect and respond to security incidents in a timely manner.
- Firmware and Software Vulnerabilities: IoT devices often run on firmware and software that may have security vulnerabilities. Manufacturers may not regularly provide security patches or updates for these devices, leaving them exposed to known vulnerabilities that can be exploited by attackers.
- Physical Security Risks: Physical access to IoT devices can lead to tampering, reverse engineering, or extraction of sensitive information. Physical security measures, such as tamper-proof packaging, secure boot processes, and device hardening, are essential to protect against physical attacks.
- Privacy issues: IoT devices collect vast amounts of data, often including personal and sensitive information. Inadequate privacy protections, such as improper data handling, lack of user consent, or unauthorized data sharing, can result in privacy breaches and compromise user trust (Ding et al., 2019).
- Scalability and Interoperability Challenges: As IoT deployments scale, managing security becomes more challenging. Ensuring interoperability and security across diverse devices, platforms, and networks requires standardized security protocols, secure APIs, and consistent security practices.
- Lack of Security Awareness: Users and organizations may lack awareness of IoT security best practices, making them more susceptible to social engineering attacks, such as phishing or credential theft. Security training and awareness programs are necessary to educate users about the risks and promote secure behaviors.
- Legacy System Integration: Integrating IoT devices with existing legacy systems introduces security challenges. Incompatibilities, lack of security measures in legacy systems, and potential vulnerabilities in integration points can expose the entire system to security risks.

In summary, addressing these security challenges and vulnerabilities requires a holistic approach that includes robust authentication and encryption mechanisms, secure communication protocols, regular security updates, strong access controls, and user awareness. Collaboration between manufacturers, service providers, policymakers, and users is important to ensuring the security and resilience of IoT and IIoT deployments.

## 3.3 Importance of Cybersecurity in IoT and IIoT

Cybersecurity is of paramount importance in the context of IoT and IIoT due to the increasing interconnectedness of devices, networks, and systems. Here are some key reasons why cybersecurity is important in IoT and IIoT:

- Protection of Sensitive Data: IoT and IIoT systems collect and transmit vast amounts of sensitive data, including personal information, operational data, and industrial trade secrets. Ensuring robust cybersecurity measures protects this data from unauthorized access, theft, tampering, or disclosure, preserving privacy and maintaining the integrity of critical information.
- Safeguarding Operational Continuity: Many IoT and IIoT deployments are used in critical infrastructure and industrial settings where disruptions can have severe consequences. Cybersecurity

safeguards prevent unauthorized access, manipulation, or disruption of IoT devices and systems, ensuring uninterrupted operations and maintaining the stability and reliability of critical services.

- Mitigating Financial Losses: Cyberattacks on IoT and IIoT deployments can lead to significant financial losses for individuals, businesses, and organizations. These losses may arise from theft of intellectual property, financial fraud, system downtime, or regulatory fines due to non-compliance. Implementing robust cybersecurity measures helps mitigate the financial impact of cyber threats.
- Protecting User Privacy: IoT devices often gather sensitive information about individuals, such as personal habits, health data, and location information (Samaniego & Yang, 2020; Yang et al., 2019). Cybersecurity measures, including encryption, secure data storage, and user authentication, are important for protecting user privacy and ensuring that personal information remains confidential.
- Preserving Trust and Reputation: Cybersecurity incidents in IoT and IIoT can undermine trust in the technology and the organizations deploying it. Demonstrating a strong commitment to cybersecurity builds trust with users, customers, and partners, enhancing reputation and ensuring continued adoption and success of IoT and IIoT solutions.
- Ensuring Safety: In certain IoT and IIoT applications, such as autonomous vehicles, healthcare devices, or industrial control systems, cybersecurity is critical for ensuring the safety of users and the public. Cybersecurity vulnerabilities in these systems can lead to physical harm, accidents, or loss of life. Robust security measures help prevent such risks and ensure the safety of individuals and communities.
- Addressing Evolving Threat Landscape: The threat landscape in the cybersecurity domain is constantly evolving, with new attack vectors, vulnerabilities, and sophisticated techniques emerging regularly. Proactive cybersecurity measures, including continuous monitoring, threat intelligence, and timely updates, are important for staying ahead of emerging threats and minimizing risks.

Hene, by giving the interconnected and complex nature of IoT and IIoT systems, we need to ensure cybersecurity requires a multi-layered approach that encompasses secure device design, strong authentication and encryption mechanisms, network security, regular updates and patching, user awareness, and industry collaboration.

## 4. ARTIFICIAL INTELLIGENCE (AI) IN CYBERSECURITY

### 4.1 Introduction to AI and Machine Learning in Cybersecurity

AI and machine learning have emerged as powerful technologies in the field of cybersecurity. They offer innovative approaches to detecting, preventing, and responding to cyber threats (Yang et al., 2019; Ylianttila & Vasilakos, 2017). Here is an introduction to AI and machine learning in cybersecurity:

- Artificial Intelligence: Artificial Intelligence refers to the development of computer systems that can perform tasks that typically require human intelligence. It involves the simulation of human intelligence processes such as learning, reasoning, problem-solving, and decision-making.
- Machine Learning: Machine Learning is a subset of AI that focuses on developing algorithms and models that enable computers to learn from data and make predictions or take actions without

explicit programming. Machine learning algorithms can automatically analyze and detect patterns in data to generate insights and make informed decisions.

- Application of AI and Machine Learning in Cybersecurity: AI and machine learning have found various applications in cybersecurity due to their ability to process large volumes of data, identify patterns, and make real-time decisions. Here are some key applications:
- Threat Detection and Prevention: Machine learning algorithms can analyze vast amounts of data to identify patterns indicative of cyber threats, such as malware, phishing attacks, or network intrusions. By learning from historical data, machine learning models can detect and prevent potential security breaches, enabling early intervention and proactive defense mechanisms.
- Anomaly Detection: AI and machine learning techniques can identify anomalous behavior or deviations from normal patterns in network traffic, user activities, or system behavior. This helps in detecting insider threats, zero-day attacks, or unusual system behavior that may indicate a security breach.
- Behavioral Analysis: Machine learning algorithms can learn and analyze user behavior to establish baseline profiles and detect deviations. This allows for the identification of suspicious activities or unauthorized access attempts, enhancing the accuracy of intrusion detection systems.
- Malware Detection: AI and machine learning techniques can be applied to identify and classify malware based on its behavior, code analysis, or signature detection. Machine learning models can continually learn from new malware samples to improve detection rates and stay updated against emerging threats.
- Vulnerability Management: AI and machine learning can assist in vulnerability management by automatically scanning and analyzing systems and applications for potential vulnerabilities. This helps in prioritizing and patching vulnerabilities, reducing the risk of exploitation by attackers.
- Security Analytics: AI and machine learning enable advanced security analytics by correlating data from multiple sources, detecting complex attack patterns, and generating actionable insights. This helps in identifying advanced persistent threats (APTs), conducting threat hunting, and improving incident response capabilities.
- Automated Response and Orchestration: AI can automate incident response by integrating with security orchestration platforms. Machine learning models can make real-time decisions, initiate automated remediation actions, or provide recommendations to security analysts, thereby accelerating incident response and reducing response times.

Note that the integration of AI and machine learning in cybersecurity brings several benefits, including enhanced threat detection capabilities, improved accuracy, faster response times, and proactive defense mechanisms. However, it is essential to address challenges such as the potential for adversarial attacks, data privacy concerns, and the need for interpretability and explainability in AI models to ensure the effective and responsible use of these technologies in cybersecurity.

## 4.2 AI-Based Threat Detection and Prevention

AI-based threat detection and prevention refers to the use of artificial intelligence techniques, such as machine learning, to identify and mitigate cybersecurity threats in real-time (Nair & Tyagi, 2023; Sk et al., 2022). These advanced systems analyze vast amounts of data, learn from patterns, and make intelligent decisions to detect and prevent various types of cyber-attacks. Here is an overview of AI-based threat detection and prevention:

Machine Learning Algorithms: Machine learning algorithms are trained on large datasets to learn patterns and characteristics of normal network behavior. These algorithms can then identify anomalies and deviations from normal behavior, indicating potential security threats. Examples of machine learning algorithms used in threat detection include decision trees, random forests, support vector machines, and neural networks.

- Anomaly Detection: Anomaly detection is a common approach in AI-based threat detection. Machine learning models are trained on historical data to establish a baseline of normal behavior. Any deviation from this baseline is flagged as an anomaly and investigated further. Anomaly detection can help identify various attacks, including network intrusions, insider threats, and unauthorized access attempts.
- Behavioral Analysis: AI-based systems can analyze user behavior to identify unusual or suspicious activities. By learning from user patterns, these systems can detect anomalies in user behavior, such as abnormal login times, excessive data access, or unauthorized privilege escalation. Behavioral analysis helps in identifying potential insider threats and detecting account compromise or misuse.
- Malware Detection: AI techniques are used for malware detection by analyzing characteristics, behaviors, or code patterns of known malware samples. Machine learning models can be trained on large datasets of malware samples to identify common features and signatures. This allows for the detection of new and previously unseen malware based on learned patterns, even if specific signatures are not available.
- Threat Intelligence and Data Analysis: AI-based systems can leverage threat intelligence feeds, vulnerability databases, and security-related news to gather information on the latest threats and attack vectors. By analyzing this data, AI models can identify emerging threats, correlate events, and provide timely alerts or recommendations to security teams.
- Automated Response and Remediation: AI-based systems can automate response and remediation actions to counter detected threats. These systems can integrate with security orchestration platforms to initiate automated actions, such as isolating compromised systems, blocking malicious IP addresses, or updating firewall rules. Automated response capabilities help in reducing response times and mitigating the impact of cyber-attacks.

Hence, AI-based threat detection and prevention systems offer the advantage of real-time monitoring, proactive defense, and the ability to handle large volumes of data. However, it is important to regularly evaluate and update these systems, as cyber threats are constantly evolving. Collaboration between AI systems and human experts is also important to ensure accurate threat identification, response prioritization, and decision-making in complex cybersecurity environments.

## 4.3 AI-Driven Anomaly Detection in IoT and IIoT

AI-driven anomaly detection plays a critical role in securing IoT and IIoT environments. It helps in identifying abnormal behaviors or events that could indicate potential security breaches, equipment malfunctions, or operational anomalies. Here's an overview of AI-driven anomaly detection in IoT and IIoT:

- Data Collection and Preprocessing: Anomaly detection starts with the collection of data from IoT and IIoT devices, which can include sensor readings, network traffic, device logs, and operational data. The collected data is preprocessed to remove noise, handle missing values, and normalize the data for further analysis.
- Feature Extraction: Relevant features are extracted from the preprocessed data to capture meaningful patterns and characteristics. These features can include statistical measures, time-series properties, frequency components, or other domain-specific metrics.
- Unsupervised Learning Techniques: AI-driven anomaly detection often utilizes unsupervised learning techniques. Unsupervised learning algorithms, such as clustering or density estimation algorithms, are applied to the feature space to identify groups or clusters of normal behavior. Instances that fall outside these clusters are flagged as potential anomalies.
- Supervised Learning Techniques: In some cases, labeled data may be available, where anomalies are already identified. Supervised learning algorithms, such as classification or regression algorithms, can be used to train models on labeled data and then apply them to detect anomalies in new, unlabeled data.
- Time-Series Analysis: IoT and IIoT data often exhibit time-dependent patterns. Time-series analysis techniques, such as autoregressive models, moving averages, or Fourier transforms, can be employed to capture temporal dependencies and identify anomalies based on deviations from expected patterns.
- Deep Learning Approaches: Deep learning, particularly recurrent neural networks (RNNs) and long short-term memory (LSTM) networks, are effective for anomaly detection in sequential data. These models can learn complex temporal patterns and detect anomalies based on deviations from learned sequences.
- Online and Real-Time Detection: Anomaly detection in IoT and IIoT requires real-time or near real-time processing to enable timely response. AI-driven approaches should be capable of analyzing streaming data, detecting anomalies in real-time, and triggering immediate alerts or automated actions.

Hence, AI-driven anomaly detection in IoT and IIoT brings several benefits, including early detection of security breaches, proactive maintenance, improved operational efficiency, and reduced downtime. However, challenges such as high-dimensional data, data quality issues, scalability, and interpretability of AI models need to be addressed to ensure the effectiveness and reliability of anomaly detection systems in IoT and IIoT environments.

## 5. BLOCKCHAIN TECHNOLOGY IN CYBERSECURITY

### 5.1 Fundamentals of Blockchain Technology and Blockchain Applications in Cybersecurity

Blockchain technology is a decentralized and distributed ledger system that enables secure and transparent record-keeping of transactions or data across multiple participants (). Understanding the fundamentals of blockchain technology is essential to explore its potential applications and implications in various industries, including cybersecurity, finance, logistics, and governance. Blockchain technology offers several applications in the field of cybersecurity, providing enhanced security, transparency, and integrity. Here are some key applications of blockchain in cybersecurity:

- Immutable Audit Trails: Blockchain provides an immutable and tamper-proof ledger, making it ideal for creating audit trails. Security events, such as system changes, access attempts, or data breaches, can be recorded on the blockchain, ensuring transparency and enabling forensic analysis.
- Secure Identity Management: Blockchain can be used for identity management, eliminating the need for centralized identity providers. By storing encrypted identity information on the blockchain, users can maintain control over their personal data, reducing the risk of identity theft and unauthorized access.
- Secure Data Storage: Blockchain can be used to securely store sensitive data. Instead of storing data on centralized servers, data can be encrypted, distributed across the blockchain network, and accessed only by authorized parties. This enhances data security and mitigates the risk of data breaches.
- Distributed Threat Intelligence: Blockchain enables the sharing of threat intelligence across multiple organizations securely. By storing and sharing threat information on the blockchain, organizations can collaborate in real-time, enhancing their ability to detect and respond to emerging threats.
- Secure Supply Chain Management: Blockchain can be utilized to enhance the security and transparency of supply chains. It enables the tracking and verification of products and components throughout the supply chain, ensuring authenticity and preventing counterfeiting or tampering.
- Decentralized DNS: Blockchain can be used to create a decentralized domain name system (DNS), reducing the risk of domain hijacking and DNS attacks. By storing domain ownership information on the blockchain, it becomes more resilient to malicious manipulation.
- Zero-Trust Networks: Blockchain can facilitate the creation of decentralized, trustless networks. By using blockchain technology, organizations can establish peer-to-peer connections and verify the integrity and authenticity of network participants without relying on a central authority.
- Secure Smart Contracts: Blockchain-based smart contracts can enhance the security of contractual agreements. Smart contracts are self-executing and self-enforcing, reducing the risk of fraud or tampering. The decentralized nature of blockchain ensures the reliability and integrity of smart contract execution.
- Incident Response and Forensics: Blockchain technology can aid in incident response and forensic investigations. By storing security event logs and digital evidence on the blockchain, investigators can ensure the integrity and non-repudiation of collected evidence.
- Threat Detection and Analytics: Blockchain can be leveraged for detecting and analyzing cyber threats. By aggregating and analyzing security-related data from multiple sources on the blockchain, patterns and anomalies can be identified, enabling proactive threat detection and response.

Note that these applications highlight the potential of blockchain technology in enhancing cybersecurity practices. However, it's important to consider the limitations of blockchain, such as scalability, privacy, and regulatory challenges, while exploring its applications in the cybersecurity domain.

## 5.2 Security and Privacy Concerns of Blockchain Based Cybersecurity

While blockchain technology offers enhanced security features, it is not without its security and privacy concerns (Jayaprakash & Tyagi, 2022; Tyagi, Agarwal, & Sreenath, 2022). Here are some of the key concerns related to blockchain-based cybersecurity:

- 51% Attack: In public blockchains that rely on consensus mechanisms like Proof of Work (PoW), there is a risk of a 51% attack. If a single entity or group controls more than 50% of the network's computing power, they can manipulate the blockchain by rewriting transaction history or excluding certain transactions from being added to the chain.
- Smart Contract Vulnerabilities: Smart contracts deployed on blockchain platforms can be vulnerable to coding errors or vulnerabilities, leading to security breaches. Exploiting these vulnerabilities can result in financial losses, unauthorized access to data, or disruption of services.
- Privacy Risks: While blockchain provides transparency, it also poses challenges to privacy. The decentralized and immutable nature of blockchain means that once data is recorded, it cannot be easily modified or erased. This can be a concern when dealing with sensitive or personally identifiable information.
- Key Management: Blockchain-based systems rely on cryptographic keys for authentication and data encryption. Poor key management practices, such as weak passwords or improper storage of keys, can lead to unauthorized access and compromise the security of the blockchain.
- Scalability and Performance: Public blockchains, like Bitcoin and Ethereum, face scalability and performance limitations. As the number of transactions increases, the network can become congested, resulting in slower transaction processing times and increased costs. These limitations can impact the overall security and effectiveness of blockchain-based systems.
- Regulatory Compliance: Blockchain-based systems may face challenges in meeting regulatory requirements, particularly in industries with strict data protection and privacy regulations. Compliance with regulations like the General Data Protection Regulation (GDPR) can be complex, considering the decentralized nature and immutability of blockchain data.
- Insider Threats: While blockchain provides security against external tampering, it may not address insider threats. Malicious actors with authorized access to the blockchain network can manipulate data or disrupt the system from within, posing a significant security risk.
- Lack of Standardization: The absence of standardized protocols and frameworks for blockchain security can lead to inconsistencies in security practices across different blockchain platforms. This lack of standardization may make it challenging to ensure consistent security measures and interoperability between blockchain networks.

Note that to address these issues, it is essential to implement additional security measures, such as secure key management practices, code audits for smart contracts, encryption of sensitive data, and access control mechanisms. It is also important to continuously monitor and update the security protocols of blockchain-based systems to mitigate emerging threats and vulnerabilities.

# 6. INTEGRATION OF BLOCKCHAIN AND AI FOR CYBERSECURITY IN IoT AND IIoT

## 6.1 Blockchain-Based Secure Communication in IoT and IIoT

Blockchain technology can be leveraged to provide secure communication in IoT and IIoT environments (Dagher et al., 2018; Fernández-Caramés & Fraga-Lamas, 2020; Jayaprakash & Tyagi, 2022; Varsha, 2022). Here's how blockchain can enhance communication security in these contexts:

- Data Integrity: Blockchain ensures the integrity of data transmitted within IoT and IIoT systems. Each data transaction is recorded as a block on the blockchain, and once added, it becomes immutable. This guarantees that the data cannot be tampered with or altered during transmission, ensuring the integrity of the information exchanged.
- Trust and Authentication: Blockchain enables trust and authentication in IoT and IIoT communication. By leveraging cryptographic algorithms, devices can securely identify and authenticate themselves on the blockchain network. This eliminates the need for a centralized authority for authentication, reducing the risk of unauthorized access or impersonation.
- Distributed and Decentralized Network: Blockchain operates as a distributed and decentralized network, making it resilient against single points of failure or attacks. IoT and IIoT devices can connect to the blockchain network, ensuring that communication occurs in a peer-to-peer manner. This decentralized architecture enhances security by minimizing the potential for a single point of vulnerability.
- Secure Device Registration: Blockchain can facilitate secure device registration and onboarding in IoT and IIoT networks. By storing device identity and registration information on the blockchain, it becomes easier to verify the authenticity and integrity of the devices joining the network. This prevents unauthorized devices from gaining access and compromising the network.
- Encrypted Communication Channels: Blockchain-based systems can leverage encryption techniques to establish secure communication channels between IoT and IIoT devices. Encryption ensures that the data transmitted between devices is protected from eavesdropping or interception, safeguarding the confidentiality of the communication.
- Consensus Mechanisms: Blockchain consensus mechanisms, such as Proof of Work (PoW) or Proof of Stake (PoS), can be used to validate and verify IoT and IIoT transactions. Consensus ensures that only legitimate and valid transactions are added to the blockchain, preventing malicious actors from injecting fraudulent or malicious data into the communication stream.
- Auditability and Traceability: Blockchain provides auditability and traceability of communication transactions. Each communication event or transaction is recorded on the blockchain, creating an immutable audit trail. This enables forensic analysis and facilitates the identification and mitigation of any security breaches or unauthorized activities within the IoT and IIoT network.

Hence, by integrating blockchain technology into IoT and IIoT communication, organizations can enhance the security and trustworthiness of their systems. Blockchain-based secure communication ensures data integrity, authentication, decentralized network architecture, and encrypted channels, addressing the security concerns associated with IoT and IIoT environments.

## 6.2 AI-Enhanced Threat Intelligence Using Blockchain

AI-enhanced threat intelligence refers to the use of artificial intelligence techniques to enhance the process of collecting, analyzing, and responding to cybersecurity threats. Blockchain technology can be integrated with AI to further enhance the effectiveness and trustworthiness of threat intelligence systems. Here's how AI and blockchain can be combined to provide AI-enhanced threat intelligence:

- Immutable Data Storage: Blockchain provides a tamper-resistant and immutable ledger to store threat intelligence data. By storing threat data on the blockchain, it becomes difficult for attackers to modify or manipulate the information. This ensures the integrity and trustworthiness of the data used by AI systems for threat analysis.
- Data Sharing and Collaboration: Blockchain allows for secure and transparent data sharing among different entities within the cybersecurity ecosystem. AI algorithms can leverage blockchain to access a shared pool of threat intelligence data from various sources, including threat feeds, security vendors, and security researchers. This enables AI systems to have a broader and more comprehensive view of the threat landscape.
- Trust and Authenticity: Blockchain's decentralized and consensus-driven nature helps establish trust and authenticity in the threat intelligence data. AI algorithms can rely on the consensus mechanism of the blockchain to validate the accuracy and reliability of the data before utilizing it for analysis. This reduces the risk of relying on inaccurate or manipulated threat data.
- Enhanced Data Privacy: Blockchain can be used to implement privacy-preserving techniques, such as zero-knowledge proofs or differential privacy, to protect sensitive threat intelligence data. AI systems can leverage these privacy-enhancing features to ensure the confidentiality of the data while still benefiting from its insights.
- Smart Contracts for Threat Sharing: Smart contracts, which are self-executing contracts with predefined conditions, can be utilized on the blockchain to automate the sharing and distribution of threat intelligence. AI systems can interact with these smart contracts to retrieve relevant threat data, ensuring a seamless and secure sharing process.
- Real-time Threat Detection and Response: AI algorithms can be trained on real-time threat intelligence data stored on the blockchain, enabling faster and more accurate threat detection and response. By leveraging AI's capabilities in pattern recognition, anomaly detection, and behavioral analysis, organizations can proactively identify and mitigate emerging threats.
- Auditable and Transparent Analysis: The transparent nature of blockchain allows for auditable and transparent analysis of AI-driven threat intelligence systems. Security auditors and regulatory bodies can examine the blockchain to verify the integrity and fairness of the AI algorithms used for threat analysis.

Hence, by combining AI and blockchain, organizations can enhance their threat intelligence capabilities with improved data integrity, enhanced collaboration, trustworthiness, and privacy protection. AI algorithms can leverage the secure and transparent nature of blockchain to provide more effective and reliable threat intelligence for proactive cybersecurity defense.

## 6.3 Privacy and Data Protection With Blockchain and AI

Privacy and data protection are essential issues when implementing blockchain and AI technologies. Here are some key aspects to consider:

- Data Encryption: Utilize encryption techniques to protect sensitive data stored on the blockchain or usedby AI systems. Encryption ensures that only authorized parties with the appropriate decryption keys can access and interpret the data, adding an extra layer of protection.
- Pseudonymization: Apply pseudonymization techniques to replace personally identifiable information (PII) with pseudonyms or identifiers. This helps protect the privacy of individuals while still allowing for data analysis and processing.
- Access Control: Implement robust access control mechanisms to restrict data access to authorized users or AI algorithms. This includes authentication and authorization protocols to ensure that only approved entities can access and interact with the data.
- Consent Management: Establish clear processes for obtaining and managing user consent regarding data collection, processing, and sharing. Blockchain can be utilized to store and enforce consent records, providing a transparent and auditable system for managing user consent preferences.
- Privacy by Design: Incorporate privacy issues into the design and development of blockchain and AI systems from the outset. By integrating privacy features and principles during the design phase, organizations can ensure that privacy is prioritized throughout the technology implementation.
- Data Minimization: Only collect and store the minimum amount of data necessary for the intended purpose. Avoid unnecessary or excessive data collection to minimize privacy risks and potential data breaches.
- Transparent Data Management: Blockchain's transparent and immutable nature can provide individuals with greater visibility and control over their data. Use blockchain to enable individuals to track and manage their data, including the ability to revoke access permissions or request data deletion.
- Privacy Policies and Documentation: Clearly communicate privacy policies and practices to users, stakeholders, and employees. Provide easily accessible documentation that outlines how personal data is collected, used, stored, and protected within the blockchain and AI systems.
- Compliance with Regulations: Ensure compliance with relevant data protection regulations such as GDPR (General Data Protection Regulation) or CCPA (California Consumer Privacy Act). Understand the legal requirements for handling personal data and implement measures to meet those requirements.
- Regular Assessments and Audits: Conduct regular privacy assessments and audits to identify any vulnerabilities or risks related to privacy and data protection. This includes assessing the security of the blockchain infrastructure, evaluating AI algorithms for potential biases or privacy implications, and reviewing data handling practices.

Hence, by integrating privacy and data protection measures into the design and implementation of blockchain and AI systems, organizations can maintain a high level of privacy assurance while benefiting from the advantages of these technologies. It is important to adopt a privacy-conscious approach and ensure ongoing monitoring and compliance with privacy regulations and best practices.

# 7. SECURE IDENTITY AND ACCESS MANAGEMENT IN IoT AND IIoT

## 7.1 Authentication and Authorization in IoT and IIoT Based Environment

Authentication and authorization are essential components of security in IoT and IIoT environments. They help ensure that only authorized devices and entities can access and interact with IoT systems. Here's an overview of authentication and authorization in IoT and IIoT:

A. Authentication:
  - Authentication verifies the identity of devices or entities attempting to access the IoT or IIoT network. It ensures that only trusted and authorized devices are granted access. Common authentication mechanisms used in IoT and IIoT include:
  - Digital Certificates: Devices are issued digital certificates that contain their unique identifiers and cryptographic keys. These certificates are used to authenticate and establish trust between devices and the IoT network.
  - Passwords and Tokens: Devices may be assigned passwords or tokens that they must provide during the authentication process. These credentials are verified against a database of authorized devices or an authentication server.
  - Biometrics: Biometric authentication, such as fingerprint or facial recognition, can be used to authenticate users or devices. Biometric data is compared against stored templates to verify identity.
  - Public/Private Key Encryption: Devices can use asymmetric encryption with public and private keys. During authentication, a device presents its public key, and the network verifies the corresponding private key to establish device identity.

B. Authorization:
  - Authorization determines the level of access and privileges granted to authenticated devices or entities within the IoT or IIoT network. It ensures that devices can only perform authorized actions based on their roles and permissions. Common authorization mechanisms used in IoT and IIoT include:
  - Role-Based Access Control (RBAC): Devices are assigned specific roles that define their access rights and privileges. Access control policies are enforced based on these roles.
  - Attribute-Based Access Control (ABAC): Access control decisions are based on specific attributes associated with devices or entities, such as their location, time of access, or environmental conditions.
  - Access Control Lists (ACL): A list of authorized devices or entities and the actions they are allowed to perform is maintained. Access requests are evaluated against this list to determine whether access should be granted or denied.
  - Policy-Based Access Control (PBAC): Access control policies define the rules and conditions for granting access. Policies may consider factors such as device type, user identity, or the state of the network.

C. Combining Authentication and Authorization:
  - Authentication and authorization work together to ensure secure access to IoT and IIoT systems. After successful authentication, devices are granted appropriate authorization based

on their identity and assigned privileges. This combination helps prevent unauthorized access, malicious activities, and data breaches within the IoT and IIoT environments.

Note that we need to implement robust authentication and authorization mechanisms that align with the specific security requirements and constraints of the IoT and IIoT applications. This includes considering factors such as device constraints, network scalability, and the need for secure communication protocols.

## 7.2 Blockchain-Based Identity Management Solutions for IoT and IIoT Based Environment

Blockchain-based identity management solutions have emerged as a promising approach to address the challenges of identity and access management in IoT and IIoT environments. These solutions leverage the decentralized and immutable nature of blockchain to enhance security, privacy, and trust in managing identities of IoT devices and entities. Here's how blockchain can be used for identity management in IoT and IIoT:

- Decentralized Identity: Blockchain enables the creation of decentralized identities for IoT devices and entities. Each device can have its unique identifier stored on the blockchain, ensuring that identity information is not controlled by a single central authority. This eliminates the need for a centralized identity management system and reduces the risk of a single point of failure.
- Immutable Identity Records: Identity information and credentials can be stored on the blockchain as immutable records. This ensures the integrity and tamper-proof nature of identity data, making it difficult for malicious actors to alter or manipulate device identities.
- Trust and Verification: Blockchain provides a trust layer for identity verification. Device identities and associated attributes can be verified through consensus mechanisms, such as proof-of-work or proof-of-stake, ensuring the authenticity and validity of the identity information.
- Self-Sovereign Identity: Blockchain allows for self-sovereign identity, where devices have control over their own identity data. Devices can selectively share identity attributes and credentials with other parties while maintaining control over their personal information. This enhances privacy and gives devices the ability to manage their own identity without relying on a centralized authority.
- Secure Authentication: Blockchain-based identity management solutions can facilitate secure authentication mechanisms for IoT devices. Digital signatures and cryptographic protocols can be used to authenticate device identities and ensure secure communication between devices and the IoT network.
- Auditability and Transparency: Blockchain's transparent and auditable nature enables the tracking and auditing of identity-related transactions. This enhances accountability and allows for the traceability of identity management activities.
- Interoperability: Blockchain-based identity management solutions can facilitate interoperability between different IoT systems and platforms. Devices from different manufacturers or ecosystems can use a common blockchain-based identity framework, simplifying the integration and interoperability of diverse IoT devices.
- Revocation and Deletion: Blockchain can provide mechanisms for revoking or deleting device identities when necessary. For example, if a device is compromised or decommissioned, its identity can be revoked, preventing further access to the network.

Hence, implementing blockchain-based identity management solutions in IoT and IIoT environments offers several benefits, including enhanced security, privacy, trust, and interoperability. However, challenges such as scalability, performance, and energy efficiency need to be considered and addressed when designing and deploying blockchain-based identity management systems for IoT and IIoT applications.

## 8. DATA INTEGRITY AND PRIVACY IN IoT AND IIoT

### 8.1 Challenges of Data Integrity and Privacy in IoT and IIoT Based Environment

Data integrity and privacy in IoT and IIoT environments present significant challenges due to the massive volume of data generated, diverse sources of data, and the distributed nature of IoT systems. Here are some key challenges related to data integrity and privacy in IoT and IIoT:

- Data Verification and Trustworthiness: Ensuring the integrity and authenticity of IoT data is challenging, as data can be tampered with or manipulated during transmission or storage. Verifying the trustworthiness of data becomes useful/ important to prevent malicious activities and maintain data integrity.
- Data Encryption and Security: IoT devices generate and transmit sensitive data, including personal information, financial data, and operational details. Protecting this data from unauthorized access requires robust encryption techniques and security measures to safeguard against data breaches and cyber-attacks.
- Data Aggregation and Fusion: IoT systems often involve multiple devices and sensors that generate vast amounts of data. Aggregating and fusing data from diverse sources while preserving data integrity and privacy is a complex task. Ensuring that data from different devices is combined accurately and securely is a challenge in maintaining data integrity.
- Data Lifecycle Management: Managing the lifecycle of IoT data, from collection to storage and disposal, presents challenges. Data must be properly handled, stored securely, and disposed of appropriately to prevent unauthorized access, data leakage, or unintended exposure.
- Consent and User Privacy: IoT devices collect data from users, and ensuring user consent and privacy becomes essential. Consent mechanisms should be transparent and allow users to have control over the data collected from them. Respecting user privacy rights while maintaining the functionality of IoT systems is a challenge.
- Data Governance and Compliance: IoT and IIoT environments may be subject to various data protection regulations and compliance requirements. Ensuring compliance with regulations such as the General Data Protection Regulation (GDPR) or industry-specific standards is a challenge due to the distributed nature of IoT systems and the complexity of data handling.
- Data Sharing and Collaboration: IoT systems often require data sharing and collaboration between different stakeholders. Balancing the need for data sharing while protecting privacy and maintaining data integrity is a challenge. Establishing secure and privacy-preserving data sharing frameworks and protocols is important.
- Edge Computing and Data Processing: With the proliferation of edge computing in IoT and IIoT, data is processed closer to the source. Ensuring data integrity and privacy during edge processing

becomes challenging due to resource-constrained devices and the need for secure communication and computation.

Hence, addressing these challenges requires a detailed approach that includes robust encryption techniques, secure data transmission protocols, privacy-preserving data management practices, and compliance with relevant regulations. Additionally, implementing secure authentication mechanisms, data access controls, and transparency in data handling can help mitigate data integrity and privacy risks in IoT and IIoT environments.

## 8.2 Blockchain-Enabled Data Integrity and Auditing in IoT and IIoT Based Environment

Blockchain technology offers a promising solution for ensuring data integrity and auditing in IoT and IIoT environments. By leveraging the decentralized and immutable nature of blockchain, it provides a transparent and tamper-proof framework for data verification and auditing. Here's how blockchain enables data integrity and auditing in IoT and IIoT:

- Immutable Data Storage: Blockchain serves as a distributed ledger where data can be stored in a decentralized manner. Each data transaction is recorded as a block, which is linked to the previous blocks using cryptographic hashes, forming a chain. Once a block is added to the blockchain, it becomes immutable, making it highly resistant to data tampering or unauthorized modifications.
- Data Integrity Verification: With blockchain, data integrity can be ensured through cryptographic techniques. IoT devices can generate digital signatures or hashes of their data, which are then stored on the blockchain. Any subsequent changes or tampering with the data can be detected by comparing the original digital signature or hash with the one stored on the blockchain.
- Consensus Mechanisms: Blockchain relies on consensus mechanisms to validate and verify data transactions. In a public blockchain, multiple nodes in the network participate in the consensus process to agree on the validity of transactions. This consensus mechanism ensures that only valid and authorized data is added to the blockchain, enhancing data integrity.
- Auditing and Traceability: Blockchain's transparent and immutable nature facilitates auditing and traceability of data in IoT and IIoT systems. Each data transaction recorded on the blockchain can be traced back to its origin, providing an auditable trail of data events. This capability enables organizations to track data flow, verify data authenticity, and ensure compliance with regulatory requirements.
- Smart Contracts for Data Validation: Smart contracts, which are self-executing agreements running on the blockchain, can be used to automate data validation and auditing processes. Smart contracts can define rules and conditions for data transactions, automatically verifying the integrity of incoming data and triggering actions based on predefined criteria.
- Data Provenance and Ownership: Blockchain can establish a clear record of data provenance and ownership. By storing transaction details on the blockchain, organizations can track the origin of data, its ownership, and any subsequent transfers or modifications. This feature enhances data transparency and accountability in IoT and IIoT environments.
- Data Access Controls: Blockchain enables fine-grained access controls to data stored on the blockchain. Access to data can be managed through cryptographic keys and permissioned blockchain

networks, ensuring that only authorized entities can view or modify data. This helps protect data privacy and prevent unauthorized access.

Note that by leveraging blockchain technology, IoT and IIoT systems can benefit from enhanced data integrity, transparency, and auditability. The decentralized and tamper-proof nature of blockchain provides a robust framework for securing and verifying data in IoT and IIoT environments, addressing the challenges of data integrity and auditing.

## 8.3 AI-Based Data Privacy Protection Techniques for IoT and IIoT Based Environment

AI-based data privacy protection techniques play an important role in safeguarding sensitive information in IoT and IIoT environments. Here are some AI-driven techniques used for data privacy protection:

- Differential Privacy: Differential privacy is a technique that adds noise or randomness to data before it is shared or analyzed. AI algorithms can apply differential privacy mechanisms to IoT and IIoT data, ensuring that individual data points cannot be re-identified, thus protecting privacy while allowing for meaningful analysis.
- Homomorphic Encryption: Homomorphic encryption allows computation on encrypted data without decrypting it. AI algorithms can be designed to perform operations on encrypted IoT data, preserving privacy throughout the data processing pipeline. This technique enables data analysis while keeping sensitive information encrypted.
- Privacy-Preserving Machine Learning: AI algorithms can be used to develop privacy-preserving machine learning models. These models are trained on decentralized or encrypted data, protecting individual data points while still providing accurate predictions. Techniques such as federated learning and secure multiparty computation enable collaborative model training without exposing raw data.
- Anonymization and Pseudonymization: AI techniques can be applied to anonymize or pseudonymize sensitive data in IoT and IIoT environments. This involves removing or obfuscating personally identifiable information (PII) to protect individual privacy. AI algorithms can be used to ensure that anonymization techniques maintain data utility while minimizing the risk of re-identification.
- Contextual Privacy Preservation: AI can be utilized to analyze the context and sensitivity of IoT data to determine appropriate privacy protection measures. By understanding the context in which data is collected and used, AI algorithms can dynamically adjust privacy settings and access controls to protect sensitive information.
- Privacy-Preserving Data Sharing: AI algorithms can facilitate secure and privacy-preserving data sharing in IoT and IIoT environments. Techniques such as secure multi-party computation and secure data federations enable collaboration while maintaining data privacy. AI can also enforce data usage policies, ensuring that shared data is only used for authorized purposes.
- Adversarial Robustness: AI techniques can be employed to enhance the robustness of IoT and IIoT systems against adversarial attacks. By training models to detect and mitigate adversarial attacks on IoT data, privacy breaches can be prevented, and the overall security of the system can be improved.

- User-Centric Privacy Control: AI-driven privacy control mechanisms can empower individuals to have more control over their IoT and IIoT data. AI algorithms can enable personalized privacy settings, allowing users to specify their privacy preferences and consent to data collection and usage. This puts individuals in control of their data and ensures compliance with privacy regulations.

Note that by leveraging AI-based data privacy protection techniques, IoT and IIoT systems can achieve a balance between data utility and privacy preservation. These techniques enable organizations to extract valuable insights from IoT data while ensuring the confidentiality and integrity of sensitive information.

## 9. BLOCKCHAIN AND AI FOR INCIDENT RESPONSE AND FORENSICS

### 9.1 Blockchain and AI Based Incident Detection and Response for IoT and IIoT based Environment

Blockchain and AI can be combined to enhance incident detection and response in IoT and IIoT environments. Here's how blockchain and AI can be leveraged for incident detection and response:

- Immutable Event Logging: Blockchain's immutable nature can be used to create a tamper-proof and transparent log of IoT and IIoT events. Each event or data transaction can be recorded on the blockchain, providing an audit trail that cannot be altered. This ensures the integrity of incident-related data and allows for reliable incident detection.
- Anomaly Detection: AI algorithms can analyze IoT and IIoT data to detect anomalies or deviations from normal behavior. By training machine learning models on historical data, AI can learn patterns and identify abnormal activities or events. This enables the early detection of potential incidents or security breaches in real-time.
- Threat Intelligence Sharing: Blockchain provides a decentralized and secure platform for sharing threat intelligence data among IoT and IIoT devices and systems. AI algorithms can analyze and correlate threat intelligence from various sources to identify emerging threats and vulnerabilities. This shared intelligence can help prevent incidents and enhance response capabilities.
- Real-time Monitoring and Alerts: AI can continuously monitor IoT and IIoT systems, analyzing data in real-time to identify potential incidents. By leveraging machine learning algorithms, AI can detect anomalies, suspicious patterns, or unauthorized activities and generate timely alerts to enable quick incident response.
- Incident Response Automation: AI can automate incident response processes in IoT and IIoT environments. Machine learning algorithms can be trained to classify and prioritize incidents based on their severity, enabling automated response actions. This can include isolating affected devices, applying security patches, or triggering alerts to security teams for further investigation.
- Decentralized Incident Management: Blockchain's decentralized nature allows for distributed incident management in IoT and IIoT environments. Incident-related information, such as event logs, response actions, and remediation steps, can be securely stored and shared across the blockchain network. This ensures that incident information is accessible to authorized parties and prevents single points of failure.

- Forensic Analysis: Blockchain's immutable and transparent nature can facilitate forensic analysis in the event of a security incident. The recorded data on the blockchain can serve as a reliable source for post-incident investigations and forensics. AI algorithms can analyze this data to reconstruct the sequence of events, identify the root cause, and gather evidence for remediation and legal purposes.

Note that by combining the strengths of blockchain and AI, IoT and IIoT environments can benefit from improved incident detection, faster response times, and enhanced overall security. The tamper-proof and decentralized nature of blockchain, along with the intelligence of AI algorithms, provides a robust framework for incident detection and response in the IoT and IIoT domain.

## 9.2 Forensic Analysis Using Blockchain and AI for IoT and IIoT Based Environment

Forensic analysis using blockchain and AI in IoT and IIoT environments can help investigate security incidents, gather evidence, and reconstruct the sequence of events. Here's how blockchain and AI can be leveraged for forensic analysis:

- Immutable and Transparent Data Storage: Blockchain's immutable nature ensures that once data is recorded on the blockchain, it cannot be altered or tampered with. This characteristic is important for preserving the integrity of forensic evidence. IoT and IIoT data, such as device logs, sensor readings, or transaction records, can be stored on the blockchain, providing a trustworthy source of evidence for forensic analysis.
- Chain of Custody: Blockchain can establish a chain of custody for forensic evidence in IoT and IIoT environments. Each transaction or event recorded on the blockchain includes a timestamp, cryptographic hash, and information about the parties involved. This creates an auditable trail of custody, ensuring the integrity and authenticity of the evidence throughout the investigation process.
- Data Attribution and Ownership: Blockchain can assist in identifying the source and ownership of IoT and IIoT data. By leveraging blockchain's decentralized architecture, AI algorithms can analyze the blockchain to trace the origin of data, identify the devices or systems responsible for generating it, and determine the ownership of the data. This information is essential for forensic investigations and establishing accountability.
- Reconstruction of Events: AI algorithms can analyze the data recorded on the blockchain to reconstruct the sequence of events leading up to a security incident. By correlating various data points, such as device logs, sensor readings, and transaction records, AI can identify the activities and interactions that occurred before, during, and after the incident. This helps investigators understand the timeline, identify potential causes, and uncover the tactics used by malicious actors.
- Pattern Recognition and Anomaly Detection: AI algorithms can leverage machine learning techniques to analyze patterns and detect anomalies in IoT and IIoT data. By comparing the current data with historical records, AI can identify abnormal behaviors, suspicious patterns, or deviations from normal operations. This aids in identifying potential security incidents and understanding the impact and scope of the incident.

- Collaborative Investigation: Blockchain's decentralized and transparent nature enables collaborative forensic analysis in IoT and IIoT environments. Multiple parties, such as investigators, analysts, or legal authorities, can access the blockchain to review and contribute to the investigation. This fosters transparency, information sharing, and coordination among stakeholders, enhancing the effectiveness of the forensic analysis process.
- Privacy-Preserving Analysis: AI algorithms can be designed to perform forensic analysis while preserving the privacy of individuals or sensitive information. Techniques such as federated learning or secure multi-party computation allow AI models to be trained on decentralized or encrypted data, ensuring that privacy is maintained during the analysis process.

Hence, by combining the immutability of blockchain with the intelligence of AI algorithms, forensic analysis in IoT and IIoT environments can become more reliable, transparent, and efficient. The integration of blockchain and AI ensures the integrity of evidence, enables thorough reconstruction of events, and supports collaborative investigation efforts, leading to effective forensic analysis in IoT and IIoT-based environments.

## 10. FUTURE DIRECTIONS AND CHALLENGES

### 10.1 Emerging Trends and Technologies for Blockchain: AI – IoT-IIoT Based Environment

The convergence of blockchain, AI, IoT, and IIoT is driving several emerging trends and technologies that hold great potential for enhancing security, efficiency, and scalability in various domains. Here are some notable trends and technologies in the blockchain-AI-IoT-IIoT-based environment:

A. Edge Computing:
    - Edge AI: Edge computing combined with AI allows for real-time data processing and decision-making at the edge devices, reducing latency and improving efficiency.
    - Edge Blockchain: Deploying blockchain technology at the edge devices enables distributed consensus and local data storage, enhancing security and privacy in IoT and IIoT applications.

B. Federated Learning:
    - Federated AI: Federated learning enables training AI models across multiple edge devices while preserving data privacy. It facilitates collaborative and decentralized AI training in IoT and IIoT environments.
    - Interoperability and Standardization:
    - Blockchain Interoperability: Efforts are underway to establish interoperability protocols and standards to enable seamless communication and data exchange between different blockchain networks.
    - IoT Standardization: Organizations are working towards developing common standards and protocols to ensure interoperability among IoT devices and platforms.

C. Hybrid Blockchain:

- Public-Private Hybrid Blockchain: Hybrid blockchain architectures combine the benefits of both public and private blockchains, offering scalability, privacy, and decentralized governance for IoT and IIoT applications.

D. Secure Hardware:
- Trusted Execution Environment (TEE): TEEs provide secure hardware environments for executing sensitive operations and protecting cryptographic keys, enhancing the security of IoT and IIoT devices.
- Hardware Security Modules (HSMs): HSMs offer secure key storage and cryptographic operations, ensuring the integrity and confidentiality of data in IoT and IIoT deployments.

E. Data Marketplace:
- Blockchain-based Data Marketplace: Decentralized data marketplaces built on blockchain enable secure and transparent data exchange, allowing individuals and organizations to monetize and trade their IoT and IIoT data.

F. AI-powered Blockchain Analytics:
- AI-driven Blockchain Analytics: AI techniques, such as machine learning and natural language processing, are employed to analyze blockchain data and detect patterns, anomalies, and potential security threats.

G. Self-Sovereign Identity (SSI):
- SSI on Blockchain: Self-sovereign identity solutions leverage blockchain to enable individuals to have control over their digital identities, enhancing privacy and security in IoT and IIoT ecosystems.

H. Quantum-Resistant Cryptography:
- Post-Quantum Cryptography: With the advent of quantum computing, post-quantum cryptography algorithms are being explored to ensure long-term security for blockchain and IoT systems against quantum attacks.

I. Energy Efficiency:
- Green Blockchain: Efforts are being made to develop energy-efficient consensus algorithms and sustainable blockchain infrastructure to minimize the environmental impact of blockchain-based systems.

These emerging trends and technologies demonstrate the continuous evolution and integration of blockchain, AI, IoT, and IIoT, creating new opportunities and addressing existing challenges in various industries. As these technologies mature, they have the potential to revolutionize multiple domains, including healthcare, supply chain, finance, energy, and smart cities, leading to increased efficiency, transparency, and security in the digital era.

## 10.2 Research Opportunities and Open Challenges for Blockchain: AI – IoT-IIoT Based Environment

The intersection of blockchain, AI, IoT, and IIoT presents numerous research opportunities and open challenges. Here are some areas where further exploration and innovation are needed:

A. Scalability and Performance:
   - Scalable Blockchain Solutions: Developing scalable blockchain architectures and consensus mechanisms to handle the increasing volume of IoT and IIoT data and transactions.
   - Efficient AI and ML Algorithms: Designing lightweight and resource-efficient AI and ML algorithms that can run on resource-constrained IoT and IIoT devices without compromising performance.

B. Privacy and Data Protection:
   - Privacy-Preserving Techniques: Investigating techniques for preserving the privacy of IoT and IIoT data while still enabling secure and efficient data sharing and analysis.
   - Differential Privacy: Exploring the application of differential privacy mechanisms to protect sensitive information in IoT and IIoT environments.

C. Trust and Security:
   - Trust Management Systems: Developing trust management frameworks and reputation systems to establish trustworthiness among IoT and IIoT devices and participants.
   - Security Auditing and Certification: Investigating mechanisms to audit and certify the security of blockchain, AI, and IoT systems to ensure their resilience against cyber threats.
   - Interoperability and Standardization:
   - Cross-Platform Interoperability: Developing standards and protocols to enable seamless integration and interoperability between different blockchain networks, AI systems, and IoT platforms.
   - Data and Device Interoperability: Addressing challenges related to data format, semantic interoperability, and device heterogeneity in IoT and IIoT environments.

D. Energy Efficiency:
   - Green Blockchain Solutions: Exploring energy-efficient consensus algorithms and sustainable blockchain infrastructures to minimize the energy consumption of blockchain-based systems.
   - AI for Energy Optimization: Utilizing AI techniques to optimize energy usage in IoT and IIoT deployments, leading to more energy-efficient operations.

E. Governance and Legal issues:
   - Legal and Regulatory Frameworks: Examining the legal implications and regulatory requirements associated with the use of blockchain, AI, IoT, and IIoT technologies, especially in sensitive domains.
   - Governance Models: Designing governance models and frameworks to ensure transparency, accountability, and fair participation in blockchain-based ecosystems.

Hence, these research opportunities and challenges highlight the multidisciplinary nature of blockchain-AI-IoT-IIoT-based environments and call for collaboration between researchers, industry practitioners, policymakers, and other stakeholders. By addressing these open issues, researchers can contribute to the advancement of these technologies and drive innovation that enables secure, efficient, and responsible applications in various domains.

## 11. CONCLUSION

The combination of blockchain, AI, IoT, and IIoT offers tremendous potential for enhancing cybersecurity in today's interconnected world. This convergence of technologies provides innovative solutions to address the complex security challenges associated with IoT and IIoT applications. Blockchain technology offers decentralized and immutable data storage, enhancing the security and integrity of IoT and IIoT data. It enables secure transactions, data sharing, and identity management, reducing vulnerabilities and ensuring trust among participants. The transparency and auditability of blockchain also facilitate incident detection, response, and forensic analysis. AI, with its advanced algorithms and machine learning capabilities, plays an essential role in cybersecurity. It enables efficient threat detection, anomaly detection, and predictive analytics, enhancing the detection and prevention of cyberattacks in real-time. AI-driven algorithms can identify patterns and anomalies in large-scale IoT and IIoT data, enabling proactive measures to mitigate potential security risks.

Hence, this chapter provides the detail explanation of the integration of blockchain, AI, and IoT/IIoT applications, with introducing a new paradigm of cybersecurity for future researchers/ scientific community. It strengthens the protection of sensitive data, safeguards critical infrastructures, and enhances the resilience of IoT and IIoT ecosystems against cyber threats. By leveraging the power of distributed ledger technology and intelligent algorithms, organizations can achieve secure and reliable communication, data integrity, access control, and incident response.

## REFERENCES

Agrawal, D., Bansal, R., Fernandez, T. F., & Tyagi, A. K. (2022). Blockchain Integrated Machine Learning for Training Autonomous Cars. In Lecture Notes in Networks and Systems, (vol 420). Springer, Cham. doi:10.1007/978-3-030-96305-7_4

Aljawarneh, S., Aldwairi, M., & Jararweh, Y. (2020). Blockchain-based Framework for Securing IoT Applications. *Journal of Ambient Intelligence and Humanized Computing*, *11*(4), 1629–1641.

Aswathy, S. U. (2021). The Future of Edge Computing with Blockchain Technology: Possibility of Threats, Opportunities and Challenges. Recent Trends in Blockchain for Information Systems Security and Privacy. CRC Press.

Biswas, K., & Koo, S. (2020). Blockchain-Based Security Framework for Industrial IoT Systems. *Future Generation Computer Systems*, *104*, 962–975.

Dagher, G. G., Mohler, J., Milojkovic, M., Marella, P. B., & Ancile, C. (2018). Ancile: Privacy-Preserving Framework for Access Control and Interoperability of Electronic Health Records Using Blockchain Technology. *Sustainable Cities and Society*, *39*, 283–297. doi:10.1016/j.scs.2018.02.014

Deshmukh, A., Patil, D., & Tyagi, A. K. (2022). Recent Trends on Blockchain for Internet of Things based Applications: Open Issues and Future Trends. In *Proceedings of the 2022 Fourteenth International Conference on Contemporary Computing (IC3-2022)*. Association for Computing Machinery. 10.1145/3549206.3549289

Deshmukh, A., Sreenath, N., Tyagi, A. K., & Eswara Abhichandan, U. V. (2022). Blockchain Enabled Cyber Security: A Comprehensive Survey. *2022 International Conference on Computer Communication and Informatics (ICCCI),* (pp. 1-6). IEEE. 10.1109/ICCCI54379.2022.9740843

Ding, S., Shao, Z., Guo, S., & Zhu, W. (2019). IoT Security Techniques Based on Machine Learning: How Do IoT Devices Use AI to Enhance Security? *IEEE Consumer Electronics Magazine, 8*(4), 22–30.

Fan, K., & Xu, H. (2019). Blockchain and Artificial Intelligence: Revolutionizing Cybersecurity and Cyberdefense. *IEEE Intelligent Systems, 34*(6), 92–96.

Fernández-Caramés, T. M., & Fraga-Lamas, P. (2020). A Review on the Use of Blockchain for the Internet of Things. *IEEE Access : Practical Innovations, Open Solutions, 8,* 154076–154130.

Gupta, V., & Sehgal, V. (2019). Blockchain and Artificial Intelligence for Secure and Privacy-Preserving Industrial IoT. *Future Generation Computer Systems, 97,* 624–637.

Jayaprakash, V., & Tyagi, A. K. (2022). Security Optimization of Resource-Constrained Internet of Healthcare Things (IoHT) Devices Using Asymmetric Cryptography for Blockchain Network. In: Giri, D., Mandal, J.K., Sakurai, K., De, D. (eds) *Proceedings of International Conference on Network Security and Blockchain Technology.* Springer, Singapore. 10.1007/978-981-19-3182-6_18

Krishna, A. M., & Tyagi, A. K. (2020). Intrusion Detection in Intelligent Transportation System and its Applications using Blockchain Technology. *2020 International Conference on Emerging Trends in Information Technology and Engineering.* IEEE. 10.1109/ic-ETITE47903.2020.332

Mishra, S., & Tyagi, A. K. (2019). Intrusion Detection in Internet of Things (IoTs) Based Applications using Blockchain Technology. *2019 Third International conference on I-SMAC (IoT in Social, Mobile, Analytics and Cloud) (I-SMAC),* (pp. 123-128). IEEE. 10.1109/I-SMAC47947.2019.9032557

Nair, M. M., & Tyagi, A. K. (2022). Preserving Privacy Using Blockchain Technology in Autonomous Vehicles. In: Giri, D., Mandal, J.K., Sakurai, K., De, D. (eds) *Proceedings of International Conference on Network Security and Blockchain Technology.* Springer, Singapore. 10.1007/978-981-19-3182-6_19

Nair, M. M., & Tyagi, A. K. (2023). Chapter 11 - AI, IoT, blockchain, and cloud computing: The necessity of the future. In Rajiv Pandey, Sam Goundar, Shahnaz Fatima (eds.), Distributed Computing to Blockchain. Academic Press. doi:10.1016/B978-0-323-96146-2.00001-2

Pandey, A. A., Fernandez, T. F., Bansal, R., & Tyagi, A. K. (2022). Maintaining Scalability in Blockchain. In A. Abraham, N. Gandhi, T. Hanne, T. P. Hong, T. Nogueira Rios, & W. Ding (Eds.), *Intelligent Systems Design and Applications. ISDA 2021. Lecture Notes in Networks and Systems* (Vol. 418). Springer. doi:10.1007/978-3-030-96308-8_4

Puthal, D., Nepal, S., Ranjan, R., Chen, J., & Mohanty, S. P. (2019). The Blockchain as a Decentralized Security Framework [Guest Editorial]. *IEEE Cloud Computing, 6*(1), 16–21.

Raza, S., Wallgren, L., & Voigt, T. (2017). Edge Computing for Industry 4.0: Scenarios and Challenges. *2017 IEEE 37th International Conference on Distributed Computing Systems (ICDCS),* (pp. 1451-1456). IEEE.

Samaniego, M., & Yang, B. (2020). A Survey on Blockchain Technology for Networking and Communications: Challenges and Opportunities. *IEEE Communications Surveys and Tutorials*, *22*(1), 713–734.

Sheth, H. S. K. (2022). Deep Learning, Blockchain based Multi-layered Authentication and Security Architectures. *2022 International Conference on Applied Artificial Intelligence and Computing (ICAAIC)*, (pp. 476-485). IEEE. 10.1109/ICAAIC53929.2022.9793179

Siddharth, M. (2021). Issues and Challenges (Privacy, Security, and Trust) in Blockchain-Based Applications, In: Opportunities and Challenges for Blockchain Technology in Autonomous Vehicles. IGI Global. doi:10.4018/978-1-7998-3295-9.ch012

Suryadevara, N. K., & Mukhopadhyay, S. C. (2018). Wireless Sensor Network Based Industrial Monitoring Applications: A Comprehensive Survey. *IEEE Communications Surveys and Tutorials*, *20*(2), 1403–1427.

Tibrewal, I., Srivastava, M., & Tyagi, A. K. (2022). Blockchain Technology for Securing Cyber-Infrastructure and Internet of Things Networks. In A. K. Tyagi, A. Abraham, & A. Kaklauskas (Eds.), *Intelligent Interactive Multimedia Systems for e-Healthcare Applications*. Springer. doi:10.1007/978-981-16-6542-4_17

Tyagi, A. (2021a). Applications of Blockchain Technologies in Digital Forensic and Threat Hunting. Recent Trends in Blockchain for Information Systems Security and Privacy. CRC Press.

Tyagi, A. (2021b). Analysis of Security and Privacy Aspects of Blockchain Technologies from Smart Era' Perspective: The Challenges and a Way Forward. In Recent Trends in Blockchain for Information Systems Security and Privacy. CRC Press.

Tyagi, A. (2021c, October). AARIN: Affordable, Accurate, Reliable and INnovative Mechanism to Protect a Medical Cyber-Physical System using Blockchain Technology. *IJIN*, *2*, 175–183.

Tyagi, A. (2022a). Preserving Privacy using Distributed Ledger Technology in Intelligent Transportation System. In *Proceedings of the 2022 Fourteenth International Conference on Contemporary Computing (IC3-2022)*. Association for Computing Machinery. 10.1145/3549206.3549306

Tyagi, A. K. (2022b). SecVT: Securing the Vehicles of Tomorrow Using Blockchain Technology. In A. A. Sk, T. Turki, T. K. Ghosh, S. Joardar, & S. Barman (Eds.), *Artificial Intelligence. ISAI 2022. Communications in Computer and Information Science* (Vol. 1695). Springer. doi:10.1109/ICCCI54379.2022.9740965

Tyagi, A. (2023). Decentralized everything: Practical use of blockchain technology in future applications. In Rajiv Pandey, Sam Goundar, Shahnaz Fatima, (eds.) Distributed Computing to Blockchain. Academic Press. doi:10.1016/B978-0-323-96146-2.00010-3

Tyagi, A. K., Agarwal, D., & Sreenath, N. (2022). SecVT: Securing the Vehicles of Tomorrow using Blockchain Technology. *2022 International Conference on Computer Communication and Informatics (ICCCI)*, (pp. 1-6). IEEE. 10.1109/ICCCI54379.2022.9740965

Tyagi, A. K., Chandrasekaran, S., & Sreenath, N. (2022). Blockchain Technology:– A New Technology for Creating Distributed and Trusted Computing Environment. *2022 International Conference on Applied Artificial Intelligence and Computing (ICAAIC)*, (pp. 1348-1354). IEEE. 10.1109/ICAAIC53929.2022.9792702

Tyagi, A. K., Fernandez, T. F., & Aswathy, S. U. (2020). Blockchain and Aadhaar based Electronic Voting System. *2020 4th International Conference on Electronics, Communication and Aerospace Technology (ICECA), Coimbatore*, (pp. 498-504). IEEE. 10.1109/ICECA49313.2020.9297655

Tyagi, A. & Kumar, S. (2022). Blockchain Technology for Securing Internet of Vehicle: Issues and Challenges. *2022 International Conference on Computer Communication and Informatics (ICCCI)*, (pp. 1-6). IEEE. 10.1109/ICCCI54379.2022.9740856

Varsha, R. (2020, January 1). 'Deep Learning Based Blockchain Solution for Preserving Privacy in Future Vehicles'. *International Journal of Hybrid Intelligent Systems*, *16*(4), 223–236.

Varsha, J. & Tyagi, A. (2022). Security Optimization of Resource-Constrained Internet of Healthcare Things (IoHT) Devices Using Lightweight Cryptography. Information Security Practices for the Internet of Things, 5G, and Next-Generation Wireless Networks. IGI Global. doi:10.4018/978-1-6684-3921-0.ch009

Wang, Y., Kumar, N., Singh, R., & Alowaisheq, E. (2018). Secure Data Sharing Framework for Industrial IoT Using Blockchain and Attribute-Based Encryption. *IEEE Transactions on Industrial Informatics*, *14*(5), 2038–2046.

Xu, L. D., He, W., & Li, S. (2018). Internet of Things in Industries: A Survey. *IEEE Transactions on Industrial Informatics*, *10*(4), 2233–2243. doi:10.1109/TII.2014.2300753

Yang, Z., Wu, S., Ning, H., & Lu, R. (2019). A Blockchain-Based Access Control Framework for Industrial IoT. *IEEE Transactions on Industrial Informatics*, *15*(1), 471–480.

Ylianttila, M., & Vasilakos, A. V. (2017). Security in Industrial IoT Networks: Approaches and Challenges. *IEEE Network*, *31*(3), 82–88.

# Chapter 8
# Application of AI and Blockchain Technologies in the Medical Domain

**Deepak Varadam**
*Ramaiah University of Applied Sciences, India*

**Sahana P. Shankar**
https://orcid.org/0000-0001-8977-9898
*Ramaiah University of Applied Sciences, India*

**Anirudh Shankar**
*Ramaiah University of Applied Sciences, India*

**Ashwath Narayan**
*Ramaiah University of Applied Sciences, India*

**Nidhi N. P**
*Ramaiah University of Applied Sciences, India*

**Tushar Narendra Kumar**
*PES University, India*

## ABSTRACT

*The rapid advancements in artificial intelligence (AI) and blockchain technologies have brought about significant transformations in various sectors, including healthcare. This proposal outlines a research paper that aims to investigate the different approaches and applications of AI and blockchain technologies in the medical domain. By exploring the integration of these technologies, the authors seek to analyze their potential benefits, challenges, and ethical implications in revolutionizing healthcare practices. The study will involve a comprehensive review of relevant literature, case studies, and interviews with experts in the field. The findings will contribute to a deeper understanding of the synergistic effects of AI and blockchain in healthcare, paving the way for enhanced decision-making, data security, interoperability, and patient-centric care.*

DOI: 10.4018/979-8-3693-0659-8.ch008

## INTRODUCTION

Blockchain technology is one of the most pioneering gizmos in the AI sector. The main focus on the facets of blockchain and artificial intelligence is to integrate these powerful technologies to bring about a meaningful distinction in healthcare by encouraging the application of a broadly analytical technology that can be involved in a more sheer risk management strategy. The required information on healthcare indices is assembled from the Web of Sciences and other Google surveys conducted by numerous governing agencies. This features has exhibited many prospects for developing dependable artificial intelligence models in e-Health employing blockchain, which is an open network for authorisation and sharing of information. The access of blockchain to view the medical reports of patients will be given to healthcare professionals and artificial intelligence uses a wide range of alleged algorithms and decision-making capabilities, as well as dealing with large quantities of information. Now, when there's an integration of the most recent breakthroughs of these technologies, there will be a medical system that is more efficient, affordable, and standardised. Blockchain makes it possible to reserve cryptographic records which is required by AI. Blockchain also assists the development of a system that prompts and manages content blocks called ledgers with secure and automated data analysis. The secure collection and evaluation of every bit of data about health will deliver rapid updates for medical professionals, payers, and healthcare workers. By adding AI algorithms to the blockchain, its functionality is advanced further (Tagde et al., 2021).

In the areas of data privacy, transparency, traceability, security, flexible access, immutability, and audit, present-day healthcare data management systems are facing serious obstacles. Additionally, the majority of the healthcare systems used for data management are centralized. This increases the risk of a single point of failure in the unfortunate occurrence of any natural disaster. To overcome these challenges blockchain technology is employed. It is a novel and decentralized technology that can drastically change the way data is being managed in healthcare systems. By integrating artificial intelligence and blockchain technology, the efficiency of data analytics could boost significantly. Blockchain helps in real-time data monitoring and AI would perform innovative analytics without any human interaction. This technology has been shown to minimize the necessity of routine medical tests in hospitals. Connected home health monitoring systems can cut the expenses of hospital stays, admission, and repeated diagnosis. It is projected that blockchain adoption can save up to \$100-\$150 billion in data breach-related expenditures by 2025, along with the reduction in fraud and counterfeit products (Yaqoob et al., 2021).

The study examines the relationship between healthcare systems and blockchain technology, focusing on the security and privacy aspects. It analyzes case studies of telecare medicine systems, Electronic Health Records (EHR), and e-Health. The study highlights potential challenges like scalability, storage capacity, and interoperability. However, it also promotes the implementation of blockchain technology in the healthcare domain, highlighting its potential to improve data security, confidentiality, and interoperability by decentralizing the healthcare network and utilizing blockchain-based approaches, schemes, algorithms, frameworks, or protocols.Overall, the importance of blockchain technology highlights the potential challenges it poses on healthcare thereafter providing apt solutions (Hussien et al., 2021).

## LITERATURE REVIEW

The current situation primarily focuses on exploring blockchain technology, remote patient monitoring, Electronic Health Records (EHR), patient-centered interoperability, clinical trial data management, and monetization aspects of the healthcare system (Fatoum et al., 2021). The sensitive nature of healthcare data presents us with major difficulties such as security, interoperability, privacy, and reliability (Hussien et al., 2021). These factors have made people stop using electronic cloud-based data. This is where decentralized and scalable technology like blockchain comes into the picture to overcome these hurdles. A vast network of nodes is used to maintain and distribute a blockchain. Each node consists of a complete replica of the data. A blockchain protocol is an assembly of predefined norms that dictate how nodes should communicate with each other, examine, verify, and update data in the ledger (Tagde et al., 2021). By integrating blockchain with other revolutionary technologies like Artificial Intelligence (AI) and the Internet of Things (IoT), there is increased efficiency and innovation in the field of healthcare management systems.

Blockchain technology is being employed to get innovative solutions to modern problems in various fields. In healthcare, blockchain network is being used to store and exchange patient records between hospital, laboratories, pharmacies, and doctors. Its applications can identify minute to severe or dangerous mistakes making it more reliable than the traditional way (Olawumi et al., 2022). The distributed blockchain technology makes it possible for the identification of fraud, cut administrative costs, get rid of overlap labor, and enforce transparency in the healthcare systems. It also promotes the idea of decentralisation, immutability and confidence (Fatoum et al., 2021). The benefit for patients is that their data is confidentially stored, their diagnostic accuracy increases and are taken care of better than usual. By integrating blockchain with IoT, the responsiveness and tracing are improved, thus helping in the proper monitoring of the patient (Fatoum et al., 2021). Blockchain also promotes new and creative research projects by exchanging EHR between many proficient laboratories and researchers (Hussien et al., 2021). This deep analysis of the reports will inspire innovative new research resulting in a remarkable collaboration between participants and researchers. Therefore, this technology could revolutionize the healthcare system for good.

In this age where there is an extensive use of electronic gadgets and social media, a large amount of data is being produced every second globally. With this growing data production, there is an increase in the risk of cyber threats. Blockchain technology is an emerging solution to these cyber threats and provides high-level security. In the past few years, the digitalisation of healthcare systems has skyrocketed. As there are advances in Information Technology (IT), the concept of smart health is gaining more perspective. Smart health is the integration of many technologies like AI, IoT, big data and cloud computing to modernize and upgrade the efficiency, security and privacy of the present healthcare system. B5G-enabled smart health framework is an application of blockchain for AI-based healthcare to battle pandemics like COVID-19. Blockchain and AI empowered telesurgery system towards 6G (BATS) is developed to address the challenges like security, throughput, trust and reliability. These are some of the real-time applications of blockchain for AI in the healthcare sector (Shinde et al., 2021). The major challenge faced by the healthcare system all around the globe is to provide the best care to the patient without increasing overall expenses. This challenge can be overcome by integrating various technologies like blockchain, IoT (Fatoum et al., 2021), AI, machine learning, and analytics into the healthcare system. The key components of blockchain technology have been discussed like immutability, decentralization, anonymity, transparent, etc. The major applications of this technology in the healthcare field are Electronic Health Records (EHR), clinical research, medical fraud detection, ensuring encryption of all transactions, and network testing. A few drawbacks of blockchain technology are data ownership, uncertainty with the overall cost of this technology, and few rules and regulation. Overall, this technology intends to make healthcare patient-centric and improve data protection, confidentiality, and interoperability (Srivastava et al., 2021).

For any decent healthcare organization, it is very challenging to balance between providing the best care to their patients and making it cost-efficient. A clinical decision support system promotes collaborative therapy with the intent of improving medical services. EHRs comprise sensitive and private information of the patients. Hence, sharing this electronically poses a lot of challenges like lack of privacy and security. Blockchain, a distributed ledger technology featuring a secured layout can be used to improve today's systems. Blockchain when paired with AI has a lot of potential to aid healthcare stakeholders and deal with crucial healthcare issues and challenges (Vijayalakshmi, 2021). Pandemics like COVID-19 showed that the application of computing communication technology can be crucial contributions to the prevention, control, and combat of diseases. With the present studies in next-generation networking (NGN), it is important to employ safe and dependable communication and networking while dealing with EHRs as they could have sensitive information. With the implementation of AI-driven NGN, healthcare systems can be improved to achieve smart and secure healthcare and thus, help in dealing with epidemics that might later become pandemics. A collaborative and distributed healthcare framework based on cutting-edge computing, communication, AI and integrating blockchain would ensure safe data exchange among medical professionals and patients. This whole technology, on combining with the low-latency and high bandwidth functionality of 5G would make it possible for widespread surveillance, monitoring, and analysis at the edge (Otoum et al., 2021). In this article, a survey of research on the most recent blockchain solutions coupled with AI technologies to enhance and develop new technical standards for the healthcare sector regarding medical diagnostics and Electronic Health Records sharing, with a special emphasis on medical imaging is delivered. In recent times, the digitalisation of the healthcare industry and EHRs has exponentially risen. Smart health improves the traditional healthcare system by integrating technologies like IoT, AI and cloud computing. The fact that smart health comes with problems like security breaches and malicious attacks cannot be ignored. Blockchain technology has recently come to prominence due to its ability in improving data management, access control, and integrity inside healthcare systems (Omrčen et al., 2021).

This article discusses the potential practical implications of using blockchain technology in healthcare settings. Some of the practical implications include creating secure and transparent e-health records and sharing patient data. Improving hospital administration and decreasing hospital costs through more efficient data integration and processing. Supporting the introduction of artificial intelligence, e-health applications and devices, and the evolution of the Internet of Medical Things. Verifying, securing, and sharing data, which can be useful for confirming identity, vaccination status, and permission to leave home, preventing health worker fatigue, and promoting targeted early intervention. Enabling secure track and trace, designing and delivering custom medical equipment, and identifying COVID-19 carriers and viral hot-spots. Providing real-time trustworthy data for analysis, planning, and modeling projections, which can aid in treatment and safety-based interventions, prepare for and manage incidents, calibrate state and national responses, and promote effective policy-making. Creating permissioned data transparency, which can supply and predict necessary resources, reduce shortfalls in equipment and vaccines, and enable improved resource-sharing. Improving citizens' relationships with governments by using secure technologies to store and share their data (Cooper, 2022). For healthcare professionals and organisations considering integrating blockchain technology, the report offers useful implications. In addition to outlining health-related blockchain products and important actors providing solutions across various applications, the paper analyses difficulties and potential for integrating blockchain technology in healthcare (Attaran, 2022). The practical implications is that the blockchain technology can provide a secure and trusted system for managing healthcare data. This technology has the ability to change how the healthcare sector provides care and make it possible for people to obtain care from anywhere in the globe. The article suggests that blockchain can be used to organize medical records and empower their propagation in a secure manner (Srivastava et al., 2022). Controlling the drug supply chain, allowing the transmission of patient medical records and facilitating the secure transfer of such records helps in the practical applications of blockchain technology in the healthcare domain (Abbas et al., 2022). Blockchain technology has the potential to completely transform the healthcare sector by resolving some of the most pressing issues that healthcare providers now face, including data security, interoperability, and ineffective administrative processes (Saeed et al., 2022). Convolutional Neural Networks (CNN), Recurrent Neural Networks (RNN), and other deep neural networks are also used in the healthcare industry and covers the applications of blockchain technology powered by AI in public health and the difficulties encountered there (Kumar, 2022).

This paper proposes an architecture for combining blockchain and artificial intelligence techniques within the privacy-preserving federated learning framework for the Internet of Medical Things. The proposed architecture protects training-specific data in the blockchain and provides new ways of handling machine learning, allowing artificial intelligence to take data for training (Połap et al., 2020). The frameworks support secure, immutable, auditable, and traceable platforms powered by blockchain that enables patients to manage their healthcare data. The system also caters to the requirements for building a trusted AI ecosystem with provenance. The Proof-of-Concept (PoC) implementation of the proposed frameworks in a permissioned blockchain shows that there is significant potential to explore further (Anoop & Asharaf, 2022). This helps to identify the challenges associated with the implementation of blockchain-based healthcare systems, such as scalability, security, and privacy issues, lack of regulations, and the need for powerful computational infrastructures (Soltanisehat et al., 2020). Natural language processing (NLP) is used to analyze opinions surrounding Alzheimer's disease. The development of machine learning algorithms to detect the early signs of Alzheimer's disease and to secure and share medical data for the improvement of participant experience in clinical studies (Pilozzi & Huang, 2020).

This systematic review explores the use of blockchain technology in the healthcare sector, focusing on its applications and implications (Merlo et al., 2023). It highlights the potential synergies and innovative approaches that emerge from the amalgamation of blockchain and deep learning (Afaq & Manocha, 2023). The review highlights the development of a blockchain-based hybrid platform for multimedia data processing in IoT healthcare (Baysal et al., 2023), offering a unique perspective on implementing blockchain in healthcare applications (Taloba et al., 2023). The review also discusses the adoption barriers of blockchain-based platforms in the healthcare industry, providing insights from a balanced scorecard perspective for informed decision-making (Govindan et al., 2023). The review also focuses on improving medical supply chain disruption management using blockchain technology, offering solutions to enhance healthcare supply chain resilience (Özcan-Top, 2023). The article explores the state of the art, challenges, and future directions in integrating blockchain with resource-constrained IoT devices for healthcare applications, discussing the evolving landscape of healthcare technology and the opportunities and challenges associated with this integration (Hegde & Maddikunta, 2023).

## AI APPLICATIONS IN THE MEDICAL DOMAIN

AI tools have significantly improved patient care, organized clinical processes, and transformed medical research. These applications include diagnostic assistance, personalized treatment, predictive analysis, drug discovery, and virtual healthcare assistants. However, challenges such as data quality, interpretability, and regulatory compliance are still present. This article explores the various types of AI applications in the medical domain and the challenges associated with implementing AI, such as data quality, interpretability, and regulatory compliance. By examining the diverse applications, opportunities, and complexities of AI in the medical field, it aims to contribute to a comprehensive discourse that informs future developments and enables responsible and impactful implementation.

### AI Applications to Medical Images: From Machine Learning to Deep Learning

AI has the potential to revolutionise medical diagnosis by making it possible to analyse medical images and other types of data more quickly and accurately. For instance, AI algorithms can be trained to spot small changes in medical images that human viewers could miss, or to find patterns in huge datasets that people might find challenging to spot. Early and accurate diagnoses may result from this, which may enhance patient outcomes and lower healthcare costs. AI has the potential to enable more personalised and precise medicine, by allowing clinicians to tailor treatments to individual patients based on their unique characteristics and medical history (Castiglioni et al., 2021).

### Machine Learning and Deep Learning in the Medical Domain

AI has the potential to significantly improve medical diagnosis and patient outcomes, and this field of study is likely to continue gaining importance in the years to come. They also point out that further study and development are required to fully realise the promise of this technology and that there are still issues that need to be resolved in the creation and application of AI tools for medical diagnosis. The researchers delve deep into utilising machine learning and deep learning to create AI applications for medical imaging (Mavropoulos et al., 2019). Table 1 also highlight some of the challenges faced by picking one design choice over the other as well as the challenges faced by each.

The authors also discussed how design decisions are made when developing AI models. Each decision block illustrates a common real-world scenario that might result in a variety of results in both classic machine learning and deep learning models as shown in the Figure 1.

Examples of how AI has been used in the medical domain is shown below:

**Medical image analysis:** AI algorithms can be taught to examine images from diagnostic tests like X-rays, CT scans, and MRIs to look for abnormalities or diagnose diseases. For instance, AI has been used to identify brain tumors from MRI scans, diagnose skin cancer from photos of skin blemishes, and detect breast cancer in mammograms.

**Disease prediction and risk stratification:** Using information from genetic data, medical history, and lifestyle factors, AI algorithms can determine the possibility of contracting a specific disease. AI has been used, for instance, to forecast the likelihood of getting diabetes, Alzheimer's illness, and heart disease.

**Personalised care:** AI algorithms can be used to create treatments that are specific to each patient, taking into account their particular traits and medical background. AI has been used, for instance, to

*Table 1. Challenges faced by ML and DL techniques*

| Challenges | Machine Learning | Deep Learning |
|---|---|---|
| Sample size | • Thorough assessments of the resilience and dependability of radiomic features.<br>• Effective feature selection methodology.<br>• Choosing a machine learning model | • Enhancement of data; transfer learning.<br>• Regularisation to increase the generalizability of the model.<br>• Pre-training that is lightly, semi-, self-supervised, or unsupervised.<br>• Change the model's architecture. |
| Medical image application domain | Avoiding reliance on the data by doing thorough evaluations of the radiomic feature robustness to prevent overfitting on the development set. | To benefit from pre-trained models or labelled instances from related domains, use transfer learning and domain adaptation. |
| Label and annotation reliability | • Data curation that takes segmentation and response characteristics into account.<br>• Multiple labels and morphological perturbations could be taken into account in the feature robustness assessments to improve reliability. | • Data management taking into account multicentric and multireader research<br>• Inexact supervision using image-level labels to infer predictions at the pixel or voxel level.<br>• Combine a few examples that are correctly labelled with occurrences that are incorrectly or incompletely labelled (weak labelling). |
| Interpretability | Some models, such as decision trees, and specific radiomic properties (in terms of relevance or importance), offer high interpretability. | Using strategies to increase model transparency during both the design and evaluation phases, such as explainability and interpretability. |

anticipate how cancer patients would likely respond to various treatments and to determine the best course of action for each patient.

**Drug discovery:** Based on their molecular structure and other characteristics, existing medications can be optimised and novel drug candidates can be found using AI algorithms. AI has been used, for instance, to discover novel antibiotics and to improve currently available medications for the treatment of cancer and other disorders. Overall, AI has the potential to revolutionise a variety of healthcare processes, including illness prevention, drug discovery, and diagnosis and therapy.

## An Overview of the Use of Chatbots in Medical and Healthcare Education

AI chatbots are revolutionizing the medical industry, particularly in rehabilitation. These intelligent devices offer personalized support to patients, providing real-time instruction, tracking progress, and suggesting fitness programs. They engage patients in meaningful conversations, encouraging them to follow rehabilitation regimens. AI chatbots can gather and analyze data to assess progress and adjust rehabilitation plans. This increased efficiency and effectiveness contribute to better patient outcomes and quality of life. A healthcare platform with multi-sensor, video, and voice data management can support medical staff and patients with mobility issues. The conversational agent can help everyone through dialogue-based activities (Frangoudes et al., 2021).

*Figure 1. A system view of medical decision-making process*

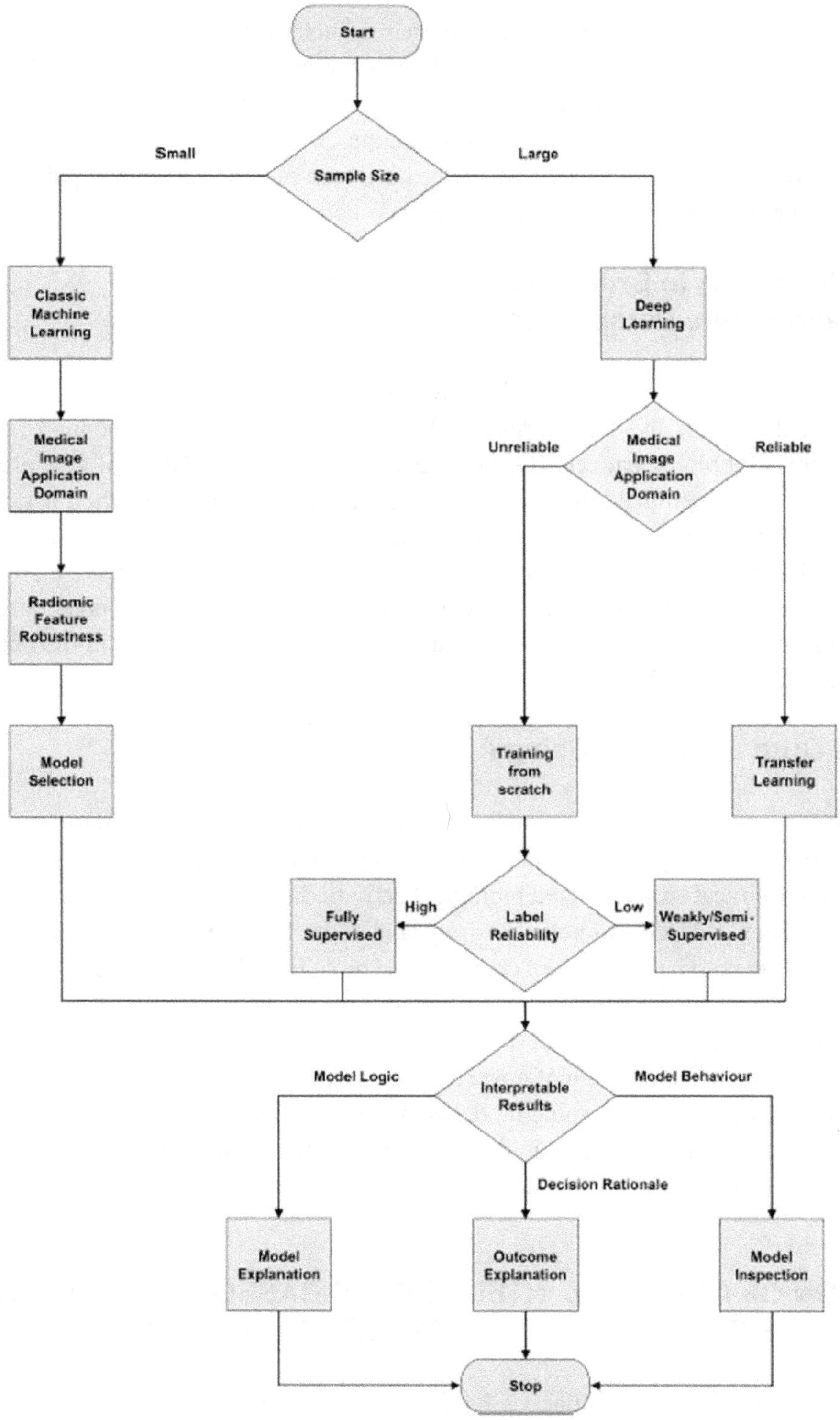

### Development of Chatbots for Healthcare Education

Chatbots for healthcare education involve several phases, including building the chatbot to understand human inquiries in natural language, using natural language processing (NLP) capabilities, and selecting the best response based on the user's query and the chatbot's knowledge base. The target audience and study objectives should be considered when creating chatbots, as a chatbot designed for medical students might require a more advanced knowledge base than one for the general public. The interface should be user-friendly and accessible to the target audience.

### Artificial Intelligence in Drug Discovery: Recent Advances and Future Perspectives

AI is being used in drug discovery to identify novel medication candidates more quickly and reliably. Machine learning algorithms and deep learning techniques are analyzing vast amounts of data, including chemical structures and biological activity, to predict the efficacy and safety of new drug candidates. AI can be used to create optimized molecules for specific activities or predict their action against specific targets. Researchers can use AI to forecast compound synthesis, resulting in faster drug discovery, lower costs, and increased success rates in medication development. However, there are also potential ethical considerations surrounding the use of AI in drug discovery, such as the potential for bias in the data used to train AI models and the need for transparency and accountability in the decision-making process (Jiménez-Luna et al., 2021).

### Methods of Drug Discovery Using AI

AI is being used in drug discovery to identify potential novel medication candidates more quickly and reliably. Machine learning algorithms and deep learning techniques are used to analyze vast amounts of data, including chemical structures and biological activity data, to find patterns and associations that can predict the efficacy and safety of potential drug candidates. One method is the creation of prediction models, which analyze massive amounts of chemical and biological data to find patterns and associations that can forecast a molecule's activity against a specific target. These models are more effective than conventional screening techniques in identifying potential therapeutic candidates from large chemical libraries. Another method is the design of novel compounds, which are tailored for specific functions or qualities. One way to train a deep learning model is to create compounds predicted to have high activity against a specific target. Numerous possible medication candidates that can be further developed and evaluated can be produced using these models (Zhavoronkov et al., 2020).

## APPLICATION OF ARTIFICIAL INTELLIGENCE-BASED TECHNOLOGIES IN THE HEALTHCARE INDUSTRY: OPPORTUNITIES AND CHALLENGES

The study on the use of artificial intelligence-based technologies in the healthcare sector is summarised by the authors. They emphasise the significance of the subject and the possible advantages of AI in the medical field. The study's goals are also outlined in the paper. They are to review the pertinent literature, present real-world examples of AI-based applications, discuss opportunities and challenges associated with their use, and conclude the study by summarising the findings and outlining their implications for both theory and practise. The report also defines concepts linked to artificial intelligence, including machine learning and deep learning. Finally, it identifies the target audience for the study, which includes hospital administrators, medical personnel, medical school curricula developers, education and training managers, human-machine roles and responsibilities specialists, privacy and cybersecurity analysts, and medical ethics professionals.

## AI Applications in the Healthcare Industry

The paper gives a thorough review of how AI-based technologies are being incorporated into both the healthcare sector and our everyday lives. The writers emphasise how AI has a positive effect on hospitals' administrative and nursing activities. For instance, AI-based solutions can assist hospitals in improving patient outcomes, reducing waiting times, and allocating resources more effectively. AI can also assist medical professionals in developing more precise diagnoses, individualised treatment regimens, and remote patient monitoring. The authors focus on the significance of looking at actual instances of how AI is now used by healthcare organisations for patient care and operational management. This is significant since it aids in determining the advantages and restrictions of AI-based healthcare technology. It also aids in determining the prerequisites and support requirements for sophisticated AI applications (Lee & Yoon, 2021).

## Diagnostic Assistance Using AI

One of the main uses of AI-based technology in the healthcare sector is addressed as diagnostic assistance. The report demonstrates how AI-supported technology may learn and diagnose from a wealth of medical research and patient treatment records, significantly assisting clinicians in making decisions about diagnosis and treatments. The study provides numerous instances of AI-based diagnostic algorithms being used to identify skin and breast cancers, acting as a "second opinion" to support radiologists' picture interpretations. AI-based diagnostic aid systems may increase the precision and effectiveness of diagnoses, moral concerns like patient privacy and safety must be taken into consideration. In order to protect patients' preferences, safety, and privacy, the article emphasises the necessity for policies and ethical principles for healthcare services that use AI and its applications.

## Opportunities With AI Applications

IBM Watson's release marked a significant shift in data-driven medical research, promoting the use of advanced technology to enhance patient care and public health. AI has significantly improved treatments for diseases like high blood pressure and lung disease, diagnosed rare DNA-based diseases, and improved patient care in rural areas. AI has also excelled in radiology, processing large amounts of medical imaging data, enhancing diagnostic precision and saving time for medical practitioners. AI-powered software aids in tasks like cancer cell mitosis analysis, reducing manual labor requirements and improving accuracy. AI systems are learning from data and research, achieving or surpassing medical experts in diagnosis. Research in AI promises to help medical staff identify overlooked areas and reduce errors during patient treatment. AI has also improved polyp detection in medical procedures, with China's doctors finding 20% more polyps overall. The University of Tokyo's AI system achieved an accuracy rate of 87.3% for patient diagnosis. Intelligent AI systems are continuously improving the standard of care, benefiting radiologists by allowing them to focus on patient engagement and preventing potential medical mistakes. The ideal healthcare system would combine patient-centered customization, compassionate medical staff, and data-driven disease prevention, diagnosis, and treatment, resulting in significant financial savings. Major US and Israeli hospitals are already using AI to prevent disease, and it is anticipated that the use of AI-supported patient education to avoid chronic illnesses like diabetes and high blood pressure will rise considerably (Vellido, 2019).

## Challenges Faced While Using AI

AI-based systems face privacy issues due to large datasets, particularly in sensitive medical records and disease-related data. The advancement of AI can be hindered by rules governing secrecy. AI's decision-making, based on accumulated data, often fails to consider specific patient situations, posing ethical, moral, and legal problems. To address these issues, discussions about ethical, legal, and societal norms are necessary. The Tesla Model S accident in 2016 highlighted the need for accountability in the healthcare industry, where AI technology is deployed. Accountability issues arise from technological, managerial, and ethical issues. While AI is rapidly developing, ethical issues have received little attention. Dr. Stephen Hawking proposed international AI regulation, while others emphasize the development of morally and ethically sound behavioral patterns. This highlights the importance of social agreements in addressing AI-related concerns.

## Societal Issues Concerning the Application of Artificial Intelligence in Medicine

Several social difficulties have been brought up by the use of machine learning and artificial intelligence in medicine. It argues that, given the industrialization of AI and its effects on society, it is important to consider human-centered perspectives, ethical considerations, and legal concerns. The report places special emphasis on issues relating to the use of AI and ML in medicine, including fairness, privacy, explainability, and interpretability (Hannan, 2023). In order to develop processes, protocols, rules, and data analysis pipelines that explicitly address these social concerns, the authors call for collaboration between the medical and healthcare communities and AI-ML.

## Legal Issues of AI and ML Applications

The development of AI and ML technologies has mainly overlooked the legal challenges that will surface as they migrate from smaller academic contexts to bigger societal contexts. Particularly, the use of AI in autonomous driving and healthcare raises complex legal questions. The development of partially autonomous vehicles and the pursuit of fully autonomous driving are putting pressure on already-existing legislation. The right to justification for judgements made by automated algorithmic systems in the medical field is now available thanks to the General Data Protection Regulation (GDPR) of the European Union. The potential for such autonomous decision-making makes AI and ML subject to legal scrutiny. Standards for GDPR compliance have been presented for ML development. The impact of GDPR on medicine and healthcare is clear, particularly for AI- or ML-based medical decision support systems (MDSS), which help medical professionals make personalised decisions that may have an impact on patient diagnoses, prognoses, and treatment recommendations, including in life-threatening circumstances.

## BLOCKCHAIN APPLICATIONS IN THE MEDICAL DOMAIN

Blockchain technology can improve the security and transparency of sharing medical data in the healthcare system. The integration of artificial intelligence can enhance the decision-making capability of healthcare professionals by providing access to large quantities of data. The use of blockchain technology and artificial intelligence can improve the overall performance of the healthcare system and strengthen the current healthcare sector. The encryption and decentralization factor of blockchain technology can improve the security of patients' electronic medical records and improve interoperability among healthcare organizations (Rao & Manvi, 2023). Blockchain technology may make decentralised data management possible outside of the banking sector. An article in the Harvard Business Review highlights several methods, including authenticating artwork and voting records. Businesses like IBM, Microsoft, Accenture, and others have teamed up to develop blockchain-based technology that clients may employ due to the diversity of interests. Several companies have begun to develop on this technology for use cases such as supply chain management, trade settlements, and identity verification.

### Blockchain's Impact on Healthcare

The Office of the National Coordinator for Health Information Technology organized an ideation challenge in 2016 to gather white papers on blockchain's potential applications in healthcare. This project produced several blockchain-based healthcare application ideas. However, potential barriers such as privacy, legal compliance, and technical challenges remain, leading to short-term ideas focusing on data validation, auditing, and permission. Despite these challenges, the potential of blockchain in healthcare IT remains promising.

Guardtime, a Dutch data security company, developed a blockchain-based system to verify patient IDs in collaboration with the Estonian government. The system uses an immutable audit trail to secure EHR data and prevent unauthorized changes. MedRec, a collaboration between MIT Media Lab and Beth Israel Deaconess Medical Centre, decentralizes permissions, authorization, and data sharing among healthcare systems, giving people ownership over their healthcare data. While most blockchain examples are conceptual, some applications have progressed to production or proof-of-concept stages (Haleem et al., 2021).

### Technical Constraints of Blockchain

Blockchain technology has potential drawbacks, including issues with sharing personally identifiable healthcare data on a public ledger, scalability, and cost-effectiveness compared to traditional methods. While blockchain data can be anonymized and encrypted, it raises the risk of identification. Scalability issues are particularly prevalent in smaller applications, while fully distributed systems must handle speed and scalability. The Bitcoin network's maximum transaction confirmation rate of 7 transactions per second may limit the throughput of larger blockchain networks. Platforms that can handle greater data volumes have not yet been tested in real settings to demonstrate their cost-effectiveness (Kumar et al., 2018). To evaluate the overall costs associated with hardware, implementation, and support, as well as the viability of this technology, is important. These difficulties highlight the fact that while blockchain provides authenticity and transparency for data transactions, converting current healthcare IT systems to blockchain-based technologies may be difficult.

## Blockchain Applications for Healthcare

Blockchain technology represents a contemporary and emerging innovation with transformative implications for healthcare. It streamlines the process of sharing and transmitting data, culminating in the creation of cost-effective therapies and advanced treatments for a range of medical conditions. Blockchain technology offers opportunities in logistics, science, logistics, and patient relationships, accelerating growth in the healthcare sector as shown in the below Table 2 (Siyal et al., 2019).

Blockchain technology enables decentralized transaction documentation in healthcare, saving time, effort, and cost. It addresses the issue of data leakage and provides access to accurate patient records. Blockchain technology contributes to healthcare by connecting therapy associates and services, ensuring everyone has access to the same data. It offers benefits for businesses in patient data, medical research, clinical trials, the medical supply chain, and the integrity of pharmaceutical products.

## Utilizing Blockchain in Healthcare Applications

The healthcare sector stands to benefit significantly from the decentralized and widely distributed attributes of blockchain technology. While initially developed for the financial sector, blockchain now holds promise for fostering secure data exchange among healthcare stakeholders. This includes patients, providers, and payers, all of whom can engage in safer data sharing within healthcare systems. The potential for blockchain in healthcare has spurred researchers to explore its practical applications as depicted in Figure 2 (Kaushik & Kumar, 2023).

# APPLICATIONS OF BLOCKCHAIN TECHNOLOGY IN MEDICINE AND FUTURE PERSPECTIVES CHALLENGES

Blockchain emerges as a novel technology making strides across various industries, offering a multitude of benefits and avenues for growth. Nonetheless, as depicted in Figure 3, this technology is not exempt from encountering distinctive challenges that necessitate resolution. The following segment delves into a couple of these significant hurdles.

## Security and Privacy of Data

Blockchain-based applications eliminate the need for third-party intermediaries in transactions, but this raises concerns about potential privacy and security vulnerabilities. The architecture relies on the entire community to validate records, leaving data exposed to potential breaches. Data privacy is compromised when all nodes can access data transmitted by a single node. In emergencies, patients designating representatives to access their information can provide access to multiple individuals, jeopardizing data security. High-security measures complicate data transfer between blocks, leading to potential access limitations for recipients. Blockchain networks are susceptible to a 51% attack, where miners control over 50% of blocks, manipulating the network and withholding consent to halt new transactions. Additionally, patient records may contain sensitive information unsuitable for blockchain inclusion.

*Table 2. Blockchain technology transaction documentation in healthcare*

| S.no | Applications | Description |
|---|---|---|
| 1 | Preserving patient data on an individual basis | Health-related information and patient data are generated at various stages, spanning both pre- and post-clinical study phases. Numerous individuals undergo blood tests, quality assessments, measurements, and health surveys. Instances where evidence of a particular record or document is needed can arise. Healthcare practitioners review the stored data, sometimes harboring doubts about its accuracy; they can easily address this by cross-referencing it with the original records preserved within the Blockchain system. Blockchain leverages established cryptographic methods, providing a secure framework for data-sharing cryptography. When it comes to patient particulars, such as name, birthdate, diagnosis, treatments, and medical history, healthcare professionals input these details into the Electronic Health Record (EHR). This data is currently housed in databases or cloud computing platforms (Panda & Satapathy, 2021). |
| 3 | Validation | Algorithms within a blockchain network authenticate transactions prior to their incorporation into the chain. The verification process involves encryption, digital signatures, and secure storage, ensuring the integrity of the content. In the realm of healthcare, both industry leaders and technological innovators are actively exploring possibilities to enhance safety and affordability. As healthcare administrators gain the ability to securely validate outcomes, blockchain holds the promise to fundamentally transform the healthcare landscape. |
| 4 | Security and transparency | This technology enables physicians to dedicate more time to patient care, enhancing safety and transparency significantly. Furthermore, it has the potential to facilitate clinical research and treatment for uncommon medical conditions. The healthcare ecosystem can experience enhanced collaboration among medical service providers through seamless data sharing, thereby promoting accurate diagnoses, successful treatments, and cost-effective operations. By leveraging blockchain, entities within the healthcare ecosystem can engage in secure and transparent communication, sharing information across a widely distributed network. This approach permits users to exchange data, monitor their own activities, and oversee the data exchanges of others without the necessity for additional measures to ensure data integrity and confidentiality (Musamih et al., 2021). |
| 6 | Trial in Medicine | Blockchain technology finds application in clinical studies to address challenges related to biased results and fragmented data that may deviate from the study's objectives. Its implementation aims to enhance transparency and integrity in clinical trials, ultimately boosting public trust. This technology also serves as a business analysis tool, providing insights into evolving market dynamics for the healthcare industry to seize opportunities. Additionally, blockchain offers a unique opportunity for tracking the medication supply chain from the manufacturer to the end-user, ensuring credibility and security in the management of medications (Ali, Abdullah, Armand, Athar, Hussain, Ali, Yaseen, Joo, & Kim, 2023). |
| 7 | Information display | The Blockchain system ensures both exceptional quality and authentic distribution of authorized medications by providing transparent information about the medicine's origins. When appropriately utilized, blockchain offers unparalleled security for sensitive data. This trend of utilizing blockchain applications to enhance data transparency and reap numerous advantages has been embraced by various industries, including finance, retail, and real estate. The healthcare sector, which is intricate and multifaceted, is another domain where this technology is making a significant impact. This assertion can lead to robust debates, as the healthcare market has undergone substantial transformations in sectors like pharmaceuticals, drug distribution, vaccine management, clinical trials, and the integration of cloud computing. The industry's intricacy and complexity contribute to these significant shifts ( Role of Chatgpt to Reduce the High Overhead Issues of Blockchain Technology in Healthcare et al., ). |
| 8 | Identifying fake material | Blockchain technology is poised to enhance transparency and enable the identification of counterfeit content. Ensuring the legitimacy of clinical trial subjects and participants becomes a straightforward process. Implementing an intelligent contract that maintains accessible records of protocol documents and their outcomes is an effective means of gaining approval. This technological advancement allows the public to closely monitor the progress of clinical trials. Designed with user convenience in mind, this system provides patients with direct and secure access to their medical and insurance information. |
| 9 | Lowers unnecessary administrative costs | Blockchain streamlines the utilization of health records by reducing unnecessary operational expenses. Furthermore, this innovative technology will mitigate the requirement for numerous intermediaries overseeing the dissemination of critical health data.<br>The foremost duty of healthcare professionals is to deliver efficient, timely, and sufficient services to their communities. With blockchain's implementation, healthcare service providers can readily access the medical history of every patient. This technology has the potential to address numerous challenges within the healthcare system, such as interoperability, accurate reporting, data security, and even the risk of severe data loss (Wenhua et al., 2023). |
| 10 | Patient observation | Blockchain technology instills confidence among medical personnel, ensuring their access to medical equipment when required. Furthermore, it enables doctors to efficiently observe patients and remotely respond to health-related incidents. The implementation of blockchain can enhance various aspects of healthcare, including monitoring patient room temperatures, bed utilization, and supply availability. A blockchain-based healthcare network combines blockchain and Internet of Things (IoT) technology to establish a dependable digital identity for healthcare organizations and providers. This integration serves to enhance the supply chain's agility and traceability, fostering transparency in healthcare logistics and facilitating accurate patient monitoring. |
| 11 | Develop research projects | Blockchains have the potential to establish a reliable and credible source of information. The traditional approach to managing membership claims and disputes can be revolutionized through the implementation of blockchain. By enabling broader exchange of patient data, blockchains can stimulate the initiation of innovative research initiatives. This deeper collaboration between patients and researchers will not only foster remarkable partnerships but also act as a catalyst for pioneering and unique research endeavors. |
| 13 | Increases security | Blockchain significantly enhances patient safety by addressing concerns related to medicine authenticity, drug traceability, and facilitating secure interoperability. It stands as the sole viable alternative to replace the existing supply chain management system, effectively thwarting the efforts of counterfeit drug manufacturers attempting to introduce counterfeit versions into the market. This holds true across all medical institutions and groups. Leveraging blockchain technology enables the consolidation of all data into a unified repository (Mozumder et al., 2023). |
| 14 | Reduce the time and expense of data transformation | Utilizing blockchain networks leads to reduced time and cost associated with data transformation. These networks hold promise in tackling the challenge of swift and precise verification of medical credentials. Furthermore, blockchain networks ensure the security and confidentiality of patient information. This, in turn, opens doors for significant breakthroughs and revelations that have the potential to revolutionize global healthcare systems. The introduction of blockchain implementation will yield valuable and privacy-centric monetized networks for data sharing. |

*Figure 2. Practical utilization of blockchain application in healthcare*

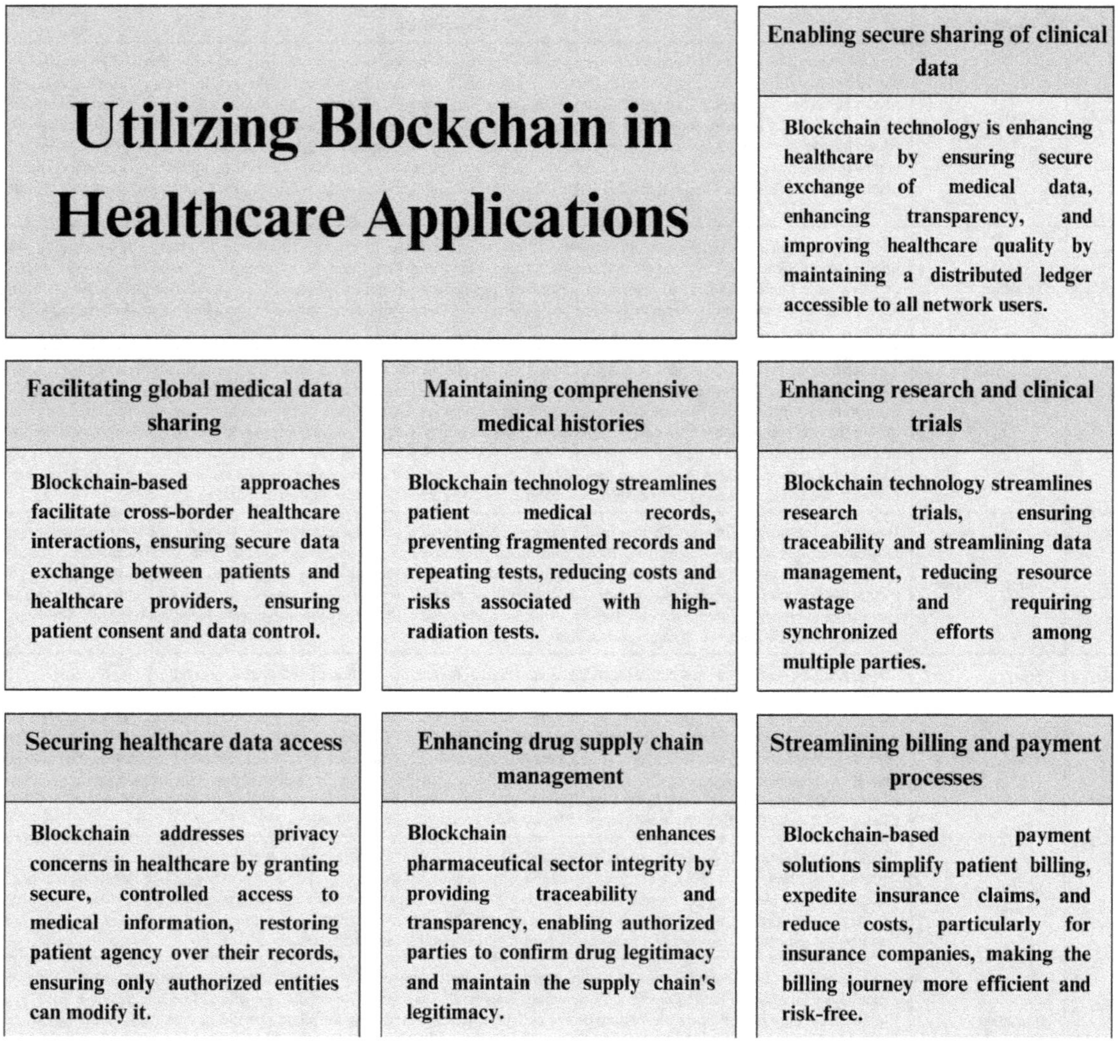


*Figure 3. Opportunities and challenges of blockchain in healthcare*

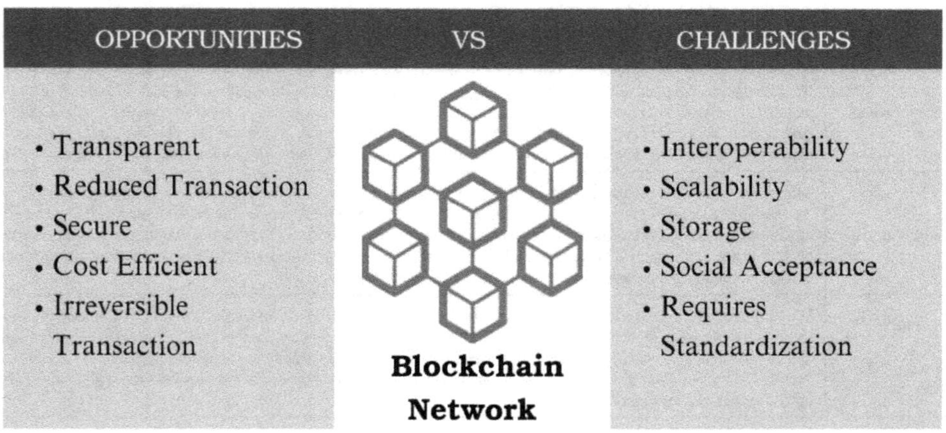

*Figure 4. SWOT analysis for blockchain in healthcare*

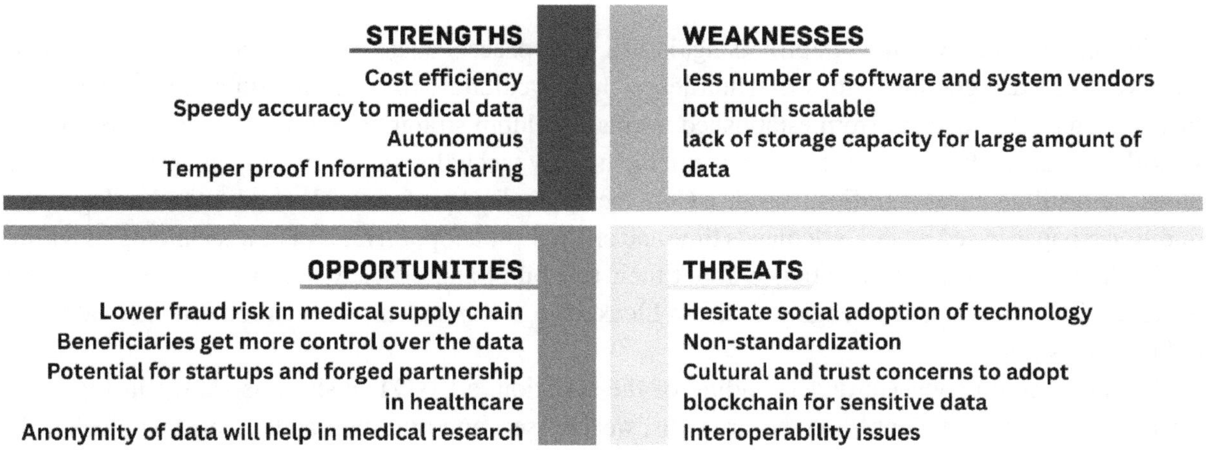


## Managing Storage Capacity

Another challenge arises in the realm of managing storage capacity. Initially designed to handle and process transaction data within a confined scope, blockchain technology demands relatively minimal storage. However, as it ventured into the healthcare sector, storage constraints gradually emerged. The healthcare industry encounters substantial volumes of data on a daily basis, encompassing patient records, health information, test results, MRI scans, X-rays, and various medical images. In a blockchain framework, all nodes within the chain gain access to this extensive array of data, necessitating a significant storage capacity. Additionally, due to the transaction-based nature of blockchain applications, the databases supporting this technology tend to rapidly expand. As these databases grow in size, the speed of data retrieval and access diminishes, becoming unsuitable for transactions that require swiftness (Ameen et al., 2023).

## Interoperability Issues

Another issue that blockchains encounter is interoperability, or making blockchains from various communicating providers and services talk to one another organically and appropriately. This difficulty hinders the effective sharing of data.

## Standardization Challenges

Blockchain technology is still in its infancy, therefore there will undoubtedly be difficulties with standardization on the road to its practical application in medicine and healthcare. International standardization organizations would be needed to provide a number of well-authenticated and approved standards. It would be useful to assess the size, data type, and structure of the information exchanged in blockchain applications using these set criteria. These requirements must act as precautionary safety measures in addition to scrutinizing the shared data.

## Social Challenges

Blockchain technology, although promising, is still in its nascent stage, and beyond the technical challenges discussed, it grapples with social hurdles, including cultural shifts. The assimilation of technology that fundamentally departs from established norms is seldom straightforward. Despite the healthcare sector's gradual shift towards digitization, embracing new technologies particularly unproven ones like blockchain still requires significant strides. Convincing medical professionals to adopt technology over traditional paper-based methods demands time and effort. Low adoption rates of new technologies within the healthcare sector foster skepticism about their reliability and efficacy. Given these obstacles and uncertainties, it remains challenging to deem blockchain a comprehensive solution for the healthcare industry's issues.

To gain a comprehensive understanding of the situation, a SWOT analysis is shown in Figure 4. This approach helps in identifying the strengths, weaknesses, opportunities, and threats that blockchain technology confronts within the healthcare sector (Kousar et al., 2023).

## Future Perspectives

Blockchain technology has the potential to revolutionize the healthcare sector by reducing monitoring costs, eliminating the need for a central server for data storage and management, and streamlining administrative tasks related to medical data. It can be likened to the internet revolutionizing healthcare and introducing telemedicine. Upon patient enrollment, the comprehensive dataset becomes instantaneously accessible, minimizing processing time. This real-time access to verifiable data allows clinicians to make more accurate diagnoses, reducing the likelihood of medical history errors and concerns about patient honesty. Transparency of data also reduces the need for second opinions from different physicians. Patients gain a global community of individuals with similar medical conditions, fostering acceptance and support, and retaining full control over their data. This new era could aptly be termed "Freedom-As-A-Service," as coined by Richie Etwaru during a book launch in 2016 (Vyas et al., 2022).

# CONVERGENCE OF AI AND BLOCKCHAIN IN HEALTHCARE

The healthcare industry is poised for a transformative era due to the convergence of AI and blockchain. AI can enhance data security, facilitate interoperability, and promote patient-centric care. Blockchain's decentralized nature presents an opportunity to reshape the healthcare landscape. The paper explores the intersection of AI and blockchain in various aspects of healthcare, including data integrity, clinical trials, personalized medicine, and supply chain management. This research paper highlights the potential of AI and blockchain in addressing healthcare challenges. It acknowledges challenges such as regulatory compliance, scalability, and standardized protocols (Mozumder et al., 2022). By analyzing current research, case studies, and industry trends, the paper comprehensively explains the benefits and hurdles associated with AI and blockchain convergence in healthcare.

## Amalgamation of AI and Blockchain Technologies in Healthcare Technologies

The healthcare sector is undergoing a transformation due to the integration of blockchain and AI technologies. Blockchain allows for the creation of healthcare indices using data from Google surveys and regulatory bodies. AI, driven by algorithms and vast datasets, enhances transparency and communication between medical professionals and patients. It can quickly identify and diagnose diseases, offer automated treatment alternatives, and facilitate comparisons. AI is also used in drug design, clinical trials, and repurposing commercial pharmaceuticals. Blockchain technology enables the development and management of secure ledgers for automated data analysis. The symbiosis of AI and blockchain extends to various healthcare dimensions, including medical supply records evaluation, health and treatment strategies implementation, and data reconciliation. This data provides real-time updates to medical experts, providers, and payers, fostering a deeper collaboration between AI and blockchain. AI also assumes a pivotal role in predictive analytics, facilitating the evaluation of health information. Research demonstrates AI's effectiveness in comparing medical supplies records, automating instructional processes through pharmacological alerts, proficiently executing health and treatment strategies, and enhancing data reconciliation (Ali, Abdullah, Armand, Athar, Hussain, Ali, Yaseen, Joo, & Kim, 2023). Through diverse methodologies rooted in blockchain and AI techniques, researchers achieve remarkable data analysis speeds without compromising precision or data security, ensuring designated experts can access information for defined periods.

## Role of Blockchain and AI in Healthcare

AI and blockchain have shown potential for growth in the healthcare sector, with AI capable of executing complex tasks and quickly analyzing patient data. However, some medical professionals remain cautious about leveraging AI for patient well-being. Blockchain has gained popularity in the medical field to address interoperability concerns and protect against data misuse. EHR-based software systems like Epic use AI to manage population health and support clinical decision-making, while MedRec allows patients to access their medical records. Scholars have combined blockchain and AI methodologies to process large data volumes effectively, maintaining data security and minimizing complexities.

This study aims to spotlight the valuable contributions of experts and offer insights into the roles and effectiveness of AI and blockchain in EHR systems. Blockchain technology has significantly enhanced the real-time processing of large datasets, leading to quicker disease detection and diagnosis, automated treatment recommendations, and improved transparency and communication between patients and healthcare providers. Figure 5 shows a comprehensive understanding of the roles of AI and blockchain in healthcare.

## Metaverse in the Medical Domain

The incorporation of augmented reality into the healthcare sector has a substantial impact on enhancing the training and skills development of aspiring medical practitioners. Surgical assistance tools, like the Microsoft HoloLens, aid surgeons in expediting and enhancing surgical procedures. Through AR headsets, real-time critical patient data such as heart rate, body temperature, blood pressure, and breathing rate can be visualized, along with pre-operative images sourced from CT, MRI, and 3D scans. Augmented reality finds current application in vein detection by healthcare professionals, addressing the challenge

*Figure 5. Roles of AI and blockchain in healthcare*

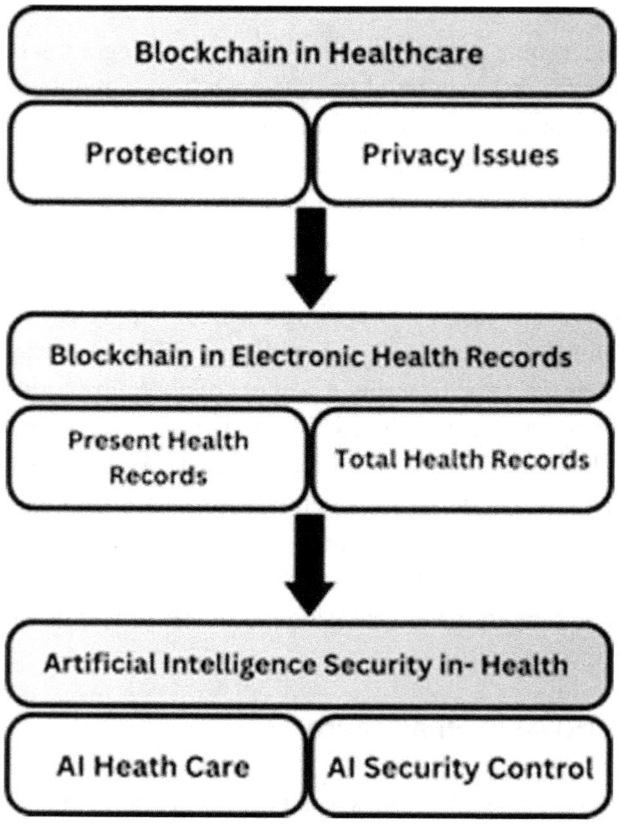


*Figure 6. Metaverse as an evolving role within the medical domain*

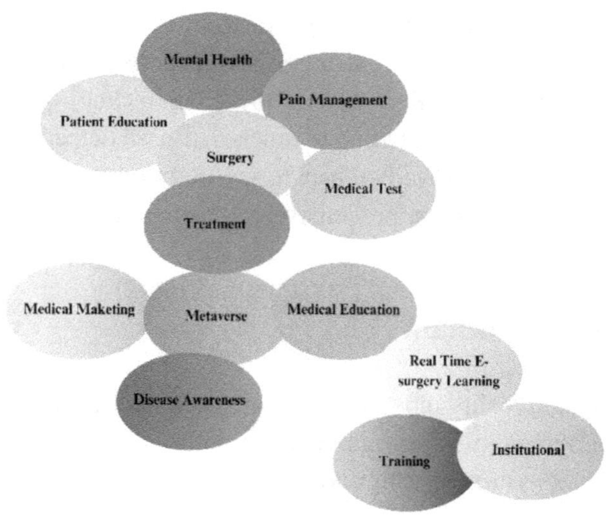

*Figure 7. Applications of blockchain in healthcare industry*

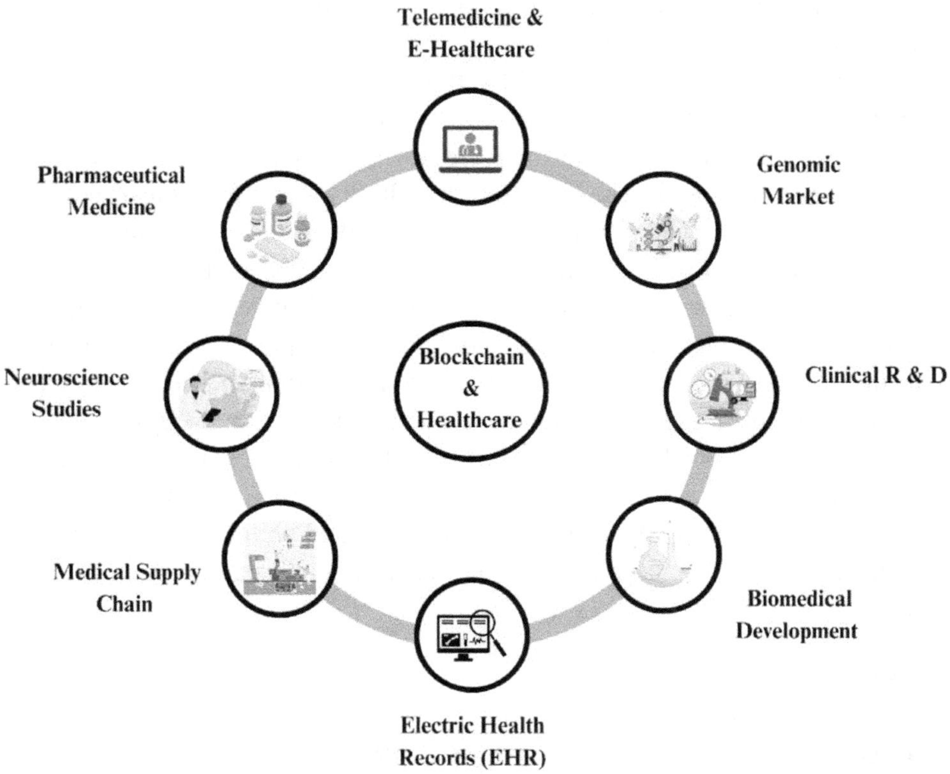


of identifying veins even in individuals with highly pigmented skin or small blood vessels (Anoop & Asharaf, 2022).

Visual-oriented technologies like X-rays and CT scans are widely used in the healthcare field, providing medical experts with insights to diagnose, detect, and treat patients by offering glimpses inside their bodies. A growing number of medical institutions are also embracing metaverse technology, as depicted in Figure 6, illustrating its evolving role within the medical domain.

Blockchain technology, initially used in banking and finance applications as cryptocurrencies, has expanded its potential in healthcare and biomedical fields. It is used in EHR, medical supply chain management, clinical research, genomic market, pharmaceutical medicine, neuroscience studies, biomedical development, and telemedicine and e-healthcare. Blockchain provides a secure mechanism for data storage and sharing in various sub-domains, enabling physicians and healthcare providers to conduct various transactions and experiments. The application of blockchain in the healthcare industry is discussed below in the light of some contemporary research in the healthcare industry as depicted in Figure 7.

Healthcare providers need to integrate technologies like blockchain and AI into their operations to optimize efficiency, ensure adherence to regulations, mitigate risks, reduce work time, and minimize overhead costs. AI is crucial for personalized drug monitoring and identifying patients, while blockchain creates secure data records for analysis. AI algorithms can discern health trends and process large volumes of patient information. However, some medical professionals have reservations about AI's impact on patient well-being. Similar to the automotive sector's use of AI for autonomous vehicles, various industries have successfully used machine learning to detect fraud and assess financial risks. While AI's maturity level is continually evolving, its potential holds great promise for enhancing healthcare systems and prioritizing patient-centric approaches.

Blockchain technology, originally designed for Bitcoin, has expanded to various sectors like data storage, financial markets, cybersecurity, IoT, and nutritional science. It has gained traction in healthcare management, potentially revolutionizing personalized, trustworthy, and secure delivery. A blockchain-based platform has been developed to assess patients' health conditions, focusing on concurrent processing and AI-driven healthcare networks. This innovative method scrutinizes patients' holistic well-being, diagnoses, recovery mechanisms, and surgical interventions. Real-world and simulated healthcare systems have undergone computational analyses to evaluate clinical decision-making effectiveness. A healthcare knowledge-sharing network rooted in blockchain principles scrutinizes the prerequisites for exchanging healthcare data, particularly personal health information and electronic medical records. The platform seamlessly integrates blockchain across diverse sources and employs both on-chain and off-chain authentication mechanisms to ensure data validity and privacy. In the domain of biomedical science and therapeutic endeavors, blockchain technology has exhibited remarkable potential. It can serve as a repository for all clinical endorsements, schedules, and protocols, resulting in more current, safeguarded, time-stamped, and transparent clinical trial-related information.

## CASE STUDIES AND USE CASES

### Utilising Blockchain and AI for Enhanced COVID-19 Surveillance in Resource-Limited Areas

The COVID-19 pandemic has significantly impacted both high-income and low-income countries, particularly in sub-Saharan Africa (SSA), where healthcare systems are struggling and resources are limited. This study proposes a low-cost self-testing and tracking system integrating AI and Blockchain technologies to reduce disease spread. The mobile application will guide patients through the self-testing process, using Blockchain and AI technology to ensure proper treatment and monitoring of positive cases. The application uses GPS for lockdown assistance, while the AI component collects patient information and analyzes it. The Blockchain component provides secure datasets for valuable insights and transactions. Localised diagnostics can help overcome financial and supply chain challenges and improve patient outcomes. The system will be particularly beneficial in areas with limited access to laboratories and poor healthcare facilities.

In conclusion, this study promotes the quick deployment of the proposed technology in resource-limited areas to combat COVID-19 more effectively. This technology is not just limited to COVID-19, but can also be used to combat similar epidemics (Mashamba-Thompson & Crayton, 2020).

### Blockchain Application in Patient Monitoring and Enhancing Healthcare 4.0

Healthcare 4.0 achieves digitisation of healthcare using advanced technologies like blockchain, AI, and IoT etc. Blockchain technology promises reliable, convenient, and transparent healthcare facilities by introducing wearable sensors for patients. These sensors allow doctors to monitor their health status through wireless communication. Blockchain technology ensures patient data security and privacy. The healthcare industry has evolved from manual records to Remote Patient Monitoring (RPM) through wearable devices. These devices track activities like blood pressure, pulse rate, and temperature, transmitting data over WiFi and ZigBee networks. Blockchain networks address security issues by providing access to patient data to authorized medical workers and patients. Smart contracts validate transactions, ensuring data exchange integrity. Integrating blockchain, AI, and Machine Learning (ML) technologies promotes patient diagnosis, monitoring, and personalized care. ML algorithms help in early disease examination and forecasting. However, blockchain deployment faces challenges like scalability, latency, and interoperability. To overcome these challenges, experts suggest using open standards to maintain interoperability, limiting validation participation to improve processing efficiency, and using blockchain models such that the privacy of patient data is maintained (Hathaliya et al., 2019).

The discussed technology has the potential to revolutionise the healthcare industry by providing accurate diagnosis and individualised treatment. Stakeholders in the healthcare industry need to collaborate and bring this technology into power so that data-driven decision-making and improved patient outcomes can come into the picture.

## CONCLUSION

The convergence of blockchain technology and artificial intelligence presents a promising avenue for enhancing patient care, medical research, and clinical trials through the amalgamation of machine learning and deep learning capabilities. Nonetheless, the reliability and precision of these predictive insights require further exploration. Challenges arise from language and cultural variations embedded in Electronic Health Record (EHR) systems, complicating the application of blockchain and AI in healthcare. Overcoming these hurdles, AI tools such as natural language processing can facilitate the translation of medical terminology across different regions. While blockchain ensures the integrity, credibility, and accessibility of medical records, it does not inherently address their interpretation. Through the utilization of aggregation and blockchain data analytics, relationships between documents can be discerned, reducing the necessity for extra storage and processing resources. This not only bolsters the authenticity of reports but also mitigates potential biases. Blockchain faces challenges in managing data storage and production due to increased transactions. To address this, researchers are exploring cloud computing's potential in healthcare environments. Distributed cloud architecture using blockchain and cloud computing can provide stable, cost-effective healthcare environments in IoT networks. This approach can improve medical decisions due to large data availability and rapid decision-making. However, few implementations have created a blockchain framework for healthcare, and further research is needed on architecture, modelling, and cost-benefit rules.

## REFERENCES

Abbas, A. F., Qureshi, N. A., Khan, N., Chandio, R., & Ali, J. (2022). The Blockchain Technologies in Healthcare: Prospects, Obstacles, and Future Recommendations; Lessons Learned from Digitalization. *International Journal of Online & Biomedical Engineering*, *18*(9).

Afaq, Y., & Manocha, A. (2023). Blockchain and Deep Learning Integration for Various Application: A Review. *Journal of Computer Information Systems*, 1–14. doi:10.1080/08874417.2023.2173330

Ali, S., Abdullah, Armand, T. P. T., Athar, A., Hussain, A., Ali, M., Yaseen, M., Joo, M.-I., & Kim, H.-C. (2023). Metaverse in healthcare integrated with explainable ai and blockchain: Enabling immersiveness, ensuring trust, and providing patient data security. *Sensors (Basel), 23*(2), 565. doi:10.339023020565 PMID:36679361

Ali, S., Abdullah, Armand, T. P. T., Athar, A., Hussain, A., Ali, M., Yaseen, M., Joo, M.-I., & Kim, H.-C. (2023). Metaverse in healthcare integrated with explainable ai and blockchain: Enabling immersiveness, ensuring trust, and providing patient data security. *Sensors (Basel), 23*(2), 565. doi:10.339023020565 PMID:36679361

Ameen, A. H., Mohammed, M. A., & Rashid, A. N. (2023). Dimensions of artificial intelligence techniques, blockchain, and cyber security in the Internet of medical things: Opportunities, challenges, and future directions. *Journal of Intelligent Systems, 32*(1), 20220267. doi:10.1515/jisys-2022-0267

Anoop, V. S., & Asharaf, S. (2022). Integrating artificial intelligence and blockchain for enabling a trusted ecosystem for healthcare sector. In *Intelligent Healthcare: Infrastructure, Algorithms and Management* (pp. 281–295). Springer Nature Singapore. doi:10.1007/978-981-16-8150-9_13

Attaran, M. (2022). Blockchain technology in healthcare: Challenges and opportunities. *International Journal of Healthcare Management, 15*(1), 70–83. doi:10.1080/20479700.2020.1843887

Baysal, M. V., Özcan-Top, Ö., & Betin-Can, A. (2023). Blockchain technology applications in the health domain: A multivocal literature review. *The Journal of Supercomputing, 79*(3), 3112–3156. doi:10.100711227-022-04772-1 PMID:36060094

Castiglioni, I., Rundo, L., Codari, M., Di Leo, G., Salvatore, C., Interlenghi, M., Gallivanone, F., Cozzi, A., D'Amico, N. C., & Sardanelli, F. (2021). AI applications to medical images: From machine learning to deep learning. *Physica Medica, 83*, 9–24. doi:10.1016/j.ejmp.2021.02.006 PMID:33662856

Cooper, N. (2022). Blockchain in Health Care. *Frontiers in Blockchain, 4*, 830459. doi:10.3389/fbloc.2021.830459

Fatoum, H., Hanna, S., Halamka, J. D., Sicker, D. C., Spangenberg, P., & Hashmi, S. K. (2021). Blockchain integration with digital technology and the future of health care ecosystems: Systematic review. *Journal of Medical Internet Research, 23*(11), e19846. doi:10.2196/19846 PMID:34726603

Frangoudes, F., Hadjiaros, M., Schiza, E. C., Matsangidou, M., Tsivitanidou, O., & Neokleous, K. 2021, July. An overview of the use of chatbots in medical and healthcare education. In *International Conference on Human-Computer Interaction* (pp. 170-184). Cham: Springer International Publishing. 10.1007/978-3-030-77943-6_11

Govindan, K., Nasr, A. K., Saeed Heidary, M., Nosrati-Abarghooee, S., & Mina, H. (2023). Prioritizing adoption barriers of platforms based on blockchain technology from balanced scorecard perspectives in healthcare industry: A structural approach. *International Journal of Production Research, 61*(11), 3512–3526. doi:10.1080/00207543.2021.2013560

Haleem, A., Javaid, M., Singh, R. P., Suman, R., & Rab, S. (2021). Blockchain technology applications in healthcare: An overview. *International Journal of Intelligent Networks, 2*, 130–139. doi:10.1016/j.ijin.2021.09.005

Hannan, S. A. (2023). Challenges of Blockchain Technology using Artificial Intelligence in Healthcare System. *International Journal of Innovative Research in Science, Engineering and Technology*, *12*(01), 64–74.

HasnainM.HussainA. Role of Chatgpt to Reduce the High Overhead Issues of Blockchain Technology in Healthcare. *S*SRN 4502924.

Hathaliya, J., Sharma, P., Tanwar, S., & Gupta, R. (2019, December). Blockchain-based remote patient monitoring in healthcare 4.0. In *2019 IEEE 9th international conference on advanced computing (IACC)* (pp. 87-91). IEEE.

Hegde, P. and Maddikunta, P.K.R., 2023. Amalgamation of Blockchain with resource-constrained IoT devices for Healthcare applications–State of Art, Challenges and Future Directions. *International Journal of Cognitive Computing in Engineering*.

Hussien, H. M., Yasin, S. M., Udzir, N. I., Ninggal, M. I. H., & Salman, S. (2021). Blockchain technology in the healthcare industry: Trends and opportunities. *Journal of Industrial Information Integration*, *22*, 100217. doi:10.1016/j.jii.2021.100217

Jiménez-Luna, J., Grisoni, F., Weskamp, N., & Schneider, G. (2021). Artificial intelligence in drug discovery: Recent advances and future perspectives. *Expert Opinion on Drug Discovery*, *16*(9), 949–959. doi:10.1080/17460441.2021.1909567 PMID:33779453

Kaushik, K., & Kumar, A. (2023). Demystifying quantum blockchain for healthcare. *Security and Privacy*, *6*(3), e284. doi:10.1002/spy2.284

Kousar, H., Sundara Rajulu Navaneethakrishnan, D. D. S. H., & Tejaswi Vuyyuru, D. D. A. M. (2023). Integrating Artificial Intelligence and Blockchain Technology for Secure and Efficient Data Sharing. *Journal of Survey in Fisheries Sciences*, *10*(1S), 5900–5911.

Kumar, R, Singh, D., Srinivasan, K. & Hu, Y.C. (2022, December. AI-powered blockchain technology for public health: A contemporary review, open challenges, and future research directions. In Healthcare (Vol. 11, No. 1, p. 81). MDPI.

Kumar, T., Ramani, V., Ahmad, I., Braeken, A., Harjula, E., & Ylianttila, M. 2018, September. Blockchain utilization in healthcare: Key requirements and challenges. In *2018 IEEE 20th International conference on e-health networking, applications and services (Healthcom)* (pp. 1-7). IEEE.

Lee, D., & Yoon, S. N. (2021). Application of artificial intelligence-based technologies in the healthcare industry: Opportunities and challenges. *International Journal of Environmental Research and Public Health*, *18*(1), 271. doi:10.3390/ijerph18010271 PMID:33401373

Mashamba-Thompson, T. P., & Crayton, E. D. (2020). Blockchain and artificial intelligence technology for novel coronavirus disease 2019 self-testing. *Diagnostics (Basel)*, *10*(4), 198. doi:10.3390/diagnostics10040198 PMID:32244841

Mavropoulos, T., Meditskos, G., Symeonidis, S., Kamateri, E., Rousi, M., Tzimikas, D., Papageorgiou, L., Eleftheriadis, C., Adamopoulos, G., Vrochidis, S., & Kompatsiaris, I. (2019). A context-aware conversational agent in the rehabilitation domain. *Future Internet*, *11*(11), 231. doi:10.3390/fi11110231

Merlo, V., Pio, G., Giusto, F., & Bilancia, M. (2023). On the exploitation of the blockchain technology in the healthcare sector: A systematic review. *Expert Systems with Applications*, *213*, 118897. doi:10.1016/j.eswa.2022.118897

Mozumder, M. A. I., Armand, T. P. T., Imtiyaj Uddin, S. M., Athar, A., Sumon, R. I., Hussain, A., & Kim, H. C. (2023). Metaverse for Digital Anti-Aging Healthcare: An Overview of Potential Use Cases Based on Artificial Intelligence, Blockchain, IoT Technologies, Its Challenges, and Future Directions. *Applied Sciences (Basel, Switzerland)*, *13*(8), 5127. doi:10.3390/app13085127

Mozumder, M. A. I., Sheeraz, M. M., Athar, A., Aich, S., & Kim, H. C. 2022, February. Overview: Technology roadmap of the future trend of metaverse based on IoT, blockchain, AI technique, and medical domain metaverse activity. In *2022 24th International Conference on Advanced Communication Technology (ICACT)* (pp. 256-261). IEEE.

Musamih, A., Salah, K., Jayaraman, R., Arshad, J., Debe, M., Al-Hammadi, Y., & Ellahham, S. (2021). A blockchain-based approach for drug traceability in healthcare supply chain. *IEEE Access : Practical Innovations, Open Solutions*, *9*, 9728–9743. doi:10.1109/ACCESS.2021.3049920

Olawumi, T. O., Chan, D. W., Ojo, S., & Yam, M. C. (2022). Automating the modular construction process: A review of digital technologies and future directions with blockchain technology. *Journal of Building Engineering*, *46*, 103720. doi:10.1016/j.jobe.2021.103720

Omrčen, L., Leventić, H., Romić, K., & Galić, I. (2021, September). Integration of Blockchain and AI in EHR sharing: A survey. In *2021 International Symposium ELMAR* (pp. 155-160). IEEE. 10.1109/ELMAR52657.2021.9550953

Otoum, S., Al Ridhawi, I., & Mouftah, H. T. (2021). Preventing and controlling epidemics through blockchain-assisted ai-enabled networks. *IEEE Network*, *35*(3), 34–41. doi:10.1109/MNET.011.2000628

Özcan-Top, Ö. (2023). Improving Medical Supply Chain Disruption Management with the Blockchain Technology. In *Supply Chain Risk and Disruption Management: Latest Tools, Techniques and Management Approaches* (pp. 217–230). Springer Nature Singapore. doi:10.1007/978-981-99-2629-9_11

Panda, S. K., & Satapathy, S. C. (2021). Drug traceability and transparency in medical supply chain using blockchain for easing the process and creating trust between stakeholders and consumers. *Personal and Ubiquitous Computing*, 1–17. doi:10.100700779-021-01588-3

Pilozzi, A., & Huang, X. (2020). Overcoming Alzheimer's disease stigma by leveraging artificial intelligence and blockchain technologies. *Brain Sciences*, *10*(3), 183. doi:10.3390/brainsci10030183 PMID:32210011

Połap, D., Srivastava, G., Jolfaei, A., & Parizi, R. M. (2020, July). Blockchain technology and neural networks for the internet of medical things. In IEEE INFOCOM 2020-IEEE conference on computer communications workshops (INFOCOM WKSHPS) (pp. 508-513). IEEE. doi:10.1109/INFOCOMWKSHPS50562.2020.9162735

Rao, K. P. N., & Manvi, S. (2023). Survey on Electronic Health Record Management Using Amalgamation of Artificial Intelligence and Blockchain Technologies. *Acta Informatica Pragensia*, *12*(1), 179–199. doi:10.18267/j.aip.194

Saeed, H., Malik, H., Bashir, U., Ahmad, A., Riaz, S., Ilyas, M., Bukhari, W. A., & Khan, M. I. A. (2022). Blockchain technology in healthcare: A systematic review. *PLoS One*, *17*(4), e0266462. doi:10.1371/journal.pone.0266462 PMID:35404955

Shinde, R., Patil, S., Kotecha, K., & Ruikar, K. (2021). Blockchain for securing ai applications and open innovations. *Journal of Open Innovation*, *7*(3), 189. doi:10.3390/joitmc7030189

Siyal, A. A., Junejo, A. Z., Zawish, M., Ahmed, K., Khalil, A., & Soursou, G. (2019). Applications of blockchain technology in medicine and healthcare: Challenges and future perspectives. *Cryptography*, *3*(1), 3. doi:10.3390/cryptography3010003

Soltanisehat, L., Alizadeh, R., Hao, H., & Choo, K. K. R. (2020). Technical, temporal, and spatial research challenges and opportunities in blockchain-based healthcare: A systematic literature review. *IEEE Transactions on Engineering Management*, *70*(1), 353–368. doi:10.1109/TEM.2020.3013507

Srivastava, S., Pant, M., Jauhar, S. K., & Nagar, A. K. (2022). Analyzing the Prospects of Blockchain in Healthcare Industry. *Computational and Mathematical Methods in Medicine*, *2022*, 2022. doi:10.1155/2022/3727389 PMID:36506597

Srivastava, S., Singh, S. V., Singh, R. B., & Shukla, H. K. (2021). Digital Transformation of Healthcare: A blockchain study. *International Journal of Innovative Science, Engineering & Technology, 8*(5).

Tagde, P., Tagde, S., Bhattacharya, T., Tagde, P., Chopra, H., Akter, R., Kaushik, D., & Rahman, M. H. (2021). Blockchain and artificial intelligence technology in e-Health. *Environmental Science and Pollution Research International*, *28*(38), 52810–52831. doi:10.100711356-021-16223-0 PMID:34476701

Taloba, A. I., Elhadad, A., Rayan, A., Abd El-Aziz, R. M., Salem, M., Alzahrani, A. A., Alharithi, F. S., & Park, C. (2023). A blockchain-based hybrid platform for multimedia data processing in IoT-Healthcare. *Alexandria Engineering Journal*, *65*, 263–274. doi:10.1016/j.aej.2022.09.031

Vellido, A. (2019). Societal issues concerning the application of artificial intelligence in medicine. *Kidney Diseases*, *5*(1), 11–17. doi:10.1159/000492428 PMID:30815459

Vijayalakshmi, S., Savita, Gayathri, S. P., & Janarthanan, S. (2021). Blockchain security for artificial intelligence-based clinical decision support tool. *Internet of things, artificial intelligence and blockchain technology*, 209-240.

Vyas, S., Shabaz, M., Pandit, P., Parvathy, L. R., & Ofori, I. (2022). Integration of AI and blockchain technology in healthcare and agriculture. *Journal of Food Quality*, 2022.

Wenhua, Z., Qamar, F., Abdali, T. A. N., Hassan, R., Jafri, S. T. A., & Nguyen, Q. N. (2023). Blockchain technology: Security issues, healthcare applications, challenges and future trends. *Electronics (Basel)*, *12*(3), 546. doi:10.3390/electronics12030546

Yaqoob, I., Salah, K., Jayaraman, R., & Al-Hammadi, Y. (2021). Blockchain for healthcare data management: Opportunities, challenges, and future recommendations. *Neural Computing & Applications*, 1–16.

Zhavoronkov, A., Vanhaelen, Q., & Oprea, T. I. (2020). Will artificial intelligence for drug discovery impact clinical pharmacology? *Clinical Pharmacology and Therapeutics*, *107*(4), 780–785. doi:10.1002/cpt.1795 PMID:31957003

# Chapter 9
# Artificial Intelligence and Blockchain Technology for Secure Smart Grid and Power Distribution Automation

**Zahira Tabassum**
https://orcid.org/0000-0002-4912-0228
*HKBK College of Engineering, India*

**Rashmi Rani Samantaray**
*HKBK College of Engineering, India*

**Syeda Husna Mohammadi**
*HKBK College of Engineering, India*

**Anees Fathima**
*HKBK College of Engineering, India*

## ABSTRACT

*A country's growth is gauged with reference to its power consumption and energy use, both of which are fast expanding. Energy consumption management improves energy generation and delivery. The smart grid is a tremendous improvement over the power system of the 20th century, using two-way electrical and communication exchanges to create a sophisticated, distributed, computerized energy delivery network. The fields of artificial intelligence (AI) and block chain distributed ledger technology (BDLT) are the most fascinating areas of study in the field of green energy and related power automation. An in-depth analysis of the most advanced automated planning, governance, optimization, confidentiality, and safety methods for the distribution of power and smart grid using integrated artificial intelligence and block chain is presented in this chapter.*

DOI: 10.4018/979-8-3693-0659-8.ch009

*Figure 1. AI and blockchain in smart grids*

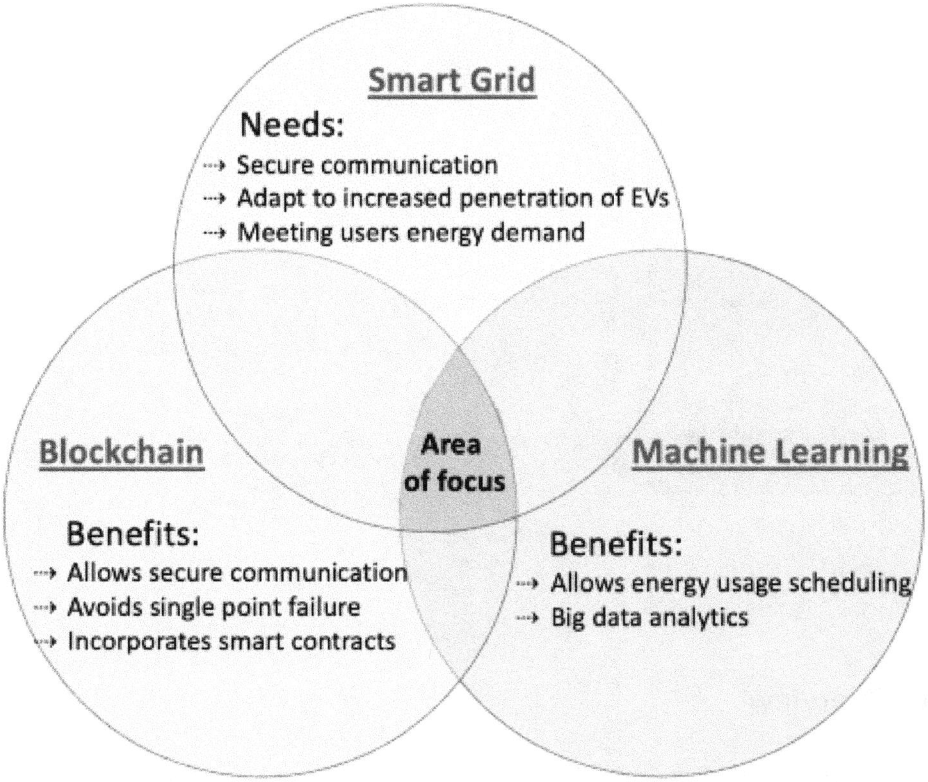


## INTRODUCTION

The smart grid concept was put forth to guarantee effective electricity distribution, maintain minimal losses and high levels of quality, and assure the security of the electrical supply. The rapid advancement of information and communication technology has led to a new paradigm in the usage of the electric power grid and associated operations. A safe control centre platform for state estimation is provided by the cooperative method of block chain, AI, and wireless sensor networks (WSN), which aids in the detection and analysis of erroneous data flow. The evaluation of technology is being seriously impacted by a number of new challenges in the field of Blockchain-AI in a green environment, such as monitoring of uncertainties, optimal flow of power, reconfiguration of network, the commitment of units with security restrictions, as well as govern auto-generation.

Artificial intelligence (AI) and blockchain technology can play significant roles in enhancing the security and efficiency of smart grids and power distribution automation systems.Figure.1 shows the correlation of AI and block chain in smart grids (Moloth,2023). This chapter covers the overview of smart grids, role of AI and block chain in smart grids followed by case study of application of AI and block chain in overall functioning of smart grids.

*Figure 2. Components of smart grid*

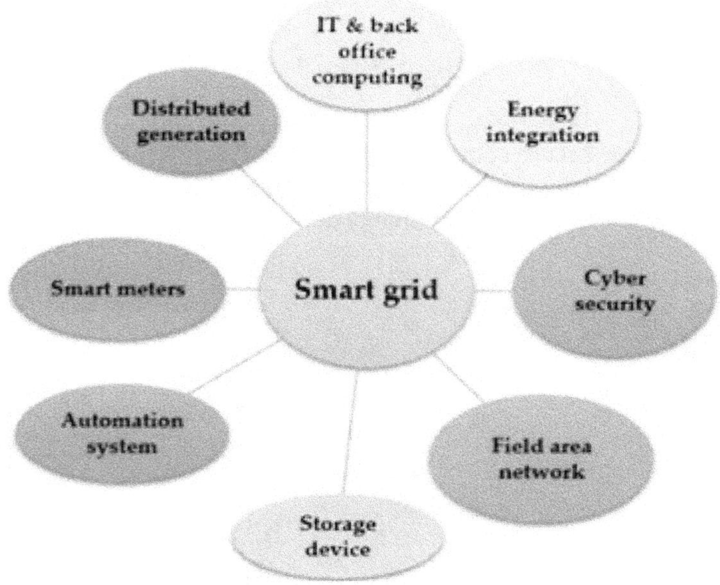


## Smart Grid Overview

A smart grid is an advanced electrical grid (Mazhar, 2023) that uses various technologies to optimize the generation, distribution, and consumption of electricity. Figure.2 shows the essential components of a smart grid. Power distribution automation involves the use of technology to monitor and control the distribution of electricity, ensuring efficient and reliable power delivery. Smart meters, intelligent devices, alternative power sources, and energy-saving measures represent few of the operational and energy strategies that a smart grid includes in its electrical system. The smart grid has several advantages over conventional grids, including enhanced power quality, self-repairing, affordability with the incorporation of green energy, flexible production of energy, environmentally conscious functioning, gathering of distributed sources of energy, real-time usage, and monitoring of energy consumption at the customer end.

## Role of AI in Smart Grid

AI (Artificial Intelligence) plays a significant role in enhancing the functionality and efficiency of smart grids, which are modernized electrical grids that leverage advanced technology for monitoring, managing, and optimizing the generation, distribution, and consumption of electricity. Here are some key roles of AI in smart grids.

- **Demand Forecasting:** AI algorithms can analyze historical consumption data, weather patterns, and other relevant factors to accurately predict electricity demand. This helps utilities better plan for generation and distribution, reducing the likelihood of blackouts and optimizing resource allocation.

- **Load Balancing:** AI can optimize the distribution of electricity across the grid in real-time by dynamically adjusting the flow of electricity to meet changing demands. This ensures that power is distributed efficiently and prevents overloads or underutilization of resources.
- **Grid Monitoring and Management:** AI-based sensors and monitoring systems can continuously assess the health and performance of grid components such as transformers, substations, and power lines. This allows for early detection of faults and proactive maintenance to minimize downtime.
- **Renewable Energy Integration:** As renewable energy sources like solar and wind power become more prevalent, AI can help predict and manage the intermittent nature of these sources. AI can optimize the integration of renewable energy into the grid while maintaining stability.
- **Energy Storage Optimization:** AI can control the charging and discharging of energy storage systems like batteries to store excess energy during periods of low demand and release it during peak demand, helping to stabilize the grid and reduce energy costs.
- **Fault Detection and Self-Healing:** AI algorithms can quickly detect faults or disturbances in the grid and autonomously reroute electricity to restore service to affected areas. This reduces downtime and improves grid reliability.
- **Energy Efficiency:** AI can analyze consumption patterns and provide recommendations to consumers and businesses on how to reduce their energy consumption during peak periods or high-price times, leading to cost savings and reduced strain on the grid.
- **Grid Security:** AI can enhance the cyber security of smart grids by identifying and responding to potential threats and vulnerabilities in real-time, protecting critical infrastructure from cyber attacks.
- **Customer Engagement:** AI-powered smart meters and home energy management systems can provide consumers with real-time information about their energy usage and costs, helping them make informed decisions to conserve energy and save money.
- **Data Analytics:** AI can process vast amounts of data generated by smart grid sensors and devices to extract valuable insights, enabling utilities to make data-driven decisions for grid optimization and future planning.
- **Predictive Maintenance:** AI can predict equipment failures and maintenance needs based on data from sensors and historical performance. This proactive approach minimizes downtime and reduces maintenance costs.

In summary, AI is a critical component of smart grid technology, helping utilities and consumers alike to optimize energy generation, distribution, and consumption, improve grid reliability, and reduce environmental impact. As the energy landscape continues to evolve, AI will play an increasingly important role in the transformation of the electrical grid (Rashmi R S, et.al, 2022).

## Role of Blockchain in Smart Grid

Blockchain technology can play a significant role in enhancing the efficiency, security, and transparency of smart grids. Smart grids are modernized electrical grids that use digital technology to optimize the generation, distribution, and consumption of electricity. Figure.3 shows a block chain enabled smart grid. Blockchain (BC) technologies produce a digital transaction record that is decentralized, secure, and impenetrable. According to Kuo Chuen (2015), a blockchain is essentially a database that is man-

*Figure 3. Blockchain enabled future smart grid*

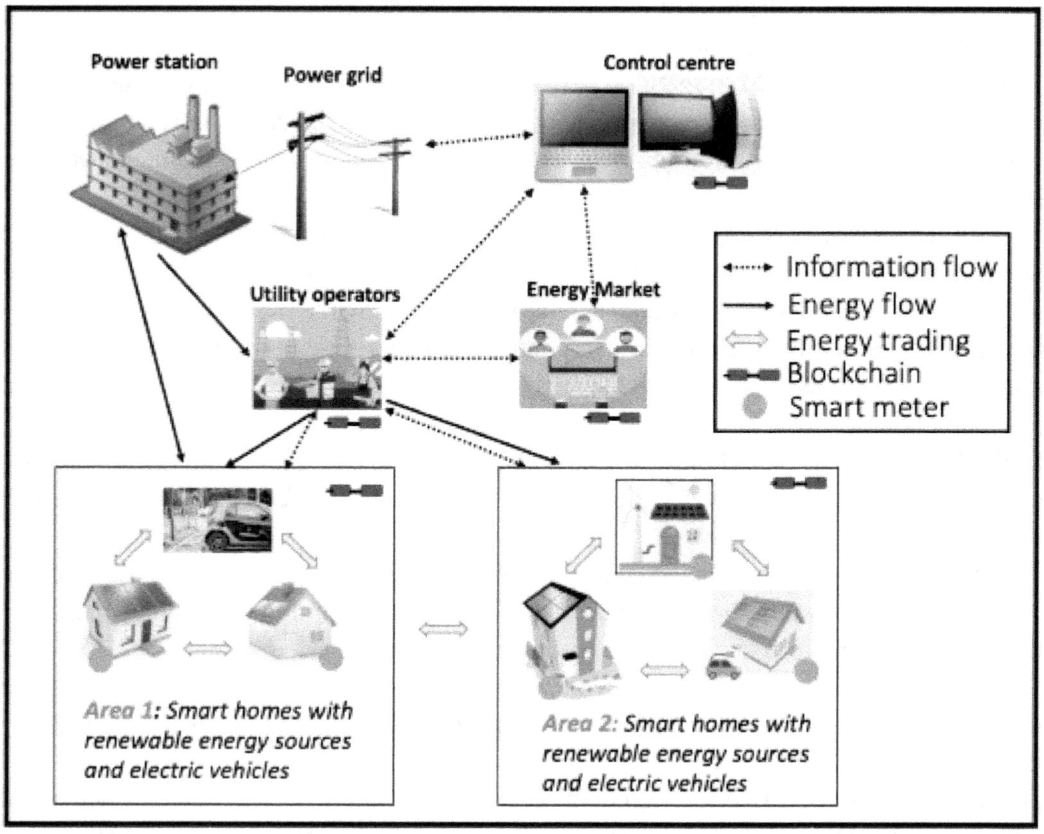


*Figure 4. Architecture of blockchain*

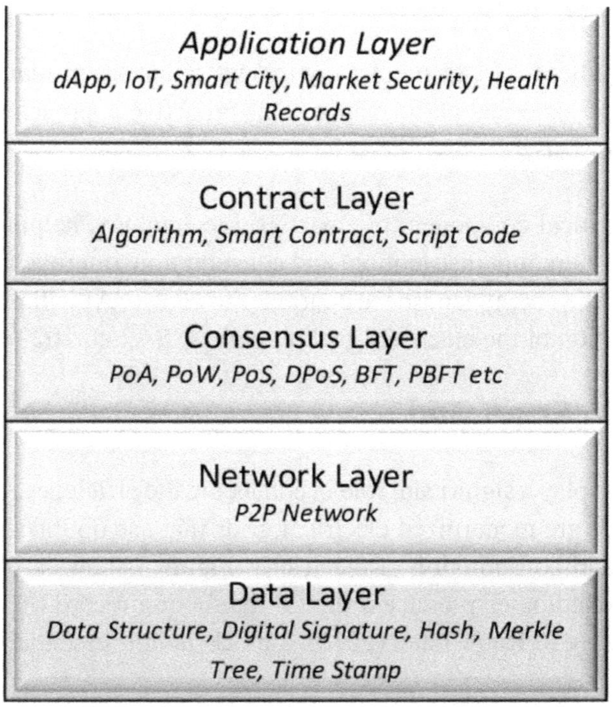

aged by a network of computers that work together to verify and record transactions. Each block in the blockchain contains a list of multiple transactions as well as a unique cryptographic signature that confirms the authenticity of the block.

As shown in Figure. 4, a block chain's design can be broken down into five layers **(Junaidi 2023)**, the network layer, the data layer, the consensus layer, the data layer, the application layer, and the contract layer.

The application layer includes decentralized applications (dApps) that run on top of the block chain. These dApps might range from straightforward applications like wallets to complex ones like supply chain management systems. The contract layer contains the smart contracts' data. By serving as self-executing contracts, smart contracts simplify the negotiation and execution of agreements.

The consensus layer, at the centre of a blockchain, is where a network of computers works together to validate and approve transactions. Making sure that only legitimate transactions are added to the blockchain is the responsibility of this layer. Different consensus mechanisms, such as Proof of Work (PoW), Proof of Stake (PoS), Delegated Proof of Stake (DPoS), and Practical Byzantine Fault Tolerance (PBFT), are used by various blockchain platforms (Sanjana Prasad, et. Al, 2022).

The network layer handles communication between the various network nodes. The nodes on a blockchain network are connected by a P2P network, enabling decentralized communication and data sharing. The network layer is also in charge of propagating transactions and blocks throughout the network because all nodes must have access to the most recent data.

Here are some key ways in which blockchain can contribute to the development and operation of smart grids:

- **Data Management and Security**: Blockchain provides a secure and tamper-resistant way to store and manage data generated by smart grid devices. This is critical for ensuring the integrity of data related to energy production, consumption, and transactions. The decentralized nature of blockchain reduces the risk of data manipulation or cyber attacks.
- **Decentralization:** Smart grids aim to decentralize power generation and distribution. Blockchain can facilitate the creation of peer-to-peer energy trading networks, allowing users to buy and sell excess energy directly to each other without the need for intermediaries. This promotes energy independence and can make the grid more resilient.
- **Smart Contracts**: Smart contracts are self-executing contracts with the terms of the agreement directly written into code. In a smart grid context, blockchain-based smart contracts can automate energy transactions, ensuring that payments are made automatically when certain conditions are met, such as when a certain amount of energy is delivered.
- **Grid Management**: Blockchain can enhance grid management by providing real-time visibility into energy production and consumption patterns. This information can help grid operators optimize energy distribution, reduce losses, and respond more effectively to demand fluctuations.
- **Energy Traceability**: Blockchain enables the tracing of the origin of energy, which can be crucial for ensuring the use of renewable sources and maintaining environmental standards. Consumers can verify the source of the electricity they consume, promoting transparency and sustainability.
- **Billing and Settlement**: Blockchain can streamline the billing and settlement process in the energy sector. With transparent and automated transactions, consumers can have more accurate and timely bills, while energy providers can reduce operational costs associated with billing.

- **Grid Resilience**: The decentralized and redundant nature of blockchain can enhance the resilience of the grid against natural disasters and cyber attacks. Even if part of the network is compromised, the rest can continue to operate.
- **Data Privacy**: Blockchain allows consumers to have more control over their energy data. They can choose to share their data securely with service providers or other parties while maintaining their privacy and control.
- **Regulatory Compliance**: Smart grids are subject to various regulations and standards. Blockchain can help ensure compliance by providing a transparent and auditable record of all transactions and activities within the grid.
- **Incentivizing Renewable Energy**: Blockchain-based incentive mechanisms, such as token rewards for energy conservation or renewable energy production, can encourage consumers to adopt eco-friendly practices and technologies.

While blockchain offers many advantages for smart grids, it's important to note that implementing this technology in the energy sector comes with challenges, such as scalability, energy consumption of blockchain networks, and interoperability issues. Nevertheless, ongoing research and development are addressing these issues, and blockchain continues to hold promise for revolutionizing the way we manage and distribute energy in smart grids.

## Related Works

## Applications of AI and Block chain in Smart Grids

### Role of AI in Renewable Energy Integration

Artificial Intelligence (AI) plays a significant role in the integration of renewable energy sources into the existing energy infrastructure. This integration is crucial for transitioning to a more sustainable and

*Table 1. Existing approaches in AI and Block chain based secure smart grid*

| Author | Title | Methodology | Limitations |
|---|---|---|---|
| (Adel Sulaiman, 2023) | Artificial Intelligence-Based Secured Power Grid Protocol for Smart City | LSTM RNN-based architecture | False data injection detection only |
| (Parminder Singh, 2021) | Blockchain and homomorphic encryption-based privacy-preserving data aggregation model in smart grid. | Deep learning and homomorphic encryption data aggregation model | Because blockchain and authentication are employed in the process, communication overhead is higher than with PDAFT and EPPDR approaches. |
| (Sunitha Basodi, 2020) | Data integrity attack detection in smart grid: a deep learning approach. | Deep-learning models using feed-forward neural network architectures | Detecting centralised data attacks only. |
| (Pradeep Kumar, 2022) | A decentralized secured grid integration system using APEBC technique with multi access AI framework. | Authenticated Privacy Ethereum based Blockchain (APEBC) technique | Secure data access |

environmentally friendly energy system. Here are some key roles that AI plays in renewable energy integration:

**Energy Forecasting**: AI can improve the accuracy of renewable energy production forecasts. By analyzing historical weather data, solar radiation, wind speed, and other factors, AI models can predict energy generation from sources like solar and wind. These forecasts help grid operators manage energy supply and demand efficiently.

**Grid Management and Optimization**: AI can optimize the distribution and transmission of renewable energy within the power grid. It can balance supply and demand, manage energy storage systems, and control the flow of electricity to ensure stability and reliability.

**Energy Storage Optimization**: AI can enhance the performance of energy storage systems (e.g., batteries). By analyzing real-time data, AI can determine the optimal times to charge and discharge these systems, maximizing their efficiency and lifespan.

**Demand Response**: AI can facilitate demand response programs, which encourage consumers to adjust their energy consumption patterns based on real-time price signals or grid conditions. This helps reduce peak demand and ensures better utilization of renewable energy.

**Grid Security**: AI can improve grid security by detecting and responding to cyber security threats. Protecting the energy infrastructure from cyber-attacks is essential for maintaining reliable energy supply from renewable sources.

**Predictive Maintenance**: AI can predict when renewable energy equipment, such as wind turbines or solar panels, is likely to fail. This allows for proactive maintenance, reducing downtime and ensuring consistent energy generation.

**Microgrid Management**: In isolated areas or during grid disruptions, AI can manage microgrids efficiently. It can balance local energy generation, storage, and consumption to ensure uninterrupted power supply.

**Energy Trading**: AI-powered platforms can facilitate peer-to-peer energy trading among renewable energy producers and consumers. Blockchain technology is often integrated with AI to ensure transparency and security in such transactions.

**Resource Assessment**: AI can aid in the selection of optimal locations for renewable energy installations. It can analyze geographical and environmental data to determine the most suitable sites for solar, wind, or hydroelectric projects.

**Policy and Regulatory Compliance**: AI can help energy companies and utilities comply with regulations related to renewable energy integration and emissions reductions by providing data-driven insights and automating reporting processes.

**Energy Efficiency**: AI-driven analytics can identify opportunities for energy efficiency improvements in buildings and industrial processes, reducing overall energy demand and increasing the effectiveness of renewable energy integration.

**Customer Engagement**: AI can enable utilities to engage with customers by providing real-time information on their energy consumption, carbon footprint, and personalized tips for reducing energy use and costs.

In summary, AI plays a pivotal role in the seamless integration of renewable energy into the energy ecosystem by improving forecasting, grid management, storage, and overall efficiency. As renewable energy sources become more prevalent, AI technologies will continue to evolve and play an increasingly important role in achieving a sustainable and resilient energy future.

## Role of AI in Grid Monitoring and Management

AI plays a crucial role in reducing and anticipating prospective grid failures and blackouts by analyzing past data on grid failures and detecting trends preceding such incidents. The integration of AI technologies into grid management helps improve efficiency, reliability, and sustainability. Here are some key roles of AI in grid management:

**Grid Optimization:** AI can optimize the operation of various grid components, such as transformers, voltage regulators, and capacitor banks. By continuously adjusting settings based on real-time data and predictive analytics, AI improves grid efficiency and reduces energy losses.

**Energy Storage Management:** AI algorithms can optimize the charging and discharging of energy storage systems, such as batteries. This helps store excess energy when supply exceeds demand and releases it when needed, improving grid resilience and supporting renewable energy integration.

**Grid Security:** AI can enhance grid security by identifying anomalies and potential threats in real-time. It can detect cyber-attacks, physical intrusions, and unusual grid behavior, allowing for a rapid response to protect critical infrastructure.

**Grid Planning and Design:** AI tools can assist in the planning and design of new grid infrastructure. By analyzing geographical and demographic data, AI helps optimize the placement of substations, power lines, and other assets to meet future demand and increase grid resilience.

**Grid Resilience:** AI-based monitoring and control systems can respond quickly to grid disturbances, rerouting power flows, and minimizing the impact of disruptions. This enhances the resilience of the grid against extreme weather events and other challenges.

In summary, AI plays a crucial role in modern grid management by optimizing operations, enhancing reliability, integrating renewable energy sources, and improving overall grid resilience. As the energy landscape continues to evolve, AI will become increasingly important in ensuring the efficient and sustainable functioning of electrical grids.

## Role of AI in Demand Response

Demand Response (DR), a practical tactic to raise the energy systems' sustainability while lowering associated costs, has grown in popularity recently. Despite this, due to the high complexity of DR-related tasks and the massive amount of data management required to make decisions with nearly real-time consequences, artificial intelligence (AI) and machine learning (ML) have recently emerged as critical technologies for demand-side management and response. Problems that can be solved with the aid of DR include choosing the best group of users to respond to, learning their priorities and consumption attitudes, optimizing prices, monitoring and controlling devices, increasing consumer participation in DR schemes, and learning how to compensate them fairly and economically (Tabassum, N, et.al, 2020).

The effective execution of DR programmes requires taking into account a variety of electrical power system parameters, such as load and energy price estimates, choosing the best customers for DR schemes, and building automated systems for managing demand-side resources. AI technologies have been applied in DR in a variety of domains, including forecasting, real-time management of networked infrastructure, making decisions that are very close to optimal, adapting to dynamic behaviour, learning from load profiles, and many others. Here, we classify the various AI-related applications for Demand Response.Figure.5 shows the role of AI in Demand Response (Khan 2023).

*Figure 5. Role of AI in demand response*

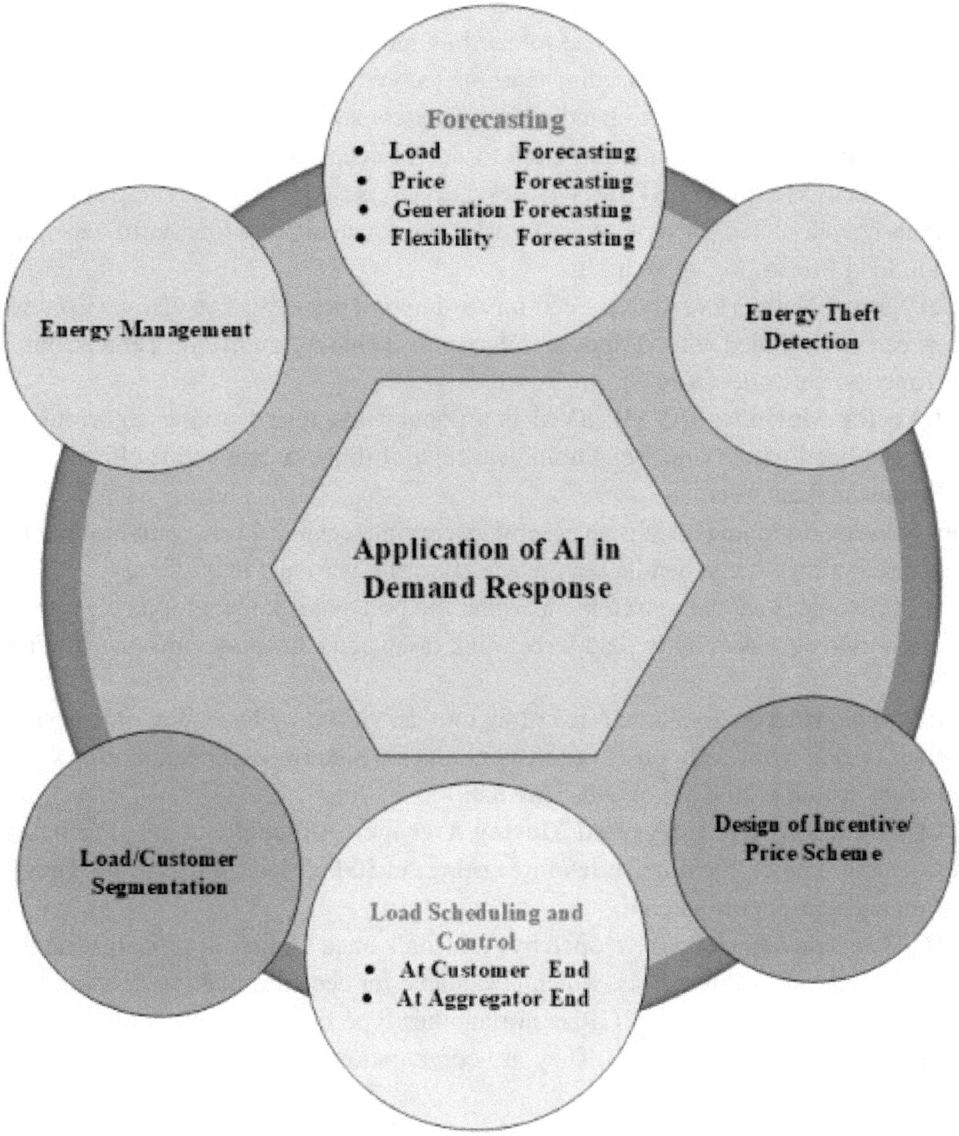


**Load Forecasting:** Load forecasting is indeed a critical component of managing and operating an electricity grid efficiently. It involves predicting the future electricity demand accurately so that grid operators can plan and allocate resources effectively. Long-term load forecasting typically covers periods ranging from months to years into the future and aids in strategic planning, while short-term forecasting (Tabassum,2022) focuses on predicting electricity demand for the next 24 hours or less and is crucial for day-to-day grid management including dispatching power generation, managing grid stability, and responding to sudden changes in demand. Accurate forecasting is also integral to the success of demand response programs, which can help balance supply and demand, reduce costs, and enhance grid reliability. Demand response programs aim to incentivize consumers to adjust their electricity consumption during periods of high demand or grid stress.

This can include shifting energy use to off-peak hours (Tabassum 2022a), reducing energy consumption temporarily, or using on-site generation or energy storage. Advances in data analytics and machine learning have improved the accuracy of load forecasting models, making them even more valuable tools for grid operators. AI programs and machine learning techniques have become increasingly popular for load forecasting due to their ability to handle large datasets and complex patterns. Here are some AI-based approaches and programs commonly used for load forecasting:

**Long Short-Term Memory (LSTM) Networks**: LSTM is a type of recurrent neural network (RNN) that is particularly well-suited for time series forecasting. It can capture long-term dependencies and is widely used in load forecasting applications.

**Artificial Neural Networks (ANN)**: Feed forward neural networks can also be used for load forecasting. They can be designed with various architectures, including multilayer perceptrons (MLPs) and radial basis function networks (RBFNs).

**Support Vector Machines (SVM)**: SVM is a popular machine learning algorithm used in load forecasting. It can handle both linear and nonlinear relationships in data and is effective when dealing with high-dimensional datasets.

**Random Forests**: Random Forest is an ensemble learning technique that combines multiple decision trees to make predictions. It can handle feature selection and is robust to outliers.

XGBoost: XGBoost is another ensemble learning method that is known for its high predictive accuracy. It has been used successfully in load forecasting tasks and can handle missing data and nonlinear relationships.

**Prophet**: Prophet is an open-source forecasting tool developed by Facebook. It is designed for forecasting time series data with daily observations that display patterns on different time scales. It can be used for load forecasting with daily or sub-daily data.

**ARIMA (Auto Regressive Integrated Moving Average)**: Although not an AI technique, ARIMA models are a classic choice for time series forecasting, including load forecasting. They can capture autocorrelation and seasonality in data.

**Deep AR**: This is an algorithm developed by Amazon's Sage Maker for probabilistic forecasting. It can model uncertainty in load forecasts, which is valuable for decision-making in energy management.

**Gated Recurrent Units (GRUs)**: GRUs are another type of recurrent neural network similar to LSTMs but with a simpler architecture. They are computationally efficient and can be used for load forecasting tasks.

**Hybrid Models**: Combining multiple forecasting methods, such as combining classical statistical models like ARIMA with machine learning techniques like neural networks, can often lead to improved accuracy.

**Custom Solutions**: Many utilities and energy companies develop custom AI programs and models tailored to their specific load forecasting needs. These solutions often involve domain expertise and feature engineering.

When implementing AI programs for load forecasting, it's essential to consider factors like data quality, feature selection, model tuning, and the specific requirements of your application. Moreover, you may need to continuously retrain and update your models to adapt to changing load patterns and external factors like weather conditions and economic trends (Tabassum, N, et.al, 2017).

## Role of AI in Price Forecasting

AI plays a crucial role in price forecasting in smart grids by leveraging advanced algorithms, data analytics, and machine learning techniques to make accurate predictions about electricity prices **(Aguiar-Pérez 2023).** Price forecasting in smart grids is essential for efficient energy management, grid stability, and cost optimization.

Supervised machine learning algorithms, such as regression, time series analysis, and neural networks, are commonly used for price forecasting. Time series forecasting models, like ARIMA (Auto Regressive Integrated Moving Average) and LSTM (Long Short-Term Memory), are well-suited for capturing temporal patterns in electricity prices. AI models can predict electricity demand patterns, including peak demand times, which help utilities optimize energy production and pricing strategies. AI models often include uncertainty analysis to provide confidence intervals or probabilistic forecasts, as electricity prices can be highly volatile and subject to unforeseen events. Smart grids require real-time updates to adapt to changing conditions. AI models can continuously update forecasts based on new data, ensuring accuracy and adaptability. AI-driven price forecasts help grid operators optimize energy distribution, manage peak loads, and reduce operational costs. The forecasts generated by AI systems can assist energy market participants, consumers, and utilities in making informed decisions about energy consumption and generation, energy trading, and grid management. AI-based price forecasts are used by energy traders to make buying and selling decisions in real-time energy markets, maximizing profit while minimizing risk. AI can also predict maintenance needs for grid infrastructure based on usage patterns and other factors, preventing costly downtime.

Overall, AI-driven price forecasting in smart grids helps create more efficient, sustainable, and reliable energy systems. These forecasts empower consumers to make informed choices about their energy consumption, support grid operators in maintaining stability, and enable utilities to optimize their operations and pricing strategies in a rapidly changing energy landscape.

## Electricity Load and Price Forecasting for Demand Side Management Using AI

Electricity load and price forecasting for demand-side management (DSM) using artificial intelligence (AI) is a crucial application in the energy sector. It helps utilities, grid operators, and consumers optimize energy consumption, reduce costs, and enhance grid reliability. The key steps and considerations for implementing AI-based load and price forecasting for DSM are:

Data Collection and Pre-processing:

Gather historical data on electricity consumption, weather conditions, market prices, and other relevant factors. Clean and pre-process the data to handle missing values, outliers, and inconsistencies.

Feature Engineering:

Create relevant features from the data, such as time of day, day of the week, holidays, temperature, and special events. Consider lagged variables, such as past load and price values, as they can be informative for forecasting.

Model Selection:

Choose appropriate AI models for load and price forecasting. Common choices include, Time series forecasting models (e.g., ARIMA, SARIMA, Prophet), Machine learning models (e.g., decision trees, random forests, gradient boosting, support vector machines), Deep learning models (e.g., recurrent neural networks (RNNs), long short-term memory networks (LSTMs), transformers)

Training Data Splitting:

Divide the data into training, validation, and test sets for model evaluation. Use cross-validation techniques to fine-tune hyper parameters and assess model performance.

Model Training:

Train the selected models using historical data. Experiment with different model architectures and hyper parameters to find the best-performing ones.

Evaluation and Validation:

Evaluate models on the validation set using appropriate metrics (e.g., Mean Absolute Error (MAE), Root Mean Square Error (RMSE), Mean Absolute Percentage Error (MAPE)). Validate the models' performance on out-of-sample data (test set) to ensure they generalize well.

Ensemble Models:

Consider using ensemble techniques like bagging or stacking to combine the predictions from multiple models, improving forecasting accuracy.

Incorporate External Factors:

Integrate external factors like market data, renewable energy generation forecasts, and demand response programs into your forecasting models for more accurate predictions.

Real-time Data Integration:

Implement mechanisms to update the forecasting models with real-time data to adapt to changing conditions.

Price Forecasting:

For electricity price forecasting, consider market-specific factors, such as supply-demand dynamics, generation mix, and regulatory changes. Explore market-specific models, like supply-demand equilibrium models or agent-based models.

Deployment:

Deploy the trained models in a production environment to provide real-time load and price forecasts. Implement monitoring and alerting systems to detect anomalies and model degradation.

Feedback Loop:

Continuously monitor model performance and retrain those periodically using fresh data to ensure accuracy and reliability.

User Interfaces and Decision Support:

Develop user-friendly interfaces or dashboards for utilities, grid operators, and consumers to access and act on the forecasts.

Demand Response Integration:

Use load forecasts to implement demand response programs effectively, optimizing energy consumption during peak periods.

Scalability and Robustness:

Design the forecasting system to scale with increasing data volume and ensure it can handle unexpected disruptions or data anomalies.

AI-based load and price forecasting for demand-side management can significantly enhance the efficiency and sustainability of the electricity grid while providing economic benefits for both utilities and consumers. However, it's essential to continuously improve and adapt the models to changing market dynamics and technological advancements.

## Role of Block Chain in Demand Response

Blockchain technology can play a significant role in enhancing the demand response capabilities of smart grids. Demand response in smart grids refers to the ability to adjust electricity consumption in response to changes in electricity prices, grid conditions, or other factors (Khan, 2022). Blockchain can facilitate this process in several ways.

Blockchain provides a secure and tamper-proof ledger for recording and sharing data related to energy consumption and demand response events. This ensures the integrity and confidentiality of sensitive information. Blockchain operates on a decentralized network of nodes, reducing the risk of a single point of failure. This ensures that demand response data and actions are more resilient and less susceptible to cyber attacks. The transparent nature of blockchain allows all parties involved in the energy ecosystem (consumers, utilities, regulators) to access a shared ledger of demand response events and verify the accuracy of transactions. This builds trust among participants. Blockchain can enable peer-to-peer (P2P) energy trading, where consumers with excess energy can sell it directly to others on the grid. This can incentivize demand response actions by providing financial benefits to consumers who reduce their consumption during peak periods. Once data is recorded on the blockchain, it cannot be altered or deleted. This feature ensures the integrity of historical demand response data, which can be valuable for auditing, compliance, and analysis. Blockchain can enable real-time monitoring of energy consumption and demand response events, allowing utilities to make more informed decisions about grid management and load balancing. Blockchain can facilitate interoperability between various devices and systems within the smart grid ecosystem, ensuring that data and commands can be exchanged seamlessly between different stakeholders and technologies. Blockchain can support tokenization of energy or incentive mechanisms such as token rewards for participating in demand response programs, encouraging more consumers to participate actively. By aggregating and analyzing data from various participants on the blockchain, utilities can gain insights into demand patterns and optimize grid operations for improved efficiency and reliability.

In summary, blockchain technology can enhance the efficiency, security, and transparency of demand response in smart grids. It enables real-time data sharing, automated smart contracts, and peer-to-peer energy trading, ultimately contributing to a more reliable and sustainable energy ecosystem. However, it's important to note that while blockchain offers many advantages, its implementation also poses technical and regulatory challenges that need to be carefully addressed.

## Role of Block Chain in Data Management and Security

Blockchain technology plays a crucial role in data management and security by providing a decentralized, immutable, and tamper-resistant ledger that can be used to store and manage data in a secure and transparent manner. Here are several ways in which blockchain contributes to data management and security:

**Data Integrity**: Blockchain uses cryptographic hashes to secure data blocks. Once a block of data is added to the chain, it becomes virtually impossible to alter or delete that data without consensus from the network. This ensures data integrity and prevents unauthorized modifications.

**Immutability**: Data stored on a blockchain is immutable, meaning it cannot be changed retroactively. This property is particularly valuable for maintaining the integrity of historical records and audit trails.

**Decentralization**: Traditional data management systems often rely on a central authority or intermediary, making them vulnerable to single points of failure and potential security breaches. In contrast, blockchain operates on a decentralized network of nodes, reducing the risk of data breaches and ensuring data availability even if some nodes fail.

**Data Ownership and Control**: Blockchain enables individuals and organizations to have greater control over their data. Users can grant permission for specific entities to access their data, reducing the risk of unauthorized access and data breaches.

**Transparency and Auditability**: Blockchain transactions are recorded on a public ledger that is visible to all participants in the network. This transparency enhances trust and accountability in data management processes. Users can easily verify the history of data transactions, making it useful for auditing purposes.

**Data Encryption**: Blockchain networks often incorporate robust encryption techniques to secure data both on and off the chain. This protects sensitive data from unauthorized access and ensures that only authorized parties can view and interact with it.

**Data Sharing and Interoperability**: Blockchain can facilitate secure data sharing and interoperability between different organizations and systems. Participants can exchange data with confidence, knowing that the block chain's security features protect it during transit and at rest.

**Supply Chain Management**: In industries like logistics and supply chain, blockchain helps trace the origin and journey of products, ensuring the authenticity of goods and reducing the risk of counterfeiting and fraud.

While blockchain offers significant advantages for data management and security, it's important to note that it is not a one-size-fits-all solution. The technology has its limitations and may not be suitable for all use cases. Additionally, the implementation of blockchain should be carefully considered, and organizations must weigh the benefits against the associated costs and complexities.

## Role of Blockchain in Energy Trading

Research and discussion on the connection between centralized energy systems and climate change are quickly gaining popularity. Numerous potential solutions, including distributed energy resources (DERs), renewable energy sources, and blockchain-based peer-to-peer (P2P) energy trade (Boumaiza, 2023), will undoubtedly be put forth for such a significant problem. Dr. Hongbiao Li and colleagues' study on blockchain energy solutions shows that the centralized electricity grid has a number of important problems. These include its limited fault tolerance and energy loss during long-distance transmissions. A decentralized smart grid serves as an alternative to a conventionally centralized system. Blockchain-based decentralized smart grids could make payments instantly and dynamically adjust to changes in supply and demand. The cost of renewable energy can be reduced, energy efficiency can be increased, climate change can be stopped, and much more precise measurements of energy generation and consumption may be made by resolving the energy transaction difficulties.

## A P2P Energy Trading System Using Blockchain

Direct electricity trading between two parties on a connected grid is known as P2P trade. Blockchain provides a safe environment for peer-to-peer trade and keeps account of the exchange of assets like energy units. Blockchain is a distributed ledger technology that uses a decentralized network to transparently store data. It is applied to asset tracking and transaction monitoring. Multiple transactions are stored in each block of the chain in a fashion that is simple to audit and verify but exceedingly difficult to amend or change. Blockchain has the potential to be a very alluring solution for distributed energy firms when combined with smart contracts, which automate the buying and selling of energy based on supply and demand.

In addition to enabling extremely localized energy production, it might substantially simplify the process of buying and selling power. Energy firms that are already in operation can also join, but they will need to modify their current business models and technological infrastructure. Energy customers could select the supplier from which they get their power, which is one potential advantage of a P2P trading system. For instance, they could designate the kind of energy source (green, etc.) and the business type (independent, big, government, etc.) from which they purchase energy. In order to make wise purchases, they can also view current prices. The consumer cost might be decreased by getting rid of "middle-men" power brokers because the system would be entirely automated.

## Block Chain Trading and Renewable Energy

One of the most prevalent and accessible DERs is solar energy. Individuals without solar panels might purchase extra green energy from their neighbours in a P2P trading system. Since it can react more quickly to local needs, a localized microgrid reduces many of the inefficiencies of a monolithic centralized power generation system. Additionally, local power production lowers the amount of power lost along transmission lines. DERs can cut expenses, lessen waste, and safeguard the environment. Renewable energy sources are regularly criticized for their reliance on particular circumstances. Solar power need sunlight, and windmills require wind. By making it much simpler for people with excess power to sell it to those who need it, blockchain trade offers a solution to these issues. Participants in a microgrid can simply buy power from other sources when renewable energy sources are unable to produce it at night or on a windless day. Although biomass, geothermal, and hydropower are also reliable sources of green energy that could profit from blockchain trade, solar and wind are the two most well-known DERs.

Energy trading using blockchain technology has the potential to transform the energy sector by enabling peer-to-peer energy transactions, increasing transparency, and reducing the need for intermediaries. However, it also faces several limitations and challenges:

**Scalability**: Blockchain networks, especially public ones like Ethereum, can struggle with scalability issues, leading to slow transaction processing times and high fees during periods of high demand. This can be a significant problem for energy trading, where real-time transactions are crucial.

**Energy Consumption**: Proof-of-work (PoW) blockchains like Bitcoin and Ethereum are energy-intensive due to the computational power required for mining. This contradicts the goal of creating a more sustainable and energy-efficient energy system.

**Regulatory Hurdles**: The energy sector is heavily regulated, and integrating blockchain technology can be challenging due to these regulations, which vary from country to country. It can be difficult to follow these rules and preserve the advantages of decentralisation.

**Lack of Standardization**: The lack of standardized protocols and formats for energy data can be a hurdle. Standardization is crucial for ensuring compatibility and smooth data exchange between different energy systems and blockchain networks.

**Data Privacy**: Energy consumption data is highly sensitive and private. Storing such data on a public blockchain may raise concerns about privacy and data protection, especially with the advent of stricter data privacy regulations like GDPR.

**User Adoption**: Convincing consumers and energy producers to adopt blockchain-based energy trading platforms can be challenging. Users may be skeptical or unwilling to change their existing habits and systems.

**Costs and Investment**: Implementing blockchain technology in the energy sector requires significant investment in infrastructure, technology, and personnel. Smaller energy players may find it cost-prohibitive.

**Smart Grid Integration**: Integrating blockchain with existing smart grid infrastructure can be challenging. Legacy systems may not be compatible with blockchain technology, requiring costly upgrades.

Despite these limitations and challenges, many organizations and governments are actively exploring blockchain solutions for energy trading. As technology evolves and these challenges are addressed, blockchain may become a more integral part of the energy sector, offering benefits such as increased efficiency, transparency, and decentralization. However, it will require careful planning, regulation, and investment to overcome these obstacles and reach its full potential.

## Role of Blockchain and AI's In the Smart Grid's Ability to Enable Consumers

The integration of energy supply, demand, and renewable sources into an autonomously controlled power grid, governed by smart software that optimizes decision-making and operations, will be made possible by a digitalized model powered by artificial intelligence and block chain technology.

The net zero emission goals set by governments (Hua, 2022) is to boost the proportion of renewable energy sources while also influencing consumer behaviour to help balance energy supply and demand in an efficient manner. The Smart grids' sophisticated information and control infrastructures, that enable interactiveness among diverse stakeholders, will be used to achieve these. Due to the fact that more customers are producing, storing, and using energy, they are now playing the role of prosumers. Prosumers integration and accommodating incurred energy and information flows depend on two essential elements: flexible energy market arrangements and intelligent power system operations. Creative technologies such as block chain and artificial intelligence (AI) can help in achieving the requirements, since the block chain provides distributed merchandise programme for the energy markets and the AI aids in the most effective control of power systems. Block chain technology which is one of the distributed ledger technology, has the potential to create a platform for decentralized power trading. By recording the ownership of these assets in the energy markets, the block chain can prevent a replay attack and double-spend attack, which happen when the identical energy is sold twice or the identical digital currency is spent twice. All participants in the energy market can hold and verify a ledger thanks to the decentralized nature of the block chain As a result; prosumers, system and market operators can access the trading platform. The block chain's disintermediation feature transforms the function of energy providers or aggregators as an impartial facilitator to encourage prosumers participation. Prosumers' personal information, including addresses, businesses, and energy profiles, are protected by the block chain's encryption. The security of trading networks is ensured by the block mining's computational difficulty and collective validation for reaching a consensus. A significant amount of relevant data is produced by the sophisticated metering infrastructures used in smart grids. The AI might use this information to enhance the circumstantial awareness and functionality of electrical systems. This is especially helpful for medium-sized and smaller prosumers that operate power systems and make decisions on their own due to the constrained budgets of their control systems. There are four stages to which the integration of AI into power system operations can proceed. The typical operations of power systems are featured in the first level, or response level, where analytical AI helps with situational awareness, problem detection, and power system restoration after getting a call about outages. Deep AI development and power system digitalization enable the move on to the next level, or anticipatory level, where more analytics based on AI and aids for decision-making are added to forecast real-time generation, demand, and uncertainty. This is done in order to sustain overall outcomes, such as stability, capacity margin, and flexibility. At the third level, or prescriptive level, with the help of AI-based software, a number of actions from the first two levels can be automatically carried out in order to prevent interruptions and failures from a systematic perspective. Future system operations would be fully automated, the system can sustain self-healing at the fourth level, or independent level, where wide-area governing decisions and network optimizations could be provided by an AI-based digitized layer without the involvement of system operators.

## Role of Block Chain and Artificial Intelligence for Cyber Attacks in Smart Grid

Among the most difficult things to prevent is a cyber attack (Moloth, 2023) on the smart grid. The main issue is brought on by the millions of sensors that are continually transmitting and receiving network packets of data. Cyber attacks might jeopardize the privacy, availability, and dependability of the smart grid. An innovative network grid is vulnerable to cyber attacks on three levels: users, the network of smart devices and sensors used for communication, and network administrators. Since the smart grid has data transfer and reception abilities, sensors are able to link to it and retrieve its data. The grid collects data packets from these sensors on a constant basis. Production-related data could be present in these data packet that gives information about generation, utilization, voltage, and frequency of energy. Current battery-integrated grids use a communication route to convey charge status, which leaves the battery

*Figure 6. AI in cyber security*

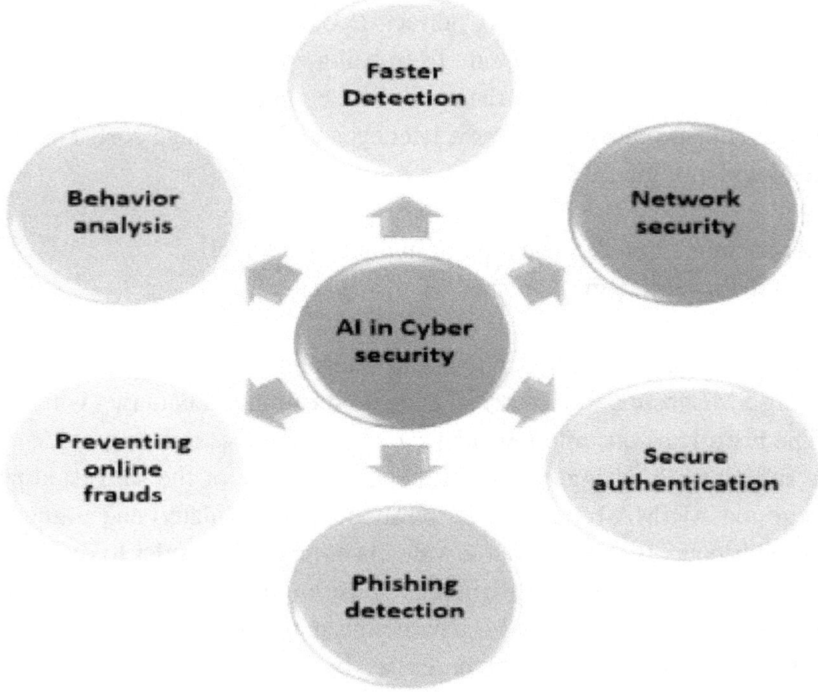

management system open to attack. Overcharged or undercharged batteries could completely lose their worth due to cyber dangers. The two main issues in smart grid are cyber security and complexity. It is more difficult to fix these errors when smart grid data is stored in the cloud.

The smart grid relies on both physical and cyber security to ensure its reliability and safety at all times. One of the most typical methods of attacking a smart grid is jamming. Communication can be obstructed by an attacker by using erratic or persistent signals. Different types of jammers, such as persistent, random, deceptive, and reactive jammers, could interfere with the working of the intelligent grid network. The application of AI in cyber security is common. Recent advances in AI have considerably increased the digital safety of intelligent grids. The defences against different dangers are strengthened by these upgrades. Security (including the detection of fraud and viruses), confidentiality, business, and IT are the top five applications of machine learning. Figure.6 shows the role of AI in cyber security.

For community detection, the data layer gathers data and sends it to the detection layer. Additionally, the community detection and transaction record are secured by the block chain layer. Smart metres operate as controller nodes, initiating all communications with the service provider. By auditing and broadcasting transactions, data is distributed and validated across a network. In a block chain setting, smart contracts and the lightning network are used to create a decentralized security paradigm. The enrolment, planning, authentication, and payment procedures are a few of the various processes.

## Case Study

The study below discusses a data analytics scheme (ArMor) to identify malicious behaviours in a block chain based smart grid system (Aparna, 2020).

Real-time data integrity problems, such as Smart Meter failure and fake data injection attacks, are identified by the Armour. In this research, the clients are categorised, and a novel harmful activity detection model based on ARIMA was suggested along with an incentive scheme for utility providers to handle the harmful activity on their end using Smart Contracts (SCs). Since transactional data is secured using SC after it is saved in BC, it stops harmful data from being entered into the SG system. To determine whether the suggested scheme is effective, the results are compared to the state-of-the-art methods in terms of characteristics like prediction accuracy, latency, and data storage cost. The two-stage Armour system model is depicted in Figure 7.

i. Customer classification and identification of malicious conduct
ii. Block Chain-based incentive mechanism

The first step is to find the harmful behaviour and determine whether or not it has affected an end-consumer ($E_C$). Using SM/sensors, the historical energy consumption data ($\psi_{hk}$) of each $E_C$ is gathered for this purpose. The historical consumption values are first pre processed to provide clean, normalised data, and they are subsequently changed to meet the requirements of the suggested prediction model. After that, the suggested ARIMA-based prediction model was validated and trained using historical data on energy use. Subsequently, the threshold value is computed in order to distinguish the attacked or non-attacked client and determine the malicious consumption behaviour. Additionally, the utility providers ($E_{UP}$) were notified by the client whose malicious data was sent to the Block Chain for storage.

In the event that the harmful data has been processed by the $E_{UP}$, the incentive mechanism is part of the second step. There are several potential causes for the harmful data to arise such as malfunctioning

*Figure 7. ArMor system model*

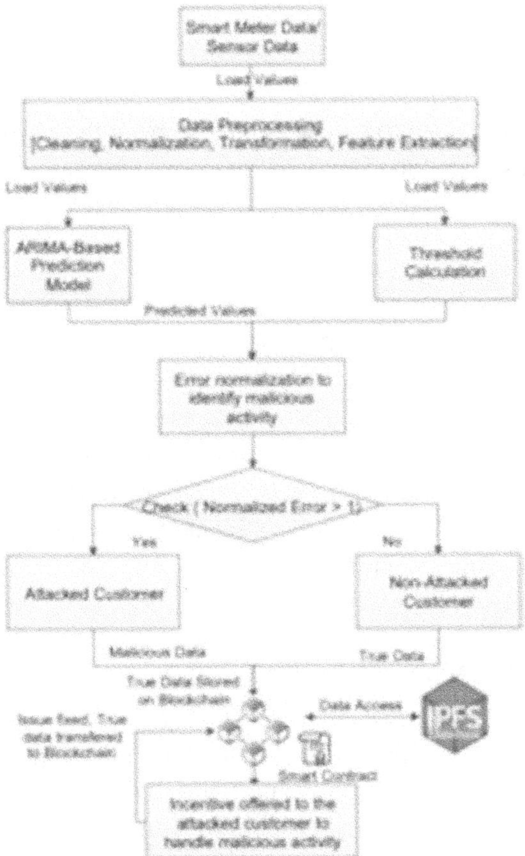


physical devices (SM, sensors, actuators), problems with the network, data assaults (such as data injection and data integrity attacks), etc. The $E_{UP}$ has several options for dealing with malicious activities, including upgrading the system software to handle data attacks, replacing or repairing malfunctioning physical components, or using other strategies. All stake holders involved like $E_{SG}$, $E_{UP}$, and $E_C$—may gain from this issue management in terms of improved Quality of Service and Quality of Experience.

Mathematical Model:

Taking into account a particular client, $E_C^i$, who possesses past consumption information $\psi^{hi}$. The goal is to identify if the received data is malicious or normal by predicting the next consumption ($\psi$) and comparing it to the actual consumption ($\psi^{ai}$).

Equation 1 is used to predict consumption using the ARIMA model.

$$\psi_t = \lambda + \theta_1\psi_{t-1} + \theta_2\psi_{t-2} + \dots + \theta_p\psi_{t-p\mu t} + \delta_1\mu_{t-1} + \delta_2\mu_{t-2} + \dots + \delta_q q\mu_{t-q} \quad (1)$$

Here, $\psi_i$ denotes the series' lag i, $\theta_i$ its coefficient, $\mu_i$ its ith error, $\delta_i$ its coefficient of error i, and $\lambda$ its intercept. The orders p and q, respectively, reflect the autoregressive term and moving average order.

ARIMA is used to forecast consumption, and prediction error is computed. Equation 2 is used to determine the error $\varepsilon_{ak}$ by comparing the actual consumption values ($\psi^{ak}$) with the anticipated values ($\psi_{pred}$).

$$\varepsilon_{ak} = |\psi - \psi^{ak}| \tag{2}$$

Additionally, incentives are provided to $E_{UP}$ by SG in the event that the hostile behaviour is resolved following its discovery and classification. The 5G-TI wireless communication channel (WCC) is used for communication between $E_{UP}$, $E_C$, and $E_{SG}$ entities.

In light of the discussion above, the ArMor's objective function is described by equation 3

$$O = \max \sum_{i=1}^{3} \text{Security}\{Wi\} + \min(\text{CostÈ}) + \min(L) \tag{3}$$

where L indicates the communication latency, Cost$\psi$ displays the cost of storing energy data, and security $\rightarrow$ {Wi} represents the wallet's security via Ethereum. Here, the optimisation goals are to reduce latency while identifying dangerous behaviours on a blockchain-based SG system, optimise security, and minimise data storage costs during analytics.

## RESULTS AND DISCUSSIONS

The Armour system, which consists of 100 buildings and incorporates historical data on residential energy consumption, is assessed. The information includes hourly energy consumption figures for a given year expressed in kilowatt-per-hour (kWh). The Armour model was used to detect harmful behaviour in the simulated malicious activities by randomly injecting synthetic data. Armour parameters, including prediction accuracy (0.9555), precision (0.7143), recall (1.0), and f1 score (0.8) were calculated and compared with the current methods.

The Remix Integrated Development Environment (IDE) compiles the SC for the incentive system and generates byte code It is then installed in the Remix testing environment in order to examine its vulnerabilities. Figure 8 depicts the Armour scheme's IDE interface and includes a number of functions

*Figure 8. Armour modules for implementing smart contracts*

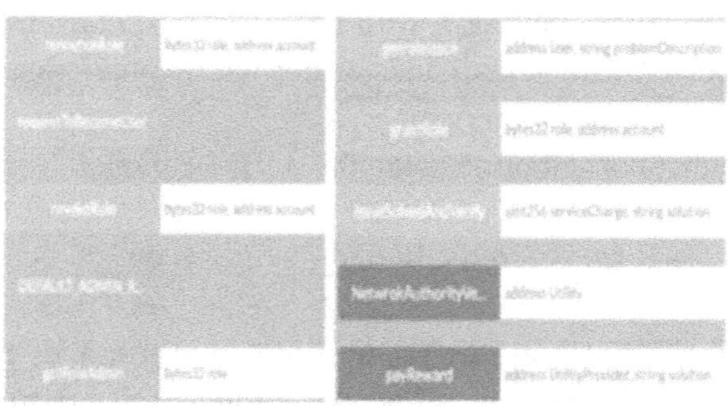

a.Registration Interface b.Rewards Interface

with corresponding input parameters. Figure 8a illustrates the SC-based registration interface for utility suppliers on the BC, while Figure 8b oversees the incentives system that offers rewards for resolving difficulties (either by updating software or replacing hardware). An extensive report is created and distributed to the SG administrator.

## SUMMARY

Due to numerous attacks, including data integrity attacks and fraudulent data injection, the SG's vital infrastructure is experiencing an increase in security problems. One of the difficult jobs to raise QoS and QoE at SG is the detection of these assaults or malicious activities. In this work, a data analytics plan called Armour to spot hostile activity in the SG system was proposed and discussed. The performance of ArMor in terms of various parameters like prediction accuracy, latency, and data storage cost that showed the efficacy of the proposed scheme.

## Challenges in Implementing AI And Block Chain in Smart Grids

Implementing AI and blockchain in smart grids presents several challenges, despite the potential benefits they can bring to the energy sector.

Limitations of block chain Technology (Younes, 2021):

- The size of the distributed ledger expands in large-scale blockchain applications, increasing the amount of chain-required data. This deficiency may cause the process of mining the blockchain to take longer each time the hash address is changed and data is saved.
- Users and producers may find it difficult to come to an agreement over how to pay for this energy use due to the significant power demand from the advanced metering infrastructure and block-chain mining facilities.
- One significant issue that poses a threat to cyber security in smart grid applications is the scalability of the blockchain.
- The interval time, which continues to be a crucially difficult problem in the blockchain application that requires time to process, is inextricably tied to the smart grid operations.
- The sharing user's ledger typically uses pseudonyms and untraceable addresses to carry out unidentified transactions.
- For the blockchain design scheme for the smart grid to be effective and reliable, several situations relating to the network system and its requirements are required.

Limitations of AI:

The complexity of the system is one of the main obstacles to adopting AI in Smart Grid management. Because Smart Grids are extremely interconnected networks, AI algorithms need to be capable of processing vast amounts of data from various sources accurately.

*Figure 9. Accountability challenges of AI*

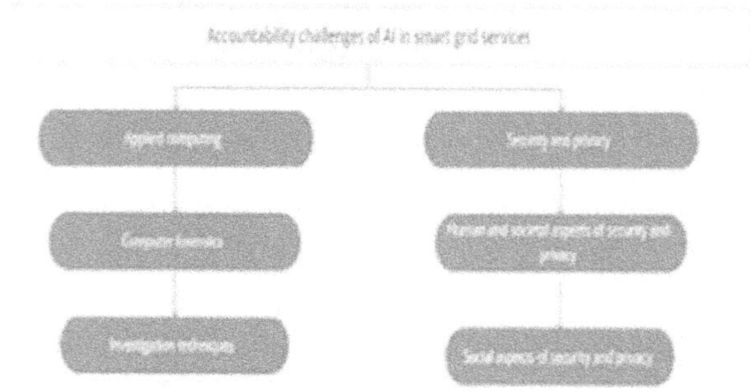


## Accountability Challenges of AI In Smart Grid Services

AI in smart grid services offers numerous benefits, such as improved efficiency, reliability, and sustainability. However, it also poses several accountability challenges **(Volkova, 2023)** that need to be addressed. Figure.9 shows the accountability challenges of AI and are discussed below:

Data Privacy and Security:

Smart grids collect vast amounts of data from various sources, including consumers' energy usage patterns. Ensuring the privacy and security of this data is a significant challenge. AI systems must be designed to protect sensitive information from breaches or unauthorized access.

Bias and Fairness:

AI algorithms can inherit biases present in their training data. In the context of smart grids, biased algorithms could lead to unfair energy distribution, disadvantaging certain communities or demographics. Ensuring fairness and equity in AI decision-making is crucial.

Transparency and Explainability:

Many AI algorithms, especially deep learning models, are often considered "black boxes" because their decision-making processes are difficult to interpret. In the smart grid, it's essential to make AI decisions transparent and explainable to gain trust and accountability.

Accountability for Errors:

AI systems can make mistakes, and when they do, it's crucial to determine who is responsible. Is it the AI developer, the utility company, or the regulator? Establishing clear lines of accountability for errors or failures is a complex issue.

Regulation and Compliance:

Regulatory bodies may struggle to keep pace with the rapid advancements in AI technology. Ensuring that AI systems in smart grid services comply with existing regulations and standards can be challenging, and there may be a need for new regulations specific to AI in the energy sector.

Dependence on AI:

As smart grid services become more reliant on AI for critical functions, there's a risk of over-dependence. In the event of AI system failures or cyber attacks, the grid's stability could be compromised. Creating backup systems and redundancy mechanisms is essential.

Algorithmic Accountability:

Establishing clear accountability for the performance and impact of AI algorithms in smart grid services is a challenge. When something goes wrong or when decisions have unintended consequences, determining who is responsible can be complex.

Ethical Considerations:

AI decisions in smart grids may involve ethical considerations, such as prioritizing energy distribution during emergencies or ensuring vulnerable populations have access to affordable energy. Balancing these ethical concerns with technical requirements can be challenging.

Human Oversight:

While AI can automate many tasks in smart grid management, there must be adequate human oversight. Defining the roles and responsibilities of human operators and AI systems and ensuring they work together effectively is essential.

Accountability in Decision-Making:

Decisions made by AI systems can have far-reaching consequences in smart grid services. Establishing accountability mechanisms for these decisions, especially in complex, interconnected systems, is a significant challenge.

Addressing these accountability challenges requires collaboration between stakeholders, including AI developers, utilities, regulators, and consumers. Additionally, ongoing research, policy development, and technology standards are essential to ensure that AI in smart grid services operates responsibly and ethically while maximizing its benefits.

## CHALLENGES OF AI AND BLOCK CHAIN IN SMART GRIDS

**Interoperability**: Smart grids often consist of diverse systems and technologies from different vendors. Ensuring interoperability between these systems, AI solutions, and blockchain networks can be complex and costly. Standards and protocols need to be established to address this challenge.

**Scalability**: As the smart grid infrastructure expands, the AI and blockchain systems need to scale accordingly. Maintaining performance and efficiency while handling a growing volume of data and transactions can be challenging.

**Security**: Smart grids are critical infrastructure, and security is a top concern. While blockchain offers security benefits through its decentralized and immutable ledger, it's not immune to attacks. Ensuring the security of AI models and blockchain networks is crucial to protect against cyber threats.

**Energy Consumption**: Blockchain, particularly proof-of-work-based systems like Bitcoin, can consume a significant amount of energy. This could be counterproductive in a smart grid context, where energy efficiency is a primary goal. Research into more energy-efficient consensus mechanisms for blockchain is ongoing.

**Regulatory and Legal Frameworks**: The integration of AI and blockchain in smart grids may face regulatory and legal hurdles. Governments and regulatory bodies need to develop frameworks that ensure compliance, data privacy, and security while enabling innovation.

**Privacy Concerns**: Smart grids collect detailed data about energy consumption patterns, which can raise privacy concerns. Balancing the need for data-driven insights with individual privacy rights is a challenge that requires careful consideration.

**Education and Workforce Skills**: Implementing AI and blockchain in smart grids requires a skilled workforce capable of developing, maintaining, and securing these systems. The shortage of qualified professionals in these fields can be a barrier to adoption.

**Costs and ROI**: The initial investment required for implementing AI and blockchain technologies in smart grids can be substantial. Organizations must carefully assess the costs and expected return on investment to justify these implementations.

**Resistance to Change**: Resistance to change within organizations and the energy industry as a whole can hinder the adoption of new technologies. Stakeholders may be hesitant to embrace AI and blockchain due to fear of job displacement or other factors.

**Reliability and Fault Tolerance**: Smart grids must operate reliably, even in the face of network disruptions or equipment failures. Ensuring the reliability and fault tolerance of AI and blockchain systems is crucial to prevent service disruptions.

Addressing these challenges requires collaboration between technology providers, utilities, regulatory bodies, and other stakeholders. It also necessitates ongoing research and development to improve the efficiency, security, and scalability of AI and blockchain solutions in smart grids.

## CONCLUSION

The integration of Artificial Intelligence (AI) and Blockchain technology represents a transformative paradigm shift in the realm of Smart Grids and Power Distribution Automation. This chapter has explored the synergies between these cutting-edge technologies, highlighting their potential to enhance the security, efficiency, and reliability of power systems.AI brings to the table advanced analytics, machine learning, and predictive capabilities that empower Smart Grids to dynamically respond to changing demand patterns, optimize energy consumption, and detect anomalies in real-time. This not only bolsters the operational efficiency of power distribution but also lays the foundation for a more resilient and adaptive grid infrastructure. On the other hand, Blockchain technology provides a secure and transparent framework for managing transactions, ensuring the integrity of data, and establishing trust in the decentralized nature of Smart Grids. The immutable and distributed ledger capabilities of Blockchain mitigate the risks of cyber threats, tampering, and unauthorized access, fostering a more robust and resilient energy ecosystem. The convergence of AI and Blockchain not only fortifies the security aspects but also enables the creation of self-executing smart contracts, facilitating automated and transparent transactions within the Smart Grid. This not only streamlines processes but also minimizes the need for intermediaries, reducing costs and enhancing overall operational efficiency. As we embark on the era of Smart Grids and Power Distribution Automation, the amalgamation of AI and Blockchain emerges as a potent force, ushering in an era of intelligent, secure, and sustainable energy management. The journey toward a more connected and resilient energy infrastructure is an ongoing process, and the advancements in AI and Blockchain technologies continue to shape the future landscape of power distribution.

In conclusion, the collaborative integration of Artificial Intelligence and Blockchain technology holds the promise of revolutionizing the energy sector, paving the way for a smarter, more secure, and efficient power grid that aligns with the evolving needs of our dynamic world.

## REFERENCES

Aguiar-Pérez, J. M., & Pérez-Juárez, M. Á. (2023). An insight of deep learning based demand forecasting in smart grids. *Sensors (Basel)*, *23*(3), 1467. doi:10.339023031467 PMID:36772509

Basodi, S., Tan, S., Song, W., & Pan, Y. (2020). Data Integrity Attack Detection in smart grid: A deep learning approach. *International Journal of Security and Networks*, *15*(1), 15. doi:10.1504/IJSN.2020.106506

Boumaiza, A., & Sanfilippo, A. (2023). Peer to peer solar energy trading demonstrator blockchain-enabled. *2023 11th International Conference on Smart Grid (icSmartGrid)*. 10.1109/icSmartGrid58556.2023.10171013

Hua, W., Chen, Y., Qadrdan, M., Jiang, J., Sun, H., & Wu, J. (2022). Applications of blockchain and Artificial Intelligence Technologies for enabling prosumers in smart grids: A Review. *Renewable & Sustainable Energy Reviews*, *161*, 112308. doi:10.1016/j.rser.2022.112308

Junaidi, N., Abdullah, M. P., Alharbi, B., & Shaaban, M. (2023). Blockchain-based management of demand response in electric energy grids: A systematic review. *Energy Reports*, *9*, 5075–5100. doi:10.1016/j.egyr.2023.04.020

Khan, H., & Masood, T. (2022). Impact of blockchain technology on Smart Grids. *Energies*, *15*(19), 7189. doi:10.3390/en15197189

Khan, M. A., Saleh, A. M., Waseem, M., & Sajjad, I. A. (2023). Artificial intelligence enabled demand response: Prospects and challenges in Smart Grid Environment. *IEEE Access : Practical Innovations, Open Solutions*, *11*, 1477–1505. doi:10.1109/ACCESS.2022.3231444

Kumar, P., Suresh, A., Anbarasu, V., Anandaraj, S. P., & Udayakumar, S. (2022). A decentralized secured grid integration system using APEBC technique with multi access AI framework. *Sustainable Computing : Informatics and Systems*, *35*, 100777. doi:10.1016/j.suscom.2022.100777

Kumari, A., Patel, M. M., Shukla, A., Tanwar, S., Kumar, N., & Rodrigues, J. J. (2020). Armor: A data analytics scheme to identify malicious behaviors on blockchain-based Smart Grid System. *GLOBECOM 2020 - 2020 IEEE Global Communications Conference*. IEEE. https://www.powerelectronicsnews.com/smart-grid-enables-quantum-improvement-in-energy-management-efficiency doi:10.1109/GLOBECOM42002.2020.9348061

Mazhar, T., Irfan, H. M., Khan, S., Haq, I., Ullah, I., Iqbal, M., & Hamam, H. (2023). Analysis of cyber security attacks and its solutions for the smart grid using machine learning and blockchain methods. *Future Internet*, *15*(2), 83. doi:10.3390/fi15020083

Mololoth, V. K., Saguna, S., & Åhlund, C. (2023). Blockchain and machine learning for future Smart grids: A review. *Energies*, *16*(1), 528. doi:10.3390/en16010528

Omitaomu, O. A., & Niu, H. (2021). Artificial intelligence techniques in Smart grid: A survey. *Smart Cities*, *4*(2), 548–568. doi:10.3390martcities4020029

Prasad, S., & Hegde, N. (2022). Information Security and Privacy in Smart Cities, Smart Agriculture, Industry 4.0, Smart Medicine, and Smart Healthcare. *IoT Based Control Networks and Intelligent Systems: Proceedings of 3rd ICICNIS 2022*. IEEE.

Samantaray, R. & Nayana Hegde, N. (2022), Smart Cart Powered by Artificial Intelligence. *2022 3rd International Conference on Communication, Computing and Industry 4.0* (C2I4). IEEE.

Siano, P. (2014). Demand response and smart grids—A survey. *Renewable & Sustainable Energy Reviews, 30*, 461–478. doi:10.1016/j.rser.2013.10.022

Singh, P., Masud, M., Hossain, M. S., & Kaur, A. (2021). Blockchain and homomorphic encryption-based privacy-preserving data aggregation model in smart grid. Computers &amp. *Computers & Electrical Engineering, 93*, 107209. doi:10.1016/j.compeleceng.2021.107209

Sulaiman, A., Nagu, B., Kaur, G., Karuppaiah, P., Alshahrani, H., Reshan, M. S., AlYami, S., & Shaikh, A. (2023). Artificial Intelligence-based secured Power Grid Protocol for Smart City. *Sensors (Basel), 23*(19), 8016. doi:10.339023198016 PMID:37836846

Tabassum, N., Devanagavi, G. D., & Biradar, R. C. (2017). Clock synchronization in wireless sensor networks using least common multiple. *AEÜ. International Journal of Electronics and Communications, 2017*(82), 446–457. doi:10.1016/j.aeue.2017.10.014

Tabassum, N., Devanagavi, G. D., Biradar, R. C., & Lazarescu, M. T. (2020). Comparative node selection-based localization technique for wireless sensor networks: A bilateration approach. *International Journal of Communication Systems, 33*(15), e4559. doi:10.1002/dac.4559

Tabassum, Z., & Shastry, B. S. (2022a). Peak power management of residential building using demand side management strategies. *International Journal of Health Sciences*, 8978–8997. doi:10.53730/ijhs.v6nS2.7333

Tabassum, Z., & Shastry, B. S. C. (2022). Short term load forecasting of residential and commercial consumers of Karnataka Electricity Board using CFNN. *International Journal of Electrical and Electronics Research, 10*(2), 347–352. doi:10.37391/ijeer.100247

Volkova, A., Patil, A. D., Javadi, S. A., & de Meer, H. (2022). Accountability challenges of AI in Smart Grid Services. *Proceedings of the Thirteenth ACM International Conference on Future Energy Systems*. ACM. 10.1145/3538637.3539636

Younes, Z., Alhamrouni, I., Mekhilef, S., & Khan, M. R. (2021). Blockchain applications and challenges in smart grid. *2021 IEEE Conference on Energy Conversion (CENCON)*. IEEE. 10.1109/CENCON51869.2021.9627297

# Chapter 10
# Blockchain Technology: A Study Applied on Intelligent Transport System

**Shridevi Kamble**
*Reva University, India*

**Manjunath Kounte**
*Reva University, India*

**Saujanya M. S.**
*Munster Technological University, Ireland*

## ABSTRACT

*Intelligent internet of vehicles is an emerging technology and has an impact on all smart vehicles on the road due to rapid development in satellite communication integrated with internet technology and enables the integration of smart vehicles with system components like inbuilt vehicle sensors and internet, surrounding vehicles, and smart devices. This integration creates a safe and better platform for vehicles to travel on the road and at the same time be connected to the surrounding environment and vehicles for better performance. Integrating blockchain technology with vehicular communication and smart vehicles is the latest emerging concept. In this chapter, the authors present the survey and latest advancement in block chain technology for intelligent transport systems. This chapter's main highlight is on the application scenario of the internet of vehicles using block chain after a decent review of the literature survey. Also, the authors focus on key challenges and explore further research paths in the intelligent internet of vehicles using block chain technology.*

## INTRODUCTION

The Internet of Vehicles (IoV) heralds a transformative era, interweaving vehicles with connectivity, data, and automation. IoV empowers seamless communication between automobiles, infrastructure, and the environment, enhancing safety, efficiency, and sustainability. By integrating sensors, Artificial

DOI: 10.4018/979-8-3693-0659-8.ch010

Intelligence (AI), and real-time data exchange, vehicles navigate smarter, foreseeing traffic patterns, optimizing routes, and preventing accidents. IoV fosters eco-friendly practices through traffic management and reduced emissions. This interconnected ecosystem engenders novel services like autonomous driving, predictive maintenance, and personalized experiences. However, concerns persist regarding data security, privacy, and infrastructure readiness. Despite challenges, IoV holds the promise of revolutionizing mobility, reshaping transportation landscapes for a connected, efficient future (Singh, Madhusudan et. al (2018)).

AI represents the development of computer systems capable of performing tasks that typically require human intelligence. It encompasses various approaches, including machine learning, neural networks, and natural language processing, among others. AI applications range from autonomous vehicles to personalized recommendations in streaming services, revolutionizing industries like healthcare, finance, and education. Its capacity to analyze vast amounts of data enables predictive insights and decision-making, optimizing processes and enhancing efficiency. Ethical considerations surrounding AI's impact on employment, privacy, and biases continue to evolve. As AI advances, the balance between innovation and responsible deployment remains crucial in shaping a future where technology augments human capabilities.

Block chain technology is a collection of information, typically structured in a table format and a very specific kind of database, mainly stores the data in the form of block that are chained together, these datasets contains huge data or information which can be accessed by hundreds and thousands. Blockchain technology is a rapidly evolving in vehicular environment due to its capability of achieving a decentralized, clear and authentic-resistant system. Intelligent Internet of Vehicles are the part of smart cities and has the capability to solve traffic problems, along with secure transmission, fast and accurate recording of data. All the traffic related, V2V, V2I and V2X communication problems issues can be managed and solved by Blockchain technology (Peng, Chunrong, et al(2020),).

The challenges in Existing Intelligent Internet of vehicles they don't collaborate with multiple entities at one time and achieving communication between V2V, V2I and V2X is done at a time (Nayana Hegde et. al (2019). In vehicular communication it is difficult to have the efficient collaboration communication between multiple entities belonging to different owner due to privacy and management issues. These issues can be solved by Blockchain technology which provide peer-t-peer network without interference of the third party. Blockchain has brought a new way of identifying the solution to the problems of intelligent transport system . The main advantage of Blockchain Technology are:

- **Decentralization:** Provides the distributed system without central controller, by using cryptograph technique.
- **Irrevocability and Traceability:** Blockchain Technology enables unchangeable records for all transaction which is Irrevocability. And by providing all the blocks to all the participants in the network and able to provide traceability.
- **Fault tolerance:** Blockchain can achieve and enable better tolerance to fault nodes in network which are under attack or disaster without being need for centralized authority (Hammoud, Ahmad, et al(2020)). Fig1. Shows the scenario for Intelligent Transport System on road consisting of V2V, V2I, V2C along with road side units. With forms the vehicular communication to monitor and improve traffic related issues (Kamble, Shridevi Jeevan. Et.al 2018).

*Figure 1. Scenario of ITS*

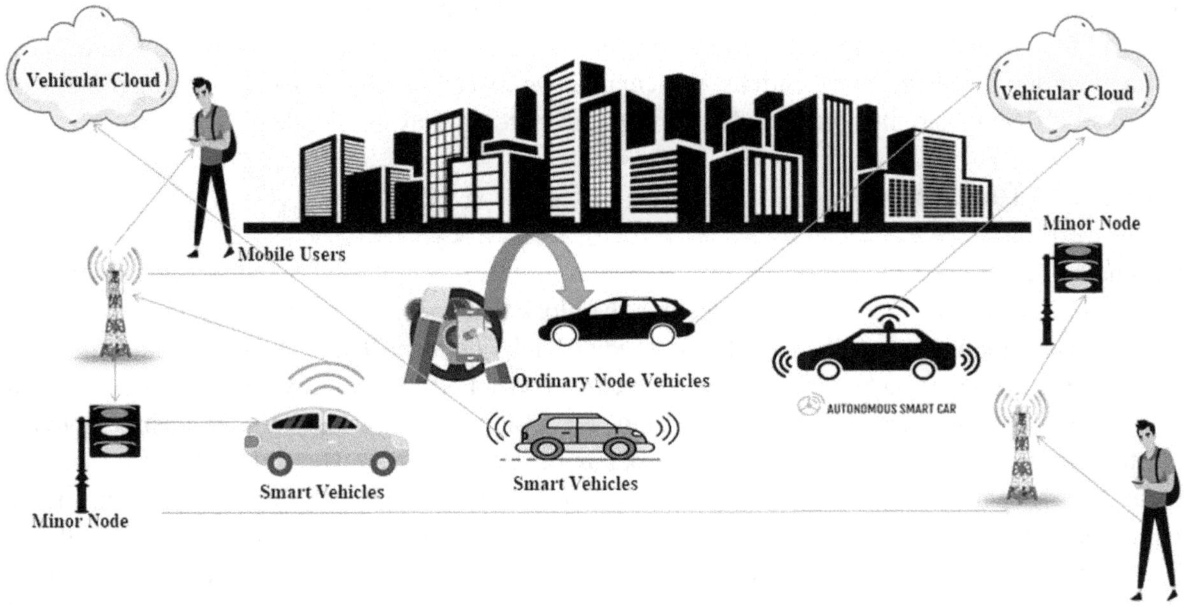


There is large volume of data generated form sensors and communication devices in ITS which can help in solving traffic related issues. Contribution of our Survey paper is summarized as [12].

- Beginning with a description of the preparatory backdrop, including blockchain technology, edge computing, ITS, and IoV, we give a detailed examination of the strategies for integrating blockchain and IoV paradigms towards developing a future ITS.
- We cover the issues connected with IoV, as well as the relevance and motives for the convergence of blockchain and IoV, and the unique challenges that blockchain can address.
- We also discusses on how Blockchain technology can have a significant improvement in existing Intelligent Transport system.
- We present current research efforts and an in-depth discussion of blockchain adoption for IoV scenarios, with an emphasis on vehicular data security, vehicle management, and on-demand transportation services.
- The major issues connected with blockchain integration with IoV, such as security and privacy, performance, IoV-specific and optimized consensus, and incentive mechanisms, are identified and investigated.
- We identify a number of outstanding questions and difficulties as future research prospects.

## AI in Vehicular Network

AI's integration into vehicular networks revolutionizes transportation by enhancing safety, efficiency, and convenience. In autonomous vehicles, AI algorithms interpret sensor data, enabling adaptive decision-making for navigation, collision avoidance, and traffic optimization. Machine learning models predict traffic patterns, aiding route planning and congestion management. AI-driven systems enable real-time analysis of road conditions, weather, and vehicle performance, ensuring proactive maintenance and safety protocols. Additionally, AI assists in developing intelligent infrastructure, such as smart traffic lights and road signs, fostering seamless vehicle-to-vehicle (V2V) and vehicle-to-infrastructure (V2I) communication. Through AI applications, vehicular networks continually evolve, promising a future of safer, more efficient, and interconnected transportation systems (Nayana Hegde et. al (2016).

## Blockchain Technology

Blockchain is distributed and public dataset in which dataset are securely stores and for any kind of data reciprocate. The very most unit in Blockchain is the block that keeps the track and records of activity made at a given amount of time. Block is like individual bank statement. These blocks are interconnected in chronological order using hash which is much secured and nobody can destroy this record book, which can be accessible to everyone. Each block contains the hash of the previous block. If a person intends to destroy must change all the hash of the block before he tries to hamper. Blockchain build-up secured system and transparent medium for record keeping. The key importance of Blockchain can help create secured, decentralized, immutable and transparent ITS. Blockchain is more secure and powerful and decentralized that the conventional datasets (Tripathi, Gautami .et.al.(2019)). The main characteristic of Block chain technology is like decentralized, secure and efficient data storage. That means untrusted and unknown parties can also operate. It is very important for other operators to agree on the data being submitted as the datasets can be operated by unknown parties. ITS built on Blockchain Technology create an entirely secure, immutable, transparent and efficient system. Figure 2 shows the Blockchain based ITS. Blockchain provides secure mean of and transparent mean of record keeping. The main purpose of Blockchain technology in ITS is it keeps the system secured, immutable, decentralized and transparent at the same time (Puri, Vikram, et al.(2020)).

*Figure 2. Blockchain based secured ITS*

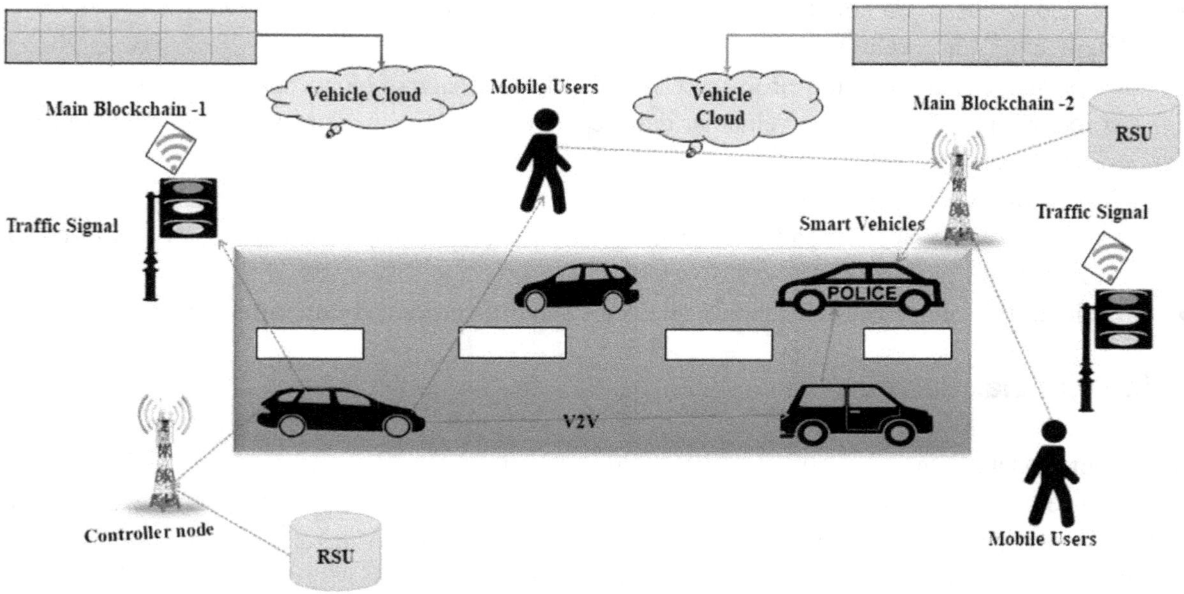

## Applications of Blockchain in Intelligent Internet of Vehicles

Blockchain is ideal for sharing and delivering the information as it provides immediate transparent information is stored on an immutable ledger that is accessible only with the help of permission network members. A blockchain network can track orders, product information and much more (Golosova, Julija et.al.(2018)). The members of blockchain share the single view of the truth, hence we can see all the details of transaction end to end and gives greater confidential approach as well as efficiencies and opportunities. The major applications of Blockchain technology I integrated with IoV are as follows and shown in figure 1. (Mollah, Muhammad Baqer, et al. (2020))

- Traffic Control Management
- Vehicle management
- Data protection and management
- Resource Management
- Content broadcasting
- Forensic Application
- Data and Resource Trading
- Resource Sharing

Divided blockchain applications used in the Internet of Vehicles into eight primary groups based on their use and industry:

*Figure 3. Integration of Block Chain with Intelligent Transport System*

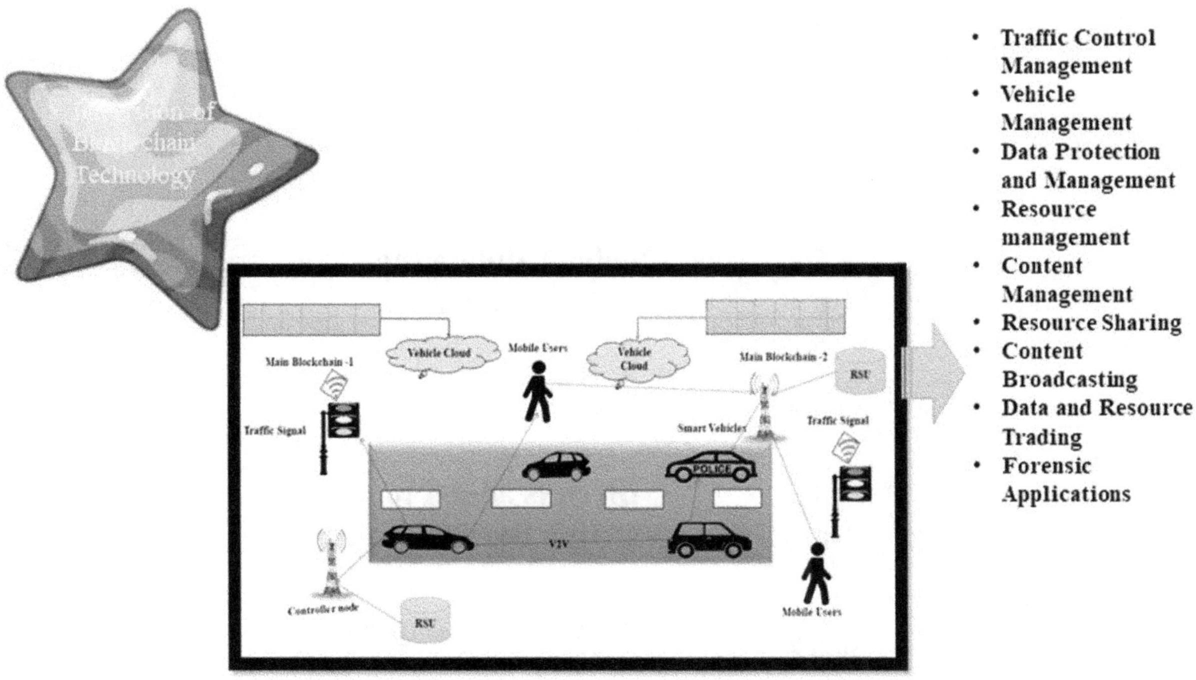

*Table 1. Vehicular cloud utilization data*

| | |
|---|---|
| Privacy and Security | 27.00% |
| IOV Management | 2.90% |
| Data protection and management; | 12.30% |
| P2P energy trading; | 6.60% |
| Microgrid management; | 9.40% |
| AI/Machine Learning (ML) and IoV; | 12.70% |
| Blockchain performance in the IoV; | 25.80% |
| General Purpose | 6.60% |

The authors found that approximately one in three applications were about Privacy & Security (27%), which included authentication, Vehicle to Everything (V2X) and identity management initiatives. The second most popular category was IoV Management (25.8%). This was followed by P2P Energy Trading and general-purpose studies, accounting for 12.7% and 12.3% of total studies, respectively. Other application areas made up around 22% of the total, as presented in Figure 4

*Figure 4. Pie chart showing vehicular cloud utilization data*

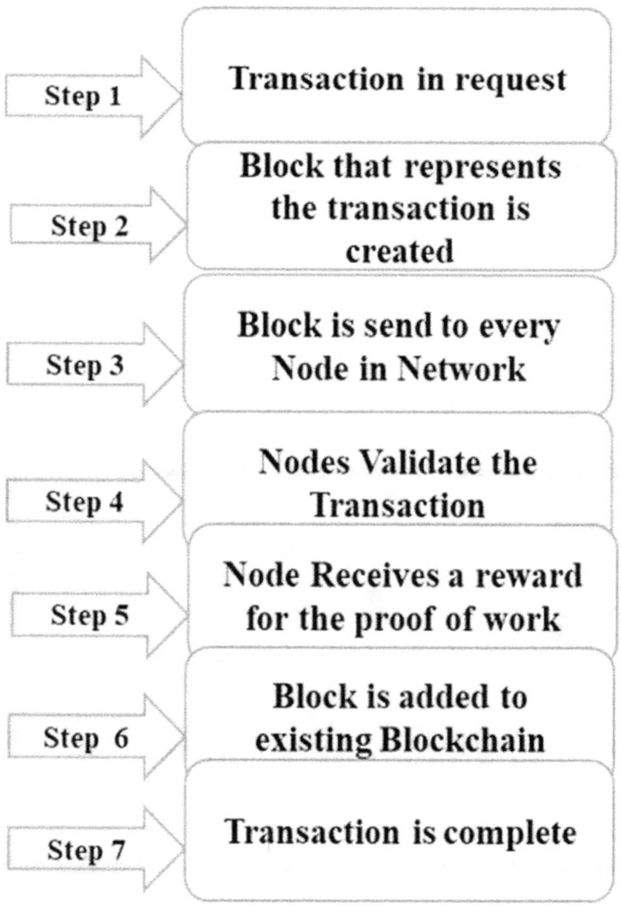

The aforementioned data makes it clear that there has been extensive evaluation of blockchain's effects on smart grids, particularly in the automotive sector. In the Internet of Vehicles (IoV) space, blockchain is starting to garner greater attention due to its tremendous potential for transformation.

## Working of Blockchain

Blockchain is a software high-end dataset in which transactions are not governed by individual party but the complete transaction history is recorded in a decentralized and distributed record book. Blockchain stores transactions. Each and every block in Blockchain includes list of transaction and metadata. Metadata is the data or information, which contain the information like when the block was created, block number, who created the block, information of previous block. Each and every block maintains hash of previous and hence creates chain of linked block. Basic features of Blockchain and the core Blockchain architecture components:

- **Transaction:** transmitting information in crypto type from node-to-node is transaction. This Blockchain keeps the track of all the transaction that are present since the beginning of the recent transmission.
- **Blocks:** Each and every transaction are grouped into blocks. Each block collects all the transaction that occurred at particular timeframe and reference to preceding block.
- **Nodes:** the entire transaction data is stored over the nodes which are also called as network members, each nodes also consists of a copy of complete Blockchain.
- **Chain:** are the sequence of block in certain specific order.
- **Miners:** these are the specific nodes that take part in maintenance of block chain called as mining process. During mining process node check previous transactions
- **Majority consensus (protocol):** The Blockchain do not have centralized database, therefore the decision in the network are take according to majority consensus. A set of rules and arrangement to carry block operations.

Always when the data comes in it enters into completely new block. Ones the block is filled into new data, these filled blocks are chained onto previous block. Thus, the data are linked together in chronological order. Different type of information can be stored on Blockchain and the most common type of information are ledger for transaction. Each block consists of cryptographic hash of the previous block as shown in fig 4. The information in the block depends on Blockchain network, the timestamp, hash and transaction are present in all the Blockchain variations. Since all the Hash information are generated automatically, it means that it is not possible to alter the information.

Ones the blocks are created with the different information. Each of the blocks in the Blockchain are related to each other and each node finds proof-of –work to broad cast to all the other nodes in the network. The other nodes validate the proof-of-work if transaction or data are valid, thus nodes expect proof –of-work by working on the next proof of work.

Proof-of-work:

Practical Byzantine fault tolerance algorithm (PBFT)

Proof-of-State:

There are several ways to build blockchain technology based on network access control mechanism blockchain is divided into three parts:

*Figure 5. Block chain components*

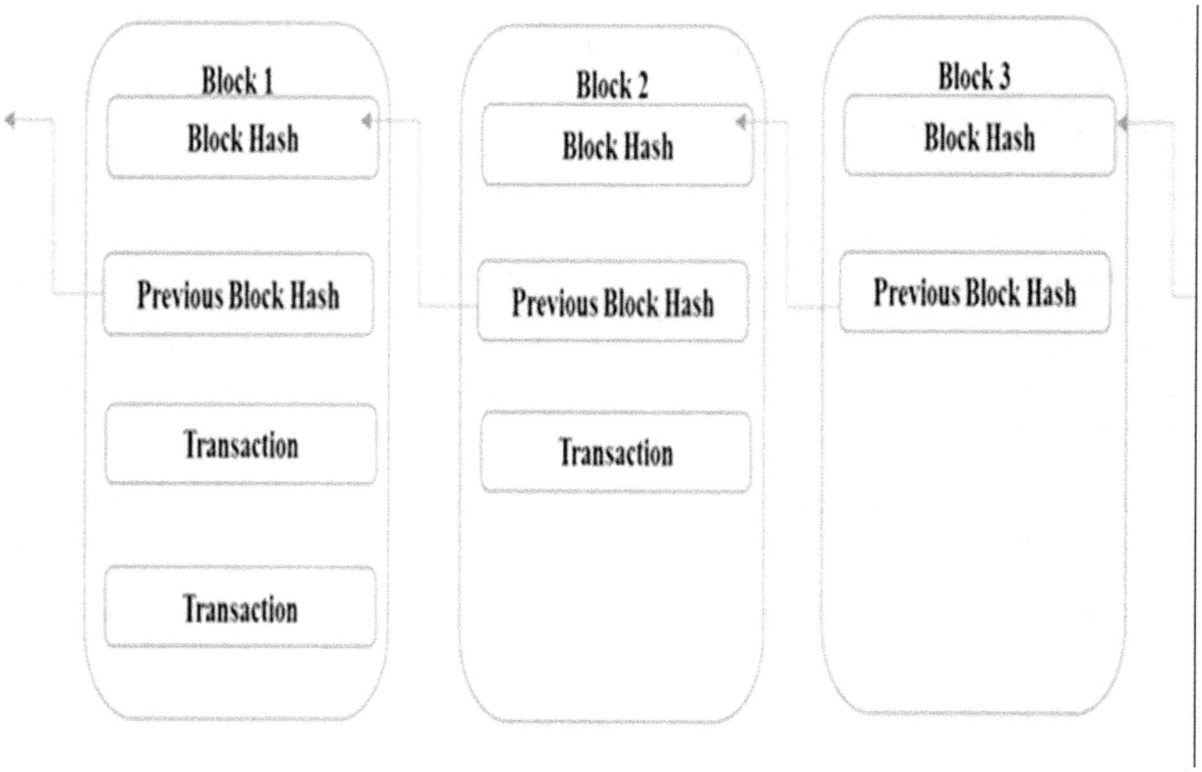


- Public Blockchain: In public blockchain network everyone can join, participate, get the access and send transactions in the chain of consent process.
- Private Blockchain Nodes: In private blockchain technology one organization governs the network, nodes will be limited and not every node can participate in blockchain
- Consortium Blockchain: In Consortium blockchain has a pre-selected organization to determine who can submit transaction or access the data. Is also known as partially decentralized.

Other blockchain platforms include Hashgraph, Corda, Hyperledger Indy, Hyperledger Sawtooth, Interchain, Cosmos and Binance. the fourth most used platform is Hyperledger.

## Blockchain Technology on Intelligent Internet of Vehicles

The main aim of the paper is to signify the benefits of blockchain technology in internet of vehicles. Blockchain technology is increasingly growing to make secure and safe connectivity of vehicles with internet technology. Duplicate record keeping and third-party validations waste a lot of time in operations. Fraud and cyberattacks can make record-keeping systems susceptible. Data verification might be slowed by a lack of openness. And, with the advent of the Internet of Things, transaction volumes have skyrocketed. All of this slows business and depletes the bottom line, indicating that we need to find a better solution. Then there's the blockchain.

## Benefits of Blockchain Technologies

It Ensures Great Trust: As a member of a members-only network, you can trust that you will get accurate and timely data from blockchain, and that your sensitive blockchain records will be shared only with network members to whom you have expressly authorized access. Every Blockchain network participants has a generated address, failing at a single point in the network cause serious issues in the network.

It Ensures Security: All network participants must agree on data accuracy, and all confirmed transactions are immutable since they are permanently recorded. A transaction cannot be deleted by anybody, not even the system administrator. It also provides security against tampering and illegal modification in network.

It Ensures Efficiencies: Time-consuming record reconciliations are removed with a distributed ledger shared across network participants. A collection of rules called a smart contract may be placed on the blockchain and implemented automatically to speed up transactions. Hence The network of Blockchain technology is transparent. All the member of the network can utilize the data and can easily identify the flaws. It is completely safe as overwriting on Blockchain network requires huge computing power.

Security Blockchain technology is well-known for providing a very secure method of transferring data. The fact that the technology is dispersed contributes to the security because no third party has control over the data. Furthermore, it implies that a transaction would be validated by several nodes. The transactions themselves are also protected by a digital signature that is encrypted. Information cannot be falsified by enemies because of its signature. Furthermore, the suitable digital key is necessary to read information. The transportation industry makes extensive use of the security that blockchain provides. Connected automobiles may now access a growing number of internet services. Threats that were previously unrelated to the transportation sector, such as tracking and remote hijacking, are now present as a result of this connection and services.

Blockchain is the decentralized, secure system and does not work with the intermediary all the parties in this Blockchain make the decisions. Blockchain technology is secure system, as the transactions have their own proof of validity and authorization to enforce constraints. Thus, the transaction can be verified and processed independently. Mainly it is new type of database creates the blocks with different type of information. Each and every block relates to other blocks in the network. And impossible to change or delete theses blocks. Some of the challenges with Block chain technology are, in addition, network needs are a source of worry. A blockchain system consumes a lot of energy, and as the technology becomes more widely adopted, network capacity will need to expand in order to maintain low latency. At decentralized junctions, high latency is unacceptable. Another issue to be addressed is the effective governance of blockchain networks (van Dijk, Jurian et. al (2020)).

## IoV Architecture

With its integration of VANET and IoT, the Internet of Vehicles is considered one of the most active ITS research topics. There are two scientific visions combined in the IoV: car intelligence and (a) car connection, which emphasize the integration of things like people, cars, objects, networks, and surrounds. In the Internet of Vehicles (IoV), communication and information technology work together to meet various traffic and driving challenges. This enhances both passenger safety and the driving experience overall. Thus, given that the Internet of Vehicles (IoV) ecosystem comprises several technologies, academics have developed an architecture that primarily consists of four levels (Kapassa, E., Themistocleous. Et. al (2021)).

Sensing Layer: This is like the "eyes and ears" of the system. It includes all the sensors in the vehicles, like cameras and detectors, that collect data. They notice things like how you're driving, what's happening around your vehicle, and the weather conditions. It helps the system understand what's happening in and around the vehicles.

Communication Layer: Think of this as the "talkative" layer. It makes sure that different vehicles can talk to each other (V2V) and to the infrastructure (V2I). It uses various wireless technologies like GSM, Wi-Fi, LTE, and Bluetooth to keep everything connected. It ensures constant communication between vehicles and the infrastructure, sharing important information for safer and more efficient driving (Nayana Hegde. Et. al. 2022).

Computing Layer: This layer is like the "brain" of the system. It stores, analyzes, and processes all the data collected by the sensors. It makes decisions based on this information and provides computing services to handle various situations in the IoV network. It processes the collected data; turning it into useful information and helping the system make smart decisions.

Application Layer: This is the "user interface" layer. It provides consumers (drivers, passengers, etc.) with different services related to their vehicles. These services could include navigation, safety alerts, or even entertainment features. It's the layer that directly interacts with users, offering them various services and making their experience with the IoV more convenient and enjoyable (Nayana Hegde. Et. al. 2023).

## Secure Message Communication in IoV

Securing message communication in the Internet of Vehicles (IoV) is crucial due to the interconnected nature of vehicles and their reliance on communication systems (Li, H., Pei, L., Liao, D., Sun. et. al, 2020). Here are some key aspects and methods for ensuring secure message communication in IoV:

- Encryption: Messages exchanged between vehicles should be encrypted using robust cryptographic algorithms. This ensures that even if intercepted, the information remains unreadable to unauthorized entities.
- Authentication and Authorization: Implement strong authentication mechanisms to verify the identity of vehicles before allowing communication. Digital certificates, biometrics, and multi-factor authentication can be used to ensure authorized access.
- Secure Communication Protocols: Use secure communication protocols such as TLS (Transport Layer Security) or DTLS (Datagram Transport Layer Security) to establish secure channels for data transmission. These protocols encrypt data and ensure integrity during transit.
- Firewalls and Intrusion Detection Systems: Deploy firewalls and intrusion detection/prevention systems to monitor and filter incoming and outgoing traffic. This helps in identifying and preventing malicious activities or unauthorized access attempts.
- Secure Software Development: Ensure that the software running on vehicles and the infrastructure is developed following secure coding practices. Regular updates and patches should be applied to address any identified vulnerabilities.
- Securing Wireless Networks: Given that IoV heavily relies on wireless communication, secure the wireless networks using strong encryption (e.g., WPA3 for Wi-Fi), change default passwords, and employ measures to prevent unauthorized access.

- Data Integrity and Non-Repudiation: Implement measures to maintain data integrity, ensuring that transmitted data remains unchanged during communication. Non-repudiation mechanisms can be used to prevent parties from denying their involvement in a communication.
- Privacy Protection: Safeguard sensitive data and user privacy. Employ techniques like data anonymization, pseudonymization, and access control to limit the exposure of personal information.
- Physical Security: Ensure the physical security of vehicle components, communication devices, and infrastructure to prevent unauthorized access or tampering.
- Regulatory Compliance: Adhere to industry standards and regulations (such as GDPR in Europe) regarding data protection, privacy, and cybersecurity in IoV deployments.
- Behavioral Analysis and AI/ML: Implement AI/ML-based systems to analyze and detect anomalous behavior within the IoV network, aiding in the identification of potential security threats.
- Continuous Monitoring and Incident Response: Continuously monitor the IoV ecosystem for potential security breaches or vulnerabilities. Develop and regularly update an incident response plan to mitigate and respond to security incidents promptly.

Secure message communication in the Internet of Vehicles demands a multi-faceted approach that encompasses encryption, authentication, secure protocols, robust infrastructure, and ongoing monitoring to ensure the safety and privacy of connected vehicles and their occupants.

## IOV on Block Chain

With increase in connectivity in IOV, the IOV devices are becoming fall victim to security vulnerabilities, such as bottleneck attacks, distributed denial of services(DDos) attacks, and insecure interferences. Blockchain is like a super secure digital ledger. It's a way of keeping records that's really, really hard to tamper with. Without BlockChain Unsecured devices gives access to cybercriminals to exploit the system. IoT used in blockchain adds a higher level of security to prevent data breaches to make IoT safer and smarter.

Blockchain technology originated from an article titled "bitcoin: a peer-to-peer electronic cash system" written by Nakamoto in 2008. From the birth of blockchain to the present, blockchain technology has experienced three generations from 1.0 to 3.0. Blockchain 1.0 focuses on the transactions of digital currency. Since the birth of bitcoin, more than 600 cryptodigital currencies have been generated correspondingly (litecoin, dogecoin, YBcoin, etc.). These cryptodigital currencies mainly be used in small payments, foreign exchange, gambling and money laundering; Blockchain 2.0 focuses on the registration, validation, and transfer of smart contract. Smart contracts rely on oracles prediction and are mainly used in the financial field, such as securities trading, bank bills, payment and clearing, and supply chain finance. Blockchain 3.0 transcends the economic field, and its application field is not limited to finance, commodity transaction. The scope of blockchain 3.0 extends to government, medical care, science, culture, transportation, etc. In recent years, a few researchers have introduced blockchain into VANET, considering with the characteristics of decentralization, redundant storage, collective maintenance and tamper-proof of blockchain. For example, Joy et al. [33] proposed the concept of blocktree. And the vehicle embedded its signature into the blockchain. Greg et al. verified the feasibility of blocktree and analyzed the end-to-end delay and the time to collect, write and update blockchain contents. To better combine the blockchain with the VANET, Dorri et al. proposed a lightweight scalable blockchain. And based on this, the reference of [36] proposed a blockchain architecture with decentralized privacy protection for intelligent vehicle systems. While Lei et al. proposed a new network topology based on blockchain structure to simplify distributed key management in heterogeneous vehicle communication systems. Furthermore, the use of the blockchain as a means for privacy-preserving proof of location has been proposed in the work of.

Blockchain system adopts distributed structure, as shown in the Fig. It consists of several decentralized nodes which cannot trust each other. Each node participates in data management. When a new block is confirmed by most of nodes (the number depends on different consensus mechanisms) in blockchain system, the new block is written into blockchain by miner. This is called as mining process. Then all nodes verify the content of that be newly added to blockchain, and have a complete copy of blockchain. Therefore, all nodes in the blockchain system jointly maintain the data information, which ensures the consistency, integrity and non-tampering of data. In this paper, blockchain technology is applied to VANET. The distributed features of the blockchain are used to remove the single trust center in traditional VANET. In this way, a novel decentralized architecture of VANET is constructed to realize the security and irreversibility of SBMs. However, it is precisely because all nodes participate in the joint maintenance of data information in blockchain. SBMs of vehicles become open and transparent, which is not conducive to identity and location privacy protection. And the two concepts of information transparency and privacy protection are contradictory. The more transparent information is, the more difficult privacy protection is. Therefore, in this paper, we adopt dynamic threshold encryption and k-anonymity unity to achieve identity and location privacy protection respectively.

## IoV Privacy Protection

In recent years, privacy protection has become a hot spot in study of VANET. According to the objects protected by researchers, we can divide the privacy protection into three types: location privacy protection, trajectory privacy protection and identity privacy protection. For example, the work in dealt with the distribution of vehicular pseudonyms using fog computing technology. The edge resources of VANET were used to effectively manage pseudonyms, which can improve the ability of location privacy protection. In, the authors proposed a policy of trajectory privacy using the multiple mix zones. By constantly changing pseudonyms, it made the pseudonyms unlikable and protected vehicular trajectory privacy. From the perspective of protection means, the privacy protection is mainly divided into two types: anonymity-based and encryption-based privacy protection. Anonymity-based privacy protection often adopts k-anonymity mechanism. K-anonymity mechanism was originally created to protect users' location privacy in LBS applications. Nowadays, many researchers use k-anonymity mechanism in VANET. By the principle of maximum entropy, we find k appropriate vehicles with the closest historical request probability. Then the real vehicle is hidden in these k vehicles, and vehicular privacy is protected. While the encryption-based privacy protection is commonly used in the authentication of VANET. Authentication is an important guarantee to receive SBMs from legitimate vehicles in VANET. It plays a vital role in privacy protection in VANET. At present the authentication of VANET generally adopts asymmetric encryption authentication methods, such as Public Key Infrastructure (PKI) and Elliptic Curve Digital Signature Algorithm (EDSA), and symmetric encryption authentication methods, such as group signature authentication mechanism [13].

Vehicular network is an emerging and advanced component in the field of transportation systems. In vehicular network, all authenticated vehicles can communicate with each other and with other components of the network. This is done to ensure the security of the information shared among vehicles.

The information may contain sensitive data like speed of vehicle, vehicle's identity, and current position of vehicle. Vehicular network provides various different advantages to drivers but at the time it is susceptible to privacy and security attacks. It is essential to address various security and privacy issues. Existing possible solutions are not feasible for providing security in the vehicular network. Therefore, blockchain has been introduced in the vehicular network that provides decentralized platform which is capable of managing and handling resources. Further, it resists various security attacks that may occur in the network[14].

With the rapid development of the Internet of Things (IoT) and its potential integration with the traditional Vehicular Ad-Hoc Networks (VANETs), we have witnessed the emergence of the Internet of Vehicles (IoV), which promises to seamlessly integrate into smart transportation systems. However, the key characteristics of IoV, such as high-speed mobility and frequent disconnections make it difficult to manage its security and privacy. The Blockchain, as a distributed tamper-resistant ledge, has been proposed as an innovative solution that guarantees privacy-preserving yet secure schemes[15].

## Research Gaps in IoV

While Blockchain technology shows promise for Intelligent Internet of Vehicles (IoV), there are several research gaps that need to be addressed to fully understand its potential and overcome associated challenges. Here are some research gaps in the application of Blockchain technology in the IoV:

- Scalable Consensus Mechanisms: Current Blockchain consensus mechanisms, such as proof-of-work and proof-of-stake, may not be suitable for the high transaction volumes and low-latency requirements of the IoV. Exploring new consensus mechanisms or optimizing existing ones to achieve scalability while maintaining security and decentralization is an important research area.
- Privacy-Preserving Techniques: Preserving privacy is crucial in the IoV, where sensitive vehicle and user data are involved. Research is needed to develop effective privacy-preserving techniques that can protect the identity and sensitive information of participants while ensuring data integrity and transparency on the Blockchain.
- Trust and Security Models: The IoV requires a high level of trust and security for reliable communication and coordination between vehicles and infrastructure. Designing trust and security models specifically tailored to the IoV ecosystem, considering the unique characteristics and requirements of vehicular networks, is an important research direction.
- Interoperability and Standardization: Achieving interoperability between different Blockchain platforms, protocols, and communication technologies is critical for seamless data exchange and communication within the IoV. Research efforts should focus on developing interoperability frameworks and standardized protocols to enable effective integration of Blockchain technology across different IoV systems.
- Energy-Efficient Blockchain Solutions: Energy consumption is a significant concern in Blockchain networks, and addressing this issue is crucial for the energy-efficient operation of the IoV. Research is needed to explore energy-efficient consensus mechanisms, optimization techniques, and sustainable architectures that can reduce the energy footprint of Blockchain in the context of the IoV.
- Real-Time Data Processing: The IoV generates large volumes of real-time data that require quick processing and decision-making. Research should focus on developing efficient mechanisms to

handle real-time data processing, storage, and retrieval on the Blockchain, ensuring timely and responsive communication between vehicles.

- Governance and Regulatory Frameworks: Blockchain-based IoV systems raise governance and regulatory challenges related to data ownership, liability, privacy regulations, and compliance. Research is needed to develop appropriate governance models and regulatory frameworks that address these challenges and provide clear guidelines for the deployment and operation of Blockchain technology in the IoV.
- Practical Implementations and Field Testing: While research on Blockchain technology in the IoV is progressing, there is a need for more practical implementations and field testing to validate the effectiveness, scalability, and security of Blockchain-based solutions in real-world scenarios. Field trials and pilot projects can help identify practical challenges and validate the potential benefits of Blockchain in the IoV.
- Addressing these research gaps will require interdisciplinary collaborations, involving experts from areas such as computer science, cryptography, transportation engineering, telecommunications, and law. Continued research and innovation are essential to advance the understanding and application of Blockchain technology in the context of Intelligent Internet of Vehicles.

## Current Status of Block Chain IOV

Proof of Concept (PoC) Projects: Several PoC projects have been conducted to explore the feasibility and potential benefits of integrating Blockchain into the IoV. These projects have demonstrated use cases such as secure vehicle-to-vehicle communication, decentralized data exchange, and transparent supply chain management in the automotive sector.

Data Security and Integrity: Blockchain provides a tamper-proof and immutable ledger, ensuring data security and integrity in the IoV. By leveraging cryptographic techniques, Blockchain enables secure vehicle identification, secure firmware updates, and secure sharing of vehicle data among stakeholders while maintaining data privacy.

Decentralized Autonomous Organizations (DAOs): Blockchain-based DAOs have the potential to revolutionize the IoV ecosystem by enabling decentralized governance and decision-making. DAOs can facilitate peer-to-peer transactions, smart contracts, and autonomous vehicle operations, enhancing efficiency and transparency in the IoV.

Tokenization and Micropayments: Blockchain enables the creation of digital tokens and supports micropayments, which can be utilized in the IoV ecosystem. Tokens can represent digital assets, such as access rights to vehicle data or charging services, and facilitate seamless transactions between vehicles, infrastructure providers, and users.

## FUTURE TRENDS

Interoperability Solutions: Efforts are underway to develop interoperability protocols and standards that enable different Blockchain networks to seamlessly communicate with each other in the IoV ecosystem. Interoperability will facilitate data sharing, smart contract execution, and interoperable tokenized ecosystems.

Privacy-Enhancing Technologies: Privacy-preserving techniques, such as zero-knowledge proofs and secure multi-party computation, will be crucial for protecting sensitive data while ensuring transparency and compliance in the IoV. Future research will focus on enhancing privacy features in Blockchain solutions.

Scalability and Performance Improvements: Enhancing the scalability and performance of Blockchain networks will be crucial to handle the high transaction volumes and low-latency requirements of the IoV. Research is focusing on developing consensus mechanisms, sharding techniques, and layer-2 scaling solutions to address these challenges.

Hybrid Blockchain Architectures: Hybrid Blockchain architectures, combining public and private Blockchain networks, offer the benefits of transparency, security, and flexibility in the IoV. Future trends involve exploring hybrid architectures that strike a balance between data privacy, scalability, and network governance.

Artificial Intelligence (AI) Integration: AI techniques, such as machine learning and intelligent data analytics, can be integrated with Blockchain in the IoV to derive valuable insights from vehicle data. AI can enhance predictive maintenance, optimize routing and traffic management, and enable intelligent decision-making in the IoV ecosystem.

Energy Efficiency and Sustainability: Research efforts are focusing on developing energy-efficient consensus mechanisms and exploring alternative Blockchain architectures to reduce the energy consumption of Blockchain networks in the IoV. Sustainable Blockchain solutions will align with the environmental goals of the automotive industry.

Regulatory Frameworks and Governance Models: As Blockchain technology matures in the IoV, there will be a need for establishing regulatory frameworks and governance models that address legal, ethical, and security concerns. Governments and industry stakeholders will collaborate to develop standards and regulations for the responsible deployment and operation of Blockchain in the IoV.

Overall, the future of Blockchain technology in the Intelligent Internet of Vehicles holds immense potential for enhancing security, privacy, efficiency, and transparency in the automotive industry. Continued research, industry collaborations, and technological advancements will drive the adoption and integration of Blockchain solutions, shaping the future of the IoV.

## PROPOSED METHOD OF IMPLEMENTATION OF BLOCKCHAIN FOR IoV

One proposed concept that can be explored for research work in the integration of Block chain technology with the Intelligent Internet of Vehicles (IoV) is "Block chain-based Autonomous Vehicle Coordination."

Autonomous vehicles are expected to play a significant role in the future of transportation. However, coordinating the movements of multiple autonomous vehicles in a decentralized manner poses various challenges. Block chain technology can provide a robust and secure platform for autonomous vehicle coordination, ensuring trust, transparency, and efficient decision-making among vehicles.

Dynamic Route Planning and Traffic Management: Explore the use of Block chain to facilitate dynamic route planning and traffic management for autonomous vehicles. Investigate how Blockchain can enable vehicles to collaborate and optimize their routes in real-time, considering factors such as traffic congestion, road conditions, and individual vehicle preferences.

Dynamic route planning and traffic management for autonomous vehicles can be enhanced by leveraging Blockchain technology. Here's an explanation of how Block chain can enable vehicles to collaborate and optimize their routes in real-time, considering factors such as traffic congestion, road conditions, and individual vehicle preferences:

Decentralized Data Sharing: Blockchain provides a decentralized platform where vehicles can securely share real-time data about their current positions, speeds, and other relevant information. By contributing their data to the Blockchain, vehicles create a transparent and immutable record of the traffic conditions.

Consensus Mechanisms: Blockchain utilizes consensus mechanisms to ensure the validity and agreement of the shared data. Through consensus algorithms like proof-of-work or proof-of-stake, vehicles in the network can collectively agree on the most accurate and up-to-date information regarding traffic congestion, road conditions, and other relevant factors.

Smart Contracts for Coordination: Smart contracts, powered by Block chain, can be employed to facilitate coordination among vehicles. These contracts contain predefined rules and conditions for route planning and traffic management. Vehicles can autonomously execute these smart contracts, adjusting their routes based on the agreed-upon rules and considering the shared data.

Real-Time Updates and Optimization: As vehicles contribute their real-time data to the Block chain, the information becomes instantly available to all other vehicles. By accessing this real-time data, vehicles can dynamically adjust their routes to optimize travel time, avoid congestion, and accommodate individual vehicle preferences. Optimization algorithms can be utilized to process the data and calculate the most efficient routes based on the shared information.

Privacy and Security: Privacy preservation mechanisms can be integrated into the Block chain system to protect sensitive information. Techniques like zero-knowledge proofs or differential privacy can ensure that individual vehicle data remains private while still contributing to the collective intelligence of the network. Security measures, such as encryption and digital signatures, can also be implemented to maintain data integrity and prevent unauthorized access.

Incentive Mechanisms: Block chain allows for the implementation of incentive mechanisms to encourage active participation and collaboration among vehicles. For instance, vehicles that share accurate and valuable data can be rewarded with tokens or other incentives. These incentives promote the sharing of reliable information, leading to a more effective and efficient traffic management system.

By utilizing Block chain in dynamic route planning and traffic management, vehicles can collaborate in real-time, optimizing their routes based on the shared data and predefined rules. This facilitates efficient utilization of road infrastructure, reduces congestion, and improves overall traffic flow. The transparency, immutability, and decentralized nature of Block chain enhance trust among vehicles, enabling better coordination and decision-making in the context of autonomous vehicles.

## FORMAL SECURITY ANALYSIS APPLIED ON BLOCKCHAIN BASED IoV

Formal security analysis in the context of blockchain-based Internet of Vehicles (IoV) involves rigorous methods to evaluate and ensure the security properties of the blockchain technology used within the IoV ecosystem. This analysis aims to identify and mitigate potential vulnerabilities or threats that could compromise the security, integrity, and functionality of the IoV.

Here are steps involved in applying formal security analysis to blockchain-based IoV:

- Threat Modeling: Begin by identifying potential threats and attack vectors specific to IoV systems. These could include attacks on the blockchain network, smart contracts, communication channels, or the interconnected vehicles themselves.
- Security Requirements: Define security requirements and objectives for the IoV system. This includes confidentiality, integrity, availability, authentication, authorization, and non-repudiation of data and transactions within the IoV.
- Formal Verification of Smart Contracts: Use formal methods to mathematically verify the correctness and security properties of smart contracts deployed on the blockchain. Formal verification techniques involve rigorous mathematical proofs to ensure the smart contracts behave as intended and are resilient against attacks like reentrancy, overflow, or other vulnerabilities.
- Consensus Mechanism Analysis: Evaluate the consensus mechanism used in the blockchain for IoV. Analyze its security properties, such as Byzantine fault tolerance, resistance to attacks, and the potential impact of different attack scenarios on the consensus mechanism.
- Network Security Analysis: Assess the security of the communication protocols and network infrastructure used in the IoV ecosystem. Ensure encryption, secure key exchange, and proper authentication mechanisms are in place to protect the communication between vehicles, IoT devices, and the blockchain network.
- Penetration Testing: Conduct simulated attacks or penetration testing on the IoV system to identify vulnerabilities that might not have been discovered through theoretical analysis. This step helps in validating the security measures and understanding the system's resilience against real-world threats.
- Continuous Monitoring and Updates: Security is an ongoing process. Implement continuous monitoring and updates to address newly discovered vulnerabilities, patch security flaws, and adapt to emerging threats in the rapidly evolving landscape of blockchain and IoT security.
- Regulatory Compliance: Ensure compliance with relevant regulatory standards and privacy laws, such as GDPR, as they pertain to the handling of sensitive data within the IoV ecosystem.

Formal security analysis in blockchain-based IoV requires a multidisciplinary approach involving cryptography, distributed systems, cybersecurity, and formal methods to comprehensively address the security challenges posed by the interconnected nature of vehicles and the underlying blockchain technology.

Here is a list of some tools commonly used for formal security analysis in the context of IoV:

- Scyther: An automatic verification tool for cryptographic protocols. It helps in analyzing the security of protocols used in the IoV.
- AVISPA (Automated Validation of Internet Security Protocols and Applications): A toolset for the analysis of Internet protocols and security-sensitive applications. It includes several tools for formal verification of security protocols.
- SPIN Model Checker: Although not specific to IoV, SPIN is a widely used tool for the formal verification of distributed systems and protocols. It can be utilized for analyzing certain aspects of security in IoV.
- ProVerif: A formal security protocol verifier for analyzing cryptographic protocols. It's useful for verifying the security properties of communication protocols within the IoV.

- SAT/SMT solvers (e.g., Z3, CVC4): These solvers are used in various formal methods to check the satisfiability of logical formulas and solve constraints. They can be employed in analyzing security properties and vulnerabilities in IoV systems.
- Uppaal: Primarily designed for modeling, simulation, and verification of real-time systems, Uppaal can be used to model and verify the behavior of protocols or systems within IoV.
- Tamarin Prover: A tool for the formal analysis of security protocols. It assists in analyzing and verifying cryptographic protocols used in the IoV.
- Attack Graph Generation Tools: These tools help in constructing and analyzing attack graphs, visualizing potential attack paths and vulnerabilities in interconnected systems like those found in IoV.
- Symbolic Model Checkers: Tools like NuSMV or SPIN can be used to formally model and verify the security properties of protocols or systems within the IoV.
- Formal Methods Frameworks (e.g., Alloy): General-purpose formal methods frameworks can be adapted to model and verify security properties in IoV systems.

Remember that the choice of tool often depends on the specific security aspects being analyzed, the nature of the IoV system, and the expertise of the security analysts. Additionally, these tools might have specific focuses or limitations, so combining multiple tools or approaches might be necessary for comprehensive security analysis in the IoV.

## INFORMAL SECURITY ANALYSIS APPLIED ON BLOCKCHAIN BASED IoV

Informal security analysis on Blockchain-based Internet of Vehicles (IoV) involves assessing the potential vulnerabilities, risks, and threats within the IoV ecosystem that utilizes blockchain technology. Here's how it can be applied:

Identifying Attack Vectors: Conduct a comprehensive analysis to identify potential attack vectors within the IoV ecosystem. This involves examining various layers, such as the communication protocols, smart contracts, consensus mechanisms, and the overall architecture of the blockchain network.

Threat Modeling: Create threat models specific to IoV applications on the blockchain. This includes considering potential threats like data tampering, Sybil attacks, 51% attacks, privacy breaches, and malicious code injection that could impact the connected vehicles and the network's integrity.

Smart Contract Audits: Perform audits of smart contracts deployed on the blockchain. Analyze the code for vulnerabilities, bugs, or loopholes that could be exploited by attackers. This step is crucial as smart contracts govern many aspects of transactions and interactions within the IoV ecosystem.

Privacy and Data Security: Assess the privacy measures and data security protocols implemented within the IoV system. Evaluate how user data is collected, stored, and utilized. Privacy-preserving techniques such as zero-knowledge proofs or homomorphic encryption can be considered to enhance security.

Consensus Mechanism Evaluation: Evaluate the consensus mechanism used in the blockchain network supporting the IoV. Assess its resilience against attacks and its ability to maintain the integrity of the network despite potential adversarial behavior.

Network Resilience: Analyze the overall resilience of the network against potential disruptions, including DDoS attacks or targeted attacks on specific nodes. Implement measures to enhance network resilience and mitigate the impact of such attacks.

Regulatory Compliance: Ensure that the IoV system complies with relevant regulatory standards and guidelines concerning data privacy, cybersecurity, and vehicle communication protocols. Compliance is essential for ensuring trust and legal adherence.

Continuous Monitoring and Updates: Security in IoV systems is an ongoing process. Implement continuous monitoring mechanisms and establish protocols for regular updates and patches to address newly identified vulnerabilities and threats.

Educating Stakeholders: Educate stakeholders involved in the IoV ecosystem, including developers, users, and manufacturers, about potential security risks and best practices to minimize these risks. Awareness is key to maintaining a secure environment.

Collaboration and Information Sharing: Foster collaboration within the industry to share insights, best practices, and emerging threat intelligence. Engaging in forums, consortiums, and information-sharing platforms can help in staying updated on evolving security challenges.

By applying these methods, an informal security analysis can help in identifying and mitigating potential risks and vulnerabilities within Blockchain-based IoV systems, ensuring a more secure and resilient environment for connected vehicles and their associated networks.

## EDGE COMPUTING IN IoV FOR BLOCKCHAIN

Edge computing in the context of blockchain and the Internet of Vehicles (IoV) plays a crucial role in enhancing the efficiency, scalability, and security of data processing within vehicular networks. The IoV, an extension of the Internet of Things (IoT), involves interconnected vehicles that communicate with each other and with infrastructure to ensure safer, more efficient transportation systems. Integrating blockchain with edge computing in IoV introduces numerous advantages and opportunities.

Reduced Latency: Edge computing brings computation and data storage closer to the source, minimizing latency. In IoV, where split-second decisions are vital for safety (e.g., autonomous driving, collision avoidance), edge nodes processing blockchain transactions locally can significantly reduce response times.

Scalability: Edge nodes in IoV can distribute the computational load associated with blockchain consensus mechanisms, enabling scalability as the number of vehicles and transactions increases. This decentralized approach prevents bottlenecks in processing power.

Enhanced Security: Edge devices can implement consensus protocols and verify transactions at the edge, securing data before it's propagated to the blockchain network. This helps in preventing potential attacks or data tampering, enhancing the overall security of vehicular communications.

Data Privacy: Edge computing allows for sensitive data processing to occur locally, preserving the privacy of users' information. With blockchain, smart contracts and encrypted transactions can be managed at the edge, ensuring that only authorized entities have access to specific data.

Optimized Resource Utilization: By offloading certain tasks to edge devices, the overall network bandwidth and cloud server load can be reduced. This optimization allows for more efficient utilization of resources and cost savings.

## USE CASES

Traffic Management: Edge computing can process traffic data from vehicles in real-time, optimizing routes and traffic flow. Blockchain can securely store this information, creating a reliable and immutable record of traffic patterns.

Vehicle-to-Vehicle (V2V) Communication: Edge nodes can facilitate direct, secure communication between vehicles for sharing data related to road conditions, potential hazards, or collaborative decision-making (e.g., forming ad hoc networks for platooning).

Payment and Transactions: With blockchain integrated at the edge, IoV can enable secure and efficient micropayments for services like tolls, parking, or energy consumption between vehicles or with infrastructure.

Fleet Management and Maintenance: Edge devices can monitor vehicle performance metrics and maintenance needs. Blockchain can securely store this data, ensuring transparency and traceability in vehicle histories.

## CHALLENGES OF EDGE COMPUTING WITH BLOCK CHAIN TECHNOLOGY

Resource Constraints: Edge devices may have limited processing power and storage capacity, posing challenges for executing complex blockchain operations.

Security Concerns: Ensuring the security of edge devices against cyber threats and attacks is crucial to maintaining the integrity of the IoV.

Interoperability: Integrating various edge devices and blockchain platforms across different vehicle manufacturers can be challenging due to interoperability issues.

Edge computing integrated with blockchain technology holds tremendous potential to revolutionize the IoV by enhancing security, reducing latency, and enabling efficient and secure data processing for connected vehicles. As technology continues to advance, addressing challenges and refining these systems will be pivotal in realizing the full potential of this integration.

## CONCLUSION

The main aim of literature study is current state of blockchain technology in ITS field as its one of the fastest developing research field. This paper we have discussed right from understanding of blockchain technology to common applications and the focusing on increasingly growing to make secure and safe connectivity in ITS. We also focused on literature survey on blockchain technology in ITS. block chain technology cab be likely adopted to overcome number of Challenges on our proposed concept of Dynamic route planning and traffic management for autonomous vehicles can be enhanced by leveraging Block chain technology.

## ACKNOWLEDGMENT

The authors would like to thank Dr P Shyamaraju, Chancellor, REVA University, for all the facilities provided to carry out the research activities.


## REFERENCES


Algarni, S., Eassa, F., & Almarhabi, K. (2021). Blockchain-Based Secured Access Control in an IoT System. *Applied Science.*

Bao, Z., Shi, W., He, D., & Chood, K.-K. R. (2018). *IoTChain: A Three-Tier Blockchain-based IoT Security Architecture.* Cryptography and Security.

Chuang, M.-C., & Lee, J.-F. (2011). PPAS: A privacy preservation authentication scheme for vehicle-to-infrastructure communication networks. *IEEE International Conference on Consumer Electronics, Communications and Networks.* IEEE. 10.1109/CECNET.2011.5768254

Dorri, A., Kanhere, S. S., Jurdak, R., & Gauravaram, P. (2017). Blockchain for IoT security and privacy: The case study of a smart home. *IEEE International Conference on Pervasive Computing and Communications.* IEEE. 10.1109/PERCOMW.2017.7917634

Fongen, A. (2012). Management and Integrity Protection in the Internet of Things. *IEEE International Conference on Emerging Security Technologies,* (pp. 111-114). IEEE.

Golosova, J., & Romanovs, A. (2018). The advantages and disadvantages of the blockchain technology. *2018 IEEE 6th workshop on advances in information, electronic and electrical engineering (AIEEE).* IEEE. 10.1109/AIEEE.2018.8592253

Hammoud, A., Sami, H., Mourad, A., Otrok, H., Mizouni, R., & Bentahar, J. (2020). AI, blockchain, and vehicular edge computing for smart and secure IoV: Challenges and directions. *IEEE Internet of Things Magazine, 3*(2), 68–73. doi:10.1109/IOTM.0001.1900109

Hegde, N. (2016). Thesis Proposal Summary: Key Management Authentication and Non Repudiation for Information Transaction in Vehicular Cloud Environments. *2016 IEEE International Conference on Cloud Computing in Emerging Markets (CCEM),* (pp. 157-160). IEEE.

Hegde, N. & Manvi, S. (2019). *A novel key management protocol for vehicular cloud security.* TELKOMNIKA (Telecommunication Computing Electronics and Control).

Hegde, N. (2023). Secure search scheme for encrypted data in the VANET cloud with random query trapdoor. Information Security Journal: A Global Perspective, 32(5), 383-400.

Hegde, N., & Manvi, S. (2022). Distributed integrity and non-repudiation scheme in the dynamic vehicular cloud environment. *International Journal of Information and Computer Security, 1*(1), 15.

Hegde, N., & Manvi, S. S. (2016). Implementation of VANET routing using computational intelligence. *2016 International Conference on Communication and Signal Processing (ICCSP),* Melmaruvathur, India. 10.1109/ICCSP.2016.7754292

Huang, J., Kong, L., Chen, G., Wu, M.-Y., Liu, X., & Zeng, P. (2019). Towards Secure Industrial IoT: Blockchain System With Credit-Based Consensus Mechanism. *IEEE Transactions on Industrial Informatics, 15*(6), 3680–3689. doi:10.1109/TII.2019.2903342

Kamble, S. J., & Kounte, M. R. (2018). Enabling Technologies for Internet of Vehicles. *International conference on Computer Networks, Big data and IoT*. Springer, Cham.

Kamble, S. J., & Kounte, M. R. (2023). Trends and Open Research Issues in Intelligent Internet of Vehicles. *Transport and Telecommunication Journal, 24*(2), 143–157. doi:10.2478/ttj-2023-0013

Kapassa, E., Themistocleous, M., Christodoulou, K., & Iosif, E. (2021). Blockchain application in internet of vehicles: Challenges, contributions and current limitations. *Future Internet, 13*(12), 313. doi:10.3390/fi13120313

Li, H., Pei, L., Liao, D., Sun, G., & Xu, D. (2019). Blockchain meets VANET: An architecture for identity and location privacy protection in VANET. *Peer-to-Peer Networking and Applications, 12*(5), 1178–1193. doi:10.100712083-019-00786-4

Lin, C., He, D., Huang, X., Choo, K.-K. R., & Vasilakos, A. V. (2018). DebiaoHe, Xinyi Huang, Kim-Kwang Raymond Choo, Athanasios V.Vasilakos, "BSeIn: A blockchain-based secure mutual authentication with fine-grained access control system for industry 4.0". *Journal of Network and Computer Applications, 116*(15), 42–52. doi:10.1016/j.jnca.2018.05.005

Lu, Z., Liu, W., Wang, Q., Qu, G., & Liu, Z. (2018). A Privacy-Preserving Trust Model Based on Blockchain for VANETs. *IEEE Access : Practical Innovations, Open Solutions, 6*, 45655–45665. doi:10.1109/ACCESS.2018.2864189

Lu, Z., Wang, Q., Qu, G., & Liu, Z. (2018). BARS: A Blockchain-Based Anonymous Reputation System for Trust Management in VANETs. IEEE International Conference On Big Data Science And Engineering. IEEE. 10.1109/TrustCom/BigDataSE.2018.00025

Mollah, M. B., Zhao, J., Niyato, D., Guan, Y. L., Yuen, C., Sun, S., Lam, K.-Y., & Koh, L. H. (2020). Blockchain for the internet of vehicles towards intelligent transportation systems: A survey. *IEEE Internet of Things Journal, 8*(6), 4157–4185. doi:10.1109/JIOT.2020.3028368

Pathak, N., & Patil, P. R. (2023, March). Securing VANET Communication Using Block Chain: A Review. In *2023 International Conference on Communication System, Computing and IT Applications (CSCITA)* (pp. 215-218). IEEE 10.1109/CSCITA55725.2023.10105098

Peng, C. (2020). Blockchain for vehicular Internet of Things: Recent advances and open issues. *Sensors, 20.*

Peng, X. (2016). A novel authentication protocol for vehicle network. IEEE International Conference on Systems and Informatics. IEEE. 10.1109/ICSAI.2016.7811036

Puri, V. (2020). A vital role of blockchain technology toward Internet of vehicles. In *Handbook of Research on Blockchain Technology* (pp. 407–416). Academic Press. doi:10.1016/B978-0-12-819816-2.00016-2

Rowan, S., Clear, M., Gerla, M., Huggard, M., & Mc Goldrick, C. (2017). *Securing Vehicle to Vehicle Communications using Blockchain through Visible Light and Acoustic Side-Channels*. Cryptography and Security.

Singh, M., & Kim, S. (2018). Branch based blockchain technology in intelligent vehicle. *Computer Networks, 145*, 219–231. doi:10.1016/j.comnet.2018.08.016

Tandon, R., Verma, A., & Gupta, P. K. (2022, November). Blockchain enabled vehicular networks: A review. In *2022 5th International Conference on Multimedia, Signal Processing and Communication Technologies (IMPACT)* (pp. 1-6). IEEE.

Tauseef, M., Kounte, M. R., Nalband, A. H., & Ahmed, M. R. (2023). Exploring the Joint Potential of Blockchain and AI for Securing Internet of Things. *International Journal of Advanced Computer Science and Applications, 14*(4). doi:10.14569/IJACSA.2023.0140498

Tripathi, G., Ahad, M. A., & Sathiyanarayanan, M. (2019). The Role of Blockchain in Internet of Vehicles (IoV): Issues, Challenges and Opportunities. *2019 International Conference on contemporary Computing and Informatics (IC3I)*. IEEE. 10.1109/IC3I46837.2019.9055613

van Dijk, J. (2020). *State-of-the-art of Blockchain Technology in Intelligent Transportation Systems: Literature review*. TU Delft Repositories.

Xie, L., Ding, Y., Yang, H., & Wang, X. (2019). Blockchain-Based Secure and Trustworthy Internet of Things in SDN-Enabled 5G-VANETs. *IEEE Access : Practical Innovations, Open Solutions, 7*, 56656–56666. doi:10.1109/ACCESS.2019.2913682

Yuan, Y., & Wang, F.-Y. (2016). Towards blockchain-based intelligent transportation systems. *IEEE International Conference on Intelligent Transportation Systems*. IEEE.

Zhao, G., Si, X., Wang, J., Long, X., & Hu, T. (2011). A novel mutual authentication scheme for Internet of Things. *IEEE International Conference on Modelling, Identification and Control*. IEEE.

# Chapter 11
# Building a Secure and Transparent Online Conversation Platform:
## A Blockchain-Based Website With Integrated Virtual Assistant

**Md Tauseef**
*Reva University, India*

**Sayak Jana**
https://orcid.org/0009-0001-3729-1433
*Reva University, India*

**Aashish R. Gouda**
*Reva University, India*

**Ashutosh Mishra**
*Reva University, India*

**Utkarsh Raj Srivastava**
*Reva University, India*

## ABSTRACT

*This chapter explores the integration of artificial intelligence (AI) and blockchain technology to create a secure and transparent online conversation platform. Advancements in AI have given rise to virtual assistants capable of human-like interactions and versatile task execution. By harnessing blockchain technology, these virtual assistants enhance security and response times, making them invaluable in applications ranging from healthcare to gaming. The chapter presents a project that enhances virtual assistant capabilities through blockchain integration, featuring a self-learning module and a Solana-based blockchain network with smart contracts. Implementation employs Vanilla JavaScript, Express.js, and Node.js, offering end-to-end customer support and adapting to modern lifestyles. The fusion of AI and blockchain showcases the potential for a robust and secure online conversation platform.*

DOI: 10.4018/979-8-3693-0659-8.ch011

*Figure 1. Current landscape of online conversation platform*

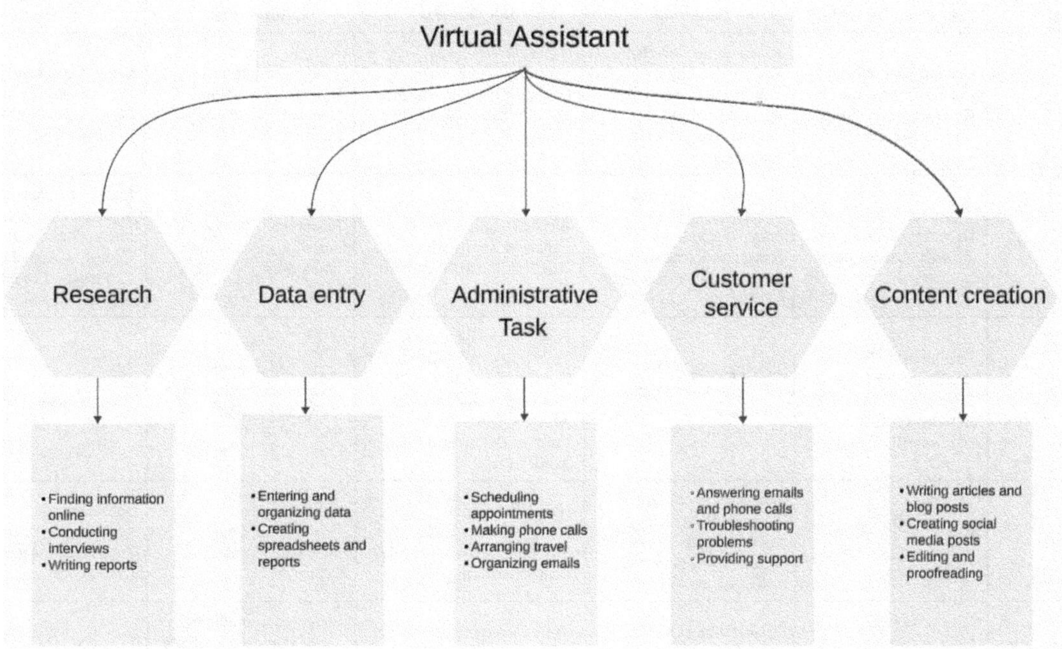


## INTRODUCTION

In today's rapidly evolving technological landscape, Artificial Intelligence (AI) is at the forefront of computer science, with a primary focus on developing intelligent machines. One of the most prominent applications of AI is the creation of virtual assistants, which utilize sophisticated algorithms to comprehend and respond to user commands, execute tasks, and provide valuable information. These virtual assistants employ an array of technologies, including natural language processing, machine learning, and voice recognition, to interpret user queries and generate contextually relevant responses. Through continuous learning and adaptation, these AI-powered virtual assistants enhance their proficiency in understanding and assisting users over time. Virtual assistants have seamlessly integrated into our daily lives, simplifying tasks, offering personalized recommendations, and bolstering overall productivity. Notable examples of such virtual assistants include Siri and Alexa, which harness AI algorithms to comprehend user commands, execute tasks, and furnish information. Figure 1 illustrates the usage of virtual assistants in research, data entry, administrative tasks, customer service, and content creation, providing a visual representation of the current landscape of online conversation platforms.

Now, let's explore how blockchain technology can be integrated into virtual assistants to augment their capabilities and confer additional advantages. By harnessing blockchain, virtual assistants can provide heightened security, trust in transactions, enhanced transparency, and data integrity. One of the methods to integrate blockchain is through the use of smart contracts. Smart contracts are self-executing agreements with the terms of the contract directly encoded into the blockchain. By implementing smart contracts within a blockchain network, virtual assistants can automate and securely execute transactions, such as making payments or transferring assets, without necessitating intermediaries. Furthermore, blockchain integration enables the creation of decentralized applications (Dapps) within the blockchain

*Table 1. Overview of existing approaches and technology*

| Ref | Approach/ Technology | Description | Key Features | Advantages | Limitation |
|---|---|---|---|---|---|
| Xie et al., 2019 | Chatbot Integration, Chatterbot Framework, Cryptocurrency Focus, Data Collection | Exploration of chatbot and cryptocurrency integration in the digital age, emphasizing automated cryptocurrency information. | Chatterbot Framework, Cryptocurrency Information, Data Collection, Real-Time Quotes | Automated Information Retrieval, Real-Time Updates, User-friendly Interface | Dependencies on APIs, Data Quality, Complex Queries, Narrow Scope, Response Time |
| Lin et al., 2023 | Hybrid Blockchain and Zero-Knowledge Proofs | Proposal of blockchain-aided semantic communication for virtual transportation networks, focusing on security and scalability. | Blockchain Framework, Enhanced Security Protocols, Real-time Data Validation | Resilience to Attacks, Scalability, Auditable Framework | Lack of Blockchain Standards for Interoperability, Privacy Concerns, Adoption Challenges, Complex Implementation |
| Zheng et al., 2023 | Decentralized Finance (DeFi) and NFT Focus, Systematic Framework, Activity, Profitability, and Security Analysis | Investigation of DeFi and NFT growth on Ethereum, offering a framework for in-depth market analysis. | Focus on DeFi and NFT - Systematic Analysis Framework - Activity, Profitability, and Security Analysis, Quantitative Insights, Dataset and Code Availability | Structured Analysis, Comprehensive Insights, Transparency and Reproducibility | Heterogeneous Data |
| Sorin, 2023 | Blockchain in Critical Infrastructure Networks | Introduction of blockchain into command and control systems of critical infrastructure networks for enhanced security. | Novel Blockchain Framework, Enhanced Security Protocols, Real-time Data Validation | Cyber Threat Fortification, Data Integrity, Decentralized Control | Self-learning, Multitasking, Natural Language Processing (NLP) |
| Abougarair et al., 2022 | Smart Voice Assistant with Academic Vocabulary Focus | Presentation of a smart voice assistant system focusing on academic vocabulary recognition. | Speech Recognition Technology, Comprehensive Academic Word Database, Real-time Feedback | Enhanced Communication Skills, Improved Learning Experiences, Speech Recognition, NLP Support | MFCC Methods Coefficients |
| Yang et al., 2022 | Blockchain and AI in the Metaverse | Exploration of blockchain and AI's potential applications in the metaverse, highlighting advantages and challenges. | Immutable Blockchain Data, Traceability, Auditability, Realism and Tailored Content for Users | Efficiency, Scalability, Immersive Experience, Trust Building | Lack of Blockchain Privacy, Regulatory Uncertainty, Technology Adoption Challenges, Complexity in Implementation |
| Kashem et al., 2022 | AI and Blockchain for Sustainable Tourism | Discussion of AI and blockchain's potential for sustainable tourism in the MENA region. | AI-driven Sustainable Tourism Products and Services, Blockchain Digital Marketplace, Environmental Impact Tracking, Resource Management | Development of Sustainable Tourism, Transparency and Accountability, Tourist Engagement | Lack of Trust in AI and Blockchain, AI Bias - Interoperability Challenges, Integration Difficulties |
| Huynh-The et al., 2022 | Machine Learning, Deep Learning, Metaverse | In-depth overview of AI's intersection with the metaverse, focusing on user experiences and content interaction. | Immersive Environment Optimization, Tailored Survey Interactions, Industry Revolutionization | Integration of AI in the Metaverse, Transparency and Examination, Content Customization | AI Integration Challenges, Lack of Metaverse Involvement in AI, Privacy Concerns, Complexity in Implementation |
| Mekni et al., 2021 | Conversational Agent, Virtual Assistant | Proposal of an AI-driven virtual assistant capable of simulating human-like interactions, enhancing user experiences. | AI-driven Virtual Assistant Implementation, Customer Service Revolution, Information Retrieval, Personalized Interaction | Seamless Interaction, Personalized Assistance, Adaptability to User Preferences | Self-learning Management System (SLMS), Open Educational Resources (OER), Voice Command Support |
| Firouzi et al. 2021 | IoT, AI, Robotics, and Blockchain for Addressing the COVID-19 Pandemic | Examination of the role of IoT, AI, Robotics, and Blockchain in addressing the COVID-19 pandemic and associated challenges. | Unified Efforts - Four Technological Domains, Comprehensive Overview, Interdisciplinary Techniques - Holistic Studies | Collective Efforts, Technological Convergence, Insights and Support, Holistic Response, Early Detection and Prediction | Technological Barriers, Data Privacy Concerns, Digital Divide, Reliability of AI and Machine Learning Models |

ecosystem. Virtual assistants can seamlessly interact with these Dapps to carry out various functions, such as securely accessing and sharing data, verifying user identities, or executing intricate transactions. This empowers users with greater control over their data and heightens the overall efficiency of virtual assistant interactions. Moreover, integrating blockchain technology can significantly bolster user privacy by granting individuals greater control over their personal information. With blockchain-based identity verification, virtual assistants can securely authenticate user identities without relying on centralized databases, thereby enhancing security and privacy. Table 1 provides a concise summary of various approaches and technologies.

It presents key features, advantages, and limitations across a range of topics, including chatbot integration, blockchain-aided semantic communication, DeFi and NFT focus, smart voice assistants, AI and blockchain in the metaverse, sustainable tourism, and COVID-19 technologies. In the context of enhancing virtual assistants through blockchain technology, Solana emerges as a significant contributor. Solana, known for its high-performance blockchain platform with exceptional scalability and rapid transaction processing capabilities, plays a crucial role in augmenting the performance and functionality of blockchain-based virtual assistants. Although Solana does not directly shape virtual assistant development, its unique characteristics provide valuable enhancements (Tabassum, N,et.al, 2020). Utilizing the Solana blockchain empowers virtual assistants to leverage its exceptional transaction speeds and cost-efficiency, resulting in swift and economical execution of smart contracts and other blockchain operations. This translates into a user-friendly experience marked by minimal delays and reduced transaction costs. Additionally, Solana's scalability is a vital asset, especially for virtual assistants handling multiple tasks and engaging with numerous users simultaneously. Its ability to manage a high volume of transactions ensures efficient and uninterrupted service, accommodating a large user base without performance bottlenecks.

Furthermore, Solana's architectural framework perfectly aligns with the development of decentralized applications, seamlessly interacting with virtual assistants. This synergy ensures secure and transparent execution for tasks like data sharing, identity verification, and financial transactions. Leveraging Solana in blockchain-based virtual assistants yields distinct advantages, including rapid transaction processing, scalability, low fees, and robust support for decentralized applications. These attributes collectively contribute to an efficient and reliable virtual assistant experience, enhancing user satisfaction and overall functionality needs (Tabassum, N et. al, 2017).

This chapter tackles data storage and security challenges within virtual assistant systems, extensively exploring aspects such as self-learning AI and blockchain implementation. Its primary objective revolves around integrating AI-powered virtual assistants with blockchain technology, emphasizing advancements in human-like interactions and versatile task execution. Moreover, the chapter investigates implementing this model using the Solana blockchain platform, highlighting its advantages in enhancing security and response times. This exploration aligns seamlessly with the book's focus on blockchain applications, particularly within industrial robotics, showcasing how blockchain augments security and task efficiency, catering directly to industrial needs (Tabassum, N,et.al, 2022).

## FUNDAMENTALS OF AI-POWERED VIRTUAL ASSISTANTS

Virtual Assistant (VA) technology has experienced rapid and exciting growth since the 19th century. From the early days of chatbots to today's AI-powered assistants, VAs have evolved to become increasingly sophisticated and capable (Soofastaei et al., 2021). Their capabilities and applications continue to expand through innovative technologies that cater to people's needs, particularly with a strong emphasis on voice command interfaces. Leading companies such as Google, Amazon, Apple, and Microsoft are competing to develop highly efficient VAs. Some of the most well-known virtual assistants in our era include Amazon Alexa, Google Assistant, and Apple Assistant (Siri), which primarily operate as cloud-based data management VAs to maximize performance. In 1922, the first voice-activated doll, named "Radio Rex," made its debut (Kiiski et al, 2020). Thirty years later, in 1952, Bell Labs introduced "Audrey," an Automatic Digit Identification device. IBM contributed to voice-activated technology with their "Shoebox" voice-activated calculator in 1961, which was a pioneering digital language identification device (Soofastaei, 2021). One of the earliest virtual assistants, "ELIZA," emerged in 1966 through the work of Joseph Weizenbaum at MIT. ELIZA was the first Natural Language Processing (NLP) web application, designed to simulate conversations with a psychotherapist. It accomplished this by using a simple pattern-matching algorithm to match user input with a set of predefined responses (Weizenbaum, 1976). Virtual Assistants (VAs) are web-based software applications capable of performing a wide range of tasks on behalf of human users, accepting both text inputs and voice commands (Yang et al. 2019). Virtual Assistants have ushered in a transformative era across industries and domains, reshaping how businesses, educational institutions, healthcare providers, and individuals manage tasks, access information, and boost productivity (Schmidt et al., 2018) These versatile digital assistants find applications in a multitude of fields, some of which are highlighted below

- **Customer Service:** Virtual assistants are invaluable for businesses in providing diverse customer service solutions. They excel at handling inquiries, offering troubleshooting guidance, facilitating purchases, and more. By doing so, they liberate human customer service representatives to tackle more intricate tasks.
- **Education:** Virtual assistants are making waves in online learning platforms, where they respond to student queries, furnish study materials, and guide learners through coursework. Additionally, they offer language learning support and tutoring services (Vishnu et al., 2022).
- **Healthcare:** In the healthcare sector, virtual assistants prove indispensable. They assist in scheduling appointments, managing medication regimens, delivering timely reminders, and even collecting crucial patient data. Their presence enhances the efficiency and quality of healthcare services (Wang et al., 2023).
- **Logistics:** In logistics operations, virtual assistants shine by overseeing tasks like shipment tracking, delivery schedules, and inventory management. These contributions not only enhance the customer experience but also drive increased sales.
- **Smart Homes and IoT:** With virtual assistants like Amazon Alexa and Google Assistant, smart homes become more convenient and efficient. They control smart home devices, adjust thermostats, play music, set alarms, and provide information—all through voice commands.
- **E-commerce:** In the realm of e-commerce, virtual assistants offer customer service in retail stores, address product inquiries, provide recommendations, and manage inventory and orders. These capabilities elevate the customer experience and contribute to sales growth.

These examples represent just a fraction of the diverse applications of virtual assistants today (Soofastaei, 2021). As technology continues to advance, we anticipate witnessing even more innovative and creative applications for virtual assistants in the future. While traditional virtual assistants have made significant strides, certain limitations and challenges persist in their development and implementation:

- **Sustainability:** Achieving real-life relevance and sustainability is paramount. Many regions lack reliable high-speed internet access, posing challenges in creating highly efficient virtual assistant models that can function without or with limited internet connectivity.
- **Security Risks:** Traditional virtual assistants can potentially pose security risks, as they may access sensitive information. The selection of a virtual assistant with robust security features is crucial to mitigate such risks.(Wang et al., 2023, Njoku et al., 2023)
- **Understanding of Natural Language:** Traditional virtual assistants may struggle to grasp natural language, particularly when it's ambiguous or open-ended. This can lead to misunderstandings and errors in interactions (Ray, 2023).
- **Limited Learning Ability:** Unlike humans, traditional virtual assistants may not adapt and learn as easily. This can result in repetitive interactions and difficulty in keeping up with new information or changes in user preferences.

## INTRODUCTION OF BLOCKCHAIN TECHNOLOGY

Blockchain, often associated with cryptocurrencies like Bitcoin, was introduced by Satoshi Nakamoto in 2008 in a whitepaper titled "Bitcoin: A Peer Electronic Cash System" (Nakamoto, 2008). At its core, blockchain is a revolutionary shift in decentralized database technology. It's essentially a chain of data blocks connected through advanced cryptographic techniques (Tapscott, 2016). Each block in this chain contains information about a batch of Bitcoin transactions, crucial for verifying data integrity and forming the next block. Think of blockchain as a unique distributed infrastructure and computational concept designed to facilitate value exchange. It acts like a digital ledger, capable of housing digital transactions, data records, and executable files. Transactions are grouped into larger units called blocks, each with a timestamp and cryptographic link to preceding blocks, forming a sequential "blockchain" of records.

Blockchain is a complex technical ecosystem, integrating multiple key technologies. Its foundational components include asymmetric encryption algorithms, distributed data storage, consensus algorithms, and smart contracts (Dasgupta et al., 2019).

As shown in Figure 2, Blockchain is a secure and transparent distributed ledger technology used to record tamper-proof transactions and data. However, it faces security threats, including:

- Double-spending attack: An attempt to spend the same digital asset twice through a fraudulent, unrecorded transaction.
- 51% attack: Control of over 50% of a blockchain network's computing power, enabling censorship, transaction reversals, and fake blocks.
- Sybil attack: Creation of numerous fake identities to control a blockchain network.
- Key management attack: Theft of a user's private keys, allowing unauthorized access and transactions.
- DDoS attack: Flooding a blockchain network with traffic to disrupt legitimate use.

To counter these threats, security mechanisms are employed:

*Figure 2. Basics of blockchain and distributed ledger technology*

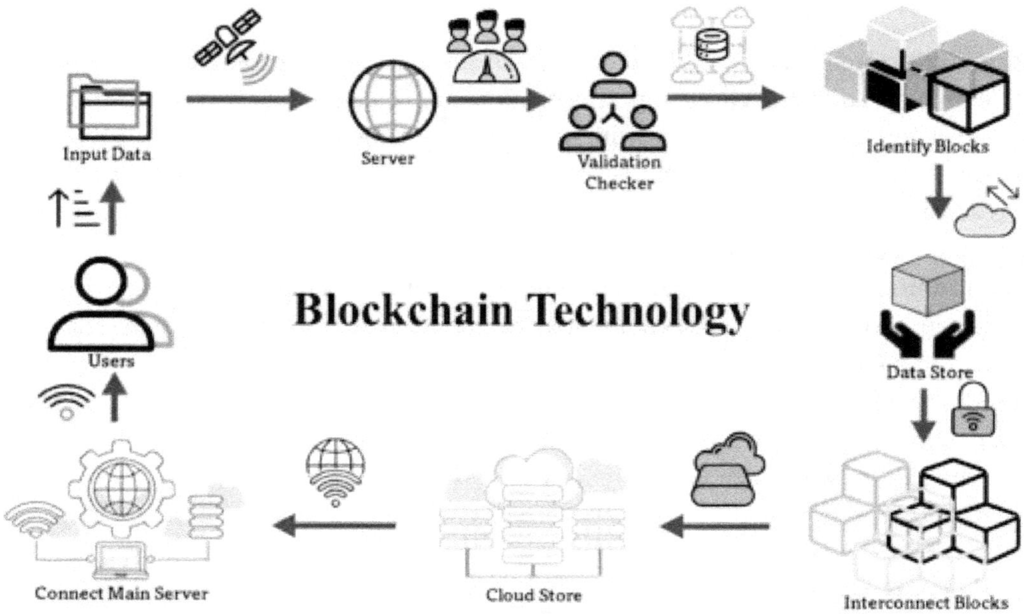


- Consensus mechanisms: Prevent double-spending and malicious activities by reaching agreement on the blockchain's state.
- Cryptography: Secures data through hashing algorithms for transaction identifiers and encryption for private keys.
- Access control: Restricts network access and transactions to authorized users.
- Monitoring and auditing: Detects suspicious activity for prevention.

Table 2 offers a comparison of various consensus mechanisms commonly used in blockchain integration for online conversation platforms. Each mechanism has strengths and limitations, and the choice depends on platform requirements. The chosen consensus mechanism significantly influences security, scalability, and efficiency in the blockchain-based virtual assistant system.

Blockchain technology, along with smart contracts, has spurred the development of numerous applications across various sectors and scenarios. Beyond its primary role in financial applications for decentralized monetary transactions, blockchain finds utility in diverse domains for recording a wide array of data types. These applications span the spectrum from public data, such as vehicle registries, to semi-public information like educational degrees, and even encompass highly private records like wills. Furthermore, blockchain's capabilities extend to recording both intangible assets, like digital coupons, and tangible ones, such as electronic hotel room keys (Wang et al., 2021).

In this section, a concise overview of both existing and envisioned applications based on blockchain technology and smart contracts is provided. These applications are categorized according to their respective sectors and contexts. The intention behind this overview is to provide the reader with a comprehensive understanding of the diverse range of applications enabled by blockchain and to illuminate the current

*Table 2. Comparison of consensus mechanisms*

| Consensus Mechanism | Description | Use Case | Advantages | Limitations |
|---|---|---|---|---|
| Proof of Work (PoW) | Miners solve complex mathematical puzzles to validate transactions and create new blocks. | Cryptocurrencies (e.g., Bitcoin) and secure decentralized networks. | Security through computational work, decentralized, established (e.g., Bitcoin). | Energy-intensive, centralization risk due to mining pools, scalability challenges. |
| Proof of Stake (PoS) | Validators lock up a certain amount of cryptocurrency as collateral to create new blocks. | Ethereum 2.0 and other blockchain platforms. | Energy-efficient, less centralization risk, faster transaction processing. | Initial wealth concentration, "Nothing at Stake" problem, validator collusion. |
| Delegated Proof of Stake (DPoS) | Token holders vote for a set of delegates to validate transactions and create blocks. | EOS and BitShares. | High throughput, energy-efficient, low latency, and governance by token holders. | Centralization risk through delegate voting, and voter apathy. |
| Practical Byzantine Fault Tolerance (PBFT) | A consensus algorithm where nodes reach an agreement through multiple rounds of voting. | Some permissioned blockchain networks. | High throughput, quick finality, robust against Byzantine failures. | Limited scalability, requires a known set of participants, and higher communication overhead. |
| Proof of History (PoH) | Uses a unique PoH (Proof of History) mechanism alongside a PoS consensus. | High-performance decentralized applications and DeFi. | Scalability, high throughput, low fees, fast confirmation times. | Less battle-tested, reliance on hardware, centralization risk. |
| Proof of Authority (PoA) | Validators are pre-approved, typically by a central authority, to create blocks. | Private and consortium blockchains. | High security, permission control, and low energy consumption. | Centralization due to trust in validators, limited decentralization. |

state of this technology and its societal impact (Al-Megren et al., 2018). Below are some of the most promising use cases of blockchain across various sectors:

1. **Financial Services:** Blockchain enables the secure and transparent recording of financial transactions, potentially reducing fraud, expediting settlement times, and simplifying asset tracking.
2. **Supply Chain Management:** Blockchain enhances efficiency, transparency, and traceability by tracking the movement of goods and materials through supply chains.
3. **Healthcare:** Blockchain offers a secure and private solution for storing and sharing medical records, potentially improving patient care and supporting medical research.
4. **Energy:** Blockchain facilitates tracking of energy production, distribution, and consumption, contributing to reduced energy consumption and enhanced efficiency.
5. **Government:** Blockchain enhances government transparency and efficiency, with applications in voting, land registry, and social welfare programs.
6. **Media and Entertainment:** Blockchain protects intellectual property rights and monitors content distribution, potentially reducing piracy and increasing revenue for content creators.
7. **Non-profit:** Blockchain improves transparency and efficiency in non-profit organizations, aiding in fundraising, grant management, and volunteer coordination.
8. **Logistics:** Similar to supply chain management, blockchain tracks the movement of goods and materials, providing efficiency gains, transparency, and traceability.

9. **Real Estate:** Blockchain securely and transparently records property ownership and transactions, potentially reducing fraud and enhancing operational efficiency.

These examples represent just a portion of the numerous potential use cases for blockchain technology. As blockchain continues to evolve, we can anticipate the emergence of even more innovative applications in the coming years (Shrimali et al., 2022).

In addition to the mentioned use cases, there are other compelling applications of blockchain technology:

1. **Smart Contracts:** These self-executing contracts, stored on the blockchain, automate various tasks, including loan agreements, insurance policies, and supply chain contracts.
2. **Decentralized Finance (DeFi):** DeFi is a financial system built on blockchain, allowing individuals to access financial services without the need for traditional intermediaries like banks.
3. **Non-fungible Tokens (NFTs):** NFTs are unique digital assets stored on the blockchain, representing a wide range of items, from art and music to in-game collectibles.

As blockchain technology continues its advancement, we can look forward to a wealth of innovative applications in the years ahead. Figure 3 represents the diverse use cases of blockchain technology across different sectors

*Figure 3. Different use cases of blockchain in different sectors*

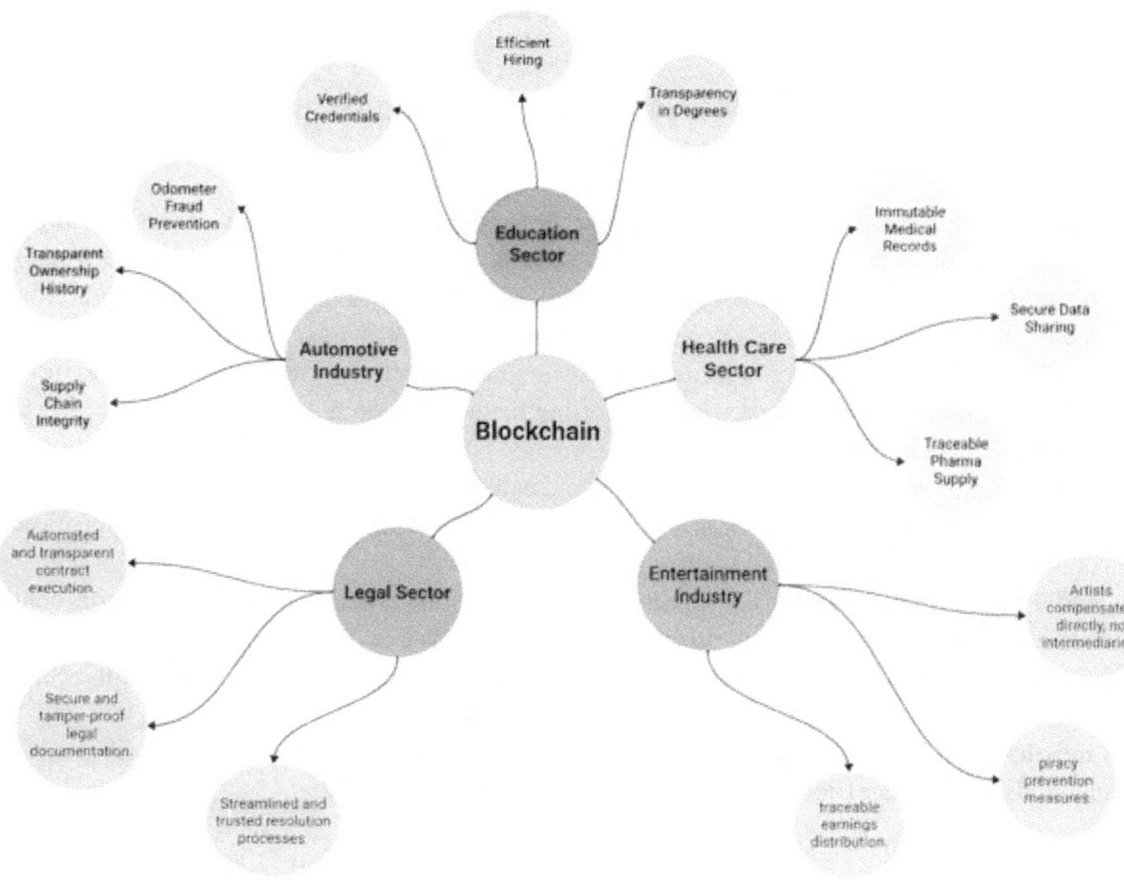

Integrating blockchain technology into virtual assistants offers a multitude of advantages. Firstly, blockchain's robust cryptographic techniques ensure the security of virtual assistant operations by encrypting user data and interactions. This security significantly reduces the risk of data breaches and unauthorized access. Moreover, users gain enhanced control over their personal data, allowing them to manage access permissions effectively. Blockchain's decentralized nature eliminates the need for a central authority, making virtual assistants resistant to censorship and manipulation. Data recorded on the blockchain becomes tamper-proof and immutable, ensuring the integrity of interactions with virtual assistants (Adel et al. 2022). Smart contracts automate tasks within virtual assistants, improving efficiency and transparency. Users can verify the accuracy of transactions and interactions, fostering trust in the system. Eliminating intermediaries reduces costs and minimizes errors or fraud. Blockchain-enabled virtual assistants are globally accessible, and micropayments using cryptocurrencies create new monetization opportunities. Users can assert true ownership of digital assets and content within virtual assistants, facilitating secure transactions. Blockchain provides robust identity verification and maintains transparent records for compliance and accountability. Its distributed nature ensures operational resilience, even in the face of failures or attacks. The trust-building properties of blockchain instill confidence in virtual assistants, enhancing their authenticity and reliability. Incentive mechanisms encourage developers to innovate and enhance virtual assistant services (Saritha et al, 2021).

## BUILDING THE BLOCKCHAIN-BASED VIRTUAL ASSISTANT PLATFORM

In the modern age, virtual assistants have become an integral part of our daily lives, seamlessly integrated into various smart devices such as smartphones and smart homes (Gatteschi et al., 2020). While AI-based virtual assistants like Personal Virtual Assistants (PVAs) hold immense potential for value creation, it's imperative to acknowledge their associated costs and the potential negative emotional impact they can have on the human workforce. Research has shed light on how malicious users can inadvertently pressure innocent users to divulge information after innocently downloading seemingly harmless data onto the PVA they are using (Hornung et al., 2022).

Within this model, as shown in Figure 4, we have developed a virtual assistant designed to address these challenges. Our solution involves a smart virtual assistant seamlessly integrated with blockchain technology. The implementation encompasses several key steps:

1. Design and Development of the Virtual Assistant using Python
2. Integration of the Virtual Assistant with the Solana Blockchain Network
3. Development of Smart Contracts in Rust to Facilitate Automated Task Execution
4. Implementation of a Frontend using Vanilla JavaScript, along with a Backend built using Express.js and Node.js
5. Rigorous Testing and Debugging of both the Virtual Assistant and Smart Contracts
6. Deployment of the Virtual Assistant on a Blockchain-Based Website

This implementation ensures that users benefit from a secure, efficient, and personalized experience, while also guaranteeing data integrity and transparency through blockchain technology. We believe that this work has the potential to revolutionize how businesses operate and provide services to their customers.

*Figure 4. Architecture of the proposed model*

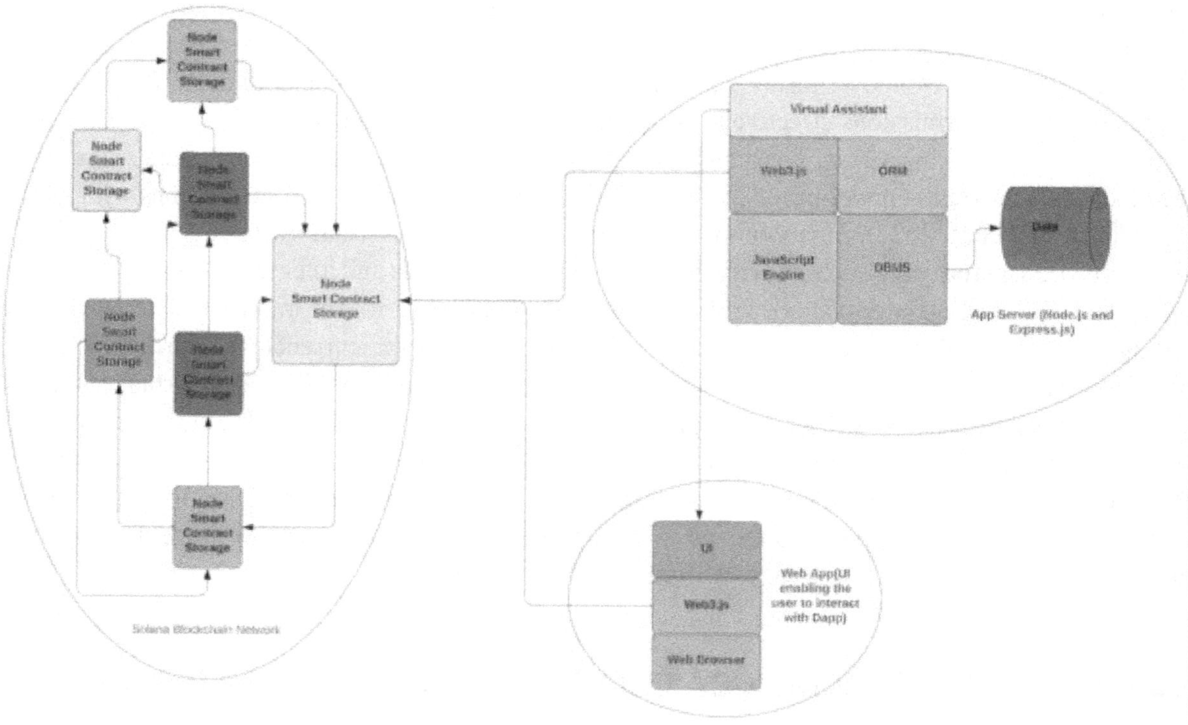


Self-learning in virtual assistant technology involves the continuous enhancement of virtual assistants' knowledge, understanding, and capabilities through machine learning algorithms and data analysis, all without human intervention. These virtual assistants, whether AI chatbots or voice assistants, can adapt and evolve based on user interactions and the data they accumulate. The synergy of machine learning algorithms and data analytics is at the core of enabling self-learning virtual assistants. This capability is pivotal in enhancing the proficiency of virtual assistants over time

Noteworthy examples of how machine learning algorithms and data analysis contribute to self-learning virtual assistants include:

1. **Natural Language Processing (NLP):** NLP is crucial in enabling virtual assistants to understand and respond to user queries in natural language. Machine learning algorithms train NLP models to recognize language patterns and generate relevant responses.
2. **Speech Recognition:** Virtual assistants use speech recognition, which involves training models with machine learning algorithms to identify and understand spoken words, allowing users to interact through voice.
3. **Recommendation Engines:** These engines recommend products or services based on user behavior, facilitated by machine learning algorithms that identify patterns in user interactions and make personalized recommendations.

The combination of machine learning and data analysis empowers virtual assistants to evolve intelligently, offering users a dynamic and tailored experience.

*Table 3. Comparison of different blockchain platforms*

| Blockchain Platform | Description | Consensus Mechanisms | Transaction Throughput | Scalability | Developer Friendly |
|---|---|---|---|---|---|
| Solana | A high-performance blockchain platform known for its scalability and fast transaction processing capabilities. | Proof of History (PoH) combined with Proof of Stake (PoS) | Thousands of transactions per second | Excellent scalability | Extensive documentation and developer support |
| Ethereum | A widely-used blockchain platform with a focus on smart contracts and decentralized applications (DApps). | Proof of Work (PoW) transitioning to Proof of Stake (PoS) for sustainability. | Around 30 transactions per second (before ETH 2.0) | Moderate scalability | Large developer community and well-established ecosystem |
| Binance Smart Chain (BSC) | A blockchain platform developed by Binance, designed for high-speed and low-cost transactions with a focus on DeFi. | Proof of Staked Authority (PoSA) | Around 1000 transactions per second | High scalability | Familiarity with Ethereum's tooling and Solidity |
| Cardano | A blockchain platform with a strong focus on sustainability, scalability, and security. | Ouroboros Proof of Stake (PoS), emphasizes security. | Around 10-15 transactions per second | Working on scalability, future upgrades | Formal methods for secure smart contract development |
| Polkadot | A multi-chain blockchain platform that enables interoperability between different blockchains. | Nominated Proof of Stake (NPoS), facilitates interoperability. | Varies based on connected chains | Highly scalable with added parachains | Substrate framework and compatibility with Ethereum Virtual Machine (EVM) |
| Avalanche | A blockchain platform known for its high throughput and fast finality. | Avalanche consensus | Thousands of transactions per second | Excellent scalability | EVM compatibility and simple deployment process |

Table 3 provides a comparison of different blockchain platforms that can be integrated into online conversation platforms to leverage their unique features and capabilities. Each platform has its strengths and weaknesses, and the selection should be based on the specific needs and requirements of the proposed blockchain-based virtual assistant system. The choice of the blockchain platform plays a crucial role in determining the overall performance, scalability, and security of the online conversation platform.

The Solana blockchain network is a high-performance, secure platform well-suited for decentralized applications. Leveraging a unique hybrid consensus mechanism that combines proof-of-stake (PoS) and proof-of-history (PoH), Solana achieves remarkable throughput and low latency. This technology finds application in various domains:

1. Decentralized Finance (DeFi): Solana supports the creation of DeFi applications, including lending platforms, exchanges, and derivatives markets.
2. Non-fungible Tokens (NFTs): Solana serves as a platform for creating and trading NFTs, unique digital assets representing ownership of diverse items.
3. Gaming: Solana facilitates the development of blockchain-based games, offering players immersive experiences.
4. Web3: Solana powers the next generation of decentralized web applications, promising enhanced security.

The advantages of using Solana include high throughput, low latency, affordability, and energy efficiency, making it an appealing choice for developers aiming to create high-performance, secure applications.

The front end of our blockchain-based AI chatbot was meticulously crafted using React.js, a well-established JavaScript toolkit renowned for its prowess in creating user interfaces. We chose React.js for its exceptional adaptability, scalability, and the abundance of tools and libraries within its ecosystem. This choice allowed us to design an exceptionally responsive and user-friendly chatbot interface. Key features of our React.js frontend include:

1. Enhanced User Experience: Our design philosophy prioritized a cutting-edge yet intuitively simple user interface. This involved the creation of user profiles, input fields, and chat windows, all meticulously designed to facilitate seamless interactions with the chatbot.
2. Real-time Updates: React.js's support for real-time updates ensures that users can engage with the chatbot without the inconvenience of webpage reloads. This feature is pivotal for sustaining dynamic and engaging conversations.

The backend architecture of our blockchain-based AI chatbot is a sophisticated system that integrates various technologies to ensure both reliability and security. To establish a scalable and efficient server infrastructure, we harnessed the power of Node.js and Express.js. Node.js's event-driven design excels in managing multiple concurrent connections, a vital requirement for real-time applications like chatbots. Express.js, a user-friendly Node.js web application framework, streamlined the development of RESTful APIs, routing, and middleware integration. Key features of our backend include:

1. Robust RESTful API: Our Node.js and Express.js backend provides RESTful endpoints that seamlessly facilitate communication between the frontend and backend. These APIs proficiently handle user requests, including chat messages and login procedures.
2. Python Powerhouse: Python played a pivotal role in various backend operations, particularly in the realms of machine learning (ML) and natural language processing (NLP). The chatbot's ability to interpret and respond effectively to natural language inquiries is attributed to the adept use of NLP libraries like NLTK (Natural Language Toolkit) and spaCy. These libraries were meticulously employed for preprocessing and analyzing user input.
3. Smart Contracts in Rust: The creation of smart contracts for the blockchain was masterfully executed using the Rust programming language. These smart contracts, intrinsic to the blockchain network, facilitate secure and transparent transactions, instilling confidence and accountability in the chatbot's administration. To handle the chatbot's transactions, user data, and tokens, we seamlessly integrated the blockchain with Rust-based smart contracts. This diverse and robust backend architecture, comprising Node.js, Python, and Rust, forms the resilient backbone of our blockchain-based AI chatbot. It guarantees the delivery of secure, efficient, and intelligent conversational experiences to our valued users.

Here, we showcase the implementation of a versatile Blockchain-Based Website with an Integrated Virtual Assistant, tailored to meet specific customer needs. While its current functionalities focus on providing answers and incorporating self-learning approaches, there's ample room for further refinements and enhancements.

*Figure 5. Model output terminal*

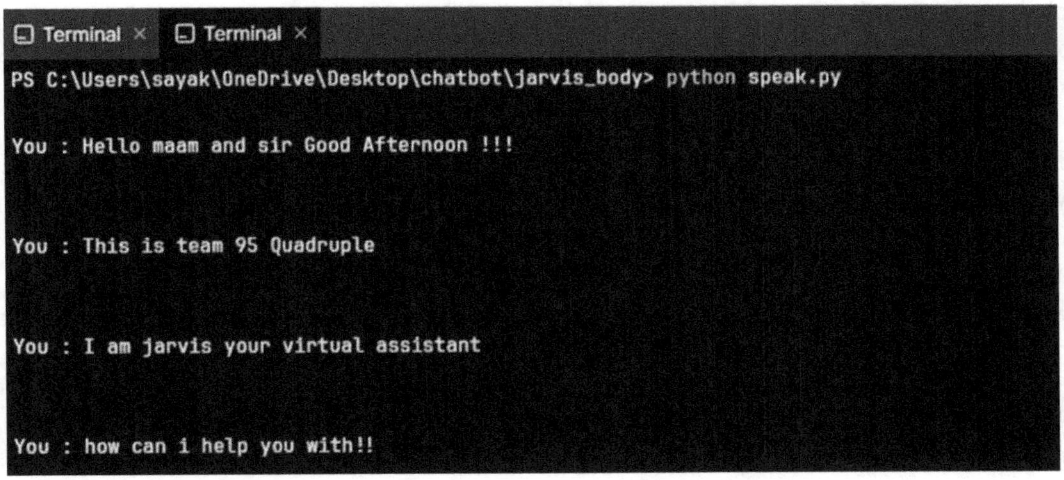


In Figure 5, the Python3-created chatbot smoothly interacts with the terminal, offering a seamless and user-friendly experience. Its sophisticated design and adeptness in addressing queries illustrate Python's capacity to craft adaptable and efficient conversational interfaces suitable for a wide array of applications.

Figure 6 showcases our Python3-based chatbot, meticulously designed to effortlessly facilitate client-side smart contract interactions. Its simplified interface for managing and querying smart contracts seamlessly integrates into existing systems, enhancing user experiences and overall functionality. This innovative fusion of chatbot engagement and smart contract capabilities extends beyond its original scope, promising transformative impacts across diverse industries. For instance, it simplifies transactional

*Figure 6. Main website*

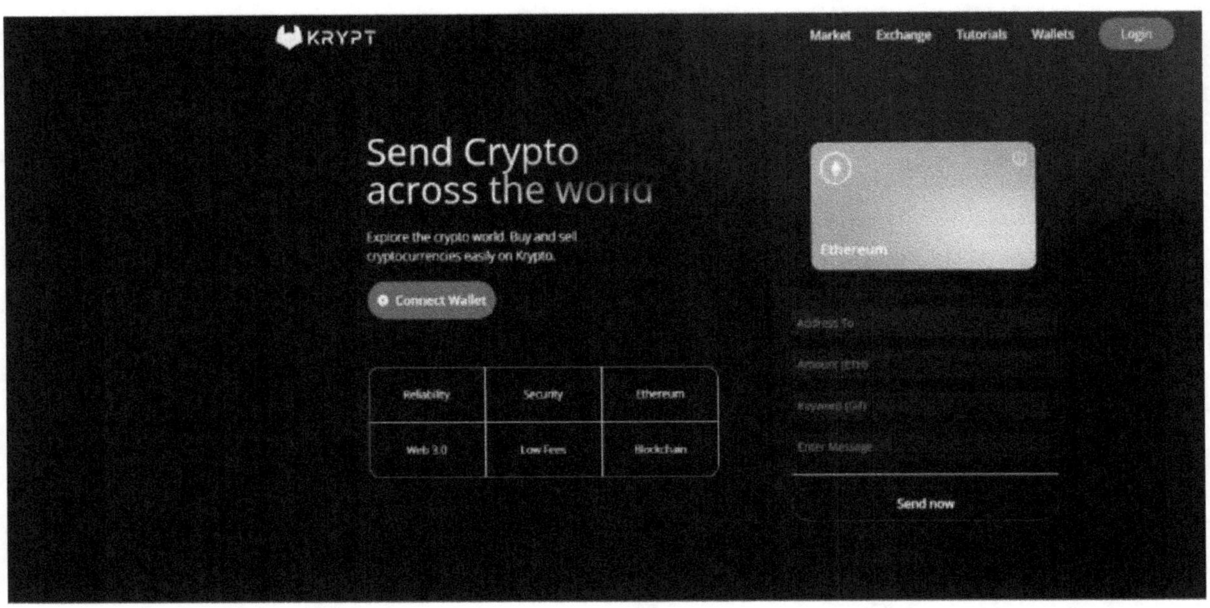

processes in finance and fortifies security measures in e-commerce, hinting at a revolutionary shift in blockchain technology's engagement across various domains.

Now, transitioning to the proposed blockchain-enabled virtual assistant system, This system offers numerous advantages over traditional virtual assistant systems:

1. **Enhanced Security:** Integrating blockchain technology, particularly the Solana-based network with smart contracts, fortifies the security aspect of your virtual assistant platform. Blockchain's inherent security features, such as immutability and a decentralized structure, ensure data integrity and protection against unauthorized access or alterations.
2. **Efficiency and Responsiveness:** Leveraging Solana's architecture ensures rapid transaction processing, enhancing the platform's efficiency. This translates to faster responses and seamless interactions between users and the virtual assistant, reducing latency and improving overall system responsiveness.
3. **User Experience:** The integration of AI and blockchain results in an end-to-end customer support system, prioritizing the user experience. By combining AI's human-like interactions with blockchain's security, the platform aims to offer a smoother and more trustworthy user interface.
4. **Adaptability and Learning:** The inclusion of a self-learning module within the virtual assistant enhances its adaptability. This AI component analyzes previous data, allowing the assistant to evolve and better cater to user needs over time. The amalgamation of AI's learning capabilities and blockchain's secure environment creates a system capable of continuous adaptation and improvement.

In essence, the integration of AI and blockchain in your work aims to fortify security, optimize efficiency, enhance user experience, and empower the virtual assistant with learning capabilities, contributing to a more robust and user-centric platform.

## COMPARING WITH EXISTING METHODS AND STUDIES

In today's modern world, virtual assistants have become indispensable, designed to automate virtual tasks seamlessly. Various machine learning algorithms are employed to achieve desired outcomes in this domain. While most virtual assistant and chatbot platforms come equipped with pre-defined rules for basic communication, a new wave of blockchain-integrated virtual assistant approaches has emerged, which we discuss below in Table 4

Presently, available virtual assistants exhibit several limitations, including their inability to learn autonomously, perform natural language processing, multitask efficiently, and more. In contrast, our proposed model addresses these limitations head-on, boasting self-learning capabilities, multitasking prowess, privacy preservation, and Natural Language Processing (NLP) expertise. Furthermore, we enhance security by seamlessly integrating the Solana-based blockchain network, ensuring not only high efficiency but also top-tier security standards. While API key-based virtual assistants exist in the market, they often fall short in terms of user-friendliness. Table 5 presents a comparison of different related works in the field of blockchain-integrated virtual assistant platforms. Each work is evaluated based on its strengths and limitations to identify key aspects that contribute to the advancement of the domain or potential challenges that need to be addressed.

*Table 4. Comparison of different blockchain-integrated virtual assistant approaches*

| Approach/ Models | Description | Block chain Platform | Consensus Mechanism | Smart Contract Integration | Security Measure | Advantages | Limitation |
|---|---|---|---|---|---|---|---|
| Proposed Model | Integration of AI-powered virtual assistant with blockchain technology for a secure and transparent online conversation platform. | Solana | Proof of History (PoH) combined with Proof of Stake (PoS) | Yes | Encryption of user data on the blockchain, Enhanced security and data integrity, Fast transaction processing | Enhanced security and data integrity, Fast transaction processing | Limited scalability with a large user base |
| Hybrid Architecture | Integrates AI chatbot with blockchain for enhanced security and data integrity. | Ethereum, Hyperledger Fabric, etc. | Varies based on the platform | Yes | Data encryption, user authentication, decentralized storage. | Improved data security, trust, and transparency. | Scalability challenges, potential latency due to blockchain transactions. |
| Decentralized AI | Employs blockchain for distributed AI model training and execution, enabling user control. | Ethereum, Binance Smart Chain, etc. | Varies based on platform | Yes | Encrypted communication, multi-party computation, federated learning. | User data privacy, user-controlled AI, trustless interactions. | Limited processing power on the blockchain, potential data leakage during training. |
| Chatbot as Smart Contract | The chatbot itself is a smart contract on the blockchain, providing AI-driven interactions. | Ethereum, EOS, etc. | Varies based on platform | N/A | On-chain validation, user permissions, data immutability. | Transparency, trust, and immutability of chatbot interactions, no central authority. | Limited AI complexity, transaction costs, scalability challenges for high-frequency use. |
| Permissioned Blockchain | Utilizes a private/ consortium blockchain for AI chatbot services with controlled access. | Hyperledger Fabric, Corda, etc. | Depends on blockchain | Yes | Identity management, access control, confidential transactions. | High privacy, fine-grained access control, regulatory compliance. | Limited decentralization, requires trust in network participants, higher setup complexity. |

## FUTURE DIRECTION AND EMERGING TRENDS

Combining artificial intelligence (AI) with blockchain offers a landscape filled with both opportunities and challenges. The inherent attributes of blockchain, with its decentralized and secure ledger system, align seamlessly with the transformative capabilities of AI. Blockchain's unique features, such as immutable peer-to-peer data and transaction sharing, significantly enhance transparency and security. In domains like healthcare, blockchain holds the potential to address interoperability hurdles and empower

*Table 5. Strengths and limitations of existing models*

| Ref | Related Work | Description | Strength | Limitations |
|---|---|---|---|---|
| Xie et al., 2019 | Chatbot Integration, Chatterbot Framework, Cryptocurrency Focus, Data Collection | This study explores the interaction between cryptocurrencies and chatbots in the digital era. It highlights chatbots' growing role as conversational agents providing automated cryptocurrency information to investors and interested parties. | Automated Information Retrieval, Real-Time Updates, User-friendly Interface | Dependencies on APIs, Narrow Scope, Response Time |
| Mekni, 2021 | Virtual Assistants that engage in conversation | The approach is to enhance user experience by combining conversational bots and artificial intelligence. The research showcases the development of a virtual assistant capable of understanding and responding to natural language queries, effectively emulating human interaction. | Seamless Interaction, Personalized Assistant, Adaptability to User Preferences | Self-learning Management System (SLMS), Open Educational Resources (OER), Voice Command |
| Huynh-The et al., 2022 | Machine Learning, Deep Learning, Metaverse | This study conducted a comprehensive analysis of the relationship between artificial intelligence and the metaverse. The study explores key aspects such as data analytics, immersive experiences, and AI-driven virtual environments. | Optimization of Immersive Environments, Tailoring Survey Interactions, Revolutionizing, Various Industries with AI | Integration of AI in Black Boxes, Greater Involvement of the Metaverse in AI, Increased Need for Transparency and Examination |

patients by granting secure access and ownership of their medical records, mobile applications, remote monitoring, and data management systems (Shah et al., 2022).

The Internet of Vehicles (IoV) has emerged as a pivotal player in modern transportation, leveraging vehicular communication to tackle traffic and safety concerns. Nevertheless, concerns related to trust and safety during data exchange have arisen. Smart blockchains step in to provide trust assurances, secure data handling, and incentives, thereby bolstering IoV advancements. Blockchain's characteristics, including transparency, anonymity, and immutability, foster trust, enhance data security, and stimulate information sharing among vehicles, individuals, and service providers (Attaran, 2022).

Blockchain technology, with its decentralization and reliability, addresses the limitations encountered in traditional supply chains. When coupled with artificial intelligence, it automates the entirety of the supply chain process, enabling predictive analysis, demand forecasting, and enhanced network management through data analysis. This integration serves as a revolutionary force in supply chain management. In the healthcare sector, integrating blockchain introduces challenges in striking a balance between data accessibility, patient privacy, and network security. Figure 7 visually outlines the landscape of challenges and opportunities arising from the integration of AI and blockchain technologies. The complexity escalates due to the integration of diverse systems like electronic health records (EHRs) with varying standards. Navigating the intricate landscape of global healthcare regulations poses implementation challenges. Encouraging user adoption necessitates substantial adjustments in behavior and workflow (Hasan et al. 2022, French et al., 2021).

The realm of smart blockchain applications grapples with the critical challenge of scalability, particularly in the context of decentralized apps (DApps). Challenges encompass consistency issues, network

*Figure 7. Challenges and opportunities in AI and blockchain integration*

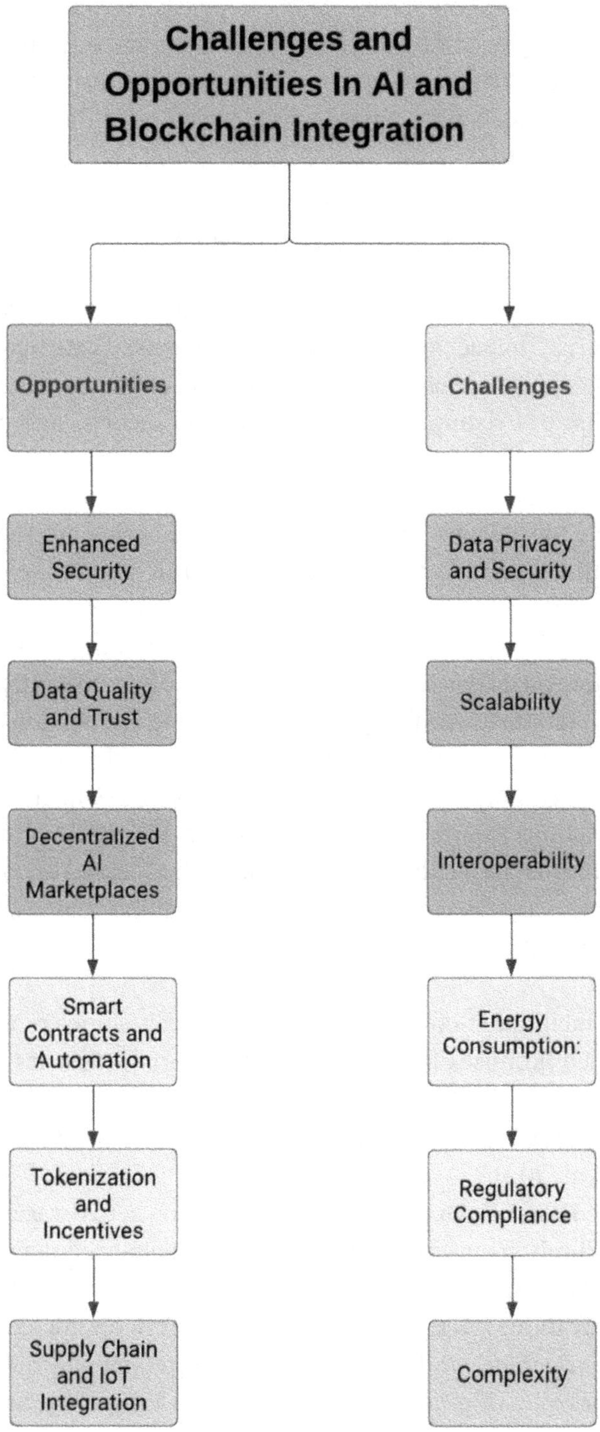

delays, and performance limitations. Balancing the facets of security and decentralization while pursuing scalability proves intricate, as achieving consensus among nodes remains pivotal for security. However, scalability efforts may inadvertently compromise network consistency, potentially leading to blockchain forks. Network delays, especially extended ones, impose constraints on scalability. Additionally, blockchain transaction performance is hampered, given that transactions cannot occur in parallel, hindering efforts to enhance throughput. Scalability emerges as a linchpin for the successful implementation of large-scale blockchain applications.

In the realm of blockchain applications, security and privacy reign supreme. Despite blockchain's inherent transparency, safeguarding sensitive user data assumes paramount importance for widespread adoption. Strategies for privacy protection encompass identity obfuscation, employing techniques like group and ring signatures, and information concealment through zero-knowledge proofs. However, the adoption of these methods may impact system efficiency and necessitate optimization. The integration of artificial intelligence algorithms to enhance efficiency represents a complex yet imperative undertaking, demanding the redesign of existing algorithms to uphold security and privacy within blockchain applications.

The Fourth Industrial Revolution unfolds as a convergence of AI, blockchain, 5G, and other innovations, reshaping industries and societies alike. This convergence brings forth opportunities, challenges, and profound shifts in work, communication, and daily life. The integration of AI, blockchain, and 5G presents extensive research possibilities spanning design and behavioral science paradigms. While each technology offers unique opportunities, their collaborative potential introduces revolutionary changes and diverse research directions. Design science research can explore innovative solutions for addressing blockchain's scalability limitations, optimizing machine learning applications, and leveraging advancements in 5G technology. Simultaneously, behavioral science research can delve into the implications of social capital theory and impression management within the context of the Fourth Industrial Revolution, particularly in social interactions and virtual identities. Table 6 illustrates Emerging Technologies and Innovations for the Virtual Assistant Platform. Furthermore, the societal impact of the Fourth Industrial Revolution necessitates research endeavors focused on understanding and resolving ethical, privacy, and security concerns associated with 5G, blockchain, and AI technologies (Salah et al., 2019; Tauseef et al., 2023).

The exploration of the integration of AI and blockchain technologies represents a dynamic frontier of inquiry. This convergence promises the potential to transform industries, fortify data security, and reshape contemporary processes. Researchers and businesses alike are at the forefront, pushing the boundaries to unlock synergies between AI and blockchain, thus paving the way for innovative solutions and applications (Zhang et al. 2021).

The integration of AI and blockchain catalyzes transformation across various sectors. It fortifies supply chain transparency, securely manages healthcare data for enhanced diagnoses, expedites financial transactions through AI-driven trading, and streamlines operations via smart contracts. Decentralized AI marketplaces foster innovation, while energy grids benefit from optimized management. Blockchain ensures secure identity verification, while AI tailors content recommendations. These advancements collectively reshape industries by offering heightened security, efficiency, and transparency, heralding a future characterized by more dependable and innovative solutions across diverse domains.

*Table 6. Emerging technologies and innovation for virtual assistant platform*

| Technology Innovation | Description | Potential Application | Advantages | Challenges |
|---|---|---|---|---|
| Natural Language Processing (NLP) | Advancements in NLP techniques, including transformer models like GPT-3, for more natural and context-aware conversations. | Virtual Assistants, Customer Support, Content Generation | Improved conversational abilities, better understanding of context and user intent, enhanced content creation. | Ethical concerns regarding biased language models, resource-intensive training, data privacy and security. |
| Multimodal AI | Integrating text, speech, and visual recognition to enable more interactive and context-rich interactions. | Virtual Shopping Assistants, Virtual Tour Guides | Enhanced user experiences, better comprehension of user needs, ability to provide visual and auditory information, improved accessibility. | Complexity in handling multiple modalities, compatibility issues across devices, privacy considerations. |
| Emotional AI | AI systems capable of recognizing and responding to user emotions through voice, text, and facial expressions. | Mental Health Support, Virtual Therapists | Improved user engagement, personalized responses, emotional support, potential for mental health applications. | Ethical concerns regarding privacy and consent, accuracy of emotion detection, potential for manipulation. |
| Conversational AI with Context Memory | AI that remembers past interactions and maintains context for more natural and consistent conversations. | Customer Service, Personal Assistants | Reduced repetition, improved user experience, smoother handoffs between human and virtual agents. | Data storage and privacy concerns, managing context over extended interactions, scalability challenges. |
| Explainable AI (XAI) | Developing AI models that can provide explanations for their decisions and recommendations. | Healthcare Diagnosis, Financial Advice | Improved trust, accountability, regulatory compliance, easier debugging, and better user understanding. | Balancing transparency with model complexity, potential trade-offs between accuracy and explainability. |

## CONCLUSION

In conclusion, this chapter has delved into the exciting integration of artificial intelligence and blockchain technology, offering a vision of a secure and transparent online conversation platform. The rise of AI, exemplified by virtual assistants capable of human-like interactions and multifaceted tasks, has paved the way for innovative applications across sectors such as healthcare, manufacturing, smart cities, and gaming. Blockchain technology emerges as a powerful companion, bolstering security and response times, thus solidifying the role of virtual assistants in future ecosystems.

The central project showcased within this chapter underscores the evolution of virtual assistants through their synergy with blockchain. Key elements, including a self-learning module, Solana-based blockchain network with smart contracts, and a tech stack comprising Vanilla JavaScript, Express.js, and Node.js, have been meticulously detailed. The resulting platform emerges as a comprehensive solution, offering end-to-end customer support and guidance, aligning seamlessly with contemporary lifestyle needs.

In summary, this chapter has illuminated the potential of merging AI and blockchain, paving the way for a robust and secure online conversation platform adaptable to diverse user requirements. It highlights the transformative power of technology to create a dynamic and efficient user experience while underlining the continuous evolution and innovation in this exciting field.

Looking forward, this work opens the door to an array of possibilities, including the integration of voice command capabilities, further enhancing self-learning aspects, and venturing into unexplored applications. The amalgamation of virtual assistants and blockchain technology remains a frontier brimming with potential, promising improved user experiences, heightened security, and far-reaching impacts across a multitude of industries. The journey towards this future is one filled with innovation, progress, and boundless opportunities.

## REFERENCES

Abougarair, A. J., Aburakhis, M., & Zaroug, M. (2022). Design and implementation of smart voice assistant and recognizing academic words. *International Robotics & Automation Journal, 8*(1), 27–32. doi:10.15406/iratj.2022.08.00240

Adel, K., Elhakeem, A., & Marzouk, M. (2022). Chatbot for construction firms using scalable blockchain network. *Automation in Construction, 141*, 104390. doi:10.1016/j.autcon.2022.104390

Attaran, M. (2022). Blockchain technology in healthcare: Challenges and opportunities. *International Journal of Healthcare Management, 15*(1), 70–83. doi:10.1080/20479700.2020.1843887

Dasgupta, D., Shrein, J. M., & Gupta, K. D. (2019). A survey of blockchain from security perspective. *Journal of Banking and Financial Technology, 3*(1), 1–17. doi:10.100742786-018-00002-6

Firouzi, F., Farahani, B., Daneshmand, M., Grise, K., Song, J., Saracco, R., Wang, L. L., Lo, K., Angelov, P., Soares, E., Loh, P.-S., Talebpour, Z., Moradi, R., Goodarzi, M., Ashraf, H., Talebpour, M., Talebpour, A., Romeo, L., Das, R., & Luo, A. (2021). Harnessing the power of smart and connected health to tackle COVID-19: IoT, AI, robotics, and blockchain for a better world. *IEEE Internet of Things Journal, 8*(16), 12826–12846. doi:10.1109/JIOT.2021.3073904 PMID:35782886

French, A., Shim, J. P., Risius, M., Larsen, K. R., & Jain, H. (2021). The 4th Industrial Revolution powered by the integration of AI, blockchain, and 5G. *Communications of the Association for Information Systems, 49*(1), 6. doi:10.17705/1CAIS.04910

Gatteschi, V., Lamberti, F., & Demartini, C. (2020). Blockchain technology use cases. *Advanced applications of Blockchain technology*, (pp. 91-114). IEEE.

Hasan, M. K., Ghazal, T. M., Saeed, R. A., Pandey, B., Gohel, H., Eshmawi, A. A., Abdel-Khalek, S., & Alkhassawneh, H. M. (2022). A review on security threats, vulnerabilities, and counter measures of 5G enabled Internet-of-Medical-Things. *IET Communications, 16*(5), 421–432. doi:10.1049/cmu2.12301

Hornung, O., & Smolnik, S. (2022). AI invading the workplace: Negative emotions towards the organizational use of personal virtual assistants. *Electronic Markets, 32*(1), 1–16. doi:10.100712525-021-00493-0

Kashem, M. A., Shamsuddoha, M., Nasir, T., & Chowdhury, A. A. (2022). The role of artificial intelligence and blockchain technologies in sustainable tourism in the Middle East. Worldwide Hospitality and Tourism Themes.

Kiiski, T. (2020). *Voice Games: The History of Voice Interaction in Digital Games.*

Lentzsch, C., Shah, S. J., Andow, B., Degeling, M., Das, A., & Enck, W. (2021). Hey Alexa, is this skill safe?: Taking a closer look at the Alexa skill ecosystem. *Network and Distributed Systems Security (NDSS) Symposium.* Springer.

Lin, Y., Du, H., Niyato, D., Nie, J., Zhang, J., Cheng, Y., & Yang, Z. (2023). Blockchain-aided secure semantic communication for ai-generated content in metaverse. *IEEE Open Journal of the Computer Society*, *4*, 72–83. doi:10.1109/OJCS.2023.3260732

Mekni, M. (2021). An artificial intelligence based virtual assistant using conversational agents. *Journal of Software Engineering and Applications, 14*(9), 455–473. doi:10.4236/jsea.2021.149027

Mekni, M. (2021). An artificial intelligence based virtual assistant using conversational agents. *Journal of Software Engineering and Applications, 14*(9), 455–47. doi:10.4236/jsea.2021.149027

Nakamoto, S. (2008). Bitcoin: A peer-to-peer electronic cash system. *Decentralized business review.*

Njoku, J. N., Nwakanma, C. I., Amaizu, G. C., & Kim, D. S. (2023). Prospects and challenges of Metaverse application in data-driven intelligent transportation systems. *IET Intelligent Transport Systems, 17*(1), 1–21. doi:10.1049/itr2.12252

Ray, P. P. (2023). *ChatGPT: A comprehensive review on background, applications, key challenges, bias, ethics, limitations and future scope.* Internet of Things and Cyber-Physical Systems.

Salah, K., Rehman, M. H. U., Nizamuddin, N., & Al-Fuqaha, A. (2019). Blockchain for AI: Review and open research challenges. *IEEE Access : Practical Innovations, Open Solutions, 7*, 10127–10149. doi:10.1109/ACCESS.2018.2890507

Saritha, K., Kurni, M., Madhavi, K., & Nagadevi, D. (2021). Integration of Artificial Intelligence and the Internet of Things with Blockchain Technology. In *Proceedings of the 2nd International Conference on Computational and Bio Engineering: CBE 2020* (pp. 449-457). Springer Singapore. 10.1007/978-981-16-1941-0_45

Schmidt, B., Borrison, R., Cohen, A., Dix, M., Gärtler, M., Hollender, M., & Siddharthan, S. (2018, October). Industrial virtual assistants: Challenges and opportunities. In *Proceedings of the 2018 ACM International Joint Conference and 2018 International Symposium on Pervasive and Ubiquitous Computing and Wearable Computers* (pp. 794-801).

Shah, K., Chadotra, S., Tanwar, S., Gupta, R., & Kumar, N. (2022). Blockchain for IoV in 6G environment: Review solutions and challenges. *Cluster Computing, 25*(3), 1927–1955. doi:10.100710586-021-03492-0

Shrimali, B., & Patel, H. B. (2022). Blockchain state-of-the-art: Architecture, use cases, consensus, challenges and opportunities. *Journal of King Saud University. Computer and Information Sciences, 34*(9), 6793–6807. doi:10.1016/j.jksuci.2021.08.005

Soofastaei, A. (2021). Introductory Chapter: Virtual Assistants. In A. Soofastaei (Ed.), *Virtual Assistant* (pp. 1–9). IntechOpen. doi:10.5772/intechopen.100248

Soofastaei, A. (2021). In A. Soofastaei (Ed.), *Introductory Chapter: Virtual Assistants. Virtual Assistant* (pp. 1–9). IntechOpen.

Sorin, T. (2023). *Innovative Approach for the Implementation of Blockchain in Command and Control Systems of Critical Infrastructure Networks.*

Tabassum, N., Devanagavi, G. D., & Biradar, R. C. (2017). Clock synchronization in wireless sensor networks using least common multiple. *AEÜ. International Journal of Electronics and Communications, 2017*(82), 446–457. doi:10.1016/j.aeue.2017.10.014

Tabassum, N., Devanagavi, G. D., Biradar, R. C., & Lazarescu, M. T. (2020). Comparative node selection-based localization technique for wireless sensor networks: A bilateration approach. *International Journal of Communication Systems, 33*(15), e4559. doi:10.1002/dac.4559

Tabassum, N., Geetha, D. D., & Biradar, R. C. (2022).. . *Joint Position Estimation and Synchronization of Clocks in WSN Lecture Notes in Networks and Systems, 2022*(235), 409–418.

Tapscott, D., & Tapscott, A. (2016). *Blockchain revolution: how the technology behind bitcoin is changing money, business, and the world.* Penguin.

Tauseef, M., Kounte, M. R., Nalband, A. H., & Ahmed, M. R. (2023). Exploring the Joint Potential of Blockchain and AI for Securing Internet of Things. *International Journal of Advanced Computer Science and Applications, 14*(4). doi:10.14569/IJACSA.2023.0140498

Vishnu, R., & Krishna Prakash, N. (2022). Mobile application-based virtual assistant using deep learning. In *Soft Computing and Signal Processing: Proceedings of 3rd ICSCSP 2020,* (pp. 609-617). Springer Singapore. 10.1007/978-981-16-1249-7_57

Wang, D. Q., Feng, L. Y., Ye, J. G., Zou, J. G., & Zheng, Y. F. (2023). Accelerating the integration of ChatGPT and other large-scale AI models into biomedical research and healthcare. *MedComm - Future Medicine, 2*(2), e43. doi:10.1002/mef2.43

Wang, Q., Li, R., & Zhan, L. (2021). Blockchain technology in the energy sector: From basic research to real world applications. *Computer Science Review, 39*, 100362.

Weizenbaum, J. (1976). Computer power and human reason: From judgment to calculation.

Xie, Q., Tan, D., Zhu, T., Zhang, Q., Xiao, S., Wang, J., & Yi, P. (2019, May). Chatbot application on cryptocurrency. In *2019 IEEE Conference on Computational Intelligence for Financial Engineering & Economics (CIFEr)* (pp. 1-8). IEEE.

Yang, H., & Lee, H. (2019). Understanding user behavior of virtual personal assistant devices. *Information Systems and e-Business Management, 17*(1), 65–87. doi:10.100710257-018-0375-1

Yang, Q., Zhao, Y., Huang, H., Xiong, Z., Kang, J., & Zheng, Z. (2022). Fusing blockchain and AI with metaverse: A survey. *IEEE Open Journal of the Computer Society, 3*, 122–136. doi:10.1109/OJCS.2022.3188249

Zhang, Z., Song, X., Liu, L., Yin, J., Wang, Y., & Lan, D. (2021). Recent advances in blockchain and artificial intelligence integration: Feasibility analysis, research issues, applications, challenges, and future work. *Security and Communication Networks, 2021*, 1–15. doi:10.1155/2021/9991535

Zheng, P., Su, B., Jiang, Z., Yang, C., Chen, J., & Wu, J. (2023, July). Exploring Heterogeneous Decentralized Markets in DeFi and NFT on Ethereum Blockchain. In *2023 IEEE 10th International Conference on Cyber Security and Cloud Computing (CSCloud)/2023 IEEE 9th International Conference on Edge Computing and Scalable Cloud (EdgeCom)* (pp. 259-267). IEEE. 10.1109/CSCloud-EdgeCom58631.2023.00052

# Chapter 12
# Enhancing Money Laundering Detection Through Machine Learning:
## A Comparative Study of Algorithms and Feature Selection Techniques

**Soumya Ranjan Mishra**
*KIIT University, India*

**Pranati**
*KIIT University, India*

**Anika**
*KIIT University, India*

**Hitesh Mohapatra**
https://orcid.org/0000-0001-8100-4860
*KIIT University, India*

## ABSTRACT

*Money laundering is a worldwide issue that jeopardizes the stability and integrity of financial institutions. Many countries have implemented anti-money laundering laws and regulations to combat this. The basics of money laundering and its influence on the financial system, as well as existing strategies for detecting and combating it, are covered in this chapter. K- nearest neighbors, random forest, naive bayes, deep neural networks, and evolution metrics are examples of machine learning techniques and algorithms used to identify suspicious transactions. Financial institutions and regulatory bodies can strengthen their ability to detect and prevent money laundering activities and help to protect the integrity of the financial system by utilizing a variety of measures. In this work, the authors have presented a deep comparative analysis among various machine learning algorithms that are used in money laundering detection.*

DOI: 10.4018/979-8-3693-0659-8.ch012

## 1. INTRODUCTION

Money laundering is a criminal activity that involves the concealment of illegally obtained funds. It is a process by which criminals attempt to make their ill-gotten gains appear legitimate. Banks have a legal and ethical obligation to ensure that their operations are conducted in a safe and sound manner, and that they do not facilitate illegal activities. Accepting ill-gotten money can expose banks to reputation risks, financial losses, and legal liabilities, as well as harm the integrity of the financial system as a whole. The term "money laundering" originated from the practice of washing dirty money through legitimate businesses to make it appear clean. The process of money laundering typically involves three stages: placement, layering, and integration (Jullum, Løland, Huseby, Ånonsen, & Lorentzen, 2020).

During the placement stage, the launderer introduces illegal funds into the financial system. In the layering stage, they move these funds through various accounts and transactions to make it difficult to trace their origin. Finally, in the integration stage, they reintroduce these funds into society as legitimate income. Money laundering poses a significant threat to global financial systems and economies. It enables criminals to profit from illegal activities such as drug trafficking, terrorism financing, and corruption while avoiding detection by law enforcement agencies (Raiter, 2021). To combat money laundering, governments around the world have implemented strict regulations and laws that require financial institutions to report suspicious activities and transactions. These measures aim to prevent criminals from using banks and other financial institutions for their illicit activities. The size of the global money laundering problem is difficult to estimate with certainty because the vast majority of illicit financial flows are not detected or reported. However, various reports and studies have provided estimates of the scale of money laundering in today's world (Yang & Wu, 2020).

According to the United Nations Office on Drugs and Crime (UNODC), it is estimated that the amount of money laundered globally in one year is between 2% to 5% of global GDP, which equates to

*Figure 1. Typical money laundering scenario*

between $800 billion and $2 trillion annually (Chohan, 2019). Another report by the Global Financial Integrity estimates that illicit financial flows, which include not just money laundering but also other forms of illicit activity such as tax evasion, corruption, and smuggling, totaled between $1.6 trillion and $4.8 trillion in 2017 alone. It is important to note that these estimates are likely to be conservative, as they are based on detected or reported cases, and much of the illicit financial flows may not be detected or reported (Firmansyah & Atmadja, 2021). According to money laundering statistics of 2020, 90% of laundered money remains undetected. Money laundering is a significant global issue that poses serious threats to the integrity of the financial system, national security, and public safety. In recent years, there has been a growing interest in the use of machine learning for money laundering detection, with many financial institutions and regulatory bodies exploring the potential of this technology (Soltani et al., 2016). Machine learning (ML) can significantly affect the rate of money laundering by improving the detection and prevention of suspicious transactions. We can leverage Machine learning algorithms to reduce money laundering. ML algorithms will help us in Fraud detection, Suspicious Activity monitoring, Effective customer profiling Automated reporting and more (). ML algorithms can monitor account activity to identify suspicious transactions or patterns of behavior that may indicate money laundering. We can also analyze customer data to create profiles that identify high-risk individuals or entities based on factors such as transaction history, location, and other risk indicators.ML algorithms can use historical data to predict future patterns of behavior, allowing financial institutions to identify potential money laundering activity before it occurs (Tertychnyi et al., 2022). It can be used to analyze large data sets to identify hidden relationships and networks between individuals and entities that may be involved in money laundering. Also, ML algorithms can automatically generate reports of suspicious activity and send alerts to compliance officers, allowing them to quickly investigate and take action (Zhang & Trubey, 2019).

Money laundering is a serious problem that has a significant impact on the global economy, funding criminal activities and undermining the integrity of the financial system. Traditional methods of combating money laundering have proved to be insufficient, which has led to the exploration of new approaches, such as machine learning. By using advanced technologies and analytical tools, financial institutions and law enforcement agencies can more effectively detect and prevent money laundering (Paula, 2016). Machine learning, a branch of artificial intelligence, has emerged as a promising tool in the fight against money laundering. This research paper proposes the use of machine learning algorithms to address this issue, highlighting the potential of this technology to significantly enhance the effectiveness of anti-money laundering efforts (Chen et al., 2018). With further research and development, machine learning has the potential to become an even more powerful tool in the fight against money laundering and other financial crimes. This research paper also aims to explore the effectiveness of machine learning algorithms in detecting and preventing money laundering. The study will focus on the application of various machine learning techniques, such as clustering, classification, and anomaly detection, to identify suspicious transactions and patterns (Lokanan, 2022).

## 1.1 Literature Reviewed

1. Money laundering: The process of disguising the proceeds of illegal activity as legitimate funds by transferring them through a complex series of transactions. This involves disguising the origin, ownership, or destination of the funds to avoid detection or prosecution. Money laundering can have serious economic and social consequences, contributing to the perpetuation of criminal enterprises and undermining the integrity of financial systems (Alotibi et al., 2022).

*Figure 2. KYC*

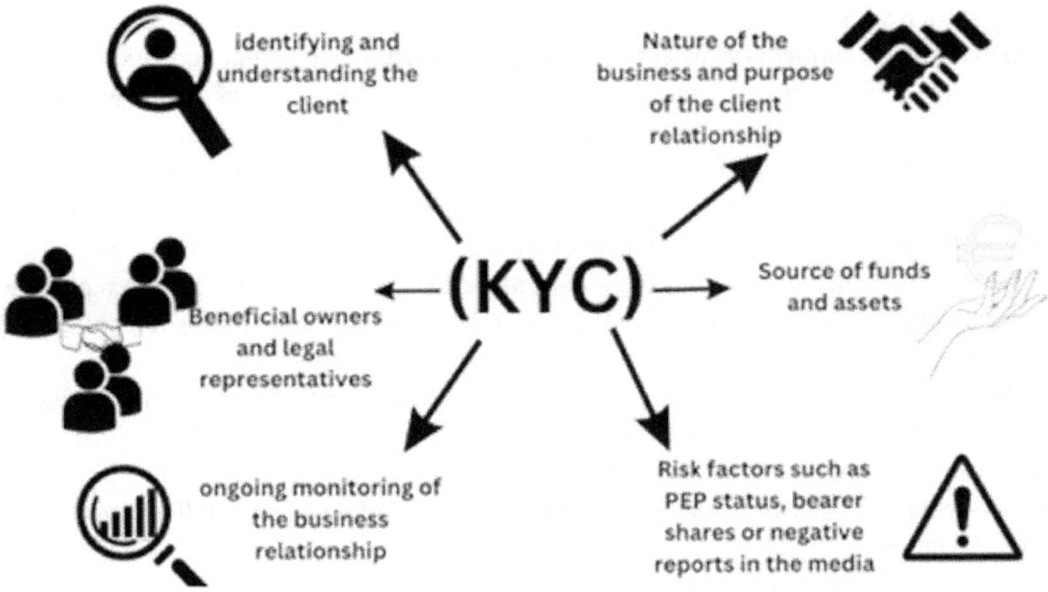


2. Mules: Individuals who are recruited to facilitate money laundering by transferring funds or opening bank accounts using their personal information. Mules are often unaware of their involvement in money laundering and may be used as a way to avoid detection by law enforcement. Cybercriminals are increasingly exploiting unsuspecting individuals as mules, capitalizing on their lack of awareness about the illicit activities they are inadvertently participating in (Pettersson Ruiz & Angelis, 2022).
3. Anti money laundering: The set of laws, regulations, and procedures established to prevent, detect, and report money laundering activities. The aim of anti-money laundering measures is to disrupt the flow of illicit funds and prevent the criminal exploitation of the financial system. Governments and international organizations collaborate to enhance anti-money laundering frameworks, adapting them to emerging techniques used by criminals in an ever-evolving landscape (Youssef et al., 2023).
4. TMS: Transaction Monitoring System, a software-based tool used to identify and report suspicious financial transactions. TMS systems analyze financial transactions, identify patterns of suspicious behavior, and alert compliance officers when they detect potential money laundering activity. Advanced TMS platforms incorporate machine learning algorithms to continuously improve their ability to detect novel money laundering patterns.
5. KYC: Know Your Customer, a process used by financial institutions to verify the identity of their customers to prevent money laundering. KYC procedures require financial institutions to obtain and verify customer identity information, assess the risk of money laundering, and monitor customer transactions for suspicious activity. In the digital age, remote and biometric-based KYC methods are gaining prominence for their convenience and enhanced security features (Kute et al., 2021).
6. Machine learning: A branch of artificial intelligence that involves training computer algorithms to learn from data and improve over time. Machine learning algorithms can automatically identify

patterns in data, classify information, and make predictions based on previous experience. As the volume of data continues to grow, machine learning is becoming increasingly essential in extracting valuable insights and supporting decision-making across various industries.

7. Deep learning: A subfield of machine learning that utilizes deep neural networks to process and analyze large amounts of data. Deep learning is used in applications such as computer vision, natural language processing, and recommendation systems. Through convolutional neural networks and recurrent neural networks, deep learning has achieved remarkable breakthroughs in tasks like image recognition and language translation (Jullum, Løland, Huseby, Ånonsen, & Lorentzen, 2020).
8. KNN: K-Nearest Neighbors, a machine learning algorithm that classifies data based on the nearest neighbors in a given dataset. KNN is a simple and effective algorithm for pattern recognition and classification, particularly in image and text data. Its simplicity often makes it a go-to choose for initial exploratory analysis and as a baseline for more complex algorithms.
9. Random Forest: A machine learning algorithm that builds multiple decision trees and combines them to improve prediction accuracy. Random forests are commonly used for classification and regression tasks in domains such as finance, healthcare, and marketing. By mitigating overfitting and improving generalization, random forests offer robust solutions for handling noisy and heterogeneous datasets.
10. Deep neural network: A type of artificial neural network composed of multiple layers that can learn complex patterns in data. Deep neural networks are used in applications such as image recognition, natural language processing, and speech recognition. Their hierarchical architecture enables them to capture intricate features, allowing for the development of sophisticated models capable of human-like perception and understanding (Canhoto, 2021).
11. Evolution metrics: A set of measures used to evaluate and optimize the performance of machine learning algorithms. Evolution metrics help data scientists assess the performance of different models and identify ways to improve their accuracy and efficiency. These metrics encompass not only traditional evaluation criteria but also novel measures that account for the dynamic nature of model performance over time and changing data distributions.

## 2. IMPLEMENTATION

In order to prevent and combat money laundering, various methods have been developed to detect and identify suspicious transactions and patterns. In this section, we will not only delve into the original content but also extend the discussion to encompass a broader historical context and the evolution of anti-money laundering efforts. The origins of money laundering can be traced back to ancient times when individuals would attempt to disguise the origin of their wealth by exchanging it for other assets or commodities, such as gold or jewelry. This practice allowed them to distance their ill-gotten gains from their criminal activities. While these early methods lacked the complexity of modern money laundering schemes, they laid the foundation for the techniques that would emerge in the modern era (Alkhalili et al., 2021).

Money laundering became an increasingly complex and sophisticated crime, often involving multiple jurisdictions, intricate financial transactions, and the exploitation of new technologies. In response to this growing threat, governments began implementing more systematic approaches to tackle money laundering. One of the pivotal strategies was the establishment of laws and regulations that required banks and financial institutions to maintain comprehensive transaction records and promptly report any suspicious activities to the authorities. These records provided law enforcement agencies with valuable insights into patterns of suspicious behavior, enabling them to initiate investigations and uncover larger criminal networks (Lorenz, 2020).

Furthermore, governments took a proactive stance by banning the use of specific types of currency known to have been stolen or counterfeit. The counterfeiting of currency has historically been a major avenue for money laundering, allowing criminals to inject illegitimate funds into the financial system. By restricting the use of such currency, authorities aimed to raise the bar for money launderers, making it harder for them to integrate their illicit gains into the legitimate economy. The battle against money laundering is an ongoing endeavor that continues to evolve alongside the tactics employed by criminals. Today, modern technology has ushered in a new era of challenges and opportunities in the fight against money laundering. From cryptocurrency to online transactions, new avenues for money laundering have emerged, necessitating innovative approaches to detection and prevention. The historical context of combating money laundering reveals a consistent effort by governments and financial institutions to thwart the activities of criminals seeking to legitimize their ill-gotten gains.

**Assembling details of possible and known networks of mules:** To prevent money laundering, authorities and financial institutions identified and tracked the flow of money through different networks. One way they did this is by assembling details of possible and known networks of mules - individuals who help to move money between different accounts or jurisdictions. By identifying these networks, authorities can work to disrupt them and prevent criminals from using them to move illicit funds. Through careful analysis of transaction patterns and the connections between various accounts, law enforcement agencies can unravel the intricate web of money laundering operations.

**Monitoring high-volume and suspicious transactions:** Monitoring high-volume and suspicious transactions is another key strategy used to prevent money laundering. Financial institutions are required to report transactions that exceed certain thresholds to the authorities, and to monitor transactions for suspicious activity. This may include monitoring for transactions that involve unusual amounts or frequency, transactions with high-risk countries, or transactions that are inconsistent with the customer's known financial profile.

**Verifying funds, including their sources and beneficiaries:** To prevent money laundering, financial institutions verified the source of funds and the identities of beneficiaries involved in transactions. This may include conducting background checks on customers, reviewing their financial history, and verifying the identities of beneficiaries. By verifying the source of funds and the identity of beneficiaries, financial institutions can ensure that transactions are legitimate and that the funds are not being laundered.

**Putting tight identification procedures:** In addition to verifying funds and monitoring transactions, financial institutions also put tight identification procedures in place before allowing certain transactions. This included requiring customers to provide additional identification documents or to answer security questions before processing the transaction. By putting these procedures in place, financial institutions could prevent criminals from using stolen identities or other fraudulent means to move illicit funds.

Today, the fight against money laundering is more complex and sophisticated than ever before. Governments and financial institutions around the world have implemented a wide range of measures to prevent, detect, and punish money laundering, including the use of advanced technologies such as artificial intelligence and blockchain. Despite these efforts, however, money laundering remains a major challenge for law enforcement agencies and financial institutions alike. Criminals are quick to adapt to changing circumstances and exploit vulnerabilities in the financial system. As a result, the battle against money laundering is ongoing, requiring continuous innovation, collaboration, and vigilance to stay one step ahead of those seeking to exploit the global financial ecosystem for their illicit gains.

While traditional methods have been effective in certain cases, they are often limited by the lack of information sharing between different agencies and jurisdictions. In addition, traditional methods are often reactive and may not detect money laundering until after it has already occurred. This has led to the development of more advanced methods and technologies, such as machine learning and data analytics, to better combat money laundering. In this section, we will delve into the methods used for detecting and preventing money laundering activities in the financial industry today. These cutting-edge approaches leverage the power of technology to analyze vast amounts of data, identify intricate patterns, and proactively flag suspicious activities. By embracing these modern techniques, authorities and financial institutions aim to stay ahead of increasingly sophisticated money laundering schemes and safeguard the integrity of the global financial system. AML (Anti-Money Laundering) and TMS (Transaction Monitoring Systems) are two of the most widely used and effective methods for detecting and preventing money laundering activities. AML stands for Anti-Money Laundering, which refers to a set of laws, regulations, and procedures designed to prevent the use of illegally obtained money or funds for legitimate purposes. Money laundering is the process of concealing the source, ownership, or destination of illegal funds to make them appear legitimate. AML laws are intended to combat money laundering and other financial crimes, such as terrorism financing and tax evasion. The primary objective of AML is to prevent money laundering by detecting and reporting suspicious financial activities. AML regulations require financial institutions, such as banks, securities dealers, and money services businesses, to implement policies and procedures to identify and verify the identities of their customers, monitor their financial transactions, and report any suspicious activities to the relevant authorities.

The origins of AML can be traced back to the 1970s when the United States passed the Bank Secrecy Act (BSA), which required financial institutions to report certain transactions to the government. Since then, AML regulations have been adopted by countries around the world, and international organizations, such as the Financial Action Task Force (FATF), have been established to promote global AML standards and provide guidance to countries on implementing effective AML regimes. The consequences of money laundering can be severe, both for individuals and for society as a whole. Money laundering can finance criminal activities, such as drug trafficking, human trafficking, and terrorism, and can undermine the integrity of the financial system. AML regulations are essential to prevent the negative consequences of money laundering and to protect the financial system and society as a whole. AML software is a type of software that helps financial institutions and other businesses comply with AML regulations. AML software typically uses data analytics, artificial intelligence, and machine learning algorithms to analyze customer data, transaction data, and other relevant information to detect suspicious activities that may indicate money laundering or other financial crimes. Here is a general overview of how AML software works:

1. Data Collection: AML software collects customer data, transaction data, and other relevant information from various sources, such as internal databases, external databases, and public records.
2. Data Analysis: Once the data has been collected, AML software uses data analytics, artificial intelligence, and machine learning algorithms to analyze the data for any suspicious activities. This involves comparing the data against various risk models and scenarios to identify any anomalies or patterns that may indicate money laundering or other financial crimes.

3. Risk Scoring: AML software assigns risk scores to customers and transactions based on the level of risk they pose. Customers and transactions that are deemed high risk are subjected to more scrutiny and monitoring.
4. Alert Generation: If the AML software identifies any suspicious activities, it generates alerts that are sent to the relevant personnel for further investigation. These alerts typically include detailed information about the suspicious activity and the risk score assigned to the customer or transaction.
5. Case Management: The relevant personnel investigate the alerts generated by the AML software and determine whether they warrant further action, such as filing a Suspicious Activity Report (SAR) with the relevant authorities. AML software typically includes case management tools that help to track and manage the investigation process.

TMS, or Transaction Monitoring Systems, play a crucial role in the AML process by monitoring and analyzing transactions in real time. These systems are designed to identify unusual or suspicious patterns of activity that may indicate money laundering or other illicit financial activities. TMS software uses a combination of rule-based and behavior-based approaches to detect potentially suspicious transactions. Rule-based monitoring involves setting up predefined rules that flag transactions that meet certain criteria, such as exceeding a specific threshold or involving high-risk countries. On the other hand, behavior-based monitoring involves using historical data and machine learning algorithms to identify deviations from normal transaction patterns for a particular customer or account. When a transaction is flagged as potentially suspicious, the TMS generates an alert, which is then reviewed by compliance officers or analysts. These professionals assess the context of the flagged transaction and decide whether further investigation is warranted. If deemed necessary, they may escalate the case for further review and, if appropriate, report the suspicious activity to the relevant authorities. In recent years, the effectiveness of AML and TMS has been enhanced by the integration of advanced technologies, including artificial intelligence (AI) and machine learning (ML). These technologies enable AML and TMS systems to analyze vast amounts of data more efficiently and accurately than traditional methods. AI and ML algorithms can identify complex patterns, trends, and anomalies in large datasets that may be difficult for human analysts to detect.

AML and TMS are pivotal components of the global effort to combat money laundering and other financial crimes. These methods utilize regulations, technology, and data analysis to detect and prevent suspicious activities that may indicate illicit financial behavior. As financial transactions become more complex and criminals continue to find new ways to exploit the system, the ongoing development and integration of advanced technologies will remain critical to staying one step ahead in the fight against money laundering. By staying vigilant and leveraging the power of modern technology, financial institutions and authorities can work together to protect the integrity of the financial system and uphold the rule of law. The KYC is an essential part of AML compliance because it is the first step in detecting and preventing money laundering and other financial crimes. KYC procedures involve collecting personal information and documentation from customers, such as their name, address, date of birth, and government issued identification documents. Financial institutions must also verify this information by conducting background checks, reviewing sanctions lists, and performing risk assessments based on the customer's profile and activities. The primary goal of KYC is to ensure that financial institutions are doing business with legitimate customers and to prevent the misuse of financial services for illegal activities, such as money laundering and terrorist financing. KYC procedures are also critical to protect financial institutions from reputational risks, regulatory sanctions, and financial losses. There are three main components of KYC procedures: customer identification, customer due diligence, and ongoing monitoring.

- *Customer Identification:* The first step of KYC procedures is to identify the customer and verify their identity using reliable and independent sources. Financial institutions must collect basic information about their customers, such as their name, address, and date of birth, and verify this information using government-issued identification documents.
- *Customer Due Diligence:* The second step of KYC procedures is to conduct a risk assessment of the customer's profile and activities to determine their level of risk. Financial institutions must assess the nature of the customer's business, the source of their wealth, and their expected transactional activity. Higher risk customers, such as politically exposed persons (PEPs) and those engaged in high-risk industries, require enhanced due diligence procedures.
- *Ongoing Monitoring:* The third step of KYC procedures is to monitor the customer's activities on an ongoing basis to ensure that their behavior remains consistent with their expected profile. Financial institutions must monitor customer transactions, including the type, frequency, and purpose of transactions, and report any unusual or suspicious activity to the relevant authorities. KYC regulations are typically enforced by government agencies, such as financial intelligence units,
- Law enforcement agencies, and regulatory authorities. These agencies may conduct investigations, impose fines and penalties, and revoke the licenses of financial institutions that fail to comply with KYC regulations' procedures are not only mandatory for financial institutions but are also essential for customers. By providing accurate and complete information about themselves, customers can gain access to a range of financial services and products, such as loans, credit cards, and investment accounts Transaction monitoring system (TMS) is a software-based solution that plays a critical role in helping organizations detect, analyze, and report on suspicious or fraudulent financial transactions. With financial crimes such as money laundering, terrorist financing, and fraud posing significant threats, TMS employs a range of tools including artificial intelligence, machine learning, and data analytics to monitor and analyze transaction patterns, making it a pivotal tool in the fight against these illicit activities. The functioning of TMS can be broken down into four main stages: data acquisition, data analysis, alert generation, and case management.
    1. Data Acquisition: TMS starts by collecting transactional data from various sources, encompassing internal databases, external data providers, and public sources. This data is drawn from diverse channels, including mobile applications, point-of-sale terminals, and online portals. TMS might also receive data feeds from other systems within the organization, like fraud detection systems or customer relationship management (CRM) platforms.
    2. Data Analysis: The strength of TMS lies in its ability to leverage advanced data analytics and machine learning algorithms to detect patterns and trends within the transactional data. This often involves configuring the system with various rules and scenarios derived from industry-specific guidelines, regulatory mandates, and internal policies. These rules and scenarios empower TMS to flag transactions that could potentially be suspicious. This can encompass transactions that deviate from established patterns or showcase unusual behavior. Employing predictive analytics further allows TMS to identify anomalies in real-time, which can be crucial in detecting fraudulent activities swiftly.
    3. Alert Generation: As TMS identifies transactions that potentially raise suspicion, it generates alerts that are then sent to the relevant personnel for review. These alerts typically include

*Figure 3. Transaction monitoring stages in anti-money laundering*

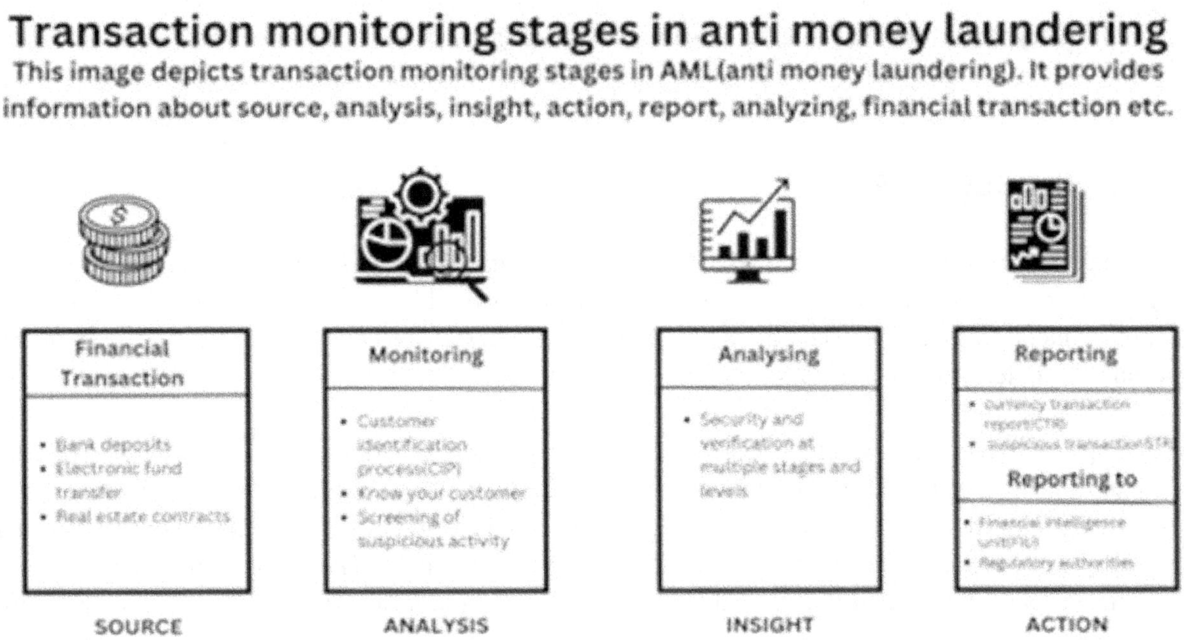


comprehensive details about the transaction, such as the involved parties, transaction amount, location, and other pertinent information. Additionally, alerts often come with a risk score assigned based on the perceived level of risk linked to the transaction. This risk score serves as a tool to prioritize alerts based on their associated risk levels, aiding organizations in focusing their attention where it matters most.

4. Case Management: Upon the generation of an alert, the designated personnel initiate an investigation into the flagged transaction to ascertain whether it is genuinely suspicious. This process, known as case management, involves a thorough examination of transaction data, supplementary research, and meticulous documentation of the findings. Should the investigation determine that the transaction is indeed suspicious, the case is escalated to the relevant authorities for further examination. Beyond its investigative role, TMS also facilitates the generation of reports and offers insights into transaction patterns and trends, which can be harnessed to continually enhance the system's effectiveness over time.

TMS emerges as an indispensable tool for organizations striving to monitor financial transactions for signs of potential suspicious activities. By harnessing advanced data analytics and machine learning algorithms, TMS is adept at real-time identification of transactions with potential red flags, actively contributing to the prevention of financial crimes such as money laundering and fraud. Furthermore, the system offers case management and reporting capabilities, providing organizations with the means to thoroughly investigate suspicious transactions and extract actionable insights that can enhance their transaction monitoring strategies. In recent times, the advent of digital technologies and the growing utilization of cryptocurrencies have introduced new complexities to Anti-Money Laundering (AML) compliance. Cryptocurrencies, in particular, have emerged as potential vehicles for money laundering and terrorism financing, given their capacity to facilitate anonymous cross-border fund transfers. Consequently, regulatory bodies have initiated the development of fresh AML frameworks specifically tailored for cryptocurrencies. Financial institutions that engage with cryptocurrencies are now subject to AML regulations in order to curb potential misuse of these technologies for illicit financial activities.

## 3. METHODOLOGIES

Machine learning possesses the remarkable potential to significantly elevate the detection of money laundering by offering more sophisticated and efficient methods to uncover patterns and anomalies within financial transactions. Here are a few illustrative examples of how machine learning can play a pivotal role in detecting money laundering:

- Improved Accuracy: Machine learning algorithms excel in learning from vast volumes of data and unearthing intricate patterns that might elude human observation. This implies that these algorithms are capable of detecting complex and sophisticated money laundering operations that may otherwise remain unnoticed when using conventional approaches. The ability to discern nuanced relationships and trends enables machine learning algorithms to enhance the accuracy of detection.
- Faster Detection: Given the capability of machine learning algorithms to process extensive datasets swiftly, they can promptly identify suspicious activities in real time. This rapid detection is particularly critical in the banking industry where timely responses are vital to detect and counteract money laundering effectively. Machine learning contributes to minimizing the time lapse between occurrence and identification, thereby mitigating potential losses and damage.
- Enhanced Automation: Machine learning algorithms possess the potential to automate numerous processes integral to money laundering detection, eliminating the necessity for manual scrutiny and intervention. This automation not only expedites the detection process but also empowers financial institutions to reduce costs and improve operational efficiency. By automating routine tasks, organizations can focus their resources on more complex analysis and decision-making tasks.
- Adaptability: A distinguishing feature of machine learning algorithms is their capacity to adapt to novel patterns and evolving trends in money laundering. This adaptability is of paramount importance as money launderers perpetually devise new strategies to evade detection. Machine learning algorithms, by perpetually learning and refining their models, remain ahead of money launderers' tactics, thereby enhancing the effectiveness of detection strategies.
- Integration with Other Technologies: To amplify the accuracy and efficiency of money laundering detection, machine learning can seamlessly integrate with other cutting-edge technologies such as blockchain and big data analytics. For instance, blockchain technology can be employed to establish an immutable ledger of financial transactions. This ledger can then be subjected to analysis using machine learning algorithms to pinpoint potentially suspicious behavior. The synergistic integration of multiple technologies bolsters the detection process and enhances the overall effectiveness of anti-money laundering efforts.

In sum, machine learning stands as a potent tool poised to revolutionize money laundering detection by furnishing methods that are more precise, streamlined, and adaptable in identifying suspicious activities. As technological advancements continue to unfold, it is anticipated that increasingly intricate and effective machine learning algorithms will be developed to counteract the constantly evolving landscape of money laundering. The potential of machine learning to bolster the efficacy of anti-money laundering endeavors holds promise for a safer and more secure financial ecosystem. Machine learning algorithms wield the potential to revolutionize the field of anti-money laundering by providing dynamic, adaptable, and highly accurate methods for detecting financial anomalies. These algorithms, ranging from Random Forest to Deep Learning, can uncover hidden patterns, enhance detection speed, and automate processes, thereby reinforcing the global fight against money laundering and bolstering the integrity of the financial sector. As technological progress continues, the synergy between innovative algorithms and comprehensive financial data is expected to yield increasingly powerful tools in the ongoing battle against financial crime. As technology advances, it is expected that more complex and effective machine learning algorithms will be developed to prevent money laundering (Soltani et al., 2016). Here are some algorithms:

*KNN:* This type of approach is regarded as a basic instance-based learning method in which a new case or instance is compared to every other instance already in existence before being classified based on a similarity measure. Accordingly, a class is consequently assigned for the new instance. The measure used is the Minkowski distance. The distance between a couple of points in a normalized vector space is represented by the Minkowski distance, which is calculated as follows:

$$d(x,y) = (\sum_{l=0}^{n=-1} |x_l - y_l|^p)^{1/p} \tag{1}$$

The KNN algorithm, short for K-Nearest Neighbors, operates through the analysis of intricate patterns within data sets, serving as a powerful tool for discerning whether a transaction bears the hallmark of fraudulence. It accomplishes this by delving into the intricacies of both fresh and historical transactions, conducting a meticulous comparison between them. If a nascent transaction bears a semblance to a prior occurrence, it raises a figurative red flag, prompting a deeper investigation into its legitimacy. In elucidating the process of implementing KNN to unmask potential instances of money laundering, the following sequential steps come to the fore:

- Data Collection: At the outset, a substantial dataset of financial transactions must be amassed, forming the bedrock of KNN's money laundering detection mechanism. Vital information encompassing transaction sums, involved parties, geographical coordinates, and assorted pertinent particulars coalesce within this dataset, engendering a comprehensive panorama.
- Feature Extraction: Post data aggregation, the onus lies on extracting salient features from the amassed corpus. Attributes intrinsic to the conundrum at hand metamorphose into features. The recurrence, quantum, and geographical context of transactions are but a few examples of features that can be harnessed to unveil concealed money laundering undertones.
- Training: Having gleaned features from the data, the KNN algorithm embarks upon a training voyage, leveraging annotated data. Data duly earmarked as indicative of potential money laundering or benign transactional behavior constitutes this annotated data. Through this symbiosis,

KNN unfurls an erudite model, adept at discerning the latent predisposition of a fresh transaction towards money laundering, imbibing knowledge from its tutored encounters.

- Prediction: Fortified with its acquired acumen, the trained KNN algorithm takes center stage, orchestrating evaluations to gauge the prospect of a transaction's entanglement in money laundering. The algorithm orchestrates a meticulous matching dance, juxtaposing the features of a new transaction against those ensconced within its training repository. This intricate dance culminates in the assignment of the new transaction to the class that prevails amongst the majority of its K closest kindred.
- Flagging: The grand denouement arrives as transactions at the brink of suspicion are unmasked. Should the KNN algorithm's calculations predicate a transaction to be a potential harbinger of money laundering, it signals the need for human intervention. An expert's discerning eye, drawn by the algorithmic flag, scrutinizes the transaction, deciding whether it warrants further inspection.

In essence, the KNN algorithm metamorphoses into an invaluable sentry, its capacity to traverse the labyrinthine realm of financial transactions facilitating the identification of money laundering machinations that might otherwise remain ensconced in obscurity. This multidimensional process, encapsulating data curation, feature discernment, algorithmic tutelage, predictive prowess, and expert appraisal, coalesces to uphold the integrity of financial landscapes against the tides of illicit monetary currents.

*Random Forest:* The Random Forest (RF) algorithm emerges as a formidable sentinel within the realm of machine learning, donning the mantle of pattern recognition and prediction with finesse. Its essence lies in the art of ensemble learning, where a collective of regression or classification trees orchestrate a harmonious symphony of insights. Unlike their solitary counterparts, these trees find strength in their diversity, collectively enhancing stability and reliability. The individuality of decision trees is a double-edged sword. Their sensitivity to training inputs, although a hallmark of their accuracy, can also render them susceptible to instability. This is where the RF algorithm takes center stage, weaving a tapestry of multiple trees into a robust ensemble.

In the dynamic domain of money laundering detection, RF unfurls its potential as a guardian of financial integrity. The algorithm's prowess to sift through immense volumes of financial data resonates with the urgency of real-time vigilance. It delves into the intricacies of transaction quantities, frequencies, and the global interplay of nations, painting a nuanced portrait of financial activities. Additionally, the RF algorithm can cast its gaze wider, encapsulating the identities of individuals or entities involved, the nature of products or services procured, and the chronology of transactions. The merits of employing RF for money laundering detection are manifold. Its rapid data processing capability stands as a beacon in the dark, enabling financial institutions to detect and flag suspicious activities in the blink of an eye.

The symbiosis of RF with the realm of money laundering detection is not without its challenges. The algorithm's potency hinges on the quality of the data that fuels it. Garbage in, garbage out – the adage resonates here, underlining the need for pristine and comprehensive datasets to extract meaningful insights. Furthermore, the complexity of RF, while a virtue in itself, demands careful parameter tuning and interpretation. In culmination, the Random Forest algorithm resonates as a multifaceted gem in the crown of machine learning. Its ensemble nature encapsulates the collective wisdom of trees, amplifying accuracy and stability. Within the arena of money laundering detection, RF shines as a beacon of vigilance, sifting through data with alacrity to flag suspicious transactions. Its adaptive capabilities ensure it stands strong against the tides of evolving threats. Yet, the alliance with RF is not devoid of considerations – data quality, parameter tuning, and interpretability are threads that must be woven into the fabric of its deployment. Through this confluence of attributes and considerations, the RF algorithm charts a course towards fortified financial security, unveiling a realm where vigilance and innovation intertwine.

*NB:* Naive Bayes (NB) algorithms is a probabilistic algorithm that is based on Bayes' theorem and assumes that the features in the data are independent of each other. The Bayesian theorem describes the likelihood that an event will occur if you have prior information about a condition associated with the specified event. The method is intended to deal with continuous attributes that are associated with each category and are distributed using a Gaussian distribution. The main advantage of the Naive Bayes is to effectively train in supervised learning, and are used for practical classification problems. The assumption that the qualities are independent, which is practically impossible to establish, is a major drawback of the Naive Bayes model. When using Naive Bayes, all features are assumed to be independent given the value of a class; this is referred to as conditional independence. In this study, there are two categories: illegal transactions, which equal zero, and legitimate transactions, which equal one. The likelihood that sample x belongs to category c is displayed in the following equation:

$$P(c|x) = \frac{p(x|c) * p(c)}{p(x)} \tag{2}$$

Employing the Naive Bayes (NB) algorithm as a sentinel against money laundering infuses the realm of financial security with a potent blend of simplicity and efficiency. This algorithm's modus operandi hinges on a foundational premise – learning from historical transactions meticulously categorized as legitimate or suspicious. This curated dataset becomes the crucible where NB imbibes the essence of financial activity, training itself to discern patterns and characteristics indicative of illicit undertakings. The dynamics of money laundering are multifaceted, encompassing transaction quantities, frequencies, and the intricate geopolitical tapestry woven by the nations involved. It's these attributes, intermingled with others that might suggest nefarious intent, that form the palette upon which NB works its magic. Armed with statistical attributes gleaned from the labeled transactions, NB operates as a probabilistic oracle, assigning the likelihood of a new transaction being lawful or dubious based on its qualities. The allure of NB rests in its swiftness and efficiency. It stands as a computational marvel, capable of sifting through immense volumes of data in the blink of an eye. This rapidity is not at the expense of complexity; rather, it emerges as a testament to the elegance of the algorithm. Its simplicity in implementation dovetails harmoniously with its computational celerity, making it an attractive choice for financial institutions grappling with the urgency of money laundering detection.

However, the moniker "Naive" is not without its implications. The algorithm assumes a level of naivety, postulating that the features within the data are independent of one another. This assumption, while often untrue in the intricate realm of finance, simplifies the computational mechanics, allowing NB to function with commendable efficacy even with limited training data. Yet, the real world seldom adheres to such naivety; correlations and interactions between features are the norm rather than the exception. This can impact the accuracy of NB's predictions, introducing an element of unpredictability. This is where the strength of synergy emerges. NB, when harmonized with other machine learning paradigms, becomes a cog in a larger machinery of detection. By uniting NB's rapidity and simplicity with algorithms capable of untangling complex interdependencies, financial institutions can unveil a more comprehensive approach to money laundering detection. This amalgamation compensates for the naive assumptions of NB, integrating it into a holistic framework where its strengths are amplified and its limitations mitigated. In the grand tapestry of financial security, NB shines as a beacon of efficiency and accessibility. Its ability to process large volumes of data with speed and grace makes it a stalwart guardian, ever vigilant against the tide of money laundering. However, its simplicity and the assumptions it carries render it more potent when juxtaposed with complementary algorithms. The symphony of machine learning algorithms, conducted by the goal of safeguarding the integrity of financial systems, unveils the potential for a brighter and more secure financial future.

*DNN:* Deep Learning, a formidable offspring of the machine learning lineage, heralds a paradigm shift by relinquishing the traditional constraints of human-engineered feature representation. Instead, it wades into the intricate labyrinth of data, wielding its innate capacity to autonomously unravel the hierarchies of information. At its core, Deep Learning's modus operandi is etched in the formidable Deep Neural Networks (DNNs), where the foundations are laid with elements like perceptron, convolutions, and nonlinear activation functions. This symphony of components, orchestrated as layers, weaves an intricate tapestry of comprehension, enabling the network to unravel the tapestry of complex concepts embedded within the raw data. A defining attribute of DNNs lies in their layered architecture, an intricate stratification that mirrors the intricacies of data hierarchies. Ranging from a modest few to an overwhelming multitude, these layers house a spectrum of insights. The bedrock layers lay the foundation by discerning low-level features, like the rudimentary edges and corners within images. As we ascend the neural strata, the complexity grows, with higher layers unraveling intricate connections that signify high-level traits. Within the expansive landscape of money laundering detection, the prowess of Deep Neural Networks assumes a commanding role. DNNs epitomize a species of artificial neural networks, pulsating with numerous interconnected nodes that possess the uncanny ability to perceive patterns within data. The modus operandi is steeped in learning - learning to discern intricate patterns, relationships, and anomalies that might evade human cognition. The journey of deploying DNNs in the battle against money laundering is marked by data-driven enlightenment. Financial institutions embarking on this quest start by nurturing the algorithm with a curated dataset, a corpus of transactions meticulously labeled as legitimate or suspicious. This forms the crucible where the DNN is molded, learning to recognize patterns that illuminate the path to suspect activity. The DNN becomes a connoisseur of statistical features, a sentinel identifying nuances that delineate suspicious transactions from their legitimate counterparts.

The palette of features used by the DNN is an eclectic mix, spanning transaction quantities, frequencies, and the intricate global web of entities involved. These attributes collectively compose a symphony of signals that might portend illicit activity. Armed with this information, the DNN becomes an oracle, making predictions about the legitimacy of a transaction as it cascades through the financial ecosystem. The allure of DNNs is underscored by their innate prowess in uncovering latent patterns within data. Human analysts might skim the surface, but DNNs, unfazed by cognitive limitations, dive deep to unearth cryptic correlations. These networks become veritable pattern detectors, unveiling concealed relationships and anomalous sequences that might evade human inspection. Moreover, DNNs are voracious learners; they consume copious amounts of data, refining their accuracy iteratively, bolstering their potential to detect nascent money laundering strategies. However, the trove of advantages is juxtaposed with a tapestry of challenges. DNNs, adorned with their complexity, present a formidable learning curve, demanding meticulous training and astute interpretability. The computational hunger of DNNs can be insatiable, often requiring vast computational resources for training and execution. The symbiotic companionship with computational expense calls for discerning resource allocation, ensuring that the benefits outweigh the costs. In the grand theater of money laundering detection, DNNs thrive when woven into a rich tapestry of complementary algorithms. Their ability to unravel complex data patterns is magnified when synchronized with other tools, each contributing a unique facet to the holistic detection strategy. In culmination, Deep Learning, through the conduit of Deep Neural Networks, emerges as a beacon of innovation in the quest to curtail money laundering. Its autonomy in identifying intricate patterns, its capacity to grasp complex relationships, and its potential to evolve with the evolving landscape of financial crime mark it as a formidable ally. While the path might be lined with challenges of interpretability and resource consumption, the alliance of DNNs with other machine learning paradigms unveils a holistic approach that charts a course towards a fortified financial future. The symphony of data-driven intelligence, in harmony with advanced algorithms, unveils the potential to reclaim transparency and integrity within the intricate tapestry of global finance.

*Evaluation Metrics:* Evaluation metrics play a pivotal role in the realm of assessing machine learning algorithms, particularly those tasked with the intricate challenge of money laundering detection. The choice of these metrics is inherently intertwined with the overarching goals of the detection system and the intrinsic characteristics of the data under scrutiny. Within the realm of money laundering detection, a diverse array of evaluation metrics has emerged to gauge algorithmic performance, each shedding light on different facets of efficacy.

1. *Precision:* The precision metric serves as a lighthouse that guides us through the turbulent waters of flagged transactions. It measures the algorithm's ability to accurately identify suspicious transactions among all those that have been flagged. A high precision value signifies not only a commendable number of correctly identified suspicious transactions but also the ability to steer clear of misidentifying valid transactions as suspect.

$$\Pr ecision = \frac{TruePositive}{TruePositive + FalsePositive} \tag{3}$$

2. *Recall:* In the intricate landscape of money laundering detection, recall stands as a sentinel, gauging the algorithm's capacity to recognize genuine suspicious transactions within the expansive sea of financial data. A robust recall metric indicates an algorithm's success in capturing a substantial portion of actual suspicious activities, even those that might be nuanced and complex in their execution.

$$\operatorname{Re} call = \frac{TruePositive}{TruePositive + FalseNegative} \tag{4}$$

3. *F1 Score:* The F1 score emerges as a harmonious symphony that melds precision and recall into a single harmonious measure. It is the harmonic mean of these two metrics and offers a comprehensive appraisal of an algorithm's accuracy. A high F1 score resonates with the successful synergy of high precision and recall, illustrating an algorithm's adeptness in striking a balance between these twin pillars of detection.

*Figure 4. Typical ROC curve*

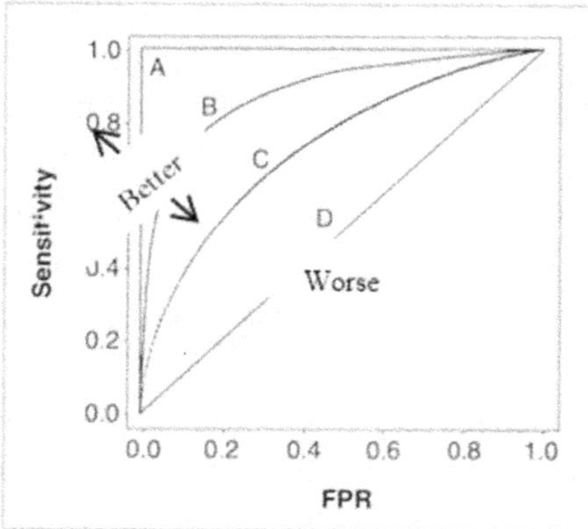


$$F1-Score = 2 * \frac{Precision * Recall}{Precision + Recall} \tag{5}$$

4. *Receiver Operating Characteristic (ROC) Curve:* The ROC curve unfurls like a navigational chart, tracing the trajectory of true positive rates against false positive rates across diverse threshold values. It paints a vivid picture of an algorithm's performance under various decision-making thresholds. A triumphant algorithm's ROC curve will ascend toward the pinnacle of the top-left corner, embodying both high recall and low false positive rates. A successful algorithm's ROC curve will be closer to the top-left corner, indicating high recall and low false positive rates.

$$FPR = \frac{FP}{FP + TN} \tag{6}$$

$$TPR = \frac{TP}{TP + FN} \tag{7}$$

- Where FPR stands for False Positive Rate
- Where TRP stands for True Positive Rate

5. *Area Under the Curve (AUC):* The AUC springs forth as a panoramic view, encapsulating an algorithm's overall prowess within a single metric. Derived from the ROC curve, a high AUC encapsulates an algorithm's capacity to consistently maintain high recall and low false positive rates across a gamut of threshold settings. In the grand tapestry of evaluating money laundering detection algorithms, it's imperative to underscore that no solitary metric can convey a holistic portrait of performance. Instead, the symphony of metrics orchestrates a chorus, harmonizing diverse perspectives to unveil a comprehensive understanding of the algorithm's capabilities. This multitudinous approach ensures that the algorithm not only identifies suspicious transactions accurately but also navigates the intricate balance between precision and recall. The realm of machine learning algorithms, however, is not devoid of challenges and limitations, particularly within the landscape

of money laundering detection. These challenges loom as potential impediments, tempering the unbridled enthusiasm for the transformative potential of these algorithms:

- *Bias:* The specter of bias can cast a long shadow over machine learning endeavors. If the training data used to fuel these algorithms is skewed or unrepresentative of the real-world population, the algorithms themselves can become skewed. This can yield predictions that mirror the biases inherent in the training data, culminating in erroneous or unjust outcomes.
- Over-fitting: The phenomenon of over fitting looms as a cautionary tale, where algorithms become hyper-specialized to the training data at hand. This specialization, while yielding high performance on the training data, might crumble when exposed to fresh data. The perpetually shifting strategies of money launderers require algorithms to swiftly adapt, necessitating the mitigation of over-fitting.
- *Data Quality:* The axiom "garbage in, garbage out" holds particular salience within the context of machine learning. Algorithms' efficacy is only as reliable as the data that fuels them. The tumultuous and labyrinthine nature of financial data renders it prone to noise, inaccuracies, and inconsistencies. This imposes a unique challenge on money laundering detection algorithms, which must grapple with the inherent chaos of financial data.
- *4.Transparency:* The opacity of machine learning algorithms presents a double-edged sword. While their predictive capabilities are formidable, deciphering the rationale behind specific predictions can be elusive. For money laundering detection, this opacity could pose difficulties for financial institutions in explaining to stakeholders why a particular transaction was deemed suspicious.
- *Price:* The journey toward harnessing machine learning's potential is not without financial tolls. Developing, implementing, and maintaining machine learning algorithms can be capital-intensive, especially for smaller financial entities. This financial burden might become a deterrent for smaller institutions, potentially hindering their access to cutting-edge technology.

In summation, the vista of machine learning algorithms for money laundering detection is rife with potential, yet threaded with challenges. These algorithms wield the power to redefine the landscape of financial security, yet their application demands circumspection. By judiciously considering data quality, biases, overfitting, transparency, and costs, stakeholders can chart a course toward a more secure financial future. This journey, underpinned by a tapestry of evaluation metrics, underscores that the pathway to effective money laundering detection must be guided by a synthesis of technological prowess and holistic wisdom.

## 4. FUTURE SCOPE

The statistics pertaining to money laundering in the year 2020 present a perplexing scenario that raises concerns within the realms of global financial security. Startlingly, it was revealed that an overwhelming proportion, accounting for nearly 90%, of illicitly laundered funds managed to slip through the cracks of vigilance, evading even the most sophisticated Anti Money Laundering software and Transaction Monitoring Systems in place. This alarming revelation underscores an undeniable urgency for the establishment of more potent and foolproof mechanisms to unearth these concealed financial transgressions. One potential avenue that holds promise in this relentless battle is the integration of advanced machine learning algorithms. It's crucial to recognize that the ramifications of money laundering extend far beyond mere statistics. The financial toll exacted by these clandestine activities is staggering, reaching an estimated 2% to 5% of the entire global Gross Domestic Product. However, what's even more disconcerting is the paltry fraction of criminal proceeds that are successfully reclaimed, standing at a meager 0.1%. This stark reality paints a dire picture of the challenges faced in the pursuit of justice against these economic offenses.

The landscape of money laundering detection has been undergoing a seismic shift, primarily due to the nefarious integration of cryptocurrencies into the playbook of criminal enterprises. This dynamic evolution has rendered the task of detection exponentially intricate and convoluted. Cryptocurrencies provide a cloak of anonymity that shields illicit transactions from traditional scrutiny, demanding a reevaluation of the tools and strategies used to counter such subversive activities. Thus, the clarion call for a reinforced arsenal of detection and monitoring software becomes all the more urgent. Against this backdrop, the trajectory of the anti-money laundering software market gains prominence as a potential catalyst for change. Projections indicate that by the year 2023, this market will burgeon into a colossal entity, boasting a value of $1.77 billion. This burgeoning market offers an opportune juncture for harnessing the prowess of machine learning algorithms and delving into the depths of deep learning. By synergizing these cutting-edge technologies with the battle against money laundering, nations can fortify their financial infrastructure and fortuitously stabilize their economic foundations. The revelations encapsulated within the 2020 money laundering statistics beckon us to confront the inadequacies of our current defenses against this global menace. The imperative to address this issue transcends numbers, striking at the heart of economic integrity and social justice. As criminal methodologies evolve, so too must our arsenal of countermeasures. The ascent of the anti-money laundering software market to a projected value of $1.77 billion by 2023 underscores a potential avenue for technological intervention. It is a clarion call for the fusion of artificial intelligence, machine learning, and deep learning into the tapestry of financial security. This convergence not only amplifies the chances of detecting money laundering activities but also bolsters nations in their quest for unwavering fiscal stability.

Amidst these challenges, international collaboration emerges as a critical linchpin in the quest to combat money laundering effectively. The fluid nature of financial transactions across borders necessitates a cohesive effort, transcending geopolitical boundaries. Forums for information exchange and joint strategies become pivotal in devising a holistic approach to stem the tide of illicit financial flows. As economies become increasingly interdependent, the need for a united front against money laundering becomes glaringly evident. Only through mutual cooperation can nations hope to harness the full potential of advanced technologies and curb the pervasive impact of money laundering on their economic well-being.

## 5. CONCLUSION

In conclusion, the utilization of machine learning algorithms as a pioneering avenue for money laundering detection emerges as a beacon of promise amid the intricate webs of financial malfeasance. In a landscape where traditional detection methods stand beleaguered by the relentless evolution of money laundering tactics, the ascendancy of machine learning algorithms ushers in a new era of hope. The very complexity that has rendered conventional methods obsolete now becomes the catalyst propelling the integration of artificial intelligence into the realm of financial security. Machine learning algorithms hold the potential to orchestrate a paradigm shift in the efficacy and precision of money laundering detection. By analyzing vast troves of transactional data with unparalleled speed and accuracy, these algorithms can uncover subtle patterns, anomalies, and hidden connections that elude human perception. This heightened level of insight empowers financial institutions to respond to potential threats promptly, thereby mitigating risks and safeguarding the integrity of their operations. Yet, within this realm of promise, lies a daunting challenge. The relentless ingenuity of money launderers is a moving target, ceaselessly adapting to circumvent detection. Thus, the efficacy of machine learning algorithms hinges upon a continuous cycle of innovation and refinement. There exists a pressing imperative for sustained research and development, birthing advanced algorithms capable of outwitting the ever-shifting stratagems of financial criminals. Crucially, the voyage toward enhanced machine learning algorithms is a journey intertwined with collaboration. Financial institutions, regulatory bodies, and the technology sector must converge in a symphony of cooperation to forge algorithms that withstand the crucible of real-world financial warfare. The sharing of insights, data, and methodologies becomes a cornerstone for these endeavors, birthing algorithms that draw strength from the collective wisdom of diverse stakeholders.

In contemplating the long-term ramifications, the symbiotic relationship between robust money laundering detection and the global financial architecture stands illuminated. A fortified defense against money laundering nurtures a more secure financial ecosystem, bolstering investor confidence and economic stability. The benefits ripple outward, resonating with a society that thrives on the tenets of transparency, accountability, and ethical financial conduct. Moreover, the advantages do not stop at financial security; they extend to the broader canvas of social prosperity. As machine learning algorithms dismantle the barriers shielding ill-gotten gains, a more equitable distribution of resources emerges. Societal resources once siphoned off into the coffers of criminals are redirected toward legitimate avenues, fostering economic growth, and driving infrastructure development. The underpinning infrastructure for this transformation is not just the algorithms themselves but also the cultivation of an ecosystem conducive to innovation. This necessitates nurturing a culture of continuous learning and adaptation, wherein financial professionals, technologists, and regulators converge to stay ahead of the curve. The dynamism of the financial landscape requires a harmonious blend of human expertise and algorithmic acumen.

In summation, the narrative that unfolds within the realm of money laundering detection paints a portrait of relentless evolution and symbiotic collaboration. The ascent of machine learning algorithms presents a beacon of promise against the backdrop of evolving money laundering tactics. With their prowess to decode complex patterns and discern hidden connections, these algorithms usher in an era of heightened vigilance, empowering financial institutions to quell threats in real-time. This journey is not solitary; it is a symphony of diverse stakeholders collaborating to outwit the innovative guile of financial criminals. The virtuous cycle thus initiated reverberates beyond financial security. It resonates with economic stability, investor trust, and societal well-being. As algorithms spearhead a more equitable distribution of resources, society at large reaps the dividends of transparency and accountability. The story that emerges is not just one of advanced algorithms but of a holistic transformation that fortifies the fabric of global finance and engenders a world where prosperity knows no bounds.

## REFERENCES

Alkhalili, M., Qutqut, M. H., & Almasalha, F. (2021). Investigation of applying machine learning for watch-list filtering in anti-money laundering. *IEEE Access : Practical Innovations, Open Solutions, 9*, 18481–18496. doi:10.1109/ACCESS.2021.3052313

Alotibi, J., Almutanni, B., Alsubait, T., Alhakami, H., & Baz, A. (2022). Money Laundering Detection using Machine Learning and Deep Learning. *International Journal of Advanced Computer Science and Applications, 13*(10), 10. doi:10.14569/IJACSA.2022.0131087

Canhoto, A. I. (2021). Leveraging machine learning in the global fight against money laundering and terrorism financing: An affordances perspective. *Journal of Business Research, 131*, 441–452. doi:10.1016/j.jbusres.2020.10.012 PMID:33100427

Chen, Z., Van Khoa, L. D., Teoh, E. N., Nazir, A., Karuppiah, E. K., & Lam, K. S. (2018). Machine learning techniques for anti-money laundering (AML) solutions in suspicious transaction detection: A review. *Knowledge and Information Systems, 57*(2), 245–285. doi:10.100710115-017-1144-z

Chohan, U. (2019). The fatf in the global financial architecture: challenges and implications. *Talking about terrorism.*

Firmansyah, W., & Atmadja, H. T. (2021). Juridical analysis awareness of profession advocacy to financial transaction reports and analysis centre (ppatk) during prevent and eradicate money laundering crime. *Journal of Multidisciplinary Academic, 5*(4), 308–314.

Jullum, M., Løland, A., Huseby, R. B., Ånonsen, G., & Lorentzen, J. (2020). Detecting money laundering transactions with machine learning. *Journal of Money Laundering Control, 23*(1), 173–186. doi:10.1108/JMLC-07-2019-0055

Kute, D. V., Pradhan, B., Shukla, N., & Alamri, A. (2021). Deep learning and explainable artificial intelligence techniques applied for detecting money laundering–a critical review. *IEEE Access : Practical Innovations, Open Solutions, 9*, 82300–82317. doi:10.1109/ACCESS.2021.3086230

Lokanan, M. E. (2022). Predicting money laundering using machine learning and artificial neural networks algorithms in banks. *Journal of Applied Security Research*, 1–25. doi:10.1080/19361610.2022.2114744

Lorenz, J. (2020). Machine learning methods to detect money laundering in the bitcoin blockchain in the presence of label scarcity. *Proceedings of the first ACM international conference on AI in finance.* IEEE. 10.1145/3383455.3422549

Paula, E. L. (2016). Deep learning anomaly detection as support fraud investigation in brazilian exports and anti-money laundering. *2016 15th ieee international conference on machine learning and applications (icmla).* IEEE. 10.1109/ICMLA.2016.0172

Pettersson Ruiz, E., & Angelis, J. (2022). Combating money laundering with machine learning–applicability of supervised-learning algorithms at cryptocurrency exchanges. *Journal of Money Laundering Control, 25*(4), 766–778. doi:10.1108/JMLC-09-2021-0106

Raiter, O. (2021). Applying supervised machine learning algorithms for fraud detection in anti-money laundering. *Journal of Modern Issues in Business Research*, *1*(1), 14–26.

Soltani, R., Nguyen, U., Yang, Y., Faghani, M., Yagoub, A., & An, A. (2016). A new algorithm for money laundering detection based on structural similarity. *2016 IEEE 7th Annual Ubiquitous Computing, Electronics & Mobile Communication Conference (UEMCON)*. IEEE.

Tertychnyi, P., Godgildieva, M., Dumas, M., & Ollikainen, M. (2022). Timeaware and interpretable predictive monitoring system for anti-money laundering. *Machine Learning with Applications*, *8*, 100306. doi:10.1016/j.mlwa.2022.100306

Yang, Y., & Wu, M. (2020, July). Supervised and unsupervised learning for fraud and money laundering detection using behavior measuring distance. In *2020 IEEE 18th International Conference on Industrial Informatics (INDIN)* (*Vol. 1*, pp. 446-451). IEEE. 10.1109/INDIN45582.2020.9442099

Youssef, B., Bouchra, F., & Brahim, O. (2023). State of the Art Literature on Anti-money Laundering Using Machine Learning and Deep Learning Techniques. *The International Conference on Artificial Intelligence and Computer Vision*. Cham: Springer Nature Switzerland. 10.1007/978-3-031-27762-7_8

Zhang, Y., & Trubey, P. (2019). Machine learning and sampling scheme: An empirical study of money laundering detection. *Computational Economics*, *54*(3), 1043–1063. doi:10.100710614-018-9864-z

# Chapter 13
# Apple Leaf Disease Identification and Segmentation Using Enhanced Learning-Driven Feature Representation Model

**Harsha Raju**
https://orcid.org/0000-0003-0159-8342
*Reva University, India*

**Veena Kalludi Narasimhaiah**
*Reva University, India*

**Mukil Alagirisamy**
https://orcid.org/0000-0003-1623-1914
*Asia Pacific University, Malaysia*

## ABSTRACT

*The field of automatic segmentation and identification of plant diseases from leaf images has become increasingly significant in recent years. There are a number of fungal diseases that affect apple quality and yield, including apple scab, cedar rust, and black rot. In order to prevent crop losses, it is essential to identify these diseases quickly as possible. Despite many approaches being discussed, segmenting the diseased part of leaves with high accuracy and low false positive rates remains a challenging task. This study suggests that by isolating the color background and highlighting the area of interest, a substantial feature set can be constructed to enhance deep learning generalization capability for disease classification. An algorithm for fitness function is developed to represent the features that are relevant to disease classes and optimally adjusts the weights and biases in the training phase. Based on visual outcome and comparative analysis in terms of precision, recall rate, and F1-scores, the efficacy of the proposed work is justified.*

DOI: 10.4018/979-8-3693-0659-8.ch013

## 1. INTRODUCTION

Apple plant diseases can have a significant impact on the productivity and profitability of fruit, and accurate identification and diagnosis of plant diseases is essential for disease management and crop protection. Traditional methods for plant disease diagnosis, such as expert manual examination, can be time-consuming, subjective, and prone to human error. The use of image processing applications and machine learning methods for plant disease classification has the potential to provide a faster, more accurate, and more objective method of disease diagnosis. The proposed research focuses on developing an automated and synchronized framework that can accelerate plant leaf disease detection and help farmers prevent diseases from spreading at early stage. The proposed system is evaluated through experiments on an apple leaf dataset to assess its accuracy and performance.

### 1.1 Background

Agriculture is a major contributor to the economies of many countries. The export and import of fruit varieties is also common, especially apple fruit, which is in high demand worldwide (Liu, J., and Wang, X,2021). Apple trees are a crucial source of food and income for many regions around the world. However, apple trees are susceptible to various diseases that can reduce their productivity and ultimately result in crop loss. There are many diseases that attack apple trees in crop fields due to their unavoidable presence in agriculture. However, apple rust and cider rust and scab are three of the most common fungal diseases that heavily impact apple crop yields and economic values (Orchi, H., Sadik, M. and Khaldoun, M., 2021.), (Ganatra, Nilay & Patel, Atul. (2018). According to estimates, diseases to plants result in a loss of 35-40% of crop yields in agriculture(Savary, Serge & Ficke, Andrea & Aubertot, Jean-Noël & Hollier, Clayton. (2012)). Due to lack of facilities and awareness, most farmers have no idea what type of disease they are dealing with. In these situations, they often use fungicides and insecticides to prevent yield loss due to fungi and pests(Nicolopoulou-Stamati, Polyxeni, Sotirios Maipas, Chrysanthi Kotampasi, Panagiotis Stamatis, and Luc Hens.). However, this practice often involves the excessive use of chemicals, which raises significant concerns over health and environmental issues, as it can be toxic to birds and many other organisms, including environmentally friendly insects (Yoon, M.Y., Cha, B. and Kim, J.C., 2013.). In addition, manual and tedious processes result in high operating costs for farmers. An appropriate remedial action can be planned to prevent fungal infections in apple plants, reduce chemical use, and ensure sustainable crop maintenance practices if diseases can be diagnosed and detected early (Vasavi, P., Punitha, A. and Rao, T.V.N., 2022). Thus, it is extremely important to develop an automated and affordable method for identifying and classifying diseases, so that farmers and agriculture experts can take effective measures.

### 1.2 Digital Image Processing

Digital Image Processing is a field of study that deals with the acquisition, representation, analysis, and interpretation of images to extract meaningful information. It utilizes mathematical and computational methods to manipulate images and extract features that are relevant to the task at hand. A typical architecture of plant disease detection using application of image processing and AI is shown in Figure 1.

In the context of plant leaf disease detection, digital image processing can help by automating the analysis of leaf images to identify and diagnose various diseases. For example, the preprocessing tech-

*Figure 1. A schematic architecture of typical plant diseases detection and classification using Application of Image processing and machine learning*

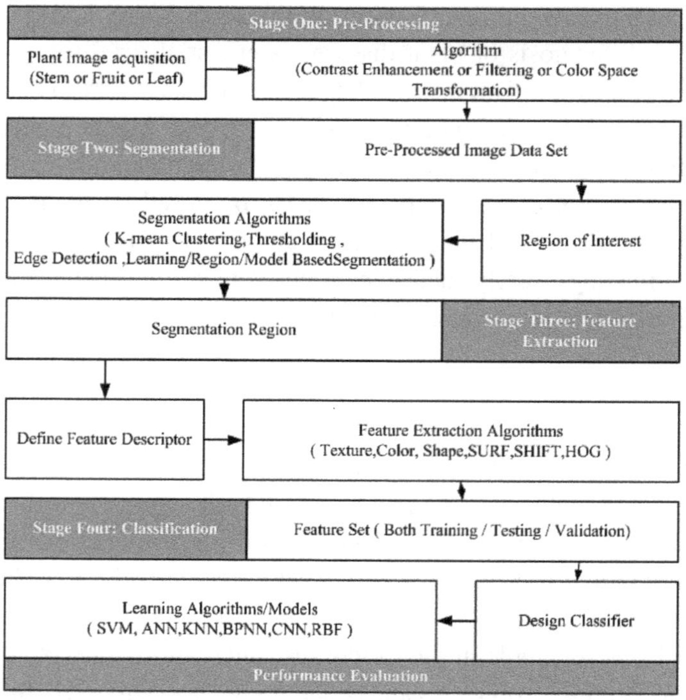


niques is the process of preparing the images for further analysis by removing or correcting any unwanted or irrelevant information. This step is important because the quality of the image can have a significant impact on the accuracy of the disease classification. The main goal of image preprocessing in plant leaf disease classification is to enhance the features of the leaves that are relevant to the classification task, and to remove any noise or artifact that may interfere with the analysis. Some common preprocessing steps includes contrast enchantment, color conversion for adjusting color and contrast variation and filtering removing any noise or artifact from the image to improve its quality and image normalization for scaling the intensity values of the image to a common range to make the image more comparable to other images. Image segmentation is a key step in plant leaf disease classification, as it divides the image into meaningful regions for further analysis. The goal of image segmentation is to separate the leaf from the background, and to identify the different regions of the leaf that are relevant to the disease classification task. In plant leaf disease classification, image segmentation can be used to separate the leaf from the background and to isolate the different regions of the leaf, such as the veins and the lamina. This information can then be used to extract relevant features, such as color, texture, and shape, that can be fed into machine learning models for disease classification. Image segmentation algorithms can be based on various approaches, such as thresholding, edge detection, region growing, and active contours. The choice of the appropriate image segmentation method will depend on the specific requirements of the classification task and the nature of the images. By properly segmenting the images, the performance of the machine learning models can be improved, leading to more accurate and reliable disease classification results. On the other hand, feature extraction is a critical step in plant leaf disease classification, as it involves selecting and extracting the relevant features of the images that are related to the disease.

The goal of feature extraction is to select the most discriminative and relevant information from the images, and to represent the images in a form that can be easily processed by machine learning algorithms.

## 1.3 Research Challenges

In addition, manual and tedious processes result in high operating costs for farmers. An appropriate remedial action can be planned to prevent fungal infections in apple plants, reduce chemical use, and ensure sustainable crop maintenance practices if diseases can be diagnosed and detected early (Vasavi, P., Punitha, A. and Rao, T.V.N., 2022). Thus, it is extremely important to develop an automated and affordable method for identifying and classifying diseases, so that farmers and experts can take effective measures. A great deal of research has been conducted over the past decades on disease detection using plant leaves. Several approaches based on image processing and learning algorithms have been explored to improve plant disease recognition (Vasavi, P., Punitha, A. and Rao, T.V.N., 2022) The learning techniques include CNN (Yang Lu, Shujuan, Nianyin Zeng, Yurong Liu, Yong Zhang, (2017),), artificial neural networks-(ANN)(Elham Omrani, Benyamin Khoshnevisan, Shahaboddin Shamshirband, HadiShaboohi, Nor Badrul Anuar, Mohd Hairul nizam md nasir, (2014),), backpropagation(Kshitij Fulsoundar, Tushar Kadlag, Sanman Bhadale, Pratik Bharvirkar, (2014), support vector machines-(SVM) (Shivaputra S.Panchal, Rutuja Sonar, (2016).) and various other image processing techniques (R.Preethi, S.Priyanka, U.Priyanka, A.Sheela, (2015),),(P.Revathi, M.Hema Latha, (2012),) For image processing, CNN is more robust because it has the ability to perform both feature extraction and classification. It is shown (P. A. Dias, A. Tabb, and H. Medeiros,)how CNN can be used to semantically classify apple flowers effectively and automatically. A comparative analysis validates the performance of the presented work, where the introduced mechanism is effective at counting total flowers. A multiclass apple disease classification system based on optimized segmentation and feature extraction techniques was presented in the study of (M. A. Khan et al.), To enhance the disease portion of leaf images, gaussian and median filtering techniques are applied first, then strong correlated pixel-based segmentation is applied. Optimal features are further extracted using a genetic algorithm. The presented system is validated on apple leaf images obtained from plant village dataset. In (Chuanlei, Z., Shanwen, Z., Jucheng, Y., Yancui, S. and Jia, C., 2017) the authors discuss enhancing the feature extraction process for apple leaf disease recognition. Segmentation was performed using histograms and region growing, and feature extraction was conducted using a genetic algorithm combined with correlation-based attribute selection. Furthermore, a custom dataset is used to classify diseases using SVM. K-means clustering based on color-based segmentation is described in(Jadhav, S.B., Udup, V.R. and Patil, S.B., 2019.). In this work, features are extracted using gray-level co-occurrence matrices (GLCMs) and SVM classifier is implemented to estimate soyabean leaf disease severity. An approach based on graph cut extraction has been presented in (Bracino, A.A., Concepcion, R.S., Bedruz, R.A.R., Dadios, E.P. and Vicerra, R.R.P., 2020,)along with a technique for selecting features based on textures. The disease classes are classified using KNN and SVM. This study, however, only examined a small set of feature maps and segmented the background portion, which may not be applicable to real-time scenarios. The application of gaussian mixture models to segment images is discussed in (Sumithra, M.G. and Saranya, N., 2021). A combination of swarm optimization and fuzzy C-means has then been used to segment disease portions. A SVM classifier is used to determine whether a leaf is healthy or diseased. The study, however, assumed uniform variations in pixel intensity, which is an unrealistic assumption and is limited to a binary classification. The authors of (Abisha, S. and Jayasree, T., 2022) combined the Delta-E and expectation maximization mechanisms to produce multiple segmented images. A single segmented image was constructed by merging wavelet coefficients retrieved from multi-segmented images. ANNs are used to detect diseases in brinjal plants based on color, texture and structural features. However, this study lacks k-fold cross validation. Using a haze-removing algorithm, the authors in (Adeel, A., Khan, M.A., Sharif, M., Azam, F., Shah, J.H., Umer, T. and Wan, S., 2019) enhance the image of grape leaves. After the enhanced image is segmented into multiple channels, a weight function is used to determine the best sub-channels, which are then subjected to mathematical morphology. Analyses of adjacent components and canonical correlation are used to determine color and structural features and the classification task is performed using SVM. The work, however, does not emphasize disease portion segmentation and may not retain all information due to the consideration of only a single channel intensity image. For enhancing diseased portions in Apple leaves, the authors of (Rehman, Z.U., Khan, M.A., Ahmed, F., Damaševičius, R., Naqvi, S.R., Nisar, W. and Javed, K., 2021) suggested a hybrid contrast stretching method. For disease segmentation, Mask R-CNN is applied, and ResNet-50 is used to learn features, while SVM is employed to classify diseases. There is, however, no ground truth for mask R-CNN in this study. The work carried out in (Abayomi-Alli, O.O., Damaševičius, R., Misra, S. and Maskeliūnas, R., 2021.) applied the deep learning architecture of MobileNetV2 for the automated extraction of features and classification of leaves diseases of cassava plants. A data augmentation approach is used to generate artificial image data to train the learning model. Various factors contribute to disease classification, including color. But artificial images have an unnatural color, which could hinder the performance of the classification model. Detection of multi-crop leaf diseases is the topic of (Gaikwad, S.S., Rumma, S.S. and Hangarge, M., 2022). In this study, the authors used color, black and white and grayscale representations of Apple, Custard apple, and Guava plants and fed them into two CNN models, AlexNet and SqueezeNet. The study results revealed that both models were equally accurate for color images, which indicates that color is an important factor in classification. A customized version of the deep CNN model ResNet with 197 layers has been proposed by (Pandian, A., Kanchanadevi, K., Rajalakshmi, N.R. and Arulkumaran, G., 2022) for detecting leaf disease in multiple crops. Data augmentation is used to generate several artificial images of similar type for the purpose of training this model. As a result, the learning model was found to be effective with a 99.58% accuracy rate, but this performance comes at a high computational cost that needs to be optimized in the design phase. Transformers are applications of attention mechanisms used in natural language processing tasks in order to comprehend complex textual patterns. Using capability transformers in CNN models, the authors in (Li, X. and Li, S., 2022)tried to improve the generalization of plant leaf features for identifying apple diseases during the training phase. The model was evaluated using a synthetic dataset and showed 96.85% accuracy. Nevertheless, this model also requires a higher level of computational complexity and may suffer from underfitting issues since it is trained on specialized datasets. Hence, it can be analyzed that the literatures are rich with variety of solution presented by different researchers. Despite many efforts, there is still a scope of improvement in various aspects like optimization in the computational complexity, unification in segmentation and feature enhancement.

The use of digital image processing in plant leaf disease detection has several advantages over traditional manual inspection methods. However, image processing in the plant diseases analysis face some unique challenges as compared to other image processing task to achieve more accurate and consistent results, making it a valuable tool for plant health management and disease control. These challenges include:

- Preprocessing: Noises in the image which exist in various disease symptoms and most are similar to background regions. Contrast mismatch variations of light condition, dimension or spread size of the disease, volume of the fruits poses significant challenges in the handling the dimension overhead.

- Segmentation: Handling the dynamics of color associated with the disease during the segmentation based on the color feature-set are another essential challenge to be considered while designing the effective segmentation algorithms.
- Feature Extraction: A large number of features can be extracted from apple leaf images, but not all of them may be relevant for disease classification. Irrelevant features do not produce good accuracy and also effects on the system execution time. Therefore, selection of most prominent features is challenging process.
- Selection of Machine Learning algorithm: There are many machine learning algorithms available for disease classification, but not all algorithms perform equally well on different types of data. Researchers need to evaluate various machine learning algorithms and determine which algorithms are best suited for apple leaf disease classification.
- Scalability: The proposed system for apple leaf disease classification should be scalable to handle large amounts of data and be able to classify new images in real-time. This requires efficient algorithms and algorithms that can handle large amounts of data.
- Generalization: The system for apple leaf disease classification should be able to generalize to new unseen data and make accurate predictions for new apple leaf images. This requires algorithms that can effectively learn from the training data and generalize to new data.

The problem statement for the proposed study based on above interpretations can be expressed as *"Designing and developing an optimal algorithm is a challenging task that can extract region of interest aware feature-map for learning model to perform adequate classification of the plant leaf diseases".*

Therefore, this chapter introduces a novel method of automating apple leaf disease segmentation and identification using an enhanced learning-driven feature representation framework. The primary consid-

*Figure 2. Proposed feature representation and enhanced learning scheme for apple leaf disease classification*

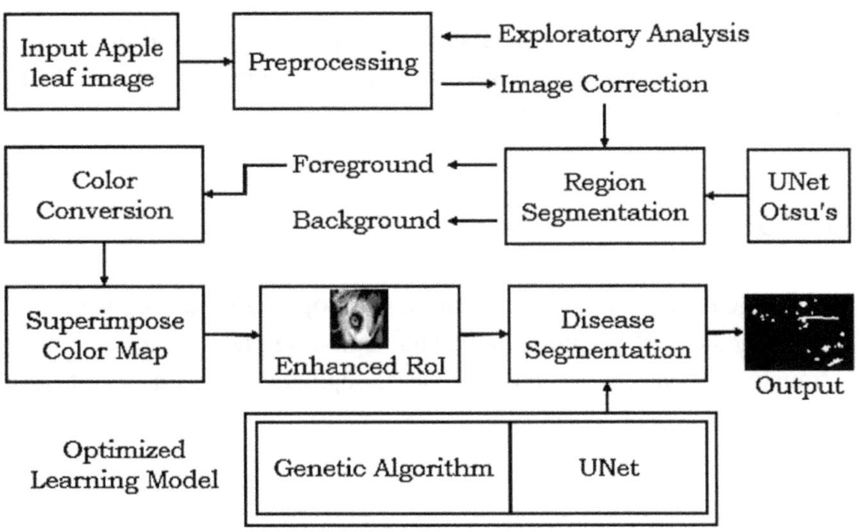

eration in the proposed feature representation model is that leaf disease can be better identified if their characteristics differ from background portions and other similar disease symptoms. For this reason, the study emphasizes on extracting regions of interest (ROIs) such as the leaf foreground region, background region, and region of disease spot. As a result of these information, the learning model will be able to generalize the underlaying pattern of data samples more effectively, since it holds vital information about which attributes are more significant and which attributes play a decisive role in identifying leaf diseases. The schematic architecture of the proposed system is illustrated in **Figure 2**.

The scheme proposed in this paper adopts analytical research methodology and implemented in phase-wise manner. The system considers an input data of apple leaf images from the dataset, which is further subjected to preprocessing and region segmentation operation. In this work, more emphasizes is given to the modelling of feature representation with an optimized learning scheme for efficient and automated leaf disease segmentation. In feature representation process, the segmented foreground image obtained after region segmentation operation is subjected to color transformation, where disease spot is transformed in HSV color map and then its superimposed to the image in RGB color map. As result, enhanced RoI is achieved which serve as a significant feature map for the learning model, which is UNet, a kind of CNN learning model developed for biomedical image segmentation. The unique contribution made in the UNet is its integration with the genetic algorithm, which will enhance its overall learning experience by optimally adjusting the learning parameters i.e., weight and bias. Also, a customized fitness function is proposed that will acts as loss function in the learning model. The overall contribution of this chapter is summarized as follows:

- The system is designed to extract RoI as a significant feature for learning based automated segmentation.
- To expose the leaf image data better to the learning model, a color conversion scheme is used to identify the influential colors significant to disease identification.
- Since the Ground Truth data isn't available for the RoI, a specialized fitness function which is aware of the type of disease is designed to aid the genetic algorithm in optimization of UNet learning model.
- The genetic algorithm measures the fitness of various population (learning parameters i.e., set of weights and biases) and chooses the best of them over several generations.
- The segmented diseased region returned as final output from the learning model. Also, the proposed learning model is then benchmarked with other different type of deep learning models and comparative analysis is also conducted.

## 2. METHODOLOGY

In this section, the implementation of the methodology adopted in the proposed system is discussed systematically which includes i) preprocessing, and region segmentation, ii) feature representation based on a color transformation scheme, and iii) disease segmentation using UNet learning model with genetic algorithm.

*Table 1. Description of apple leaf dataset*

| SI. No | Apple Leaf Classes | Total Number |
|---|---|---|
| 1 | Apple_scap | 2,016 |
| 2 | Apple Black_rot | 1,987 |
| 3 | Cedar_apple_rust | 1,760 |
| 4 | normal | 2,008 |

## 2.1 Preprocessing and Region Segmentation

The exploratory analysis is conducted as first step in the preprocessing phase to get insight about the dataset and leaf image characteristics. The exploratory findings reveal following significant observations:

- The dataset consists of apple leaf images which are subjected to most common diseas as shown in Table I
- There is total of 7771 images of apple leaves in the dataset. The total number of images is sufficient for deep learning model hence no data augmentation procedure needs to be applied.
- Dataset is nearly uniform and does not associate with class imbalance problem
- Outer shape is similar and greener color indicates healthier leaf
- The stem is always present in center of leaf
- Edge of the leaf is similar.
- As a preprocessing step, dataset split operation is performing, where out of 7771 images, 6216 leaf images are considered in training set and remaining 1555 images are considered in testing set.
- Since the dataset size is quite big, the images are resized.

The region segmentation is performed using the proposed Augmented Otsu's based thresholding technique which is a joint approach of conventional Otsu and multi-channel CNN i.e., UNet.

**Algorithm-1:** *Proposed Augmented Otsu's based thresholding technique*

```
begin
Define HMCRmin, HMCRmax, PARmin,PARmax, HMS, BWmin,BWmax
Define Maximum number of iterations (NI)
Define fitness function
Define maximum number of clusters (Kmax)
Generate the initial population of harmony vectors HVi, (i= 1,2,..,HMS)
while (iter < NI) do
    for i=1 to HMS do
        for j=1 t0 Kmax,
        if (Thi,j > Thcutoff;) then
            J™ cluster center of i″ harmony vector is active
        Else
```

```
                J™ cluster center of i* harmony vector is inactive
            endif
        end for
    for each data points in the given dataset do
    Compute the weighted Euclidean distance from all the active centers
    Recompute the active cluster centers
        end for
    end for
        Generate the new harmony vector using PAHS
        if (fitness(new_harmony_vector) < worst(fitness(HM))) then
        Replace the worst in HM with the new harmony vector
        endif
    end while
        Return the best harmony vector
    end
```

Since, the color plays an important role in the detection, the leaf images are not converted to gray scale representation and UNet is employed to perform region segmentation without loosing vital aspects of color leaf image. Overall usage of this algorithm is for enhancing further operation of feature representation and disease segmentation making the model robust.

## 2.3 Feature Representation

The obtained segmented color leaf images are in the RGB space, which may not provide better feature set or represent color attributes significant to different diseases in apple leaf because multiple diseases may occur on the same leaf and share similar color. Therefore, the proposed feature representation scheme adopts a color transformation where RGB image is converted into HSV color map to better comprehend the colors variation among the diseases in the leaf image. The HSV stands for Hue Saturation and value, where Hue component plays an important role in color detection, and the value determines the darkness of the leaf in order to detect diseases like scab. The algorithm for RGB to HSV is represented as follows:

**Algorithm-2:** *Color transformation from RGB to HSV*

```
Input: I_FS Segmented foreground color leaf image
Output: HSV Image
Start
    For ∀ i_x,y in I_FS
    Compute C_max = f_max(R,G,B)
      Compute C_min  = f_min(R,G,B)
      Compute Diff = C_max - C_min
        if C_max==C_min
          Value of H = 0
        else if C_max==R
```

Value of H = $\left(60\times\left(\dfrac{G-B}{Diff}\right)+360\%360\right)$

else if $C_{max}$==G

Value of H = $\left(60\times\left(\dfrac{B-R}{Diff}\right)+120\%360\right)$

else if $C_{max}$==B

Value of H = $\left(60\times\left(\dfrac{R-G}{Diff}\right)+240\%360\right)$

end of if

if $C_{max}$==$C_{min}$

Value of S = 0

else

Value of S = $(Diff/C_{max})\times100$

end of if

Value of V = $C_{max}\times100$

**end** of **For**

Return HSV color map

**End**

The algorithm takes input of the segmented foreground image $I_{FS}$ obtained from region segmentation operation using Augmented Otsu's technique. After successful execution of all the steps it returns the image in HSV color map from the RGB color space. The algorithm initially computes the maximum $C_{max}$ and minimum $C_{min}$ color value of the RGB components for pixels ($i_{x,y}$) in the input image. Further it computes a chroma (strength) by computing difference (*Diff*) of $C_{max}$ and $C_{min}$. Afterwards, the algorithm executes an operation of conditional check to compute the individual value of H and S and V. Here, the RGB components, saturation (s) and value (v) lies in the range [0,1], while the hue (h) should

*Figure 3. Frequency distribution of HSV normal leaves*

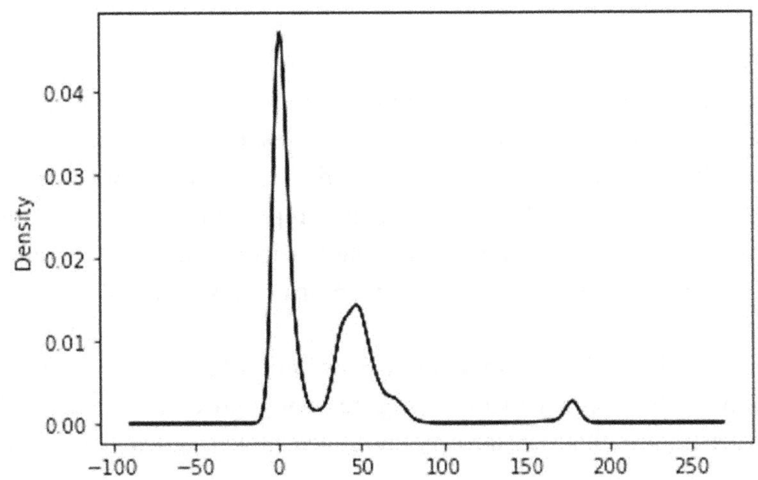

*Figure 4. Frequency distribution of HSV normal leaves*

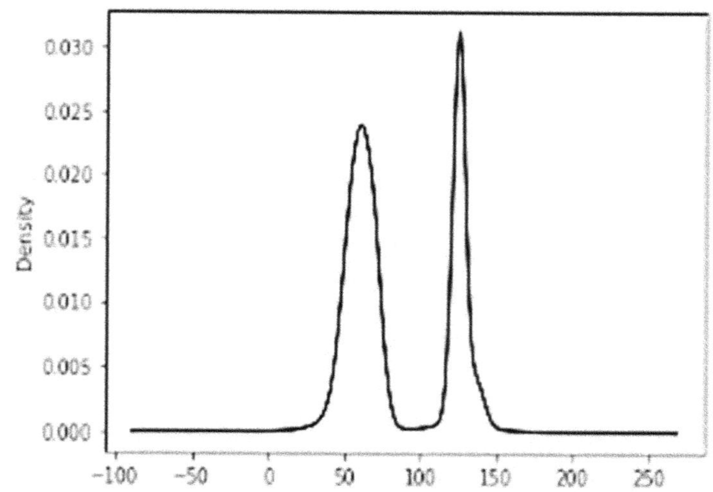


*Figure 5. Frequency distribution of HSV scab diseased leaves*

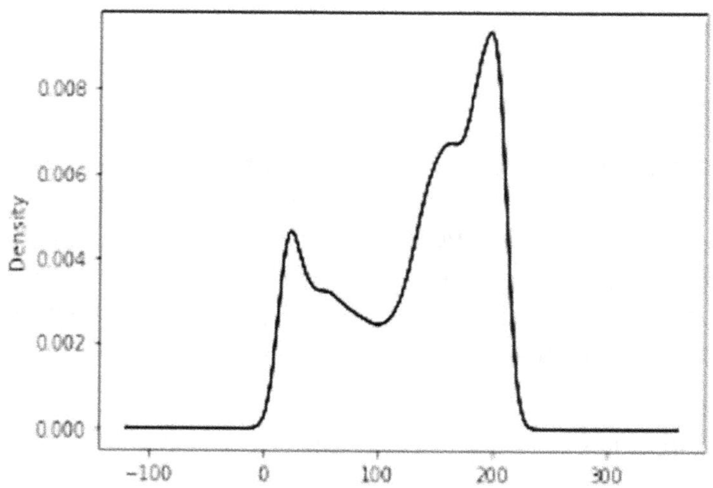


have values in the range [0,360]. For this a modulo operation is performed in each step of computing HVS value. After execution, the algorithm returns leaf images in HSV color map, which is then used to highlight the RoI (i.e., disease spot) in the leaf image and further superimposed to the leaf image in the RGB color map. This operation provides a better feature representation of leaf image part along with the enhanced diseased portion towards achieving better performance by the genetic-driven learning model. Figure 2 illustrates the average HSV color map of healthy or normal leaf image and diseased (Apple_Scab) leaf image.

The above-mentioned Figure 3 is the Frequency distribution of HSV normal leaves, Figure 4 is the Frequency distribution of HSV apple scab image, Figure 5 and Figure 6 is Frequency distribution of HSV scab diseased leaves illustrating statistical information concerning frequency distribution in the apple leaf

*Figure 6. Frequency distribution of HSV scab diseased leaves*

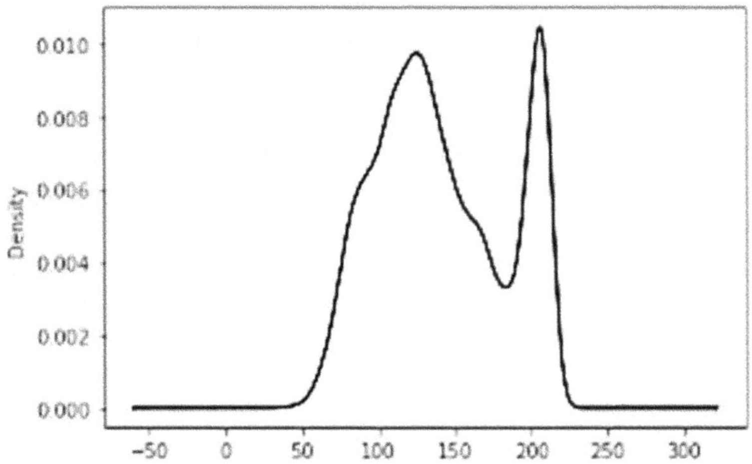


image concerning HSV color map for normal leaf and scab image. This analysis helps in deciding the responsible factors to derive optimal for fitness value in genetic algorithm based UNet learning model.

## 2.5 Leaf Disease Segmentation

The proposed study adopts UNet learning model for the segmentation of disease portion from the leaf image. The UNet takes input of feature map which has enhanced RoI part for the segmentation operation. However, the Ground Truth (GT) data or mask is not available for the obtained enhanced RoI features, so training the UNet model for disease spot segmentation without GT is not a feasible task and its output cannot effectively be evaluated against the actual problem. Therefore, the UNet is integrated with genetic algorithm and a specialized fitness function is devised which is aware of the type of disease. Here, the genetic algorithm acts as an optimizer and the proposed fitness function acts like a loss function that will to aid the genetic algorithm in the proper adjustment of learning parameters of UNet in the training phase. The schematic illustration of the proposed enhanced learning model is shown in Figure 17.

*Figure 17. Enhancement and optimization of UNet Learning model with genetic algorithm*

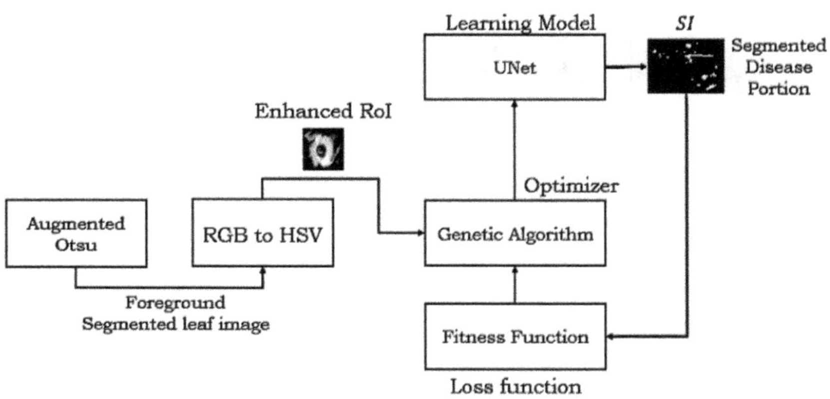

The genetic algorithm measures the fitness of various population (set of weights and biases) and chooses the best of them over several generations. Then the segmentation is carried out with the trained UNet and is compared with the other learning model for the benchmarking. The implementation of the fitness function is discussed in algorithm 3.

**Algorithm-3:** *Fitness function for Genetic Algorithm for UNet Optimization*

**Input:** $I_{FS}$ (Segmented foreground leaf image), $SI$ (Segmented Image or mask), $Y$ (Leaf Class)

**Output:** Fitness ($f$)

Start

Initialize $f = 0$

**if** $Y==0$

compute $f = 100 - \left( \frac{\sum_{i=1}^{n} \sum_{j=1}^{m} SI_{i,j}}{m \times n} \right) \times 100$

**end** of **if**

**if** $Y==1$

$I_{FS}$ = HSV ($I_{FS}$)

$X = I_{FS} \times SI$

s = $\sum_{i=1}^{n} \sum_{j=1}^{m} SI_{i,j}$

compute $f = 100 * \left( \frac{s - \sum_{i-1}^{n} \sum_{j=1}^{m} T1 > Hue(X_{i,j}) > T2}{s} \right)$

**end** of **if**

**if** $Y==2$

$I_{FS}$ = HSV ($I_{FS}$)

$X = I_{FS} \times SI$

s = $\sum_{i=1}^{n} \sum_{j=1}^{m} SI_{i,j}$

compute $f = 100 * \left( \frac{s - \sum_{i-1}^{n} \sum_{j=1}^{m} D1 > Volume(X_{i,j}) > D2}{s} \right)$

**end** of **if**

**if** $Y==3$

$I_{FS}$ = HSV ($I_{FS}$)

$X = I_{FS} \times SI$

s = $\sum_{i=1}^{n} \sum_{j=1}^{m} SI_{i,j}$

```
        compute f = 100*( (s - Σ_{i-1}^{n} Σ_{j=1}^{m} M1 > Hue(X_{i,j}) > M2) / s )
else
    Y==4
End of if
Return output f
End
```

$$f = 100 * \left( \frac{s - \sum_{i-1}^{n} \sum_{j=1}^{m} M1 > Hue(X_{i,j}) > M2}{s} \right)$$

The genetic algorithm is inspired by the natural evolution process for solving both constrained and unconstrained optimization problems that is based on natural selection, the process that drives biological evolution. The genetic algorithm repeatedly modifies a population of individual solutions. The fitness function of the genetic algorithm is based on fuzzy logic and is used as loss function in the learning model. There are two threshold values are selected for each of the different types of the images. The threshold values are chosen based on the training images where certain set of training images are selected and then their color values in HSV are averaged out. The threshold vales are selected based on the comparative central tendencies of both the color spectrums. Based on the threshold values the fitness algorithm is built. In the fitness algorithm, if the class of the object is 0 then fitness is decided by total number of unmasked pixels as in a normal image all pixels must be masked as it contains no region of interest. If class type is 1 in that case, the region of interest in the leaf are reddish in color and hence the hue of the image is considered and measured if it is under threshold in the masked image. In this way, the algorithm computes fitness value for all generated segmented image of type mask from the UNet learning model and accordingly the genetic algorithm as an optimizer will adjust the weight and bias through the training process. The next section discusses the outcome and performance of the proposed system for disease segmentation.

## 3. RESULTS AND DISCUSSION

The design and the development of the proposed system is carried out using Python programming language executed in the Anaconda distribution, installed on the Windows 10 OS system. This section presents the outcome of the study and discusses the performance of the system based on the comparative analysis. The model is validated against the Apple leaf dataset obtained from plant village which consists of total 7771 number of data

samples concerning four different disease classes. The training of the learning model is done with 80% (i.e., 6216) of the data samples and while the trained model is tested over 20% (i.e.,1555) of the dataset samples. The statistical information about the testing dataset along with class label is given in table II. The overall result and performance discussion are presented with respect to visual outcome and numerical outcomes concerning accuracy, precision recall rate and F1-score.

*Table 2. Labels of disease class*

| Disease class | No. of Samples | Label |
|---|---|---|
| Apple scab | 431 | 0 |
| Black Rot | 391 | 1 |
| Cider Apple Rust | 367 | 2 |
| Healthy | 365 | 3 |

## 3.1 Visual Outcome

Figure 7, Figure 8 and Figure 9 presents the sample visualization of the feature representation process. Using Otsu's approach segmented image is transformed in the HSV color map. Afterward, superimposition of image is done over the input image to highlight RoI as the disease spot. As a result, learning model during its training will get better generalization of the latent attributes and underlaying pattern that are identical to the particular leaf disease class. The next Figure 10. and Figure 11 provides the analysis of mask or segmented disease spot obtained from the trained UNet learning model after introducing it with unknow input leaf image of type Cedar apple rust disease.

Figure 10, Figure 11 and figure 12 shows the flow diagram highlighting the main steps to elaborate the methodology and implementation of the proposed system for disease classification. As already discussed, the input apple leaf images are first preprocessed to enhance the visual characteristics, making them suitable for region-based segmentation. Afterward, a primary segmentation is done using the proposed augmented Otsu's algorithm to segment the background and foreground region of the input leaf image without compromising the contextual aspects of the image. HSV based color space contributes towards improving the ROI and hence the scheme can be also adopted for leaf-image based disease detection of other crops too. The segmented foreground leaf images are then subjected to feature representation, where a color conversion algorithm is used to generate a feature map. Afterward, secondary segmentation operation is executed, which takes the feature map as its input. U-Net, with the genetic algorithm, is trained to generate a segmented mask for all color apple leaf images in the dataset. The obtained segmented masks serve as a ground truth or basis for training supervised classification models to predictive tasks. The study then split the dataset into training and testing sets with a ratio of 80:20. For the disease prediction and classification, the study trained the CNN model with the obtained mask and input leaf images in the training set. For benchmarking, the proposed work focuses on implementing two pre-trained CNN models, ResNet and Mobile Net. The following section discusses the outcome and performance of the proposed system for disease detection and classification.

## 3.2 Model Analysis

The performance of the proposed classification model CNN is analyzed concerning the multiple performance metrics of confusion plot, true positive, and false positive. The confusion matrix is exhibited in Figure 13 followed by classification outcomes in Table 2 to 5.Figure 13 presents the confusion plot for the classified disease obtained while testing the proposed learning model. Since the proposed system is a multiclass classification system, the outcome is shown for each disease class, including healthy leaf. Further, numerical outcomes of classification for i) apple scab is shown in Table 2, ii) black rot disease

*Figure 7. Visualization of feature representation of HSV color map*

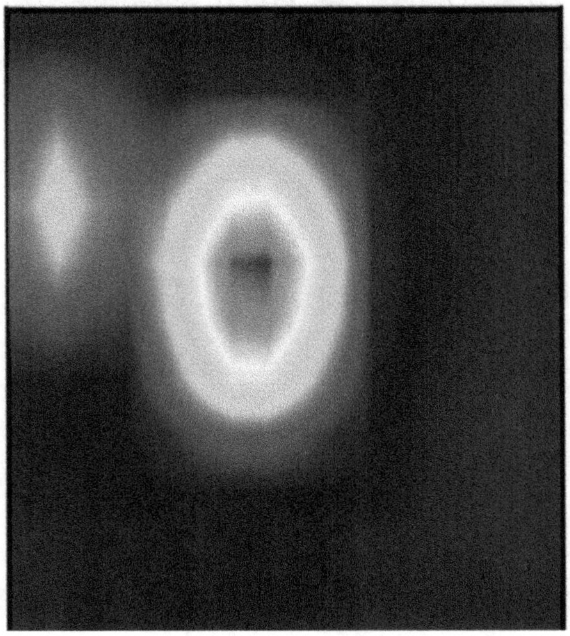

*Figure 8. Visualization of feature representation of RGB color map*

*Figure 9. Visualization of feature representation of Enhanced RoI*

is shown in Table3, iii) cider apple rust is shown in Table 4, and iv) healthy leaf is shown in Table 5. The complete classification outcomes is shown in form of correct and false prediction for better accuracy analysis.

*Figure 10. Visualization of input leaf image of apple leaf disease segmentation*

*Figure 11. visualization of system outcome as apple leaf disease segmentation*

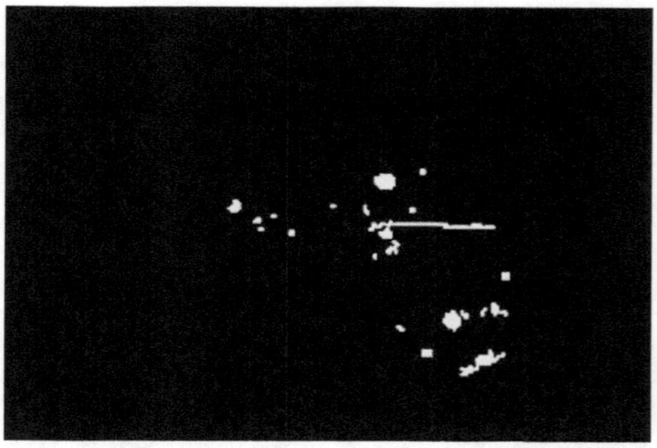

*Figure 12. Workflow for the proposed system*

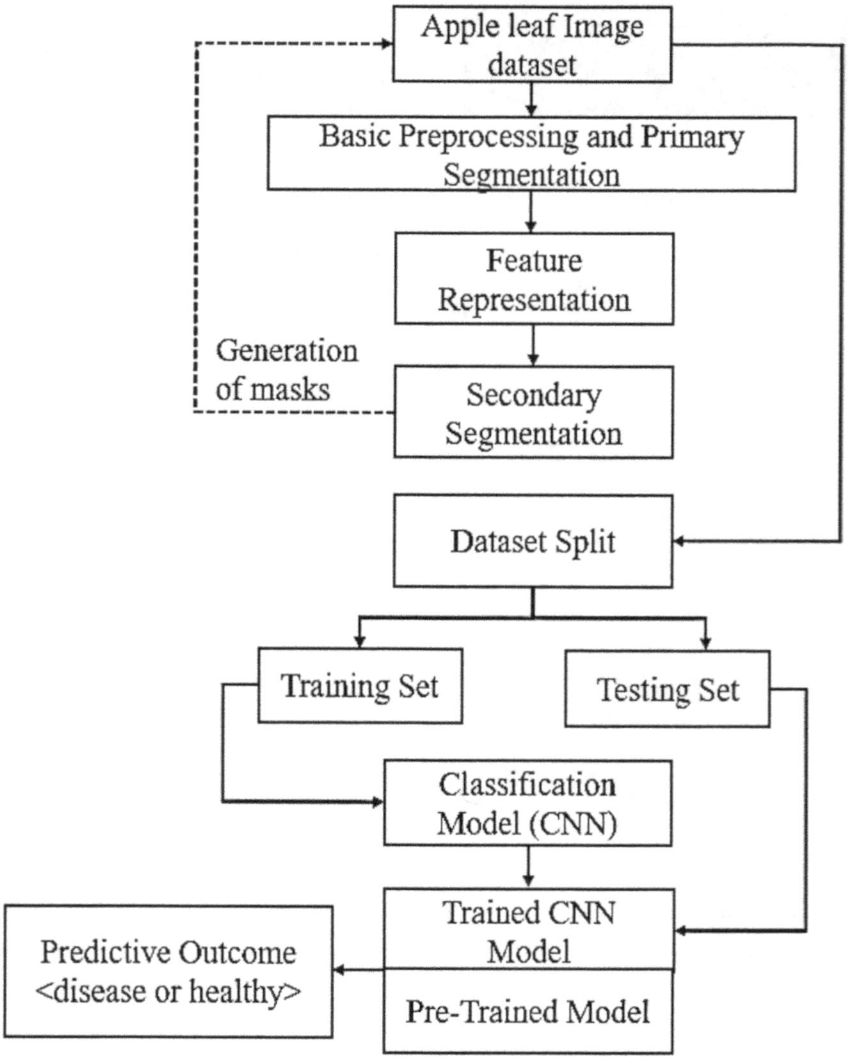

*Figure 13. Confusion plot for classification model CNN*

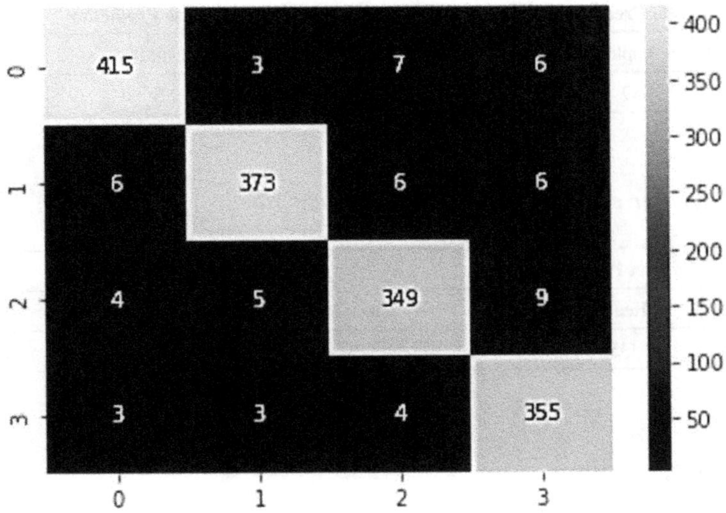


*Table 3. Outcome statists for classification of apple scab*

| Total Apple Scab | Correct Prediction | False Prediction | | |
|---|---|---|---|---|
| | Apple Scab | Black Rot | Cider Apple Rust | Healthy |
| 431 | 415 | 3 | 7 | 6 |

*Table 4. Outcome statists for classification of black rot*

| Total Apple Scab | Correct Prediction | False Prediction | | |
|---|---|---|---|---|
| | Black Rot | Apple Scab | Cider Apple Rust | Healthy |
| 391 | 373 | 6 | 6 | 6 |

A closer analysis of the confusion plot reveals a total of 431 Apple Scab leaves, out of which the proposed system correctly predicted 415 samples belonging to Scab disease, whereas 16 samples were falsely predicted, as shown in Table 3.

An analysis of black rot disease showed 391 black rot leaves, out of which the proposed system correctly predicted 373 samples, and 18 were misclassified, as shown in Table 4.

An analysis of cider apple rust disease showed 367 black rot leaves, out of which the proposed system correctly predicted 349 samples, and 18 were misclassified, as shown in Table 5.

An analysis of the confusion matrix for healthy leaves shows a total of 365 healthy leaves, out of which the proposed system correctly predicted 355 samples, and ten samples were misclassified, as shown in Table 6.

*Table 5. Outcome statists for classification of cider apple rust*

| Total Apple Scab | Correct Prediction | False Prediction | | |
|---|---|---|---|---|
| | Cider Apple Rust | Apple Scab | Black Rot | Healthy |
| 367 | 349 | 4 | 5 | 9 |

*Table 6. Outcome statists for classification of healthy leaf*

| Total Apple Scab | Correct Prediction | False Prediction | | |
|---|---|---|---|---|
| | Healthy | Black Rot | Cider Apple Rust | Black Rot |
| 365 | 355 | 3 | 3 | 4 |

## 3.3. Comparative Analysis

This section presents the comparative analysis to justify the effectiveness of the proposed methodology. The comparative analysis is conducted with another similar type of learning model, i.e., pre-trained CNN models, namely ResNet and MobileNet, which are implemented in the same test-bed towards performing a uniform comparison.

Figure-14, figure 15 and figure 16 shows that the proposed CNN learning model effectively outperforms both ResNet and MobileNet, achieving higher true positive outcomes. The comparative analysis of other matrices, such as precision, recall, and F1-score, is illustrated in Figure 14 Figure 15 and figure 16. A closer analysis of all presented graphs proves the effectiveness of the CNN learning model. The reason behind achieving good performance by CNN is its design according to the problem context and its training on the dataset, which has segmented masks that serve as a significant feature map to generalize

*Figure 14. Analysis of confusion plots for different CNN classification models*

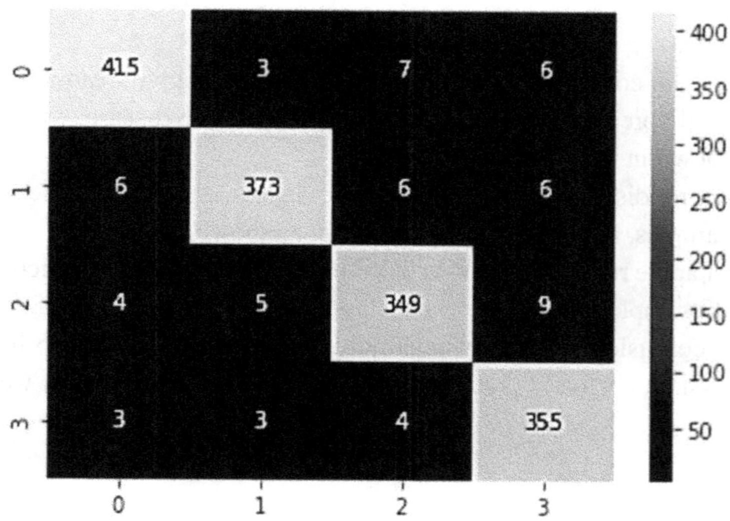

*Figure 15. Analysis of confusion plots for different for ResNet classification models*

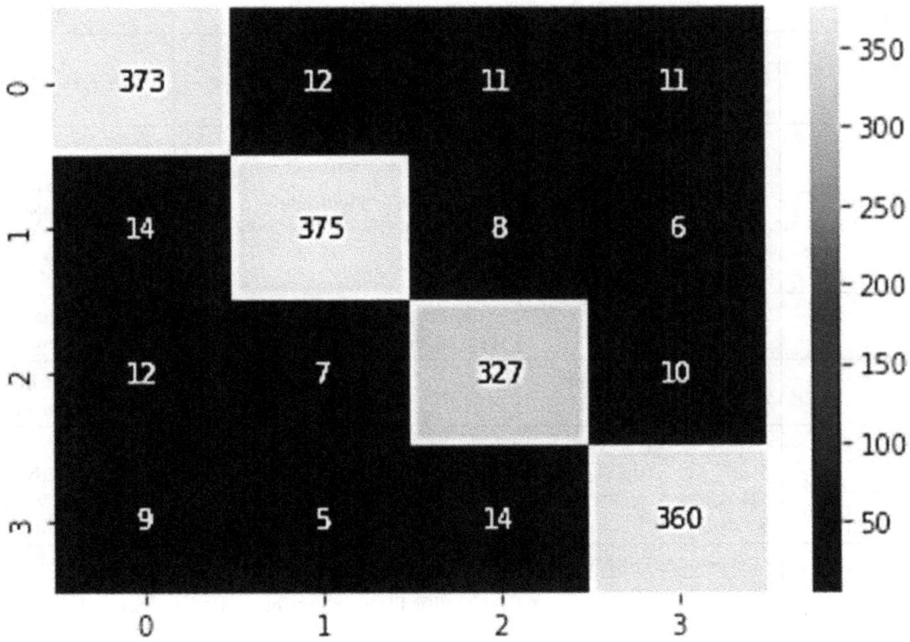


*Figure 16. Analysis of confusion plots for different for Mobile Net classification models*

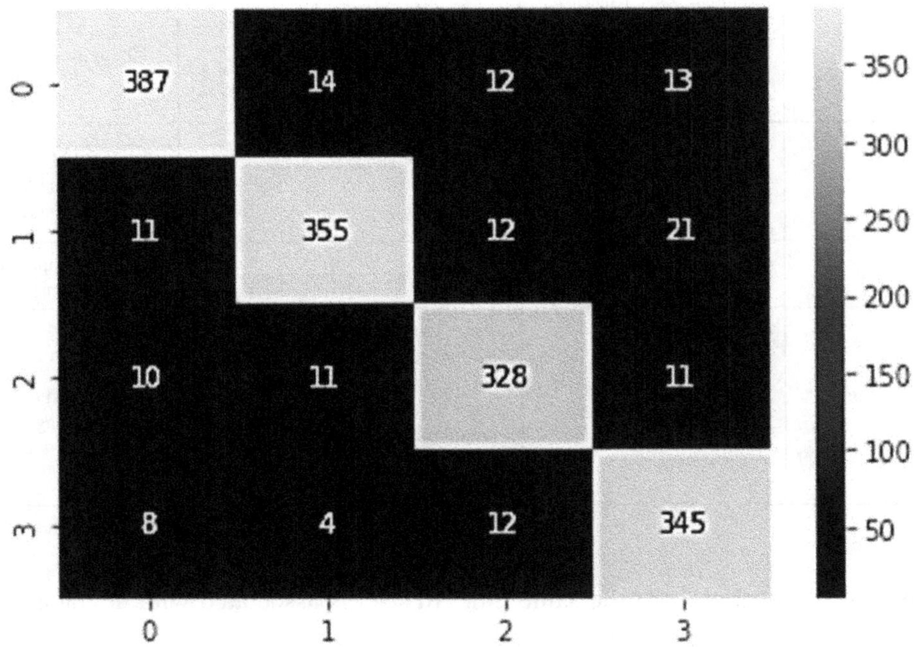

*Table 7. Comparison towards state-of-art method*

| Approaches | Author Reference | Advantage | Limitation |
|---|---|---|---|
| CNN | Chuanlei, Z., Shanwen, Z., Jucheng, Y., Yancui, S. and Jia, C., 2017 | Higher accuracy | Higher resource dependencies |
| Genetic algorithm, correlation SVM | Jadhav, S.B., Udup, V.R. and Patil, S.B., 2019. Bracino, A.A., Concepcion, R.S., Bedruz, R.A.R., Dadios, E.P. and Vicerra, R.R.P., 2020, | Search-based optimization, simplified and faster processing | Fitness function cannot adapt to potential change when test dataset is increased |
| K-Means clustering, color segmentatio | Sumithra, M.G. and Saranya, N., 2021 | Simplified approach, effective segmentation | Narrowed analysis to offer higher precision |
| SVM, KNN | Abisha, S. and Jayasree, T., 2022 | Supports instance-based learning | Biased prediction not addressed; outcomes have outliers too |
| Swarm Optimization, FCM | Adeel, A., Khan, M.A., Sharif, M., Azam, F., Shah, J.H., Umer, T. and Wan, S., 2019 | Applicable for varied forms of images even with noisy | Highly sensitive during initialization & parameters |
| Wavelet, ANN | Rehman, Z.U., Khan, M.A., Ahmed, F., Damaševičius, R., Naqvi, S.R., Nisar, W. and Javed, K., 2021 | -Simplified feature learning for leaf disease | -demands large number of parameters, leading to overfitting |
| SVM | Abayomi-Alli, O.O., Damaševičius, R., Misra, S. and Maskeliūnas, R., 2021 | -Capable of processing high-dimensional problem space<br>-Less Overfitting issues | -computationally intensive operation when exposed to multiple classes of leaf disease |
| CNN, ResNet-50, | Rehman, Z.U., Khan, M.A., Ahmed, F., Damaševičius, R., Naqvi, S.R., Nisar, W. and Javed, K., 2021 | improved | -Highly Iterative Operation |
| MobileNet | Pandian, A., Kanchanadevi, K., Rajalakshmi, N.R. and Arulkumaran, G., 2022 | -Faster inference<br>-Effective Transfer Learning & finetuning | -Reduced Model Capacity<br>-Sensitive to input quality<br>-Lower Accuracy |
| CNN with ResN | Li, X. and Li, S., 2022. | Optimal Resource dependencies | -Higher computational burden<br>-Higher Accuracy |
| Proposed (CNN, Genetic Algorithm, U-Net) | | -lower computational burden<br>-Higher Accuracy<br>-Highly optimized performance<br>-Low Iteration & highly Progressive<br>-Supports generalization<br>-Efficient use of training data<br>-Highly adaptable and Flexible | -not assessed over high resolution images |

spatially coherent data attributes. At the same time, ResNet is associated with degradation problems. However, MobileNet performs better than ResNet due to its fast convergence features.

Table 7 highlights the comparison with the state of art methods which shows that proposed scheme offers significant edge of advantage towards addressing the identified research problem. The outcome in Table VI also showcases that these existing studies with varied approaches introduces both beneficial aspect as well as limiting factors. A closer look into the limiting factors of existing system will infer that majority of them are highly case specific towards disease detection, whereas the proposed scheme offers better flexibility, adaptability, and generalization towards disease detection and classification. Integrating genetic algorithm in the form of an optimizer significant assists in generating a masked image that can be further analyzed in training and testing with respect to its feature map. Hence, the developed fitness function as well as loss function using fuzzy logic over genetic algorithm significantly improved the classification performance of CNN. The limitation of the proposed scheme is that it has been implemented and assessed over standard dataset and not on high resolution image. Hence, adoption of high-resolution images may incur higher memory demands too, which could be a part of future research work direction.

## 4. CONCLUSION

This paper has presented enhanced learning driven feature representation framework for apple leaf disease segmentation and mask generation. The proposed study has performed optimization in the learning process where fitness function of the genetic algorithm is used as optimizer of the UNet learning model. Using this scheme, a ROI based segmentation is achieved, with better training performance. This ROI aware segmentation model served as a support system for a more accurate leaf disease identification and its treatment at early stage. The generated mask which is segmented diseased portion can be effectively utilized to any supervised learning scheme to localize the disease spot and its class. The experimental results verified scope of the proposed system that is has higher precision, recall rate and F1-score. The result analysis demonstrates its scope that it can be applied to real-world application towards benefitting farmers by an effective decision-making process. Using the proposed system, disease can be detected at early stage and suitable treatment can be applied. Hence, leading to quality aware fruits, higher production and provide economic benefits to the producer.

## REFERENCES

Abayomi-Alli, O. O., Damaševičius, R., Misra, S., & Maskeliūnas, R. (2021). Cassava disease recognition from low-quality images using enhanced data augmentation model and deep learning. *Expert Systems: International Journal of Knowledge Engineering and Neural Networks*, *38*(7), e12746. doi:10.1111/exsy.12746

Abisha, S., & Jayasree, T. (2022). Application of image processing techniques and artificial neural network for detection of diseases on brinjal leaf. *Journal of the Institution of Electronics and Telecommunication Engineers*, *68*(3), 2246–2258. doi:10.1080/03772063.2019.1696716

Adeel, A., Khan, M. A., Sharif, M., Azam, F., Shah, J. H., Umer, T., & Wan, S. (2019). Diagnosis and recognition of grape leaf diseases: An automated system based on a novel saliency approach and canonical correlation analysis based multiple features fusion. *Sustainable Computing : Informatics and Systems*, *24*, 100349. doi:10.1016/j.suscom.2019.08.002

Bracino, A. A., Concepcion, R. S., Bedruz, R. A. R., Dadios, E. P., & Vicerra, R. R. P. (2020). Development of a Hybrid Machine Learning Model for Apple (Malus domestica) Health Detection and Disease Classification. In *2020 IEEE 12th International Conference on Humanoid, Nanotechnology, Information Technology, Communication and Control, Environment, and Management (HNICEM)* (pp. 1-6). IEEE.

Chuanlei, Z., Shanwen, Z., Jucheng, Y., Yancui, S., & Jia, C. (2017). Apple leaf disease identification using genetic algorithm and correlation based feature selection method. *International Journal of Agricultural and Biological Engineering, 10*(2), 74–83.

Dias, P. A., Tabb, A., & Medeiros, H. (2018, October). Multispecies fruit flower detection using a refined semantic segmentation network. *IEEE Robotics and Automation Letters, 3*(4), 3003–3010. doi:10.1109/LRA.2018.2849498

Fulsoundar, K., Kadlag, T., Bhadale, S., & Bharvirkar, P. (2014). Detection and classification of plant leaf diseases. *International Journal of Engineering Research and General Science, 2*(6).

Gaikwad, S. S., Rumma, S. S., & Hangarge, M. (2022). Fungi affected fruit leaf disease classification using deep CNN architecture. *International Journal of Information Technology : an Official Journal of Bharati Vidyapeeth's Institute of Computer Applications and Management, 14*(7), 1–10. doi:10.100741870-022-00860-w

Ganatra, N., & Patel, A. (2018). A Survey on Diseases Detection and Classification of Agriculture Products using Image Processing and Machine Learning. *International Journal of Computer Applications, 180*(13), 7–12. doi:10.5120/ijca2018916249

Jadhav, S. B., Udup, V. R., & Patil, S. B. (2019). Soybean leaf disease detection and severity measurement using multiclass SVM and KNN classifier. *Iranian Journal of Electrical and Computer Engineering, 9*(5), 4092. doi:10.11591/ijece.v9i5.pp4077-4091

Khan, M. A., Lali, M. I. U., Sharif, M., Javed, K., Aurangzeb, K., Haider, S. I., Altamrah, A. S., & Akram, T. (2019). An Optimized Method for Segmentation and Classification of Apple Diseases Based on Strong Correlation and Genetic Algorithm Based Feature Selection. *IEEE Access : Practical Innovations, Open Solutions, 7*, 46261–46277. doi:10.1109/ACCESS.2019.2908040

Li, X., & Li, S. (2022). Transformer Help CNN See Better: A Lightweight Hybrid Apple Disease Identification Model Based on Transformers. *Agriculture, 12*(6), 884. doi:10.3390/agriculture12060884

Liu, J., & Wang, X. (2021). Plant diseases and pests detection based on deep learning: A review. *Plant Methods, 17*(1), 22. doi:10.118613007-021-00722-9 PMID:33627131

Lu, Y., Shujuan, N. Z., Liu, Y., & Zhang, Y. (2017). *Identification of Rice diseases using Deep Convolutional Neural Networks*. Elsevier Journal on Neurocomputing. doi:10.1016/j.neucom.2017.06.023

Nicolopoulou-Stamati, P., Maipas, S., Kotampasi, C., Stamatis, P., & Hens, L. (2016). Chemical pesticides and human health: The urgent need for a new concept in agriculture. *Frontiers in Public Health, 4*, 148. doi:10.3389/fpubh.2016.00148 PMID:27486573

Omrani, E., Khoshnevisan, B., & Shamshirband, S. (2014). Potential of redial basis function- based support vector regression for apple disease detection. Measurement, 55.

Orchi, H., Sadik, M., & Khaldoun, M. (2021). On using artificial intelligence and the internet of things for crop disease detection: A contemporary survey. *Agriculture*, *12*(1), 9. doi:10.3390/agriculture12010009

Pandian, A., Kanchanadevi, K., Rajalakshmi, N. R., & Arulkumaran, G. (2022). An Improved Deep Residual Convolutional Neural Network for Plant Leaf Disease Detection. *Computational Intelligence and Neuroscience*, 2022. doi:10.1155/2022/5102290 PMID:36156945

R.Preethi, S.Priyanka, U.Priyanka, A.Sheela, (2015), Efficient knowledge based system for leaf disease detection and classifcation, *International Journal of Advance Research In Science And Engineering, 4*(1).

Rehman, Z. U., Khan, M. A., Ahmed, F., Damaševičius, R., Naqvi, S. R., Nisar, W., & Javed, K. (2021). Recognizing apple leaf diseases using a novel parallel real-time processing framework based on MASK RCNN and transfer learning: An application for smart agriculture. *IET Image Processing*, *15*(10), 2157–2168. doi:10.1049/ipr2.12183

Revathi, P., & .Hema Latha, M. (2012). Classification Of Cotton Leaf Spot Diseases Using Image Processing Edge Detection Techniques. *International Journal of Applied Research.*

Savary, S., Ficke, A., Aubertot, J.-N., & Hollier, C. (2012). Crop losses due to diseases and their implications for global food production losses and food security. *Food Security*, *4*(4), 519–537. Advance online publication. doi:10.100712571-012-0200-5

Shivaputra, S. (2016). Pomegranate Leaf Disease Detection Using Support Vector Machine. *International Journal Of Engineering And Computer Science, 5*(6).

Sumithra, M. G., & Saranya, N. (2021). Particle Swarm Optimization (PSO) with fuzzy c means (PSO-FCM)–based segmentation and machine learning classifier for leaf diseases prediction. *Concurrency and Computation*, *33*(3), e5312. doi:10.1002/cpe.5312

Vasavi, P., Punitha, A., & Rao, T. V. N. (2022). Crop leaf disease detection and classification using machine learning and deep learning algorithms by visual symptoms: A review. *Iranian Journal of Electrical and Computer Engineering*, *12*(2), 2079. doi:10.11591/ijece.v12i2.pp2079-2086

Yoon, M. Y., Cha, B., & Kim, J. C. (2013). Recent trends in studies on botanical fungicides in agriculture. *The Plant Pathology Journal*, *29*(1), 1–9. doi:10.5423/PPJ.RW.05.2012.0072 PMID:25288923

# Chapter 14
# Machine Learning and Blockchain Integration in Industrial Robotics:
## Challenges and Opportunities

**Dankan Gowda V.**
https://orcid.org/0000-0003-0724-0333
*BMS Institute of Technology and Management, Bangalore, India*

**Manojkumar S. B.**
https://orcid.org/0000-0002-3037-2564
*BGS Institute of Technology, Adichunchanagiri University, India*

**Pullela S. V. V. S. R. Kumar**
https://orcid.org/0000-0003-0621-132X
*Aditya College of Engineering, Surampalem, India*

**Aparna Atul Junnarkar**
*Vishwakarma Institute of Information Technology, India*

**Sheetalrani Rukmaji Kawale**
https://orcid.org/0000-0001-7887-0346
*Karnataka State Akkamahadevi Women University, India*

## ABSTRACT

*Machine learning and blockchain have the potential to completely change the landscape of industrial robots. Learning and adaptation give robots an edge in functionality and independence. Blockchain, on the other hand, provides a decentralized and secure platform for information exchange and transaction verification. In this chapter, the authors look at the potential benefits and drawbacks of integrating machine learning and blockchain technology for use in manufacturing robots. The study begins with a brief introduction to machine learning, blockchain, and their respective industrial robot applications. Some of the possibilities and advantages that will be covered in the next sections include better data security, more transparent and auditable decision making, and the use of decentralized control systems. The importance of collaboration between academic institutions, businesses, and government agencies is emphasized in order to speed up the process of mainstreaming machine learning and blockchain integration in industrial robots.*

DOI: 10.4018/979-8-3693-0659-8.ch014

## INTRODUCTION

In recent years, industrial automation has entered a new phase thanks to the confluence of cutting-edge technology. Machine learning and blockchain in particular have the potential to radically alter the landscape of industrial robots. This chapter discusses the benefits, drawbacks, and possibilities of combining machine learning and blockchain technology in industrial robots systems. Robots' capacity to make decisions is improved by machine learning because of the data it can sift through and the way it can adjust on the fly. The blockchain, on the other hand, is a decentralized digital ledger that is both immutable and transparent, making it an ideal system upon which to build a trustworthy data and transaction management infrastructure. Together, they form a formidable duo, promising heightened efficiency, transparency, and security across various industrial domains.

A robot is a sophisticated, programmable entity designed to mimic the functions of living organisms. It's adept at executing intricate and recurrent tasks. Robots are primarily classified based on their movement capability into two groups: stationary robots and mobile robots. Mobile robots further branch out into three types: those that move on wheels or legs, those that fly, and those that navigate underwater. The latter utilize water thrust for movement. In this section, we'll focus solely on robots with wheels(Verma, 2022). In contemporary robotics, the term "agency" is prevalent. An agent is either software or hardware that enables a robot to achieve a specific objective. Just as humans pool resources when confronted with complex challenges, robots sometimes require the combined intelligence of multiple agents to reach their goal. These agents can detect and act logically. At times, agents might share data or decisions with their counterparts either via direct communication systems or through distinct gestures, serving as communication cues to their peers.

Communication is essential for agents to create plans. Yet, due to the time-intensive nature of communication, it's often sidelined in real-world robotic applications. In this thesis, we explore a learning-based approach to discern agent behavior patterns, reducing communication lags in real-time planning.

The field of computer science has witnessed a constant evolution, encompassing everything from lightning-fast computations to the emergence of intelligent, self-sustaining systems. These advancements have rippled through numerous related domains, including biomedical engineering, aerospace engineering, and more(V. Kumar, B. Pant, G. Elkady, 2022). Among these interdisciplinary fields, robotics has recently taken center stage in the scientific community. Robotics draws upon a multitude of technological resources from diverse disciplines, and one such rapidly evolving and transformative asset is blockchain technology.

When it comes to transactions, blockchain technology, a P2P electronic record management system, enables authentication and validation. It does this by functioning as a decentralized ledger and doing away with the necessity for a trusted third party. The ledger of past trades is a growing set of documents called "blocks," which are linked cryptographically. Blockchain's overwhelming acceptance may be traced back to its resilient byzantine fault-tolerant consensus process and distributed ledger design(R. Kavitha, K. Prasad, 2023). Any fraudulent transactions may be quickly identified and removed from the system since every node has a complete and accurate copy of the block chain.

In parallel, researchers have extended the applicability of blockchain beyond its original purpose in Bitcoin and finance sectors, now embracing its potential within the realm of robotics. This integration aims to create a more dependable, secure, and adaptable environment for robots, whether they function collaboratively or independently.

The progress of robotics in the previous decade may be attributed to developments in related fields including electrical engineering, mechanical engineering, computer science, and Artificial Intelligence (AI). These advancements have allowed for greater in-depth study and work in robotics, opening up enormous potential applications of these technologies. Concepts formerly confined to science fiction are quickly becoming concrete reality, as seen by the fascinating dancing robots built by Boston Dynamics and the semi-autonomous rovers investigating the surface of Mars.

Robots confined to controlled laboratory environments can often surpass basic expectations in terms of efficiency, resilience, and cost-effectiveness. However, their performance can take a significant hit when deployed in real-world work settings(Grau, M. Indri, L. L. Bello, 2017). Recent years have seen an uptick in robotics-related research and development thanks to the rising need for automation and improved process throughput across a wide range of industries. These developments have given rise to a wide variety of sectors where robots may be put to use.

To facilitate the comprehension of research efforts within these application areas, we propose a straightforward and pertinent classification of robots. It's important to acknowledge that robots can be classified in various ways due to their wide-ranging applications. In this context, we opt to focus on the primary application areas that are most relevant in today's world. Broadly speaking, robots are predominantly categorized into two main groups: industry-oriented and service-oriented robots(Ranjan, K. K. Hegge, 2023). These categories are further subdivided into specific groupings tailored to each sector. For instance, industrial robots encompass those used in manufacturing and agriculture, while service-oriented robots encompass professional and domestic applications. Within the realm of professional and domestic service robots, further categorization is based on their specific areas of expertise, such as medical, workplace, and defense.

There's been a lot written on planning algorithms throughout the years. Nilsson's early robot planning work at Stanford's AI lab, based on reasoning research, paved the way for algorithms that would eventually be included into STRIPs. From the late 1980s through the early 1990s, a number of planning algorithms based on static settings, such as A Voronoi diagrams, Quad tree, and prospective field, appeared.

By the early 1990s, Michalewicz introduced genetic operations in his notable work, paving the way for dynamic planning with local adaptations using specialized mutation operators. The decade from 1990 to 2000 marked considerable advancements in planning algorithms(H. A. F. Almurib, H. F. Al-Qrimli, 2012). This era introduced neural learning methods that operated in both static and dynamic settings. In dynamic contexts, these neural algorithms could forecast movement direction and speed to avoid collisions. However, their limited learning scope meant challenges in planning amidst unpredictable movements of dynamic obstacles or individuals.

Around the early 1990s, Sutton presented the reinforcement learning (RL) algorithm, offering robots the capability to comprehend their environment via semi-supervised learning.

Coordination and planning go hand in hand in the field of multi-agent robots. Coordination involves the strategic interaction of agents to achieve both short-term and long-term goals, whereas planning is concerned with specifying a sequence of activities to obtain a goal. Coordination is essential for carrying out the various phases of planning.

Centralized planning doesn't necessitate coordination among agents since a primary controller manages all agent states, crafting a planning cycle that incorporates their collective states and goals. However, this approach has downsides. Centralized planning is inappropriate for real-time systems with many agents because it may be slow and has the potential for a single point of failure(Avinash, Parismita, 2023). In contrast, distributed planning requires every agent to independently come up with a planning step while simultaneously coordinating with the other agents involved.

Coordination manifests in two primary forms: cooperation and competition. Cooperation compels agents to collaboratively achieve the team's shared goal. Competition, conversely, pits one team's success against another's failure. Take robot soccer as an example: within a team, agents cooperate seamlessly, but when it comes to opposing teams, they compete, each vying for victory at the other's expense.

This amalgamation, however, is not without of difficulties and obstacles. In this chapter, we explore some of the difficulties that academics and practitioners have on the road to integrating machine learning and blockchain into industrial robots(Gite, Pratik, Premkumar, S., 2023). When these challenges are resolved, a bright future of intelligent, trust-driven automation open up, and we'll examine all the possibilities that come with it.

***An abbreviated history of industrial robots:*** A fascinating journey spanning several decades, the development of industrial robots is marked by technical progress, growing automation, and revolutionary shifts in production methods. Here is a quick recap of the major turning points in the history of industrial robots.

The First Robots Appear in the 1950s The development of the first crude industrial robots in the 1950s marks the beginning of industrial robotics. The Unimate, invented by George Devol and Joseph. Engelberger, was utilized in the General Motors assembly line to do simple, repetitive operations like loading and unloading parts.

Pioneering Uses from the '60s The first widespread use of industrial robots occurred in the 1960s, initially in the production of automobiles in the United States and Japan. Spot welding, spray painting, and other assembly line duties were performed by robots.

Significant progress was made in the 1970s in the areas of robot control and programming. The accuracy and adaptability of industrial robots were greatly boosted by the development of computer numerical control (CNC) technology.

In the 1980s, advancements in sensors, end-of-arm tooling, and vision systems allowed industrial robots to become more flexible and adaptive. In addition to welding and assembly, these robots may also do material handling and quality control inspections.

In the 1990s, we saw the introduction of collaborative robots, often known as cobots, which were built to operate securely alongside people without the need for elaborate safety barriers(Grau, M. Indri, L. Lo Bello and T. Sauter, 2021). The development of mobile robots made it possible for industries to carry their own materials without human intervention.

Precision sectors such as electronics and semiconductors were early adopters of robotics technology in the 2000s. The assembly of microelectronics and other cleanroom procedures require smaller, more accurate robots.

In the 2010s, industrial robots were connected to the Internet of Things (IoT) and data analytics thanks to the idea of Industry 4.0. To provide constant monitoring and flexible production, robotics and automation quickly rose to the forefront of smart manufacturing.

2020s - Artificial Intelligence and Sensing Technology The combination of sophisticated sensors, AI, and machine learning is a current trend in industrial robots. Autonomous robots may now respond to new information in real time and modify their behavior accordingly.

The future of industrial robotics is projected to be inextricably linked to artificial intelligence, allowing robots to learn and adapt to new and different jobs. Improved energy efficiency and less environmental impact will motivate the creation of future robotic systems.

Industrial robots has been a driving force in the modernization of the manufacturing sector, contributing to gains in productivity, quality, and worker safety. Industrial robots are already changing the production and distribution of commodities in many sectors, and this trend is only anticipated to accelerate as technology develops.

***The introduction of cutting-edge technologies like machine learning (ML) and blockchain:*** The emergence of ML and blockchain technologies has had far-reaching consequences for many sectors, ushering in novel capabilities, opportunities, and difficulties. Let's look at how these technologies entered the market and what effect they had there.

With the advent of ML technology, we can... Machine learning is a subfield of AI that focuses on teaching computers to analyze data for patterns and make inferences or predictions as a result. Businesses and other organizations' use of data for decision making and automation has undergone a sea change since the advent of ML technology. Here are a few highlights from the advent of ML technology:

With the help of ML algorithms, businesses can tap into the potential of data to make better, more well-informed decisions. Companies may increase their efficiency, process optimization, and insight by analyzing massive databases.

Predictive maintenance is one area where ML has been widely adopted, especially in the manufacturing and logistics sectors(Ranjan, K. K. Hegge and R. Agrawal, 2023). Businesses may save money and effort by using sensor data to determine when maintenance is required for machinery and other pieces of equipment. Manufacturing operations now employ ML systems for real-time quality control. Defects may be found and quality standards maintained with the use of computer vision automation.

Marketing and e-commerce use ML extensively for tailoring the shopping and browsing experiences they provide to each individual client. The goal of recommendation systems is to provide consumers with relevant and personalized suggestions based on an analysis of their actions and interests.

Medical Diagnosis and Treatment: Medical diagnosis, medication discovery, and patient care all benefit greatly from the use of ML. Medical pictures may be analyzed by algorithms, illness outbreaks predicted, and treatment strategies optimized.

Chatbots for customer support, translation, sentiment analysis, and content production are just some of the NLP applications that make use of ML-enhanced NLP models(K. Prasad, V. Srinivas, T. De 2023). The development of autonomous cars relies heavily on ML since it allows these vehicles to understand their surroundings and move freely through them.

***Introduction of Blockchain Technology:*** Transactions made using blockchain, a distributed ledger technology, are recorded reliably and openly across a network of computers. As a result of its widespread adoption, trust, openness, and data security have all seen revolutionary changes. Here we look at how the spread of blockchain technology has altered certain markets.

*Supply Chain Management:* Blockchain technology improves supply chain visibility and tracking. Many businesses in the food, pharmaceutical, and luxury goods industries utilize blockchain to monitor where their products come from and where they go.

*Financial Services:* In the financial industry, blockchain caused a disruption by eliminating the need for middlemen via the facilitation of transparent and trustworthy digital transactions. Bitcoin and other cryptocurrencies and blockchain platforms have become more popular.

*Smart Contracts:* Blockchain facilitates the execution of smart contracts, self-executing agreements with predefined rules. These contracts automate and secure various processes, from real estate transactions to insurance claims.

*Healthcare Data Management:* Patients' medical records may be kept private while also being easily shared among healthcare practitioners thanks to blockchain technology.

*Intellectual Property:* Creative industries use blockchain to protect intellectual property rights, proving the ownership and provenance of digital assets.

*Voting and Governance*: Some countries are exploring blockchain for secure and transparent voting systems and government record-keeping.

*Energy and Sustainability*: Blockchain is used in energy trading, enabling peer-to-peer transactions and the monitoring of renewable energy production and consumption. *Tokenization*: Assets such as real estate, art, and commodities are being tokenized on blockchain platforms, making them more accessible to investors.

***Impact on Industry:*** The introduction of ML and blockchain technologies has had a transformative impact on industries.

Blockchain has simplified complicated, multi-party transactions by eliminating middlemen and lowering transaction costs, while machine learning has increased the efficiency of processes through automation. Due to its increased security and trust, Blockchain is gaining popularity in sectors where data integrity is paramount.

These technological advancements encourage creativity since they help companies create cutting-edge services and merchandise. Successful businesses that use ML and blockchain will have a leg up on the competition by providing safer and more efficient services to their customers. Data privacy worries, legal roadblocks, and a lack of available experienced personnel are just a few of the issues plaguing both technologies. As a conclusion, the advent of machine learning and blockchain technology has transformed sectors by facilitating data-driven decision making, enhancing security, and introducing fresh avenues for innovation(G. Singh, V. K. Banga and T. Yingthawornsuk, 2022). The continued development of these technologies is anticipated to have an ever-increasing influence across a wide range of industries, causing greater disruption and upheaval.

## BASICS OF MACHINE LEARNING IN INDUSTRIAL ROBOTICS

The use of ML to industrial robotics has unleashed previously unimaginable levels of robot intelligence, flexibility, and productivity. In this article, we will cover the fundamentals of machine learning as they apply to industrial robots. Knowing How Machines Learn: Fundamentally, machine learning is a branch of AI concerned with creating programs that can analyze data and provide conclusions or suggestions based on that analysis. Industrial robots can evaluate data, adjust to new circumstances, and complete jobs more efficiently thanks to ML.

The first step in using machine learning (ML) in industrial robots is collecting and preprocessing the necessary data. There are a number of sensors and cameras included into the robotic system that can provide this information. It is necessary to sort, clean, and arrange the data before analyzing it. In the field of industrial robotics, supervised and unsupervised learning are the two most popular forms of machine learning. In supervised learning, the robot is taught to make predictions or classifications by being exposed to an already-labeled dataset. In the manufacturing industry, for instance, robots may be taught to spot flaws in goods by comparing photographs of both faulty and flawless components(K. H. Tantawi, A. Sokolov and O. Tantawi, 2019). Learning without labeled examples, or unsupervised learning, is the process of discovering patterns and structures in data. Unsupervised learning frequently employs clustering and dimensionality reduction strategies for applications like anomaly detection and data compression. The field of industrial robotics relies heavily on the reinforcement learning (RL) ML paradigm. Training robots to make successive choices that maximize a reward signal is the primary goal of RL. If a robot is rewarded for behaviors that bring it closer to its objective, it can learn to traverse a complicated environment. The goal of feature engineering is to enhance the predictive power of machine learning models by picking and manipulating useful properties (features) from the data. To make sensor data more amenable to training ML algorithms, this may involve identifying significant properties in industrial robots.

Selecting the Appropriate Model Neural networks, decision trees, support vector machines, and other ML models are used by industrial robots. The job at hand and data characteristics should guide the selection of a suitable model. When all the information is ready, the chosen ML model is trained on a portion of it, while the remaining data is used to test and evaluate the model. The objective is to test the model's ability to adapt to novel inputs. Decision Making and Adaptation in Real Time: ML models are frequently used on the robotic system in an industrial scenario for real-time inference (Tellaeche, I. Maurtua and A. Ibarguren, 2015). A robot on a manufacturing line, for instance, may use ML to spot defective items as they roll by.

One common application of ML in industrial robotics is a feedback loop, in which the robot iteratively refines its model in light of new information and experiences.

The robot's performance can evolve to meet changing conditions thanks to this type of adaptive learning.

*Industrial Robotics Use Cases:* Many areas of industrial robotics make use of machine learning. Quality Assurance Robots can do thorough flaw inspections of manufactured goods.

*Predictive Maintenance:* ML can predict when robots and machinery require maintenance, reducing downtime.

*Path Planning:* Robots can use ML to optimize their paths for tasks like material handling and logistics.

*Object Recognition:* ML models enable robots to recognize and manipulate objects with varying shapes and sizes.

*Human-Robot Collaboration:* ML allows robots to work safely alongside humans, adapting to human behavior and needs.

Machine learning is a rapidly evolving field, and its applications in industrial robotics continue to expand. As the technology matures, we can expect even more intelligent and capable industrial robots that enhance productivity and efficiency across various industries.

***Definition and fundamental concepts:*** Machine learning is a branch of AI that focuses on teaching computers to learn from and make predictions or judgments based on data on their own, without being specifically programmed to do so. Fundamentally, it involves training computers to learn from previous experiences and make adjustments to new circumstances.

Machine learning cannot function without data (S. P. M, A. M. Reddy and R. Tanguturi, 2022). Data consists of information gathered from numerous sources, such as records, measurements, or observations. Training ML models need data, which may be either organized (like tables) or unstructured (like text, photos, and audio).

Features are the distinct qualities or aspects of data that are employed by ML models for classification and prediction. In image recognition, for instance, features might consist of things like pixel values, colors, and forms.

*Classifications and Aims:* Labeling data is a common practice in supervised learning. Labels are the proper or desired results for each data point. Labels are used to represent categories ("cat" or "dog") in a classification job, and numerical values (such as property prices) in a regression work. The heart of each machine learning system is its model. It's a mathematical model that recognizes trends in data and uses that information to draw conclusions or make forecasts(D. Palanikkumar, and N. Patwari, 2023). Decision trees, neural networks, and support vector machines are just some of the models that may be used.

*Training:* Training is the process of teaching a machine learning model to recognize patterns in data. During training, the model learns to map input features to output labels by adjusting its internal parameters.

*Testing and Validation:* After training, the model's performance is evaluated using a separate dataset not used during training. This is called testing or validation. It assesses how well the model generalizes to new, unseen data.

*Supervised Learning:* In supervised learning, algorithms are trained using data for which the desired outcome has been labeled. The aim is for the model to be able to generate predictions on fresh, unlabeled data by learning a mapping from inputs to outputs.

*Unsupervised Learning:* With unlabeled data, the goal of the model is to unearth previously unseen patterns, structures, or correlations. Common techniques include clustering and dimensionality reduction.

*Reinforcement Learning:* Reinforcement learning focuses on decision-making in dynamic environments. Learning occurs when an agent engages with its environment, receives feedback on its activities in the form of rewards or punishments, and adjusts its behaviors to maximize the sum of those rewards (S. Venkatakiran, B. Ashreetha and N. S. Reddy, 2023). Mechanical Design: Selecting, manipulating, or fabricating important characteristics from raw data is what feature engineering is all about. It's a vital part of ML, but it demands specialized training and understanding.

*Overfitting and Underfitting*: When a model is too complicated and fits the training data too closely, it loses the ability to generalize to new data, a phenomenon known as overfitting. If a model is overly simplistic, it will underfit the data and fail to adequately represent the data's structure. Finding a happy medium is crucial.

*Bias and Fairness:* Biases in the training data may be passed on to ML models, which can then provide biased predictions. Ethical concerns in machine learning center on ensuring justice and reducing bias in models.

*Hyperparameters:* Hyperparameters are settings that control a model's learning process, such as the learning rate or the depth of a decision tree. Tuning hyperparameters is an important aspect of model optimization.

*Evaluation Metrics:* Metrics like accuracy, precision, recall, and F1-score are used to quantify a model's performance on classification tasks. For regression, metrics like mean squared error (MSE) and R-squared are commonly used.

*Deployment:* Once a model is trained and validated, it can be deployed in production environments to make real-time predictions or automate decision-making processes.

These basic ideas are the building blocks for learning and using machine learning. Applications of machine learning may be found in many different sectors, including healthcare, finance, manufacturing, and more. A solid understanding of these concepts is necessary for effectively using ML in practical settings.

Use-cases: predictive maintenance, anomaly detection, adaptive processes, etc.

Machine learning (ML) has several practical uses in industry, including enhancing output, decreasing downtime, and refining processes. Here are only a few real-world applications of advanced robotics and manufacturing techniques in the industrial sector.

Predictive maintenance makes use of machine learning (ML) models to monitor the health of vital equipment in real time(Shivashankar, M. Sahana, 2017). Algorithms trained on data obtained by sensors embedded in machines may use this information to predict when maintenance is required. The equipment will need less maintenance and will last longer as a result.

*Anomaly Detection:* A Real-World Example The process of actively searching for anomalies in operational environments is known as anomaly detection. Exactly How It Functions: Models trained using machine learning data may detect deviations from the norm in a system's functioning and provide alerts in response. There will be fewer defects, less waste, and a more consistent final product.

Adaptive processes in manufacturing and assembly lines may benefit from ML for real-time optimization.

*Exactly How It Functions:* Models trained using machine learning data analyze sensor information and adjust parameters like machine speed or process flow to boost output and quality(Grau, M. Indri, L. Lo Bello and T. Sauter, 2017). Increased productivity, decreased energy use, and the flexibility to meet ever-changing production demands are just a few of the benefits.

*Use-Case:* During production, ML is put to use for automated quality control, examining items for flaws. How It Works: Defects in manufactured goods may be detected using computer vision models by analyzing photos or sensor data. Product quality is increased, rework is decreased, and customer satisfaction is boosted.

*Use-Case:* ML models improve inventory management at distribution centers and factories. How It Works: Machine learning algorithms make predictions about future demand based on existing data and then modify supply levels accordingly(F. G. Pereira, R. F. Vassallo, 2013). Advantages include lower storage costs, less waste, and guaranteed product availability.

*Case Study:* Human-Robot Collaboration ML allows robots to coexist with human employees in a safe environment and learn from their actions. Methodology: Safe and effective human interaction is maintained by robots equipped with sensors and machine learning algorithms. Benefits include increased productivity, safer working conditions, and the ability to produce variations on a theme.

*Use-Case:* Energy Management ML models reduce wasteful energy use in factories. MethodologyML algorithms examine information gathered by sensors and control systems to maximize the efficiency of equipment that consumes a lot of power. Advantages include progress toward environmental targets and cheaper energy bills.

*Case Study:* Using ML to optimize intricate manufacturing processes for greater productivity. Methodology Machine learning algorithms examine process data to single out slowdowns, inefficiencies, and room for growth. Gains include higher output, less waste, and better overall efficiency.

*Case Study in Autonomous Robotics:* Using Machine Learning and sensors, autonomous robots can move about and complete tasks without human guidance. Here's How It Works: By using ML models, robots may learn to interpret and respond to their surroundings in real time. Benefits: Automates tasks, reduces labor costs, and increases flexibility in material handling and logistics.

These use-cases demonstrate how machine learning is revolutionizing industrial robotics and manufacturing by enabling smarter, more adaptive, and efficient processes. ML-driven solutions have the potential to significantly reduce costs, improve product quality, and drive innovation in various industries.

## BASICS OF BLOCKCHAIN IN INDUSTRIAL ROBOTICS

Blockchain technology is gaining traction in various industries, including industrial robotics, due to its ability to enhance security, transparency, and trust in data and transactions. Here are the basics of blockchain in the context of industrial robotics.

Data immutability, transparency, and security are the hallmarks of blockchain, a decentralized and distributed digital ledger system that records transactions across several computers. Blockchain is a distributed ledger system, meaning that all participants (nodes) have a copy of the complete database.

***Key Concepts***: Blocks: Blocks are containers that store a set of transactions. Each block is linked to the previous block, creating a chain of blocks, hence the term "blockchain." The entries or acts that are recorded on the blockchain are referred to as "transactions." Information on machine status, maintenance, and supply chain events are all fair game for transactions in the realm of industrial robots. Each block includes a cryptographic hash of the prior block, which verifies the authenticity of the whole chain. Because altering one block would necessitate altering all future blocks, tampering is very difficult. Blockchain networks require consensus techniques to verify and agree on the data included within each block. Proof-of-Work (PoW) and Proof-of-Stake (PoS) are two common techniques.

***Advantages of Blockchain Technology for Industrial Robotics:*** Blockchain's cryptographic properties make the data recorded on the distributed ledger very safe and difficult to alter. All users on the network can see and confirm all transactions, eliminating any doubt about their legitimacy and minimizing the potential for fraud. Data stored on a blockchain cannot be edited or removed once it has been recorded. The ability to accurately record machine performance, maintenance, and supply chain events relies heavily on this function. When it comes to legal agreements, "smart contracts" are agreements that may carry out their own terms and conditions automatically. In the field of industrial robots, contracts like maintenance agreements and supply chain contracts may be automated and secured. Blockchain technology can trace where goods originated, where they traveled, and who handled them throughout the supply chain. This is especially helpful for sectors that must adhere to rigorous standards of quality assurance and documentability.

***Industrial Robotics Case Studies:*** Blockchain technology may be used for supply chain management, allowing for the monitoring of robotic parts and components as they are sent from one location to another. Predictive maintenance and auditing can benefit from storing maintenance records and data on the blockchain for industrial robots.

*Quality Assurance:* The blockchain can record quality control data and inspections during the manufacturing process, enhancing product quality.

*Asset Ownership:* In cases where multiple parties share ownership of industrial robots, blockchain can establish clear ownership records and automate revenue sharing through smart contracts.

*Robot-as-a-Service (RaaS):* Blockchain can facilitate secure and transparent transactions for RaaS providers and customers, ensuring efficient billing and usage tracking.

Challenges and Considerations

*Scalability:* Handling a large volume of transactions in real-time can be challenging for some blockchain networks.

*Energy Consumption:* Some blockchain networks, especially those using PoW, can be energy-intensive, which may not be suitable for resource-constrained robotic systems.

*Regulatory Compliance:* Compliance with industry-specific regulations and data privacy laws must be considered when implementing blockchain in industrial robotics. The difficulty and perhaps need for specialist knowledge in integrating blockchain technology into existing industrial systems and procedures.

Enhanced security, transparency, and automation through smart contracts are just a few of the benefits that blockchain technology provides in the field of industrial robots. Improved reliability and productivity in industrial robotic systems are possible outcomes of this technology's continued development and widespread use.

## TECHNICAL CHALLENGES IN ROBOTICS

There are several technical obstacles in robotics that scientists and engineers are working hard to solve. Hardware and software development, autonomy, and human-robot interaction are only few of the areas that face difficulties. Here are some of the key technical challenges in robotics.

Despite the significant strides made in technology, there remain substantial challenges that pose a potential threat to both robots and the individuals who use them. Even robots equipped with advanced features can have vulnerabilities that compromise their performance and reliability. This section provides an overview of the broad spectrum of challenges encountered in the field of robotics.

Security stands out as a paramount concern in the realm of robotics, given its central role in technology. The multidisciplinary nature of robotics gives rise to a multitude of security challenges spanning firmware, communication protocols, data storage, cryptography, control systems, and mechanical designs. In a referenced study, the author highlights the deficiencies present in widely employed robotic software architectures, both in research and industrial robot development. Notably, open-source middleware frameworks like Robot Operating System (ROS) and Yet Another Robot Platform (YARP) are commonly used. These frameworks facilitate the orchestration of multiple intercommunicating programs hosted on various nodes (processors), enabling interaction with the robot's sensors, actuators, and other peripherals.

ROS, for instance, transmits its messages without encryption via Transmission Control Protocol/Internet Protocol (TCP/IP) or User Datagram Protocol/Internet Protocol (UDP/IP) across nodes, leaving it susceptible to interception by unauthorized entities. While newer versions have introduced enhancements like Message Digest 5 (MD5) algorithms and other authentication mechanisms, overall system security remains incomplete as the data itself remains unencrypted. Similarly, YARP employs unencrypted TCP handshakes when communicating with sensors, exposing the infrastructure to potential data corruption, thus rendering it vulnerable.

Shifting our attention to the external environment, Starship Technologies Ltd. has tackled a fundamental issue confronting Autonomous Delivery Vehicles (ADVs) when operating in public spaces: the vulnerability to theft or tampering of the goods they carry. Moreover, these robots themselves can be captured and reprogrammed for illicit purposes. Beyond hardware misuse, there are also concerns related to the valuable data continually collected by the robot's sensors, such as cameras, which is used for mapping routes and identifying obstacles along the way. This data holds significant commercial value, making data protection not only a matter of security but also a legal imperative.

In a separate study, researcher's have expressed apprehensions about the security of Autonomous Social Robots (ASRs) concerning the information they gather on user interaction patterns. These users span a wide range, from vulnerable young children who may find the robot's adaptations intimidating to adults who serve as valuable human resources for various organizations. The improper use of their personal and behavioral data could have far-reaching consequences. Consequently, ensuring data privacy in ASRs becomes a paramount necessity.

A field diverging from traditional robotics is known as soft robotics, where robots are constructed using flexible composite materials with minimal reliance on rigid actuators. In stark contrast to conventional rigid robots, soft robots utilize their outer membranes and semi-rigid structures within these membranes, controlled through pneumatic or hydraulic pressure fluctuations, cables, and thermal variations. While soft robotics offers numerous advantages over its traditional counterpart, it is important to acknowledge significant limitations.

Whether mobile or stationary, robots must be designed with careful consideration of energy source characteristics at the early stages of development. Factors such as availability, reusability (or renewability), capacity, and lifespan of the energy sources are critical. In practice, mobile robots, particularly autonomous and tele-operated ones, need to carry their energy sources, typically batteries, on their frames as they navigate and perform tasks. In contrast, factory or stationary robots often tap into virtually unlimited power supplies. Unlike living organisms that efficiently harvest energy from carbohydrates through food ingestion, mobile robots tend to be less energy-optimized and less efficient in their operations, losing a significant portion of energy to undesired mechanical vibrations, thermal dissipation, and audible noise generation.

The field of robotics has experienced significant growth in recent times, particularly with the emergence of the COVID-19 pandemic. Robotics, when integrated with the Internet of Things (IoT), have found widespread adoption across various sectors, including healthcare, modern societies, industry, agriculture, military, and transportation. However, like many other technological advancements, robotics are not immune to security concerns, presenting numerous risks and vulnerabilities.

The field of robotics has a number of difficulties. Security threats to robots might mirror those faced by PCs for decades. Cyberattacks and hacks leave both industrial and service robots open to exploitation if they lack proper security safeguards. Connecting robots to networks through IP addresses makes them vulnerable to hacking, which, if successful, can lead to severe interruptions and damage. Many different types of vulnerabilities can affect the connection, productivity, operations, and general dependability of robotic systems.

Creating sensors and vision systems that give robots a thorough understanding of their surroundings is a difficult task. Sensor fusion, depth sensing, object identification, and computer vision all fall under this category. Robots require localization and mapping services so that they can find their way around and comprehend their surroundings. Especially in uncertain or changing surroundings, solving the challenge of simultaneous localization and mapping (SLAM) is difficult. High levels of autonomy are an ongoing pursuit in robotics. It entails giving robots the ability to think for themselves, form plans, and respond to changing circumstances. Making robots as dexterous and precise in their object manipulation as humans is a difficult task. Advanced hardware and control algorithms are needed for fine motor abilities, gripping, and object manipulation. Saving energy is essential, especially for robots that need to move about or be placed in remote locations. New energy-efficient technologies and more sustainable energy sources are urgently required.

Interacting with people in a safe and natural way is a complex task for robot developers. Speech synthesis, NLP, and secure physical contact are all a part of it. The capacity to adjust to new circumstances quickly is essential for robots. Making robots that can effectively learn and adjust to novel conditions is a formidable problem. It is crucial that robots be able to move around in many settings, some of which may be dangerous or difficult to navigate due to obstacles like clutter or uneven ground. Search and rescue operations and autonomous vehicles both depend critically on mobility. Maintaining the safety of robots is of the utmost importance, especially when they are used in close proximity to people. It is crucial to have fail-safes, solid control systems, and dependable hardware and software. The broad use of robots across a wide range of businesses depends on their becoming more cost-effective and cheap. The emergence of low-cost manufacturing techniques and components contributes to this difficulty. Downtime and expenses associated with robotics systems may be mitigated by improving both the longevity of the systems and the ease with which maintenance can be performed.

Ethical and Legal Considerations It is becoming increasingly vital to address ethical and legal difficulties associated with robots, such as liability, privacy, and ethical decision-making. Challenges in communication, coordination, and job allocation arise when a large number of robots, referred to as a "robotic swarm," must work together. Integrating cutting-edge AI and ML methods with robots to improve perception, decision making, and flexibility is an ongoing issue. Standardization and Interoperability: Establishing common standards and protocols for robotics hardware and software can improve interoperability and ease of integration in various applications.

These challenges are driving innovation and research in the field of robotics, with the goal of creating more capable, reliable, and versatile robotic systems that can address a wide range of applications and industries.

## CLASSIFICATION OF MULTI-AGENT SYSTEM

Various cutting-edge attributes define the classification of a multi-agent system (MAS). For the context of this thesis, we'll use specific attributes to categorize MAS. At its core, MAS falls into two categories: cooperative and competitive. In a cooperative MAS, agents collaborate in a manner reminiscent of social living beings. Conversely, in a competitive MAS, agents vie against each other for limited essential resources. This chapter will solely focus on cooperative MAS.

***Classification by Cooperation in MAS:*** MAS can be classified by evaluating an agent's capability to cooperate with other agents during task execution. Agents that work in tandem with others are termed "cooperative agents". In contrast, those that don't cooperate and instead compete with others are labeled "non-cooperative agents". The primary aim of cooperative agents is to fulfill a shared objective, whereas non-cooperative agents typically have opposing goals. Figure.1 offers an in-depth categorization of only cooperative agents, segmented by their knowledge levels.

***Classification by Knowledge Level in Cooperative MAS:*** Cooperative MAS can be further categorized according to an agent's awareness of its teammates. As depicted in Figure 1, we can distinguish between "aware agents" and "unaware agents." An aware agent possesses knowledge about its teammates, while an unaware agent lacks such information about other agents in the environment.

***Classification by Coordination in Aware Agents:*** Aware agents in a multi-agent system can be further sorted based on the coordination methods they employ. There are three primary types of coordination. In strong coordination, agents rigorously stick to the established coordination protocols. Conversely, in weak coordination, agents don't consistently adhere to these protocols. Lastly, there's a category where agents simply don't engage in coordination with their counterparts. This categorization, which captures the nuances of coordination among aware agents, is visually delineated in Figure 1.

***Classification by Organizational Structure:*** *Strongly* coordinated agents can be further categorized based on their roles within a team or organizational structure, as illustrated in Figure 1. This perspec-

*Figure 1. Categorization of multi-robot systems*

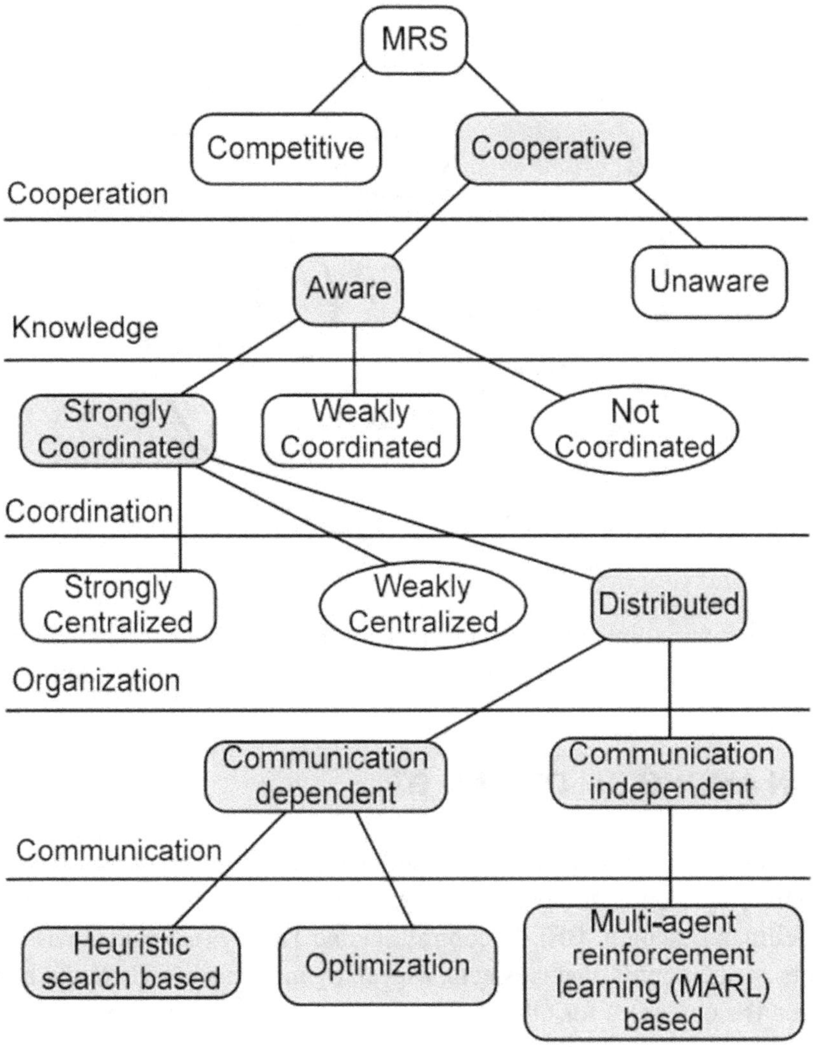


tive distinguishes between centralized and distributed approaches. In the centralized model, one agent is designated as the leader for the entire group. This leader is tasked with delegating responsibilities to the other agents. The other agents, or followers, operate based on directives from the leader. Conversely, in the distributed approach, all agents operate autonomously without a specific leader guiding them. Furthermore, the centralized structure can be sub-divided based on the leadership model. A "strongly centralized" system is where a single robot takes the lead for the entire operation. In contrast, a "weakly centralized" system allows multiple agents to assume leadership roles throughout the mission's duration.

***Classification by Communication Dependency:*** Distributed robots can be differentiated based on their reliance on communication between agents, as depicted in Figure.1. Two primary types emerge in this categorization: those that are communication-dependent and those that are communication-independent. Beyond this, multi-agent systems (MAS) can also be categorized by other criteria, such as "Team Com-

Figure 2. Interrelationship between MARL, DP, and GT

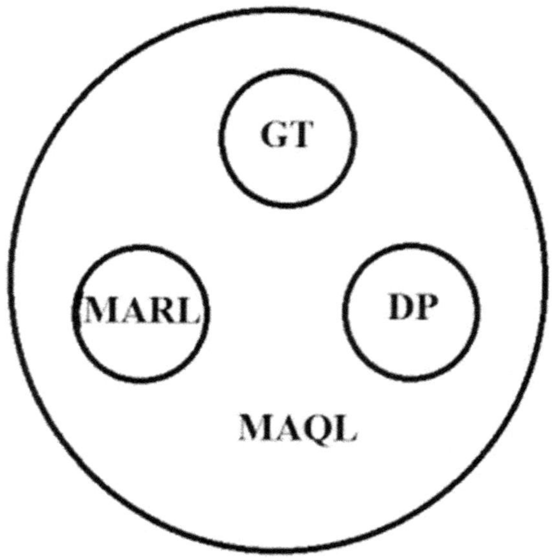


position", which looks at mixtures of heterogeneous and homogeneous robots; "System Architecture", which examines the structural design; and "Team Size", which considers the number of robots involved.

## CORRELATION AMONG RL, DP, AND GT

Reinforcement Learning (RL) operates based on the principle of rewards or penalties that agents receive as feedback from their environment. Dynamic Programming (DP) serves as an optimization technique that refines the Bellman Equation (BE). In contrast, Game Theory (GT) aids in assessing the strategic positions of agents within a multi-agent system, especially in scenarios where the actions of one agent markedly influence the outcomes for other agents in the system. Figure 2 highlights how multi-agent Q-learning (MAQL) integrates elements of Multi-Agent Reinforcement Learning (MARL), GT, and DP. However, in academic and industry literature, MARL is commonly referred to as MAQL for the sake of simplicity.

Figure 3 illustrates the coordination and planning strategy used by multiple robots for the stick-carrying challenge. This challenge involves moving a stick from its starting position to a specified destination. Currently, two robots are positioned at a joint state <4, 7>, holding a stick as depicted in Figure 3(c). Since each robot has access to both robots' Q-tables in a shared state-action space, they can determine the optimal joint action or the pure strategy Nash Equilibrium (PSNE) at <4, 7> by assessing equilibrium. To achieve this, they gather data from the joint state <4, 7> (as shown in Figures 3(a) and 3(b)) and determine the PSNE as "FL" based on the Nash Equilibrium definition, illustrated in Figure 3. Notably, both robots come to the same PSNE conclusion. This shared understanding allows them to coordinate without direct communication, and as a result of their combined action "FL", they move the stick to the next joint state <5, 4>.

The stick-carrying problem is a classic challenge in the field of robotic coordination. In essence, it tests how multiple robots can work together to transport an object, in this case, a stick, from a starting point to a desired destination. Addressing this problem not only requires individual robot proficiency but also efficient inter-robot collaboration. Here's an overview of the coordination mechanisms and considerations involved.

*Initial State and Destination:* The problem begins with robots at an initial joint state, holding the stick. The primary objective is to move the stick to a specified end point.

*Shared Knowledge Base:* It's crucial for the robots to have a shared knowledge base, such as joint Q-tables in a common state-action space. This shared information allows robots to predict and understand each other's actions and tendencies. *Decision-making Process:* The robots must make decisions based on Nash Equilibrium principles, specifically the pure strategy Nash Equilibrium (PSNE). By assessing equilibrium, each robot can determine the optimal joint action. *Communication:* One of the hallmarks of this problem is that robots often need to operate without direct communication. Instead, they rely on the shared knowledge base and the predictable behaviors derived from it. However, advanced solutions might incorporate limited or non-verbal communication mechanisms to further optimize coordination.

*Action Execution:* Once a decision is made, both robots execute their joint action. If done correctly, their combined efforts will move the stick closer to the desired destination.

*Re-evaluation:* The robots continually assess their position and the stick's position relative to the end point. If adjustments are needed, the robots will consult the shared knowledge base and make new decisions based on the current context.

*End Goal:* The problem is considered solved when the robots successfully transport the stick to the predetermined destination.

The stick-carrying problem serves as a foundational exercise for understanding multi-agent coordination. Lessons learned from this problem can be applied to more complex tasks and environments, further advancing the capabilities of multi-robot systems.

## SYNCHRONIZATION THROUGH OPTIMIZATION TECHNIQUES.

Coordination by optimization refers to the process of harmonizing or adjusting different elements or agents in a system to achieve a desired outcome. In the context of algorithms, it typically means using computational techniques to find the best possible solution to a problem considering the constraints and objectives provided.

Some key points about Coordination by Optimization are:

*Problem Formulation:* Before optimization can be done, the problem needs to be formulated mathematically. This involves defining the objective function (what you're trying to maximize or minimize) and the constraints (limitations or boundaries within which solutions must lie).

*Multi-agent Systems:* In systems where multiple agents or entities are involved, coordination ensures that all agents' actions lead to globally optimal or near-optimal outcomes. Each agent can use local information, but the global outcome should be optimized.

*Techniques:* There are various techniques used for optimization such as linear programming, integer programming, genetic algorithms, simulated annealing, and gradient-based methods, to name a few.

*Decentralized Optimization:* In some scenarios, especially in large-scale systems like power grids or traffic systems, centralized optimization might be infeasible. In such cases, decentralized optimization algorithms, where each agent optimizes its action based on local information and some coordination mechanism, can be used. *Dynamic Systems:* In cases where the system is dynamic and changes over

*Figure 3. Cooperative robot interaction in the stick-transport challenge*

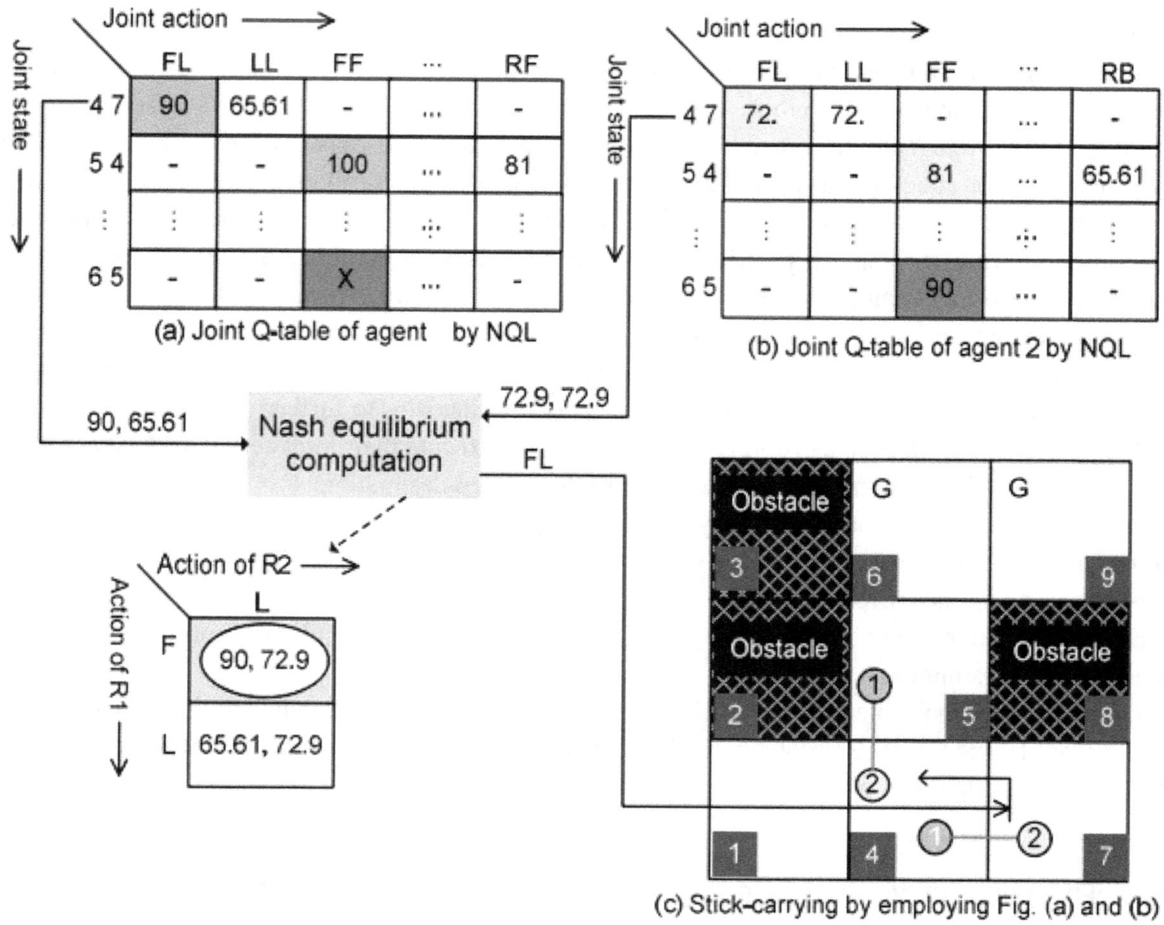


time, adaptive and online optimization techniques are used. These algorithms adjust to the changing conditions to ensure that the system remains optimized.

*Coordination vs. Cooperation:* While both terms are often used interchangeably, coordination focuses on orchestrating the actions of agents to achieve a global objective. In contrast, cooperation is about agents working together, sharing resources, and collaborating for a common goal. Coordination can be seen as a method to achieve effective cooperation. In essence, coordination by optimization is a powerful approach to manage and harmonize complex systems, ensuring that they operate at their best given the circumstances and constraints.

Swarm Intelligence (SI) and Evolutionary Algorithms (EAs) offer compelling solutions to address common bottlenecks encountered in search-based coordination and Multi-Agent Reinforcement Learning (MARL) algorithms. One of the significant challenges in these approaches is the high memory requirement and the potential for suboptimal solutions. SI algorithms, exemplified by Particle Swarm Optimization (PSO) and Firefly algorithm (FA), stand out for their scalability, adaptability, collective robustness, and individual simplicity. They can efficiently handle large swarms of agents, adapt swiftly to dynamic environments, exhibit resilience against single-point failures, and achieve sophisticated

*Figure 4. Local planning for multiple robots using swarm and evolutionary algorithms*

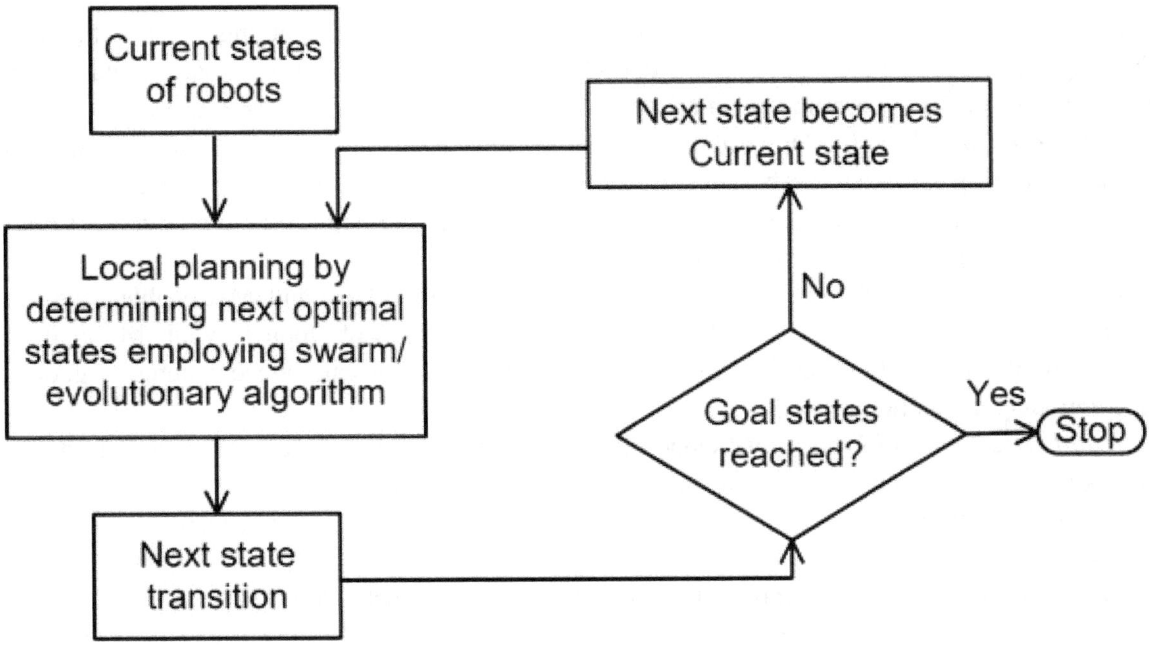


group behavior despite the simplicity of individual swarm members. PSO, for instance, excels in high-dimensional problems with non-differentiable fitness functions, while FA excels in global optimization. On the other hand, EAs excel in coping with complex optimization challenges characterized by discontinuities, non-linear constraints, multi-modalities, and multi-objective objectives. The choice between SI and EA depends on the specific problem characteristics and requirements, allowing researchers and practitioners to select the most suitable algorithm for their particular application.

Evolutionary Algorithms (EAs), such as Differential Evolution (DE) and Imperialist Competitive Algorithm (ICA), while versatile, do not guarantee optimal solutions within a finite amount of time. DE is known for its stability compared to Genetic Algorithms (GA). ICA, a socio-political-based algorithm, incorporates neighborhood movements in both continuous and discrete search spaces, but it also lacks a guarantee of finding optimal solutions and often requires tuning a larger number of parameters compared to algorithms like PSO, FA, and DE. In such scenarios, hybridization emerges as a promising approach. By combining the strengths of two or more algorithms, hybrid algorithms can be created to harness their efficient attributes and produce more powerful optimization methods. For the multi-robot stick-carrying problem, for instance, hybridization has been investigated. One method, described in reference, merges the firefly motion dynamics from the Firefly Algorithm (FA) with a meta-heuristic search algorithm based on social and political evolution. The resulting algorithm is called the Imperialist Competitive Firefly Algorithm (ICFA). Using the best features of FA and ICA, this hybridization strategy aims to improve multi-robot cooperation. Figure 4 depicts how combining these algorithms into a coordination scheme might be a useful approach to solving difficult coordination problems. In conclusion, hybridization provides a way to integrate the benefits of several algorithms, such as FA, to develop more resilient and efficient solutions to complicated problems like multi-robot coordination, where algorithms like DE and ICA may not guarantee optimality.

## OPPORTUNITIES ARISING FROM ML AND BLOCKCHAIN INTEGRATION

Many sectors might benefit from the synergies that would result from combining ML with blockchain. This might lead to synergies that improve data-driven processes' effectiveness, safety, and openness. Here are some of the opportunities that arise from the integration of ML and blockchain

***Enhanced Data Security:*** Opportunity: Combining blockchain's cryptographic security with ML's anomaly detection can create a robust defense against data breaches and cyberattacks. How: Blockchain ensures data immutability and tamper resistance, while ML can continuously monitor data for unusual patterns, signaling potential security threats.

***Transparent Supply Chains:*** Opportunity: Integrating blockchain and ML enables end-to-end transparency in supply chains. How: Blockchain records the journey of products, while ML algorithms analyze supply chain data for optimization and fraud detection. This can reduce counterfeit goods and ensure product authenticity.

***Fraud Detection and Prevention:*** Opportunity: ML models can analyze transaction data stored on a blockchain to detect and prevent fraudulent activities. How: ML algorithms can identify suspicious patterns or deviations in transactions, such as anomalies in financial records, leading to early fraud detection and prevention.

***Smart Contracts and Automation:*** Opportunity: Smart contracts on blockchain can be enhanced with ML to enable more complex, self-executing agreements. How: ML models can analyze real-time data to trigger smart contract actions automatically. For example, insurance claims can be processed and verified using ML, and payments can be released through smart contracts.

***Improved Compliance and Auditing:*** Opportunity: Integration of ML and blockchain can streamline compliance processes and auditing in various industries. How: ML algorithms can continuously monitor transactions and data for compliance violations, and the immutable blockchain ledger provides a transparent audit trail.

***Predictive Maintenance and Optimization:*** Opportunity: Combining ML predictive maintenance with blockchain traceability can optimize industrial processes. How: ML models predict maintenance needs based on sensor data, while blockchain records maintenance history, facilitating efficient scheduling and optimization.

***Data Monetization and Ownership:*** Opportunity: Blockchain can enable individuals to have more control over their data and its monetization. How: ML algorithms can analyze user data to provide personalized services, and blockchain can record data ownership and usage, allowing individuals to monetize their data while maintaining control.

***Decentralized AI and Federated Learning:*** Opportunity: Decentralized AI models can be trained on distributed data without exposing sensitive information. How: Blockchain can facilitate secure data sharing and federated learning, where ML models are trained collaboratively across multiple parties without sharing raw data.

***Energy Efficiency in Blockchain:*** Opportunity: ML can optimize energy-intensive blockchain networks, making them more sustainable. How: ML algorithms can analyze network traffic and adjust consensus mechanisms and energy consumption strategies to minimize environmental impact.

***Customized Healthcare:*** Opportunity: Integration of ML and blockchain can enable personalized healthcare and research. How: Patients can securely share health data on a blockchain, and ML models can analyze this data to provide personalized treatment recommendations.

The integration of ML and blockchain technologies opens doors to innovative solutions across a wide range of industries, enhancing security, transparency, and efficiency. Organizations that seize these opportunities can gain a competitive edge, deliver better services, and unlock new business models.

## CHALLENGES IN ML AND BLOCKCHAIN INTEGRATION

While the integration of machine learning (ML) and blockchain technologies presents significant opportunities, it also comes with several challenges. Addressing these challenges is crucial to ensure the successful implementation of integrated solutions. Here are some of the key challenges in ML and blockchain integration.

Scalability:

Challenge: Blockchain networks, especially public ones, can face scalability issues when processing a large number of transactions in real-time. This can hinder the integration of ML models that require frequent data updates.

Solution: Implementing efficient consensus mechanisms, exploring off-chain solutions, or using permissioned blockchains can help alleviate scalability concerns.

Data Privacy:

Challenge: Blockchain's transparency may conflict with the need to protect sensitive or confidential data used in ML models. Sharing private data on a blockchain can pose privacy risks. Solution: Techniques such as zero-knowledge proofs, secure multi-party computation, and federated learning can help protect data privacy while allowing ML to be performed on blockchain data.

Energy Consumption:

Challenge: Numerous blockchain networks need a significant amount of energy, particularly those that use the Proof of Work (PoW) consensus mechanism. This may be a cause for worry in settings where energy efficiency is of the utmost importance, such as in resource-constrained systems or with Internet of Things devices.

Solution: Think about utilizing consensus processes that use less energy, such as Proof of Stake (PoS), or investigate other blockchain technologies that are meant to be more energy efficient.

Latency and Speed:

Challenge: Blockchain transactions may have latency, making real-time decision-making in ML applications challenging. Solution: Combining blockchain with off-chain solutions or sidechains can help reduce transaction confirmation times and improve speed.

Interoperability:

Challenge: When dealing with varying blockchain platforms and ML frameworks, it might be difficult to integrate ML systems with blockchain networks.

Solution: The gap between ML and blockchain technology can be closed through standardization initiatives and the use of middleware solutions.

Skill Gap:

Challenge: Organizations may find it difficult to obtain the specialized skills in both ML and blockchain needed to implement integrated solutions.

Solution: It is possible to close the skills gap by investment in training and development programs or through collaboration with specialists in both disciplines.

Regulatory Compliance:

Challenge: Integrating blockchain and ML requires meeting regulatory standards specific to various businesses and geographical locations. It might be difficult to ensure that data stored on a blockchain complies with all applicable laws and regulations.

Solution: Compliance issues can be mitigated by consulting with legal professionals and maintaining awareness of industry-specific legislation.

Data Quality and Accuracy:

Challenge: Machine learning models require precise and reliable inputs. Even if data is stored accurately on a blockchain, it may still contain flaws or errors.

Solution: Data validation and quality assurance techniques implemented prior to data being recorded on the blockchain can assist in reducing this risk.

Cost of Implementation:

Challenge: Costs associated with setting up and running systems that combine machine learning with blockchain technology might be significant.

Solution: In order to gain a return on the money spent for integration, a thorough cost-benefit analysis and financial strategy are required.

To overcome these obstacles, we need to take a comprehensive strategy that incorporates technical knowledge, strategic forethought, and an in-depth familiarity with machine learning (ML) and blockchain technology. The full potential of combined ML and blockchain solutions will only be available to organizations who can effectively manage these difficulties.

## CASE STUDIES

### Case Study One: Secure Data Sharing in Robotic Manufacturing

*Challenge:* The production process at a robotic factory often involves a number of robots from various manufacturers. Safe data exchange and verification are required to guarantee that robots are making decisions based on reliable information.

*Solution:* Combining machine learning with blockchain technology for trustworthy data exchange and verification.

How It Works:

*Data Recording:* Robot sensors capture data on performance indicators, maintenance records, and operational data on a distributed ledger called the blockchain.

*Immutable Records:* The immutability of blockchain assures that information cannot be changed once it has been recorded.

*Data Verification:* Data on the blockchain is constantly being analyzed by machine learning models. Anomalies, out-of-the-ordinary behavior, and malfunction indicators are what they scan for.

*Optimization:* Optimizing robot operations with ML models via validated data. In the case of robots, the ML model may plan predictive maintenance to avoid failures if performance data suggest wear and tear.

*Secure Data Sharing:* Robot makers and service providers may use the blockchain to confirm the accuracy of the data and make sound judgments.

Benefits:

*Data Integrity:* Blockchain ensures that data is secure and tamper-proof, enhancing trust in the recorded information.

*Predictive Maintenance:* ML models optimize robot maintenance schedules, reducing downtime and increasing efficiency.

*Streamlined Data Sharing:* Authorized stakeholders can access verified data easily, leading to informed decisions and improved collaboration.

## Case Study Two: Decentralized Robot Coordination in Warehousing

***Challenge:*** In a large warehousing facility with numerous autonomous robots, coordinating their tasks efficiently without human intervention can be complex.

***Solution:*** Integration of smart contracts on a blockchain to facilitate decentralized robot coordination.

How It Works:

*Smart Contracts:* Smart contracts are stored in a blockchain. Robots can be hired on a contract basis for a variety of tasks, including product picking, inventory management, and delivery.

*Task Assignment*: The smart contract initiates a bidding procedure whenever work has to be done. Depending on their skills, location, and availability, robots can place bids for jobs. *Consensus and Execution*: When a robot is the highest bidder, the smart contract immediately puts it to work. Robots carry out the agreed-upon work.

*Task Completion Verification*: Robotic sensors and data verify the work has been done as agreed upon in the contract.

*Payment and Reputation*: Payments to the robots are released automatically under the terms of the smart contracts. The blockchain keeps a public record of each robot's history, including its successes and failures.

Benefits:

*Efficiency:* In order to improve work allocation and execution, decentralized coordination decreases the requirement for centralized control and human involvement.

*Transparency:* Transparency and trust are maintained by making the smart contracts and task history publicly viewable on the blockchain.

*Automation:* Automation eliminates the need for human interaction and supervision while getting work done.

These examples illustrate how data security, job coordination, and operational optimization may all be improved by using machine learning and blockchain technology with industrial robots. These coordinated approaches enhance productivity, visibility, and reliability in automated production and storage facilities.

## CONCLUSION

In conclusion, the convergence of ML and blockchain technologies in industrial robots marks a revolutionary step toward a safer, more transparent, and more efficient tomorrow. There are several advantages to combining ML with blockchain, such as increased data security, simplified procedures, and data-driven decisions. By safeguarding data, automating hard procedures, and improving operations, it might cause a major shift in many fields. It is impossible to overestimate the importance of this merger. It allows businesses to keep up with technology advancements, respond to shifting market conditions, and provide customers with superior goods and services. Predictive maintenance, transparent supply chains, and decentralized coordination of robotic systems are all made possible by this combination, as demonstrated in the use cases and case studies, increasing productivity while decreasing overhead costs. It's time for businesses to take note and implement the necessary changes, such as adopting innovative technologies, funding R&D, encouraging a culture of innovation, and working together with experts and peers. In this way, businesses can take use of the full power of ML and blockchain integration in today's data-driven economy. Those that are able to adapt to this shifting environment and take use of the possibilities offered by ML and blockchain will not only be able to compete successfully, but will also help shape the future of industrial robots and set new benchmarks for excellence. The path forward is one of innovation, collaboration, and continuous adaptation to unlock the immense possibilities that lie at the intersection of machine learning and blockchain.

## REFERENCES


Almurib, H. A. F., & Al-Qrimli, H. F. (2012). A review of application industrial robotic design. *2011 Ninth International Conference on ICT and Knowledge Engineering*, Bangkok, Thailand. 10.1109/ICTKE.2012.6152387

Avinash, P. (2023). A novel approach of unsupervised feature selection using iterative shrinking and expansion algorithm. *Journal of Interdisciplinary Mathematics*, *26*(3), 519–530. doi:10.47974/JIM-1678

Gite, P. (2023). A novel RF-SMOTE model to enhance the definite apprehensions for IoT security attacks. *Journal of Discrete Mathematical Sciences and Cryptography*, *26*(3), 861–873. doi:10.47974/JDMSC-1766

Grau, A., Indri, M., Lo Bello, L., & Sauter, T. (2017). Industrial robotics in factory automation: From the early stage to the Internet of Things. *Proc. IECON 2017–43rd Annu. Conf. IEEE Ind. Electron. Soc.*, (pp. 6159-6164). IEEE. 10.1109/IECON.2017.8217070

Grau, M. (2017). Industrial robotics in factory automation: From the early stage to the Internet of Things. *IECON 2017 - 43rd Annual Conference of the IEEE Industrial Electronics Society*, (pp. 6159-6164). IEEE.

Grau, M. I., Bello, L. L., & Sauter, T. (2021). Robots in Industry: The Past, Present, and Future of a Growing Collaboration With Humans. *IEEE Industrial Electronics Magazine*, *15*(1), 50–61. doi:10.1109/MIE.2020.3008136

Kavitha, R., & Prasad, K. (2023). *A Novel Method of Identification of Delirium in Patients from Electronic Health Records Using Machine Learning*. 2023 World Conference on Communication & Computing (WCONF), Raipur, India. 10.1109/WCONF58270.2023.10235245

Kumar, V., Pant, B., & Elkady, G. (2022*). Examining the Role of Block Chain to Secure Identity in IOT for Industry 4.0.* 2022 5th International Conference on Contemporary Computing and Informatics (IC3I), Uttar Pradesh, India. 10.1109/IC3I56241.2022.10072516

Palanikkumar, D., & Patwari, N. (2023). *Symmetrized Feature Selection with Stacked Generalization based Machine Learning Algorithm for the Early Diagnosis of Chronic Diseases*. 2023 5th International Conference on Smart Systems and Inventive Technology (ICSSIT), Tirunelveli, India.

Pereira, F. G., Vassallo, R. F., & Salles, E. O. T. (2013). Human-robot interaction and cooperation through people detection and gesture recognition. *J. Control Automat. Elect. Syst.*, *24*(3), 187–198. doi:10.100740313-013-0040-3

Prasad, K., Srinivas, V., & De, T. (2023). Wi-Fi Router Signal Coverage Position Prediction System using Machine Learning Algorithms. *2023 International Conference on Sustainable Computing and Smart Systems (ICSCSS)*, Coimbatore, India.

Ranjan, K. K. (2023). *Wireless and Solar-Powered Multipurpose Robot for Industrial Automation: A Sustainable Solution*. 2023 8th International Conference on Communication and Electronics Systems (ICCES), Coimbatore, India.

Ranjan, K. K. (2023). *Industrial Automated Multipurpose Robot Using WIFI*. 2023 4th International Conference for Emerging Technology (INCET), Belgaum, India.

Shivashankar, M. S. (2017). Implementation of swarm intelligence in obstacle avoidance. in *RTEICT 2017 - 2nd IEEE International Conference on Recent Trends in Electronics, Information and Communication Technology*. IEEE.

Singh, G., Banga, V. K., & Yingthawornsuk, T. (2022). *Artificial Intelligence and Industrial Robot*. 2022 16th International Conference on Signal-Image Technology & Internet-Based Systems (SITIS), Dijon, France. 10.1109/SITIS57111.2022.00098

Tanguturi, R. (2022). Improved Secure Communication Mechanism for IoT Platform Devices. *2022 International Conference on Smart Generation Computing, Communication and Networking (SMART GENCON)*, Bangalore, India.

Tantawi, K. H., Sokolov, A., & Tantawi, O. (2019). *Advances in Industrial Robotics: From Industry 3.0 Automation to Industry 4.0 Collaboration*. 2019 4th Technology Innovation Management and Engineering Science International Conference (TIMES-iCON), Bangkok, Thailand. 10.1109/TIMES-iCON47539.2019.9024658

Tellaeche, I. (2015). Human robot interaction in industrial robotics. Examples from research centers to industry. *2015 IEEE 20th Conference on Emerging Technologies & Factory Automation (ETFA)*. IEEE.

Venkatakiran, S., Ashreetha, B., & Reddy, N. S. (2023). Implementation of a Machine Learning-based Model for Cardiovascular Disease Post Exposure prophylaxis. *2023 International Conference for Advancement in Technology (ICONAT)*, Goa, India.

Verma. (2022). Blockchain for Industry 5.0: Vision, Opportunities, Key Enablers, and Future Directions. *IEEE Access*, *10*, 69160-69199.

# Compilation of References

Abayomi-Alli, O. O., Damaševičius, R., Misra, S., & Maskeliūnas, R. (2021). Cassava disease recognition from low-quality images using enhanced data augmentation model and deep learning. *Expert Systems: International Journal of Knowledge Engineering and Neural Networks*, *38*(7), e12746. doi:10.1111/exsy.12746

Abbas, A. F., Qureshi, N. A., Khan, N., Chandio, R., & Ali, J. (2022). The Blockchain Technologies in Healthcare: Prospects, Obstacles, and Future Recommendations; Lessons Learned from Digitalization. *International Journal of Online & Biomedical Engineering*, *18*(9).

Abinaya, B., Gurupriya, S., & Pooja, M. (2017). IoT based smart and adaptive lighting in street lights. In *Second International Conference on Computing and Communications Technologies, ICCCT17)* (pp. 195-198). IEEE. 10.1109/ICCCT2.2017.7972267

Abisha, S., & Jayasree, T. (2022). Application of image processing techniques and artificial neural network for detection of diseases on brinjal leaf. *Journal of the Institution of Electronics and Telecommunication Engineers*, *68*(3), 2246–2258. doi:10.1080/03772063.2019.1696716

Abomhara, M., & Koien, G. M. (2014). Security and Privacy in the Internet of Things: Current Status and Open Issues. *International Conference on Privacy and Security in Mobile Systems (PRISMS)*, (pp. 1-8). IEEE. 10.1109/PRISMS.2014.6970594

Abougarair, A. J., Aburakhis, M., & Zaroug, M. (2022). Design and implementation of smart voice assistant and recognizing academic words. *International Robotics & Automation Journal*, *8*(1), 27–32. doi:10.15406/iratj.2022.08.00240

Adeel, A., Khan, M. A., Sharif, M., Azam, F., Shah, J. H., Umer, T., & Wan, S. (2019). Diagnosis and recognition of grape leaf diseases: An automated system based on a novel saliency approach and canonical correlation analysis based multiple features fusion. *Sustainable Computing : Informatics and Systems*, *24*, 100349. doi:10.1016/j.suscom.2019.08.002

Adel, K., Elhakeem, A., & Marzouk, M. (2022). Chatbot for construction firms using scalable blockchain network. *Automation in Construction*, *141*, 104390. doi:10.1016/j.autcon.2022.104390

Aditya, M. S., & Gupta, S. C. (2018). An Internet of Things Based Smart Surveillance and Monitoring System using Arduino. *International Conference on Advances in Computing and Communication Engineering(ICACCE)*, (pp. 428–433). IEEE. 10.1109/ICACCE.2018.8441725

Aditya, U. S., Singh, R., Singh, P. K., & Kalla, A. (2021). A survey on blockchain in robotics: Issues, opportunities, challenges and future directions. *Journal of Network and Computer Applications*, *196*, 103245. doi:10.1016/j.jnca.2021.103245

Afaq, Y., & Manocha, A. (2023). Blockchain and Deep Learning Integration for Various Application: A Review. *Journal of Computer Information Systems*, 1–14. doi:10.1080/08874417.2023.2173330

Agrawal, D., Bansal, R., Fernandez, T. F., & Tyagi, A. K. (2022). Blockchain Integrated Machine Learning for Training Autonomous Cars. In Lecture Notes in Networks and Systems, (vol 420). Springer, Cham. doi:10.1007/978-3-030-96305-7_4

Aguiar-Pérez, J. M., & Pérez-Juárez, M. Á. (2023). An insight of deep learning based demand forecasting in smart grids. *Sensors (Basel)*, *23*(3), 1467. doi:10.339023031467 PMID:36772509

Alanzi, H., Alshammri, A. S., Alzu'bi, R., & Hamed, E. (2019). Internet of Things Objectives, Benefits, and Applications. *Oral Science International, 31*(1), 95-100.

Alferidah, D. K. (2020). A Review on Security and Privacy Issues and Challenges in Internet of Things. *International Journal of Computer Science and Network Security IJCSNS*, *20*(4), 263–286.

Algarni, S., Eassa, F., & Almarhabi, K. (2021). Blockchain-Based Secured Access Control in an IoT System. *Applied Science.*

Ali, S., Abdullah, Armand, T. P. T., Athar, A., Hussain, A., Ali, M., Yaseen, M., Joo, M.-I., & Kim, H.-C. (2023). Metaverse in healthcare integrated with explainable ai and blockchain: Enabling immersiveness, ensuring trust, and providing patient data security. *Sensors (Basel)*, *23*(2), 565. doi:10.339023020565 PMID:36679361

Alja'afreh, M., El Ahmar, W., Hoda, M., & Karime, A. (2022), A Review on Industrial Blockchain. In *2022 Fourth International Conference on Blockchain Computing and Applications (BCCA)*. IEEE. 10.1109/BCCA55292.2022.9922219

Al-Jaroodi, J., & Mohamed, N. (2019b, January). Industrial applications of blockchain. In *2019 IEEE 9th Annual Computing and Communication Workshop and Conference (CCWC)* (pp. 0550-0555). IEEE. 10.1109/CCWC.2019.8666530

Al-Jaroodi, J., & Mohamed, N. (2019a). Blockchain in industries: A survey. *IEEE Access : Practical Innovations, Open Solutions*, *7*, 36500–36515. doi:10.1109/ACCESS.2019.2903554

Aljawarneh, S., Aldwairi, M., & Jararweh, Y. (2020). Blockchain-based Framework for Securing IoT Applications. *Journal of Ambient Intelligence and Humanized Computing*, *11*(4), 1629–1641.

Alkhalili, M., Qutqut, M. H., & Almasalha, F. (2021). Investigation of applying machine learning for watch-list filtering in anti-money laundering. *IEEE Access : Practical Innovations, Open Solutions*, *9*, 18481–18496. doi:10.1109/ACCESS.2021.3052313

Alladi, T., Chamola, V., Sahu, N., & Guizani, M. (2020). Applications of blockchain in unmanned aerial vehicles: A review. *Vehicular Communications*, *23*, 100249. doi:10.1016/j.vehcom.2020.100249

Almurib, H. A. F., & Al-Qrimli, H. F. (2012). A review of application industrial robotic design. *2011 Ninth International Conference on ICT and Knowledge Engineering*, Bangkok, Thailand. 10.1109/ICTKE.2012.6152387

Alotibi, J., Almutanni, B., Alsubait, T., Alhakami, H., & Baz, A. (2022). Money Laundering Detection using Machine Learning and Deep Learning. *International Journal of Advanced Computer Science and Applications*, *13*(10), 10. doi:10.14569/IJACSA.2022.0131087

Al-Qaseemi, S. A., Almulhim, H. A., Almulhim, M. F., & Chaudhry, S. R. (2016). IoT architecture challenges and issues: Lack of standardization. *2016 Future Technologies Conference (FTC)*, San Francisco, CA, USA. 10.1109/FTC.2016.7821686

AlQerm, I., & Pan, J. (2022). I-HARF: Intelligent and Hierarchical Framework for Adaptive Resource Facilitation in Edge-IoT Systems. *IEEE Internet of Things Journal*, *10*(5), 3954–3967. doi:10.1109/JIOT.2022.3151667

Alsamhi, S. H., Lee, B., Guizani, M., Kumar, N., Qiao, Y., & Liu, X. (2021). Blockchain for decentralized multi-drone to combat COVID-19 and future pandemics: Framework and proposed solutions. *Transactions on Emerging Telecommunications Technologies*, *32*(9), e4255. doi:10.1002/ett.4255

AlZubi, A. A. (2022). IoT-Based Automated Water Pollution Treatment Using Machine Learning Classifiers. *Environmental Technology*, 1–9. doi:10.1080/09593330.2022.2034978 PMID:35083949

Ameen, A. H., Mohammed, M. A., & Rashid, A. N. (2023). Dimensions of artificial intelligence techniques, blockchain, and cyber security in the Internet of medical things: Opportunities, challenges, and future directions. *Journal of Intelligent Systems, 32*(1), 20220267. doi:10.1515/jisys-2022-0267

Ammar, M., Russello, G., & Crispo, B. (2018). Internet of Things: A survey on the security of IoT frameworks. *Journal of Information Security and Applications, 38*, 8–27. doi:10.1016/j.jisa.2017.11.002

Andriansyah, M., Subali, M., Purwanto, I., Irianto, S. A., & Pramono, R. A. (2018). e-KTP as the basis of Home Security System using Arduino UNO. *4th International Conference on Computer Applications and Information Processing Technology (CAIPT).* IEEE.

Anoop, V. S., & Asharaf, S. (2022). Integrating artificial intelligence and blockchain for enabling a trusted ecosystem for healthcare sector. In *Intelligent Healthcare: Infrastructure, Algorithms and Management* (pp. 281–295). Springer Nature Singapore. doi:10.1007/978-981-16-8150-9_13

Antunes, L. (2022). IoT security certifications: Challenges and Potential Approaches. *Computers and Security, 116.* doi:10.1016/j.cose.2022.102669

Aoun, A., Ilinca, A., Ghandour, M., & Ibrahim, H. (2021). A review of Industry 4.0 characteristics and challenges, with potential improvements using blockchain technology. *Computers & Industrial Engineering, 162*, 107746. doi:10.1016/j.cie.2021.107746

Asham, A. D., Hanaa, M., Alyoubi, B., Badawood, A. M., & Alharbi, I. (2018). A Simple Integrated Smart Green Home Design. *Intelligent Systems Conference (IntelliSys),* (pp. 194–197).

Aswathy, S. U. (2021). The Future of Edge Computing with Blockchain Technology: Possibility of Threats, Opportunities and Challenges. Recent Trends in Blockchain for Information Systems Security and Privacy. CRC Press.

Atta, A., Abbas, S., Khan, M., Ahmed, G., & Farooq, U. (2018). An Adaptive Approach: Smart Traffic Congestion Control System. *Journal of King Saud University- Computer and Information Science, 32*(9), 1012-1019.

Attaran, M. (2022). Blockchain technology in healthcare: Challenges and opportunities. *International Journal of Healthcare Management, 15*(1), 70–83. doi:10.1080/20479700.2020.1843887

Avinash, P. (2023). A novel approach of unsupervised feature selection using iterative shrinking and expansion algorithm. *Journal of Interdisciplinary Mathematics, 26*(3), 519–530. doi:10.47974/JIM-1678

Azaria, A., Ekblaw, A., Vieira, T., & Lippman, A. (2016). Medrec: Using blockchain for medical data access and permission management. In *Proceedings of the 2016 2nd International Conference on Open and Big Data* (pp. 25–30). IEEE. 10.1109/OBD.2016.11

Azorin-Molina, C., Asin, J., McVicar, T. R., Minola, L., Lopez-Moreno, J. I., Vicente-Serrano, S. M., & Chen, D. (2018). Evaluating Anemometer Drift: A Statistical Approach to Correct Biases in Wind Speed Measurement. *Atmospheric Research, 203*, 175-188. doi:10.1016/j.atmosres.2017.12.010

Babu, S. M., Lakshmi, A. J., & Rao, B. T. (2015). A Study on Cloud Based Internet of Things: Cloudiot, *Global Conference on Communication Technologies (GCCT)*, Thuckalay, India. 10.1109/GCCT.2015.7342624

Bai, Y., & Lu, L. (2020). Acoustic-Based Sensing and Applications: A Survey. *Computer Networks, 181.* doi:10.1016/j.comnet.2020.107447

Bajpai, S., & Radha, D. (2019). Smart Phone as a Controlling Device for Smart Home using Speech Recognition. *IEEE International Conference on Communication and Signal Processing, ICCSP,* (pp. 701–705). IEEE. 10.1109/ICCSP.2019.8697923

Bala Anand Muthu, C. B. (2020). IOT Based Wearable Sensor for Diseases Prediction and Symptom Analysis in Healthcare Sector. *Peer-to-Peer Networking and Applications, 13*(6), 2123–2134. doi:10.100712083-019-00823-2

Bang, J., & Choi, M. J. (2019, September). Design and implementation of storage system for real-time blockchain network monitoring system. In *2019 20th Asia-Pacific Network Operations and Management Symposium (APNOMS)* (pp. 1-4). IEEE. 10.23919/APNOMS.2019.8892967

Bao, Z., Shi, W., He, D., & Chood, K.-K. R. (2018). *IoTChain: A Three-Tier Blockchain-based IoT Security Architecture.* Cryptography and Security.

Baqari, M., & Barka, E. (2020). Biometric-Based Blockchain EHR System (BBEHR). In *Proceedings of the 2020 International Wireless Communications and Mobile Computing* (pp. 2228–2234). IEEE. 10.1109/IWCMC48107.2020.9148357

Barka, E., Dahmane, S., Kerrache, C. A., Khayat, M., & Sallabi, F. (2021). STHM: A Secured and Trusted Healthcare Monitoring Architecture Using SDN and Blockchain. *Electronics, 10,* 1787.

Basodi, S., Tan, S., Song, W., & Pan, Y. (2020). Data Integrity Attack Detection in smart grid: A deep learning approach. *International Journal of Security and Networks, 15*(1), 15. doi:10.1504/IJSN.2020.106506

Baysal, M. V., Özcan-Top, Ö., & Betin-Can, A. (2023). Blockchain technology applications in the health domain: A multivocal literature review. *The Journal of Supercomputing, 79*(3), 3112–3156. doi:10.100711227-022-04772-1 PMID:36060094

Bello, O., Zeadally, S., & Badra, M. (2017). Network layer Inter-operation of Device-to-Device Communication Technologies in Internet of Things (IoT). *Ad Hoc Networks, 57.* doi:10.1016/j.adhoc.2016.06.010

Benotsmane, R., Dudás, L., & Kovács, G. (2020). Survey on artificial intelligence algorithms used in industrial robotics. *Multidiszciplináris Tudományok: A Miskolci Egyetem Közleménye, 10*(4), 194-205.

Bernieri, A., Ferrigno, L., Laracca, M., & Rasile, A. (2017). An AMR-Based Three-Phase Current Sensor for Smart Grid Applications. *IEEE Sensors Journal, 17*(23), 7704–7712. doi:10.1109/JSEN.2017.2725983

Bhatt, H., Bahuguna, R., Singh, R., Gehlot, A., Akram, S. V., Priyadarshi, N., & Twala, B. (2022). Artificial Intelligence and Robotics Led Technological Tremors: A Seismic Shift towards Digitizing the Legal Ecosystem. *Applied Sciences (Basel, Switzerland), 12*(22), 11687. doi:10.3390/app122211687

Bhavke, A., & Pai, S. (2017). Advance Automatic Toll Collection and Vehicle Detection During Collision Using RFID. *International Conference on Nascent Technologies in Engineering (ICNTE),* Vashi, India. 10.1109/ICNTE.2017.7947958

Bilancia, P., Schmidt, J., Raffaeli, R., Peruzzini, M., & Pellicciari, M. (2023). An Overview of Industrial Robots Control and Programming Approaches. *Applied Sciences (Basel, Switzerland), 13*(4), 2582. doi:10.3390/app13042582

Biswas, K., & Koo, S. (2020). Blockchain-Based Security Framework for Industrial IoT Systems. *Future Generation Computer Systems, 104,* 962–975.

Bluetooth. (n.d.). *Home.* Bluetooth. https://www.bluetooth.com/learn-about-bluetooth/tech-overview/

Boumaiza, A., & Sanfilippo, A. (2023). Peer to peer solar energy trading demonstrator blockchain-enabled. *2023 11th International Conference on Smart Grid (icSmartGrid).* 10.1109/icSmartGrid58556.2023.10171013

Bracino, A. A., Concepcion, R. S., Bedruz, R. A. R., Dadios, E. P., & Vicerra, R. R. P. (2020). Development of a Hybrid Machine Learning Model for Apple (Malus domestica) Health Detection and Disease Classification. In *2020 IEEE 12th International Conference on Humanoid, Nanotechnology, Information Technology, Communication and Control, Environment, and Management (HNICEM)* (pp. 1-6). IEEE.

Bt Aripin, N., & Othman, M. B. (2014). *Voice Control of Home Appliances using Android, Electrical Power, Electronics, Communications, Controls, and Informatics Seminar*. EECCIS.

Cachin, C. (2016). *The architecture of the hyper ledger blockchain fabric*. In *Workshop on Distributed Cryptocurrencies and Consensus Ledgers*. Chicago, IL, USA.

Caliper, H. (2019). Hyperledger Caliper Architecture. *Hyperledger.* (https://hyperledger.github.io/caliper/docs/2_Architecture) (accessed on 15 September 2020).

Camarinha-Matos, L. M., & Katkoori, S. (2020). Challenges in IoT Applications and Research. In L. M. Camarinha-Matos, G. Heijenk, S. Katkoori, & L. Strous (Eds.), *IFIP Advances in Information and Communication Technology* (Vol. 641, pp. 3–10). Springer. doi:10.1007/978-3-030-96466-5_1

Canhoto, A. I. (2021). Leveraging machine learning in the global fight against money laundering and terrorism financing: An affordances perspective. *Journal of Business Research, 131*, 441–452. doi:10.1016/j.jbusres.2020.10.012 PMID:33100427

Castiglioni, I., Rundo, L., Codari, M., Di Leo, G., Salvatore, C., Interlenghi, M., Gallivanone, F., Cozzi, A., D'Amico, N. C., & Sardanelli, F. (2021). AI applications to medical images: From machine learning to deep learning. *Physica Medica, 83*, 9–24. doi:10.1016/j.ejmp.2021.02.006 PMID:33662856

Chaka, C. (2023). Fourth industrial revolution—A review of applications, prospects, and challenges for artificial intelligence, robotics and blockchain in higher education. *Research and Practice in Technology Enhanced Learning, 18.*

Chakraborti, T., Isahagian, V., Khalaf, R., Khazaeni, Y., Muthusamy, V., Rizk, Y., & Unuvar, M. (2020). From Robotic Process Automation to Intelligent Process Automation: –Emerging Trends–.*Business Process Management: Blockchain and Robotic Process Automation Forum: BPM 2020 Blockchain and RPA Forum, Seville, Spain, September 13–18, 2020 Proceedings*, *18*, 215–228.

Chanal, P. M., & Kakkasageri, M. S. (2021). Preserving Data Confidentiality in Internet of Things. *SN Computer Science*, *02*(53), 53. doi:10.100742979-020-00429-z

Chandu, Y., Rakesh Kumar, K. S., Khanolkar, N. V., Anish, A. N., & Rawal, S. (2017). Design and Implementation of Hybrid Encryption for Security of IoT Data. *International Conference on Smart Technologies for Smart Nation (Smarttechcon'17)*, Bangalore. 10.1109/SmartTechCon.2017.8358562

Chen, G., Jin, Z., & Chen, J. (2023). A Review: Magneto-Optical Sensor Based on Magnetostrictive Materials and Magneto-Optical Material. *Sensors and Actuators Reports, 5.* doi:10.1016/j.snr.2023.100152

Chen, S., Xu, H., Liu, D., Hu, B., & Wang, H. (2014, August). A Vision of IoT: Applications, Challenges, and Opportunities With China Perspective. *IEEE Internet of Things Journal, 1*(4), 349–359. doi:10.1109/JIOT.2014.2337336

Chen, Z., Van Khoa, L. D., Teoh, E. N., Nazir, A., Karuppiah, E. K., & Lam, K. S. (2018). Machine learning techniques for anti-money laundering (AML) solutions in suspicious transaction detection: A review. *Knowledge and Information Systems, 57*(2), 245–285. doi:10.100710115-017-1144-z

Chohan, U. (2019). The fatf in the global financial architecture: challenges and implications. *Talking about terrorism.*

Chowdhry, D., Paranjape, R., & Laforge, P. (2015). Smart Home Automation System for Intrusion Detection. *IEEE 14th Canadian Workshop on Information Theory (CWIT)*, (pp. 75–78). IEEE. 10.1109/CWIT.2015.7255156

Chuang, M.-C., & Lee, J.-F. (2011). PPAS: A privacy preservation authentication scheme for vehicle-to-infrastructure communication networks. *IEEE International Conference on Consumer Electronics, Communications and Networks*. IEEE. 10.1109/CECNET.2011.5768254

Chuanlei, Z., Shanwen, Z., Jucheng, Y., Yancui, S., & Jia, C. (2017). Apple leaf disease identification using genetic algorithm and correlation based feature selection method. *International Journal of Agricultural and Biological Engineering*, *10*(2), 74–83.

Cognizant. (2016). *How the Internet of things will overcome a lack of standards*. Cognizant. https://www.cognizant.com/perspectives/how-the-internet-of-thingswill-overcome-a-lack-..........mpaign=ThoughtLeadership.

Computing. (2016). *Internet of things undermined by a lack of standards, warns Pentaho VP EMEA Paul Scholey*. Computing.Co. http://www.computing.co.uk.

Cooper, N. (2022). Blockchain in Health Care. *Frontiers in Blockchain*, *4*, 830459. doi:10.3389/fbloc.2021.830459

Cornelius, K., Kumar, N. K., Pradhan, S., Patel, P., & Vinay, N. (2020). An Efficient Tracking System for Air and Sound Pollution using IoT. *6th International Conference on Advanced Computing and Communication Systems (ICACCS)*, Coimbatore, India. 10.1109/ICACCS48705.2020.9074301

CSA. (n.d.). *Governing Documents and Support from the Alliance*. CSA. https://csa-iot.org/resources/security/

Dagher, G. G., Mohler, J., Milojkovic, M., & Marella, P. B. (2018). Ancile: Privacy-preserving framework for access control and interoperability of electronic health records using blockchain technology. *Sustainable Cities and Society*, *39*, 283–297.

Dagher, G. G., Mohler, J., Milojkovic, M., Marella, P. B., & Ancile, C. (2018). Ancile: Privacy-Preserving Framework for Access Control and Interoperability of Electronic Health Records Using Blockchain Technology. *Sustainable Cities and Society*, *39*, 283–297. doi:10.1016/j.scs.2018.02.014

Dasgupta, D., Shrein, J. M., & Gupta, K. D. (2019). A survey of blockchain from security perspective. *Journal of Banking and Financial Technology*, *3*(1), 1–17. doi:10.100742786-018-00002-6

De Angelis, S., Aniello, L., Baldoni, R., Lombardi, F., Margheri, A., & Sassone, V. (2018). PBFT vs. proof-of-authority: Applying the CAP theorem to permissioned blockchain. In *Italian Conference on Cyber Security*. Milan, Italy.

Debrah, C., Chan, A. P., & Darko, A. (2022). Artificial intelligence in green building. *Automation in Construction*, *137*, 104192. doi:10.1016/j.autcon.2022.104192

Deshmukh, A., Patil, D., & Tyagi, A. K. (2022). Recent Trends on Blockchain for Internet of Things based Applications: Open Issues and Future Trends. In *Proceedings of the 2022 Fourteenth International Conference on Contemporary Computing (IC3-2022)*. Association for Computing Machinery. 10.1145/3549206.3549289

Deshmukh, A., Sreenath, N., Tyagi, A. K., & Eswara Abhichandan, U. V. (2022). Blockchain Enabled Cyber Security: A Comprehensive Survey. *2022 International Conference on Computer Communication and Informatics (ICCCI)*, (pp. 1-6). IEEE. 10.1109/ICCCI54379.2022.9740843

Dewasiri, N. J., Karunarathne, K. S. S. N., Menon, S., Jayarathne, P. G. S. A., & Rathnasiri, M. S. H. (2023). Fusion of Artificial Intelligence and Blockchain in the Banking Industry: Current Application, Adoption, and Future Challenges. In Transformation for Sustainable Business and Management Practices: Exploring the Spectrum of Industry 5.0 (pp. 293-307). Emerald Publishing Limited.

Dhirani, L. L., Mukhtiar, N., Chowdhry, B. S., & Newe, T. (2023). Ethical dilemmas and privacy issues in emerging technologies: A review. *Sensors (Basel), 23*(3), 1151. doi:10.339023031151 PMID:36772190

Dias, P. A., Tabb, A., & Medeiros, H. (2018, October). Multispecies fruit flower detection using a refined semantic segmentation network. *IEEE Robotics and Automation Letters, 3*(4), 3003–3010. doi:10.1109/LRA.2018.2849498

Dileep, G. J. R. E. (2020). A Survey on Smart Grid Technologies and Applications. *Renewable Energy, 146*, 2589–2625. doi:10.1016/j.renene.2019.08.092

Ding, S., Shao, Z., Guo, S., & Zhu, W. (2019). IoT Security Techniques Based on Machine Learning: How Do IoT Devices Use AI to Enhance Security? *IEEE Consumer Electronics Magazine, 8*(4), 22–30.

Divatankar, S., & Hivarkar, U. N. (2019). Survey and Comparative Study of Various Approaches to Monitor the Road Traffic. *Journal of Emerging Technologies and Innovative Research, 6*(4), 248–253.

Docker, Inc. (2017). What is Docker? Docker, Inc. https://www.docker.com/what-docker

Dorri, A., Kanhere, S. S., Jurdak, R., & Gauravaram, P. (2017). Blockchain for IoT security and privacy: The case study of a smart home. *IEEE International Conference on Pervasive Computing and Communications.* IEEE. 10.1109/PERCOMW.2017.7917634

Duan, R., Chen, X., & Xing, T. (2011). *A QoS Architecture for IoT.* International Conference on Internet of Things and 4th International Conference on Cyber, Physical and Social Computing, Dalian, China. 10.1109/iThings/CPSCom.2011.125

Duc, T. L., Leiva, R. G., Casari, P., & Östberg, P. O. (2019). Machine learning methods for reliable resource provisioning in edge-cloud computing: A survey. *ACM Computing Surveys, 52*(5), 1–39. doi:10.1145/3341145

Du, J., Zhao, L., Feng, J., & Chu, X. (2017). Computation offloading and resource allocation in mixed fog/cloud computing systems with min-max fairness guarantee. *IEEE Transactions on Communications, 66*(4), 1594–1608. doi:10.1109/TCOMM.2017.2787700

Du, K. L., & Swamy, M. (2016). Search and optimization by metaheuristics. In *Techniques and Algorithms Inspired by Nature*. Birkhauser. doi:10.1007/978-3-319-41192-7

El Ouadghiri, M., Aghoutane, B., & El Farissi, N. (2020). Communication model in the Internet of Things. *Procedia Computer Science, 177*, 72-77. doi:10.1016/j.procs.2020.10.013

Ericsson Working Group. (2016). *Cellular Networks for Massive IoT.* (Ericsson White Paper). Ericcson. https://www.ericsson.com/en/white-papers/cellular-networks-for-massive-iot& ndash;enabling-low-power-wide-area-applications

Fadhil, J. A., Omar, O. A., & Sarhan, Q. I. (2020). A Survey on the Applications of Smart Home Systems. *International Conference on Computer Science and Software Engineering (CSASE)*, Duhok, Iraq. 10.1109/CSASE48920.2020.9142103

Fadi, Al-Turjman. (2019). Smart Parking in IoT-Enabled Cities: A Survey. *Sustainable Cities and Society, 49*. doi:10.1016/j.scs.2019.101608

Fahim, A., Hasan, M., & Chowdhury, M. A. (2021). *Smart Parking Systems: Comprehensive Review Based on Various Aspects, Heliyon, 7*(5). doi:10.1016/j.heliyon.2021.e07050

Fakih, A., Naik, R., Rais, U., & Maniar, S. (2018). A Green Energy System for Smart abode. *International Conference on Smart City and Emerging Technology (ICSCET),* (pp. 1–5). IEEE. 10.1109/ICSCET.2018.8537365

Falayi, A., Wang, Q., Liao, W., & Yu, W. (2023). Survey of Distributed and Decentralized IoT Securities: Approaches Using Deep Learning and Blockchain Technology. *Future Internet, 15*(5), 178. doi:10.3390/fi15050178

Fan, K., Wang, S., Ren, Y., Li, H., & Yang, Y. (2018). Medblock: Efficient and secure medical data sharing via blockchain. *Journal of Medical Systems, 42*, 136.

Fan, K., & Xu, H. (2019). Blockchain and Artificial Intelligence: Revolutionizing Cybersecurity and Cyberdefense. *IEEE Intelligent Systems, 34*(6), 92–96.

Farahani, B., Firouzi, F., & Chakrabarty, K. (2020). *Healthcare IoT, Intelligent Internet of Things*. Springer. doi:10.1007/978-3-030-30367-9_11

Fareed, M. M. S., Pradhan, B., & Bhattacharyya, S. P. (2021). Kunal, IoT-Based Applications in Healthcare Devices. *Journal of Healthcare Engineering, 2021*, 1–18. doi:10.1155/2021/6632599

Fatima, N., Agarwal, P., & Sohail, S. S. (2022). Security and Privacy Issues of Blockchain Technology in Health Care—A Review. In ICT. *Analysis and Applications*, 193–201.

Fatoum, H., Hanna, S., Halamka, J. D., Sicker, D. C., Spangenberg, P., & Hashmi, S. K. (2021). Blockchain integration with digital technology and the future of health care ecosystems: Systematic review. *Journal of Medical Internet Research, 23*(11), e19846. doi:10.2196/19846 PMID:34726603

Feng, Y., Mao, G., & Cheng, B. (2020). MagMonitor: Vehicle Speed Estimation and Vehicle Classification Through a Magnetic Sensor. *IEEE Transactions on Intelligent Transportation Systems, 23*(2), 1311–1322. doi:10.1109/TITS.2020.3024652

Fernandez-Carames, T. M., & Fraga-Lamas, P. (2019). A review on the application of blockchain to the next generation of cybersecure industry 4.0 smart factories. *IEEE Access : Practical Innovations, Open Solutions, 7*, 45201–45218. doi:10.1109/ACCESS.2019.2908780

Fernández-Caramés, T. M., & Fraga-Lamas, P. (2020). A Review on the Use of Blockchain for the Internet of Things. *IEEE Access : Practical Innovations, Open Solutions, 8*, 154076–154130.

Ferrag, M. A., Derdour, M., Mukherjee, M., Derhab, A., Maglaras, L., & Janicke, H. (2019). Blockchain Technologies for the Internet of Things: Research Issues and Challenges. *IEEE Internet of Things Journal, 6*, 2188–2204.

Firmansyah, W., & Atmadja, H. T. (2021). Juridical analysis awareness of profession advocacy to financial transaction reports and analysis centre (ppatk) during prevent and eradicate money laundering crime. *Journal of Multidisciplinary Academic, 5*(4), 308–314.

Firouzi, F., Farahani, B., Daneshmand, M., Grise, K., Song, J., Saracco, R., Wang, L. L., Lo, K., Angelov, P., Soares, E., Loh, P.-S., Talebpour, Z., Moradi, R., Goodarzi, M., Ashraf, H., Talebpour, M., Talebpour, A., Romeo, L., Das, R., & Luo, A. (2021). Harnessing the power of smart and connected health to tackle COVID-19: IoT, AI, robotics, and blockchain for a better world. *IEEE Internet of Things Journal, 8*(16), 12826–12846. doi:10.1109/JIOT.2021.3073904 PMID:35782886

Floris, A., Girau, R., Porcu, S., Pettorru, G., & Atzori, L. (2020). *Implementation of a Magnetometer based Vehicle Detection System for Smart Parking* applications, IEEE International Smart Cities Conference (ISC2), Piscataway, NJ, USA. 10.1109/ISC251055.2020.9239005

Fongen, A. (2012). Management and Integrity Protection in the Internet of Things. *IEEE International Conference on Emerging Security Technologies,* (pp. 111-114). IEEE.

Fontes, J. (2005). CHAPTER 12 - Humidity Sensors. In J. S. Wilson (Ed.), *Sensor Technology Handbook* (pp. 271–284). Newnes. doi:10.1016/B978-075067729-5/50052-5

Fox, S., & Griffy-Brown, C. (2023). Robotics in society: Technology in Society Briefing. *Technology in Society, 72*, 102174. doi:10.1016/j.techsoc.2022.102174

Frangoudes, F., Hadjiaros, M., Schiza, E. C., Matsangidou, M., Tsivitanidou, O., & Neokleous, K. 2021, July. An overview of the use of chatbots in medical and healthcare education. In *International Conference on Human-Computer Interaction* (pp. 170-184). Cham: Springer International Publishing. 10.1007/978-3-030-77943-6_11

French, A., Shim, J. P., Risius, M., Larsen, K. R., & Jain, H. (2021). The 4th Industrial Revolution powered by the integration of AI, blockchain, and 5G. *Communications of the Association for Information Systems*, *49*(1), 6. doi:10.17705/1CAIS.04910

Fulsoundar, K., Kadlag, T., Bhadale, S., & Bharvirkar, P. (2014). Detection and classification of plant leaf diseases. *International Journal of Engineering Research and General Science, 2*(6).

Fuzi, M. F. M., Ibrahim, A. F., Ismail, M. H., & Halim, N. S. A. (2014). HOME FADS: A Dedicated Fire Alert Detection System using Zigbee Wireless Network. *IEEE 5th Control and System Graduate Research Colloquium,* (pp. 53–58). IEEE. 10.1109/ICSGRC.2014.6908695

Gaikwad, S. S., Rumma, S. S., & Hangarge, M. (2022). Fungi affected fruit leaf disease classification using deep CNN architecture. *International Journal of Information Technology : an Official Journal of Bharati Vidyapeeth's Institute of Computer Applications and Management*, *14*(7), 1–10. doi:10.100741870-022-00860-w

Ganatra, N., & Patel, A. (2018). A Survey on Diseases Detection and Classification of Agriculture Products using Image Processing and Machine Learning. *International Journal of Computer Applications, 180*(13), 7–12. doi:10.5120/ijca2018916249

García, L., Parra, L., Jimenez, J. M., Lloret, J., & Lorenz, P. (2020). IoT-based Smart Irrigation Systems: An Overview on the Recent Trends on Sensors and IoT Systems for Irrigation in Precision Agriculture. *Sensors (Basel)*, *20*(4), 1042. doi:10.339020041042 PMID:32075172

Gatteschi, V., Lamberti, F., & Demartini, C. (2020). Blockchain technology use cases. *Advanced applications of Blockchain technology*, (pp. 91-114). IEEE.

Ghosh, P., & Dhar, P. K. (2019). GSM Based Low-cost Gas Leakage, Explosion and Fire Alert System with Advanced Security. *International Conference on Electrical, Computer and Communication Engineering (ECCE)*, (pp. 1–5). IEEE. 10.1109/ECACE.2019.8679411

Gigli, M., & Koo, S. (2011). Internet of Things: Services and Applications Categorization. *Advances in Internet of Things*, *1*(2), 27–31. doi:10.4236/ait.2011.12004

Gite, P. (2023). A novel RF-SMOTE model to enhance the definite apprehensions for IoT security attacks. *Journal of Discrete Mathematical Sciences and Cryptography*, *26*(3), 861–873. doi:10.47974/JDMSC-1766

Golosova, J., & Romanovs, A. (2018). The advantages and disadvantages of the blockchain technology. *2018 IEEE 6th workshop on advances in information, electronic and electrical engineering (AIEEE)*. IEEE. 10.1109/AIEEE.2018.8592253

Govindan, K., Nasr, A. K., Saeed Heidary, M., Nosrati-Abarghooee, S., & Mina, H. (2023). Prioritizing adoption barriers of platforms based on blockchain technology from balanced scorecard perspectives in healthcare industry: A structural approach. *International Journal of Production Research*, *61*(11), 3512–3526. doi:10.1080/00207543.2021.2013560

Grau, M. (2017). Industrial robotics in factory automation: From the early stage to the Internet of Things. *IECON 2017 - 43rd Annual Conference of the IEEE Industrial Electronics Society*, (pp. 6159-6164). IEEE.

Grau, A., Indri, M., Lo Bello, L., & Sauter, T. (2017). Industrial robotics in factory automation: From the early stage to the Internet of Things. *Proc. IECON 2017–43rd Annu. Conf. IEEE Ind. Electron. Soc.*, (pp. 6159-6164). IEEE. 10.1109/IECON.2017.8217070

Grau, M. I., Bello, L. L., & Sauter, T. (2021). Robots in Industry: The Past, Present, and Future of a Growing Collaboration With Humans. *IEEE Industrial Electronics Magazine*, *15*(1), 50–61. doi:10.1109/MIE.2020.3008136

Gubbi, J., Buyya, R., Marusic, S., & Palaniswami, M. (2013). Internet of things (IoT): A Vision, Architectural Elements, and Future Directions. *Future Generation Computer Systems*, *29*(7), 1645–1660. doi:10.1016/j.future.2013.01.010

Guerbouj, S. S. E., Gharsellaoui, H., & Bouamama, S. (2019). A Comprehensive Survey on Privacy and Security Issues in Cloud Computing, Internet of Things and Cloud of Things. *International Journal of Service Science, Management, Engineering, and Technology, 10,* 32–44.

Guerrero-Ibáñez, J., Zeadally, S., & Contreras-Castillo, J. (2018). Sensor Technologies for Intelligent Transportation Systems,. *Sensors (Basel)*, *18*(4), 1–24. doi:10.339018041212

Gunatilaka, D. (2021). *An IoT-Enabled Acoustic Sensing Platform for Noise Pollution Monitoring.* IEEE 12th Annual Ubiquitous Computing, Electronics & Mobile Communication Conference (UEMCON), New York, NY, USA. doi:10.1109/UEMCON53757.2021.9666534

Gupta, R., Shukla, A., & Tanwar, S. (2020). BATS: A blockchain and AI-empowered drone-assisted telesurgery system towards 6G. *IEEE Transactions on Network Science and Engineering*, *8*(4), 2958–2967. doi:10.1109/TNSE.2020.3043262

Gupta, V., & Sehgal, V. (2019). Blockchain and Artificial Intelligence for Secure and Privacy-Preserving Industrial IoT. *Future Generation Computer Systems*, *97*, 624–637.

Haleem, A., Javaid, M., Singh, R. P., Suman, R., & Rab, S. (2021). Blockchain technology applications in healthcare: An overview. *International Journal of Intelligent Networks*, *2*, 130–139. doi:10.1016/j.ijin.2021.09.005

Hammoud, A., Sami, H., Mourad, A., Otrok, H., Mizouni, R., & Bentahar, J. (2020). AI, blockchain, and vehicular edge computing for smart and secure IoV: Challenges and directions. *IEEE Internet of Things Magazine*, *3*(2), 68–73. doi:10.1109/IOTM.0001.1900109

Hanes, D., Salgueiro, G., Grossetete, P., Barton, R., & Henry, J. (2017). *IoT Fundamentals: Networking Technologies, Protocols, and Use Cases for the Internet-of-Things*. Cisco Press.

Hannan, S. A. (2023). Challenges of Blockchain Technology using Artificial Intelligence in Healthcare System. *International Journal of Innovative Research in Science, Engineering and Technology*, *12*(01), 64–74.

Hans, S., Chohan, J. S., Kumar, R., & Pareek, S. 2023, March, Artificial Intelligence with Robotics-A Bibliometric Study, In *2023 10th International Conference on Computing for Sustainable Global Development (INDIACom)*. IEEE.

Hasan, M. K., Ghazal, T. M., Saeed, R. A., Pandey, B., Gohel, H., Eshmawi, A. A., Abdel-Khalek, S., & Alkhassawneh, H. M. (2022). A review on security threats, vulnerabilities, and counter measures of 5G enabled Internet-of-Medical-Things. *IET Communications*, *16*(5), 421–432. doi:10.1049/cmu2.12301

Hasan, M., Anik, M. H., & Islam, S. (2018). *Microcontroller Based Smart Home System with Enhanced Appliance Switching Capacity, 5th HCT Information Technology Trends.* ITT.

HasnainM.HussainA. Role of Chatgpt to Reduce the High Overhead Issues of Blockchain Technology in Healthcare. SSRN 4502924.

Hathaliya, J., Sharma, P., Tanwar, S., & Gupta, R. (2019, December). Blockchain-based remote patient monitoring in healthcare 4.0. In *2019 IEEE 9th international conference on advanced computing (IACC)* (pp. 87-91). IEEE.

Hegde, N. & Manvi, S. (2019). *A novel key management protocol for vehicular cloud security.* TELKOMNIKA (Telecommunication Computing Electronics and Control).

Hegde, N. (2016). Thesis Proposal Summary: Key Management Authentication and Non Repudiation for Information Transaction in Vehicular Cloud Environments. *2016 IEEE International Conference on Cloud Computing in Emerging Markets (CCEM)*, (pp. 157-160). IEEE.

Hegde, N. (2023). Secure search scheme for encrypted data in the VANET cloud with random query trapdoor. Information Security Journal: A Global Perspective, 32(5), 383-400.

Hegde, P. and Maddikunta, P.K.R., 2023. Amalgamation of Blockchain with resource-constrained IoT devices for Healthcare applications–State of Art, Challenges and Future Directions. *International Journal of Cognitive Computing in Engineering*.

Hegde, N., & Manvi, S. (2022). Distributed integrity and non-repudiation scheme in the dynamic vehicular cloud environment. *International Journal of Information and Computer Security*, *1*(1), 15.

Hegde, N., & Manvi, S. S. (2016). Implementation of VANET routing using computational intelligence. *2016 International Conference on Communication and Signal Processing (ICCSP)*, Melmaruvathur, India. 10.1109/ICCSP.2016.7754292

Hemdan, E. E. D., El-Shafai, W., & Sayed, A. (2023). Integrating Digital Twins with IoT-Based Blockchain: Concept, Architecture, Challenges, and Future Scope. *Wireless Personal Communications*, *131*(3), 1–24. doi:10.100711277-023-10538-6 PMID:37360142

He, Y., Wang, Y., Qiu, C., Lin, Q., Li, J., & Ming, Z. (2020). Blockchain-based edge computing resource allocation in IoT: A deep reinforcement learning approach. *IEEE Internet of Things Journal*, *8*(4), 2226–2237. doi:10.1109/JIOT.2020.3035437

Ho, G., Leung, D., Mishra, P., Hosseini, A., Song, D., & Wagner, D. (2016). Smart locks: Lessons for securing commodity Internet of things devices. In *Proceedings of the 11th ACM on Asia Conference on Computer and Communications Security* (pp. 461–472). ACM. 10.1145/2897845.2897886

Hornung, O., & Smolnik, S. (2022). AI invading the workplace: Negative emotions towards the organizational use of personal virtual assistants. *Electronic Markets*, *32*(1), 1–16. doi:10.100712525-021-00493-0

Huang, Y. C., & Huang, S. C. C. (2014). A Personalized Smart Living Room. Lecture Notes in Computer Science. Springer. doi:10.1007/978-3-319-07788-8_4

Huang, J., Kong, L., Chen, G., Wu, M.-Y., Liu, X., & Zeng, P. (2019). Towards Secure Industrial IoT: Blockchain System With Credit-Based Consensus Mechanism. *IEEE Transactions on Industrial Informatics*, *15*(6), 3680–3689. doi:10.1109/TII.2019.2903342

Huang, Y., Zhang, J., Duan, J., Xiao, B., Ye, F., & Yang, Y. (2021). Resource allocation and consensus of blockchains in pervasive edge computing environments. *IEEE Transactions on Mobile Computing*, *21*(9), 3298–3311. doi:10.1109/TMC.2021.3053230

Hua, W., Chen, Y., Qadrdan, M., Jiang, J., Sun, H., & Wu, J. (2022). Applications of blockchain and Artificial Intelligence Technologies for enabling prosumers in smart grids: A Review. *Renewable & Sustainable Energy Reviews*, *161*, 112308. doi:10.1016/j.rser.2022.112308

Humayun, M., Afsar, S., Almufareh, M. F., Jhanjhi, N. Z., & AlSuwailem, M. (2022). N. Z. Jhanjhi, Mashayel AlSuwailem, Smart Traffic Management System for Metropolitan Cities of Kingdom Using Cutting Edge Technologies, Journal of Advanced Transportation, Vol. 2022. *Journal of Advanced Transportation*, *4687319*, 1–13. doi:10.1155/2022/4687319

Hussain, A. A., & Al-Turjman, F. (2021). Artificial intelligence and blockchain: A review. *Transactions on Emerging Telecommunications Technologies*, *32*(9), e4268. doi:10.1002/ett.4268

Hussien, H. M., Yasin, S. M., Udzir, N. I., Ninggal, M. I. H., & Salman, S. (2021). Blockchain technology in the healthcare industry: Trends and opportunities. *Journal of Industrial Information Integration*, *22*, 100217. doi:10.1016/j.jii.2021.100217

Hyperledger Blockchain. (2018). *Performance Metrics.* (White paper, Vol. 1). https://www.hyperledger.org/wpcontent/uploads/2018/10/HL_Whitepaper_Metrics_PDF_V1

IBM Think Academy. (2015). *How it works: Internet of things.* [Video]. Youtube. https://www.youtube.com/watch?v=QSIPNhOiMoE

Idris, M., Tamil, E. M., Noor, N. M., Razak, Z., & Fong, K. W. (2009). Parking Guidance System Utilizing Wireless Sensor Network and Ultrasonic Sensor. *Information Technology Journal*, *8*(2), 138–146. doi:10.3923/itj.2009.138.146

Imario, A., Sudiharto, D. W., & Ariyanto, E. (2018). The Validated Voice Recognition Measurement Of Several Tribes In Indonesia Using Easy VR 3.0. Case Study: The Prototype Of Automated Doors. *International Seminar on Application for Technology of Information and Communication (iSemantic).* IEEE.

Islam, M. A. (2023). *AI & Blockchain as sustainable teaching and learning tools to cope with the 4IR.* arXiv preprint arXiv:2305.01088.

Islam, M. J., Rahman, A., Kabir, S., Karim, M. R., Acharjee, U. K., Nasir, M. K., Band, S. S., Sookhak, M., & Wu, S. (2021). Blockchain-SDN-based energy-aware and distributed secure architecture for IoT in smart cities. *IEEE Internet of Things Journal*, *9*(5), 3850–3864. doi:10.1109/JIOT.2021.3100797

Ismail, Y. (2019). I*nternet of Things (IoT) for Automated and Smart Applications.* IntechOpen, London. http://dx.doi.org/ doi:10.5772/intechopen/77404

Ismail, W. H. W., Husny, H. R. M., & Abdullah, N. Y. (2016). Smoke Detection Alert System via Mobile Application. *10th International Conference on Ubiquitous Information Management and Communication (IMCOM)*, (pp. 1–6). ACM. 10.1145/2857546.2857581

Jadhav, S. B., Udup, V. R., & Patil, S. B. (2019). Soybean leaf disease detection and severity measurement using multiclass SVM and KNN classifier. *Iranian Journal of Electrical and Computer Engineering*, *9*(5), 4092. doi:10.11591/ijece.v9i5.pp4077-4091

Jain, N. K., Saini, R. K., & Mittal, P. (2019). A Review on Traffic Monitoring System Techniques. In K. Ray, T. Sharma, S. Rawat, R. Saini, & A. Bandyopadhyay (Eds.), *Soft Computing: Theories and Applications. Advances in Intelligent Systems and Computing* (Vol. 742, pp. 569–577). Springer., doi:10.1007/978-981-13-0589-4_53

Javaid, M., & Khan, I. H. (2021). Internet of Things (IoT) Enabled Healthcare Helps to Take the Challenges of COVID-19 Pandemic. *Journal of Oral Biology and Craniofacial Research, 11*(2), 209-214. doi:10.1016/j.jobcr.2021.01.015

Javaid, M., Haleem, A., Singh, R. P., Khan, S., & Suman, R. (2021). Blockchain technology applications for Industry 4.0: A literature-based review. *Blockchain: Research and Applications*, *2*(4), 100027.

Jayaprakash, V., & Tyagi, A. K. (2022). Security Optimization of Resource-Constrained Internet of Healthcare Things (IoHT) Devices Using Asymmetric Cryptography for Blockchain Network. In: Giri, D., Mandal, J.K., Sakurai, K., De, D. (eds) *Proceedings of International Conference on Network Security and Blockchain Technology.* Springer, Singapore. 10.1007/978-981-19-3182-6_18

Jiménez-Luna, J., Grisoni, F., Weskamp, N., & Schneider, G. (2021). Artificial intelligence in drug discovery: Recent advances and future perspectives. *Expert Opinion on Drug Discovery*, *16*(9), 949–959. doi:10.1080/17460441.2021.1909567 PMID:33779453

Joglekar, J., Bhutani, S., Patel, N., & Soman, P. (2019). Lightweight Elliptical Curve Cryptography (ECC) for Data Integrity and User Authentication in Smart Transportation IoT Systems. In Sustainable Communication Networks and Application (pp. 270–278).

José, M. (2018). Environmental Perception For Intelligent Vehicles. Intelligent Vehicles: Enabling Technologies and Future Developments (pp. 23–101). Elsevier.

Juhana, T., & Anggraini, V. G. (2016). Design and Implementation of Smart Home Surveillance System, *10th International Conference on Telecommunication Systems Services and Applications (TSSA)*, (pp. 1-5). IEEE. 10.1109/TSSA.2016.7871101

Jullum, M., Løland, A., Huseby, R. B., Ånonsen, G., & Lorentzen, J. (2020). Detecting money laundering transactions with machine learning. *Journal of Money Laundering Control, 23*(1), 173–186. doi:10.1108/JMLC-07-2019-0055

Junaidi, N., Abdullah, M. P., Alharbi, B., & Shaaban, M. (2023). Blockchain-based management of demand response in electric energy grids: A systematic review. *Energy Reports, 9*, 5075–5100. doi:10.1016/j.egyr.2023.04.020

Kakkar, L., Gupta, D., Saxena, S., & Tanwar, S. (2021). IoT Architectures and Its Security: A Review. In: Goyal, D., Gupta, A.K., Piuri, V., Ganzha, M., Paprzycki, M. (eds.) *Proceedings of the Second International Conference on Information Management and Machine Intelligence.* Springer, Singapore. 10.1007/978-981-15-9689-6_10

Kalla, A. (2019). Pawani Prombage, Madhusanka Liyanage, Introduction to IoT. In M. Liyanage, A. Braeken, P. Kumar, & M. Ylianttila (Eds.), *IoT Security: Advances in Authentication* (pp. 1–25). Wiley. doi:10.1002/9781119527978.ch1

Kamble, S. J., & Kounte, M. R. (2018). Enabling Technologies for Internet of Vehicles. *International conference on Computer Networks, Big data and IoT*. Springer, Cham.

Kamble, S. J., & Kounte, M. R. (2023). Trends and Open Research Issues in Intelligent Internet of Vehicles. *Transport and Telecommunication Journal, 24*(2), 143–157. doi:10.2478/ttj-2023-0013

Kamelia, L., Effendi, M. R., & Pratama, D. F. (2018). Integrated Smart House Security System Using Sensors and RFID. *4th International Conference on Wireless and Telematics (ICWT)*, (Vol. 91, pp. 1–5). IEEE. 10.1109/ICWT.2018.8527803

Kamruzzaman, J., Wang, G., Karmakar, G., Ahmad, I., & Md, Z. A. B. (2018). Acoustic sensor networks in the Internet of Things applications. *Future Generation Computer Systems, 86,* 1167-1169. doi:10.1016/j.future.2018.05.019

Kapassa, E., Themistocleous, M., Christodoulou, K., & Iosif, E. (2021). Blockchain application in internet of vehicles: Challenges, contributions and current limitations. *Future Internet, 13*(12), 313. doi:10.3390/fi13120313

Kasgar, A. K., Dhariwal, M. K., Tantubay, N., & Malviya, H. (2013). A Review Paper of Message Digest 5 (MD5). *International Journal of Modern Engineering and Management Research, 1*(4).

Kashani, M. H., Madanipour, M., Nikravan, M., Asghari, P., & Mahdipour, E. (2021). A Systematic Review of IoT In Healthcare: Applications, Techniques, and Trends. *Journal of Network and Computer Applications, 192*. doi:10.1016/j.jnca.2021.103164

Kashem, M. A., Shamsuddoha, M., Nasir, T., & Chowdhury, A. A. (2022). The role of artificial intelligence and blockchain technologies in sustainable tourism in the Middle East. Worldwide Hospitality and Tourism Themes.

Kaur, A., Saini, S. S., Singh, L., Sharma, A., & Sidhu, E. (2016). Efficient Arduino UNO Driven Smart Highway/Bridge/Tunnel Lighting System Employing Rochelle Piezoelectric Sensor. *International Conference on Control, Computing, Communication and Materials (ICCCCM)*, Allahbad, India. 10.1109/ICCCCM.2016.7918247

Kaur, N., & Sood, S. K. (2017). An Energy-Efficient Architecture for the Internet of Things (IoT). *IEEE Systems Journal, 11*(2), 796–805. doi:10.1109/JSYST.2015.2469676

Kaushik, K., & Kumar, A. (2023). Demystifying quantum blockchain for healthcare. *Security and Privacy, 6*(3), e284. doi:10.1002py2.284

Kavitha, R., & Prasad, K. (2023). *A Novel Method of Identification of Delirium in Patients from Electronic Health Records Using Machine Learning*. 2023 World Conference on Communication & Computing (WCONF), Raipur, India. 10.1109/WCONF58270.2023.10235245

Kavitha, S., & Alphonse, P. J. A. (2019). A Hybrid Cryptosystem to Enhance Security in IoT Health Care System. *International Journal of Wireless and Microwave Technologies, 9*(1), 1–10.

Kesavan, G., Sanjeevi, P., & Viswanathan, P. (2016). A 24 Hour IoT Framework for Monitoring and Managing Home Automation *International Conference on Inventive Computation Technologies (ICICT),* (Vol. 1, pp. 1–5). IEEE. 10.1109/INVENTIVE.2016.7823205

Khan, M. A., & Salah, K. (2018). IoT security: Review, blockchain solutions, and open challenges. *Future Generation Computer Systems, 82,* 395–411.

Khanfar, A. A., Iranmanesh, M., Ghobakhloo, M., Senali, M. G., & Fathi, M. (2021). Applications of blockchain technology in sustainable manufacturing and supply chain management: A systematic review. *Sustainability (Basel), 13*(14), 7870. doi:10.3390u13147870

Khan, H., & Masood, T. (2022). Impact of blockchain technology on Smart Grids. *Energies, 15*(19), 7189. doi:10.3390/en15197189

Khan, M. A., & Khaled, S. (2018). IoT security: Review, Blockchain Solutions, And Open Challenges. *Future Generation Computer Systems, 82*(May), 395–411. doi:10.1016/j.future.2017.11.022

Khan, M. A., Lali, M. I. U., Sharif, M., Javed, K., Aurangzeb, K., Haider, S. I., Altamrah, A. S., & Akram, T. (2019). An Optimized Method for Segmentation and Classification of Apple Diseases Based on Strong Correlation and Genetic Algorithm Based Feature Selection. *IEEE Access : Practical Innovations, Open Solutions, 7*, 46261–46277. doi:10.1109/ACCESS.2019.2908040

Khan, M. A., Saleh, A. M., Waseem, M., & Sajjad, I. A. (2023). Artificial intelligence enabled demand response: Prospects and challenges in Smart Grid Environment. *IEEE Access : Practical Innovations, Open Solutions, 11*, 1477–1505. doi:10.1109/ACCESS.2022.3231444

Khanna, A., & Kaur, S. (2020). Internet of Things (IoT), Applications and Challenges: A Comprehensive Review. *Wireless Personal Communications, 114*(2), 1–76. doi:10.100711277-020-07446-4

Khattab, A., & Youssry, N. (2020). Machine Learning for IoT Systems. In M. Alam, K. Shakil, & S. Khan (Eds.), *Internet of Things*. Springer. doi:10.1007/978-3-030-37468-6_6

Khujamatov, K. E., Toshtemirov, T. K., & Lazarev, A. P. (2021). IoT and 5G Technology In Agriculture. *International Conference on Information Science and Communications Technologies (ICISCT)*, Tashkent, Uzbekistan. 10.1109/ICISCT52966.2021.9670037

Kiiski, T. (2020). *Voice Games: The History of Voice Interaction in Digital Games*.

Kory-Westlund, J. M. (2023). Implications of children's social, emotional, and relational interactions with robots for human–robot empathy. *Conversations on Empathy: Interdisciplinary Perspectives on Empathy, Imagination and Othering*, (pp. 256-279). Taylor & Francis.

Kousar, H., Sundara Rajulu Navaneethakrishnan, D. D. S. H., & Tejaswi Vuyyuru, D. D. A. M. (2023). Integrating Artificial Intelligence and Blockchain Technology for Secure and Efficient Data Sharing. *Journal of Survey in Fisheries Sciences, 10*(1S), 5900–5911.

Krishna, A. M., & Tyagi, A. K. (2020). Intrusion Detection in Intelligent Transportation System and its Applications using Blockchain Technology. *2020 International Conference on Emerging Trends in Information Technology and Engineering.* IEEE. 10.1109/ic-ETITE47903.2020.332

Krishnendhu, S. P., & Mohandas, P. (2023). SAD: Sensor-based Anomaly Detection System for Smart Junctions. *IEEE Sensors Journal, 23*(17), 20368–20378. doi:10.1109/JSEN.2023.3297205

Kumar, A. (2022, October). A Broad Survey on AI Integration in Blockchain: A Forward-Looking Approach. In *Proceedings of the National Conference on Recent Trends of Engineering & Technologies,(RTET-2022) Ramgovind Group of Colleges, Koderma, Jharkhand, India* (pp. 5-7). Academic Press.

Kumar, J. & Patel, D. (2014). A Survey on Internet of Things: Security and Privacy Issues. *International Journal of Computer Applications, 90(*11).

Kumar, R, Singh, D., Srinivasan, K. & Hu, Y.C. (2022, December. AI-powered blockchain technology for public health: A contemporary review, open challenges, and future research directions. In Healthcare (Vol. 11, No. 1, p. 81). MDPI.

Kumar, S., & Saha, G. K. (2020). *Blockchain Technology Testing Challenges and Approaches.* Academic Press.

Kumar, T., Ramani, V., Ahmad, I., Braeken, A., Harjula, E., & Ylianttila, M. 2018, September. Blockchain utilization in healthcare: Key requirements and challenges. In *2018 IEEE 20th International conference on e-health networking, applications and services (Healthcom)* (pp. 1-7). IEEE.

Kumar, V., Pant, B., & Elkady, G. (2022*). Examining the Role of Block Chain to Secure Identity in IOT for Industry 4.0.* 2022 5th International Conference on Contemporary Computing and Informatics (IC3I), Uttar Pradesh, India. 10.1109/IC3I56241.2022.10072516

Kumar, G., Saha, R., Buchanan, W. J., Geetha, G., Thomas, R., Rai, M. K., Kim, T.-H., & Alazab, M. (2020). Decentralized accessibility of e-commerce products through blockchain technology. *Sustainable Cities and Society, 62*, 102361. doi:10.1016/j.scs.2020.102361

Kumari, A., Patel, M. M., Shukla, A., Tanwar, S., Kumar, N., & Rodrigues, J. J. (2020). Armor: A data analytics scheme to identify malicious behaviors on blockchain-based Smart Grid System. *GLOBECOM 2020 - 2020 IEEE Global Communications Conference.* IEEE. https://www.powerelectronicsnews.com/smart-grid-enables-quantum-improvement-in-energy-management-efficiency doi:10.1109/GLOBECOM42002.2020.9348061

Kumar, P., Suresh, A., Anbarasu, V., Anandaraj, S. P., & Udayakumar, S. (2022). A decentralized secured grid integration system using APEBC technique with multi access AI framework. *Sustainable Computing : Informatics and Systems, 35*, 100777. doi:10.1016/j.suscom.2022.100777

Kumar, R. L., Khan, F., Kadry, S., & Rho, S. (2022). A survey on blockchain for industrial internet of things. *Alexandria Engineering Journal, 61*(8), 6001–6022. doi:10.1016/j.aej.2021.11.023

Kumar, R. L., Wang, Y., Poongodi, T., & Imoize, A. L. (Eds.). (2021). *Internet of Things, artificial intelligence and blockchain technology.* Springer. doi:10.1007/978-3-030-74150-1

Kute, D. V., Pradhan, B., Shukla, N., & Alamri, A. (2021). Deep learning and explainable artificial intelligence techniques applied for detecting money laundering–a critical review. *IEEE Access : Practical Innovations, Open Solutions, 9*, 82300–82317. doi:10.1109/ACCESS.2021.3086230

Lakshmikantha, V., Hiriyannagowda, A., Manjunath, A., Patted, A., Basavaiah, J., & Anthony, A. A. (2021). IoT Based Smart Water Quality Monitoring System. *Global Transitions Proceedings*. Science Direct. doi:10.1016/j.gltp.2021.08.062

Lartigue, J. W., McKinney, C., Phelps, R., Rhodes, R., Rice, A. D., & Ryder, A. (2014). A Tablet-Controlled, Mesh-Network Security System. *ACM Southeast Regional Conference.* ACM. 10.1145/2638404.2638500

Latif, S., Idrees, Z., Ahmad, J., Zheng, L., & Zou, Z. (2021). A blockchain-based architecture for secure and trustworthy operations in the industrial Internet of Things. *Journal of Industrial Information Integration, 21*, 100190. doi:10.1016/j.jii.2020.100190

Latif, S., Idrees, Z., Huma, Z., & Ahmad, J. (2021). Blockchain technology for the industrial Internet of Things: A comprehensive survey on security challenges, architectures, applications, and future research directions. *Transactions on Emerging Telecommunications Technologies, 32*(11), e4337. doi:10.1002/ett.4337

Lee, D., & Yoon, S. N. (2021). Application of artificial intelligence-based technologies in the healthcare industry: Opportunities and challenges. *International Journal of Environmental Research and Public Health, 18*(1), 271. doi:10.3390/ijerph18010271 PMID:33401373

Lee, W.-H., & Chiu, C.-Y. (2020). Design and Implementation of a Smart Traffic Signal Control System for Smart City Applications. *Sensors (Basel), 20*(2), 508–526. doi:10.339020020508 PMID:31963229

Lentzsch, C., Shah, S. J., Andow, B., Degeling, M., Das, A., & Enck, W. (2021). Hey Alexa, is this skill safe?: Taking a closer look at the Alexa skill ecosystem. *Network and Distributed Systems Security (NDSS) Symposium.* Springer.

Leung, D. Y. C. (2015). Outdoor-Indoor Air Pollution in Urban Environment: Challenges and Opportunity. *Frontiers in Environmental Science, 2*(69), 1–7. doi:10.3389/fenvs.2014.00069

Lian, K.-Y., Hsiao, S. J., & Sung, W. T. (2014). Smart Home Safety Handwriting Pattern Recognition with Innovative Technology. *Computers & Electrical Engineering, 40*(4), 1123–1142. doi:10.1016/j.compeleceng.2014.02.010

Li, H., Pei, L., Liao, D., Sun, G., & Xu, D. (2019). Blockchain meets VANET: An architecture for identity and location privacy protection in VANET. *Peer-to-Peer Networking and Applications, 12*(5), 1178–1193. doi:10.100712083-019-00786-4

Lin, C., He, D., Huang, X., Choo, K.-K. R., & Vasilakos, A. V. (2018). DebiaoHe, Xinyi Huang, Kim-Kwang Raymond Choo, Athanasios V. Vasilakos, "BSeIn: A blockchain-based secure mutual authentication with fine-grained access control system for industry 4.0". *Journal of Network and Computer Applications, 116*(15), 42–52. doi:10.1016/j.jnca.2018.05.005

Lin, H.-E., Zito, R., & Taylor, M. (2005). A review of travel-time prediction in transport and logistics. *Proceedings of the Eastern Asia Society for Transportation Studies, 5*, 1433–1448.

Lin, Y., Du, H., Niyato, D., Nie, J., Zhang, J., Cheng, Y., & Yang, Z. (2023). Blockchain-aided secure semantic communication for ai-generated content in metaverse. *IEEE Open Journal of the Computer Society, 4*, 72–83. doi:10.1109/OJCS.2023.3260732

Li, S., Xu, L. D., & Zhao, S. (2015). The Internet of Things: A Survey. *Information Systems Frontiers, 17*(2), 243–259. doi:10.100710796-014-9492-7

Liu, J., Wu, J., & Sun, L. (2020). Control Method of Urban Intelligent Parking Guidance System Based on Internet of Things. *Computer Communications, 153*, 279-285. doi:10.1016/j.comcom.2020.01.063

Liu, J., & Wang, X. (2021). Plant diseases and pests detection based on deep learning: A review. *Plant Methods, 17*(1), 22. doi:10.118613007-021-00722-9 PMID:33627131

Li, X., & Li, S. (2022). Transformer Help CNN See Better: A Lightweight Hybrid Apple Disease Identification Model Based on Transformers. *Agriculture*, *12*(6), 884. doi:10.3390/agriculture12060884

Lokanan, M. E. (2022). Predicting money laundering using machine learning and artificial neural networks algorithms in banks. *Journal of Applied Security Research*, 1–25. doi:10.1080/19361610.2022.2114744

Lopes, V., & Alexandre, L. A. (2018). *An overview of blockchain integration with robotics and artificial intelligence.* arXiv preprint arXiv:1810.00329

Lopes, V., & Alexandre, L. A. (2018). *An overview of blockchain integration with robotics and artificial intelligence.* arXiv preprint arXiv:1810.00329.

Lopes, V., Alexandre, L. A., & Pereira, N. (2019). *Controlling robots using artificial intelligence and a consortium blockchain.* arXiv preprint arXiv:1903.00660

Lorenz, J. (2020). Machine learning methods to detect money laundering in the bitcoin blockchain in the presence of label scarcity. *Proceedings of the first ACM international conference on AI in finance*. IEEE. 10.1145/3383455.3422549

Lu, Z., Wang, Q., Qu, G., & Liu, Z. (2018). BARS: A Blockchain-Based Anonymous Reputation System for Trust Management in VANETs. IEEE International Conference On Big Data Science And Engineering. IEEE. 10.1109/TrustCom/BigDataSE.2018.00025

Luan, F., Yang, X., Chen, Y., & Regis, P. J. (2022). Industrial robots and air environment: A moderated mediation model of population density and energy consumption. *Sustainable Production and Consumption*, *30*, 870–888. doi:10.1016/j.spc.2022.01.015

Lu, Y., Shujuan, N. Z., Liu, Y., & Zhang, Y. (2017). *Identification of Rice diseases using Deep Convolutional Neural Networks*. Elsevier Journal on Neurocomputing. doi:10.1016/j.neucom.2017.06.023

Lu, Z., Liu, W., Wang, Q., Qu, G., & Liu, Z. (2018). A Privacy-Preserving Trust Model Based on Blockchain for VANETs. *IEEE Access : Practical Innovations, Open Solutions*, *6*, 45655–45665. doi:10.1109/ACCESS.2018.2864189

Magyar, G. (2017). Blockchain: Solving the privacy and research availability tradeoff for EHR data: A new disruptive technology in health data management. In *Proceedings of the 2017 IEEE 30th Neumann Colloquium* (pp. 135–140). IEEE. 10.1109/NC.2017.8263269

Mahmoud, R., Yousuf, T., Aloul, F., & Zualkernan, I. (2015). Internet of things (IoT) security: Current status, challenges and prospective measures. *10th International Conference for Internet Technology and Secured Transactions (ICITST)*, London, UK. 10.1109/ICITST.2015.7412116

Mahyuni, L. P., Adrian, R., Darma, G. S., Krisnawijaya, N. N. K., Dewi, I. G. A. A. P., & Permana, G. P. L. (2020). Mapping the potentials of blockchain in improving supply chain performance. *Cogent Business & Management*, *7*(1), 1788329. doi:10.1080/23311975.2020.1788329

Makori, E. O. (2020). Blockchain applications and trends that promote information management. In *Emerging trends and impacts of the internet of things in libraries* (pp. 34–51). IGI Global. doi:10.4018/978-1-7998-4742-7.ch002

Ma, R., Gong, J., Liu, G., & Hao, Q. (2020). Enabling Cognitive Pyroelectric Infrared Sensing: From Reconfigurable Signal Conditioning to Sensor Mask Design. *IEEE Transactions on Industrial Informatics*, *16*(7), 4436–4446. doi:10.1109/TII.2019.2944700

Mashamba-Thompson, T. P., & Crayton, E. D. (2020). Blockchain and artificial intelligence technology for novel coronavirus disease 2019 self-testing. *Diagnostics (Basel)*, *10*(4), 198. doi:10.3390/diagnostics10040198 PMID:32244841

Mathur, S., Jin, T., & Kasturirangan, N. (2010). Parknet: Drive-By Sensing of Road-Side Parking Statistics. *Proc. of the 8th Int. Conf. on Mobile Systems, Applications, and Services*, ACM. 10.1145/1814433.1814448

Mavropoulos, T., Meditskos, G., Symeonidis, S., Kamateri, E., Rousi, M., Tzimikas, D., Papageorgiou, L., Eleftheriadis, C., Adamopoulos, G., Vrochidis, S., & Kompatsiaris, I. (2019). A context-aware conversational agent in the rehabilitation domain. *Future Internet, 11*(11), 231. doi:10.3390/fi11110231

Mazhar, T., Irfan, H. M., Khan, S., Haq, I., Ullah, I., Iqbal, M., & Hamam, H. (2023). Analysis of cyber security attacks and its solutions for the smart grid using machine learning and blockchain methods. *Future Internet, 15*(2), 83. doi:10.3390/fi15020083

McGhin, T., Choo, K. K. R., Liu, C. Z., & He, D. (2019). Blockchain in healthcare applications: Research challenges and opportunities. *Journal of Network and Computer Applications, 135*, 62–75.

Mekni, M. (2021). An artificial intelligence based virtual assistant using conversational agents. *Journal of Software Engineering and Applications, 14*(9), 455–473. doi:10.4236/jsea.2021.149027

Merlo, V., Pio, G., Giusto, F., & Bilancia, M. (2023). On the exploitation of the blockchain technology in the healthcare sector: A systematic review. *Expert Systems with Applications, 213*, 118897. doi:10.1016/j.eswa.2022.118897

Microsoft Azure. (2020). *What is Azure Files?* Microsoft Azure. https://azure.microsoft.com/en-in/products/storage/files/

Miorandi, D., Sicari, S., De Pellegrini, F., & Chlamtac, I. (2012). Internet of Things: Vision, Applications and Research Challenges. *Ad Hoc Networks, 10*(7), 1497-1516. doi:10.1016/j.adhoc.2012.02.016

Mishra, S., & Tyagi, A. K. (2019). Intrusion Detection in Internet of Things (IoTs) Based Applications using Blockchain Technology. *2019 Third International conference on I-SMAC (IoT in Social, Mobile, Analytics and Cloud) (I-SMAC)*, (pp. 123-128). IEEE. 10.1109/I-SMAC47947.2019.9032557

Mistry, I., Tanwar, S., Tyagi, S., & Kumar, N. (2020). Blockchain for 5G-enabled IoT for industrial automation: A systematic review, solutions, and challenges. *Mechanical Systems and Signal Processing, 135,* 106382.

Moin, S., Karim, A., Safdar, Z., Safdar, K., Ahmed, E., & Imran, M. (2019). Securing IoTs in the distributed blockchain: Analysis, requirements, and open issues. *Future Generation Computer Systems, 100*, 325–343.

Mollah, M. B., Zhao, J., Niyato, D., Guan, Y. L., Yuen, C., Sun, S., Lam, K.-Y., & Koh, L. H. (2020). Blockchain for the internet of vehicles towards intelligent transportation systems: A survey. *IEEE Internet of Things Journal, 8*(6), 4157–4185. doi:10.1109/JIOT.2020.3028368

Mololoth, V. K., Saguna, S., & Åhlund, C. (2023). Blockchain and machine learning for future Smart grids: A review. *Energies, 16*(1), 528. doi:10.3390/en16010528

Mondal, M. A., & Rehena, Z. (2021). An IoT-Based Congestion Control Framework for Intelligent Traffic Management System. In N. N. Chiplunkar & T. Fukao (Eds.), *Advances in Artificial Intelligence and Data Engineering* (Vol. 1133, pp. 1287–1297). Springer. doi:10.1007/978-981-15-3514-7_96

Mouskos, K., Boile, M., & Parker, N. A. (2007). *Technical Solutions to Overcrowded Park and Ride Facilities*. New Jersey Department of Transportation.

Mozumder, M. A. I., Sheeraz, M. M., Athar, A., Aich, S., & Kim, H. C. 2022, February. Overview: Technology roadmap of the future trend of metaverse based on IoT, blockchain, AI technique, and medical domain metaverse activity. In *2022 24th International Conference on Advanced Communication Technology (ICACT)* (pp. 256-261). IEEE.

Mozumder, M. A. I., Armand, T. P. T., Imtiyaj Uddin, S. M., Athar, A., Sumon, R. I., Hussain, A., & Kim, H. C. (2023). Metaverse for Digital Anti-Aging Healthcare: An Overview of Potential Use Cases Based on Artificial Intelligence, Blockchain, IoT Technologies, Its Challenges, and Future Directions. *Applied Sciences (Basel, Switzerland)*, *13*(8), 5127. doi:10.3390/app13085127

Muheidat, F., & Tawalbeh, L. A. (2021). Artificial intelligence and blockchain for cybersecurity applications. In *Artificial Intelligence and Blockchain for Future Cybersecurity Applications* (pp. 3–29). Springer International Publishing. doi:10.1007/978-3-030-74575-2_1

Mukta, M. Y., Rahman, M. A., Asyhari, A. T., & Md, Z. A. B. (2020). IoT for energy efficient green highway lighting systems: Challenges and issues. *Journal of Network and Computer Applications, 158.* doi:10.1016/j.jnca.2020.102575

Musamih, A., Salah, K., Jayaraman, R., Arshad, J., Debe, M., Al-Hammadi, Y., & Ellahham, S. (2021). A blockchain-based approach for drug traceability in healthcare supply chain. *IEEE Access : Practical Innovations, Open Solutions, 9*, 9728–9743. doi:10.1109/ACCESS.2021.3049920

Nair, M. M., & Tyagi, A. K. (2022). Preserving Privacy Using Blockchain Technology in Autonomous Vehicles. In: Giri, D., Mandal, J.K., Sakurai, K., De, D. (eds) *Proceedings of International Conference on Network Security and Blockchain Technology*. Springer, Singapore. 10.1007/978-981-19-3182-6_19

Nair, M. M., & Tyagi, A. K. (2023). Chapter 11 - AI, IoT, blockchain, and cloud computing: The necessity of the future. In Rajiv Pandey, Sam Goundar, Shahnaz Fatima (eds.), Distributed Computing to Blockchain. Academic Press. doi:10.1016/B978-0-323-96146-2.00001-2

Nakamoto, S. (2008). Bitcoin: A peer-to-peer electronic cash system. *Decentralized business review.*

Nakamoto, S. (2019). Bitcoin: A Peer-to-Peer Electronic Cash System. Decentralized Business Review. https://manubot.org/

Naseri, F., Qaedi, R., Karimi, S., Ghanbari, T., & Farjah, E. (2019). An effective Arc Fault Detection Approach for Smart Grid Solar Farms Using Rogowski Coil Sensor. *Smart Grid Conference (SGC)*, Tehran, Iran. 10.1109/SGC49328.2019.9056622

Nath, P., & Pati, U. C. (2018). Low-Cost Android App Based Voice Operated Room Automation System. 3rd International Conference for Convergence in Technology (I2CT), (pp. 1–4). IEEE. 10.1109/I2CT.2018.8529157

Nayak, S., & Vijayalakshmi, M. N. (2016). Analysis on IoT Challenges, Opportunities, Applications and Communication Models. *International Journal of Advanced Engineering, Management and Science, 2*(4).

Nest Labs. (2015). Meet the Nest Thermostat. Nest Labs.

Nicolopoulou-Stamati, P., Maipas, S., Kotampasi, C., Stamatis, P., & Hens, L. (2016). Chemical pesticides and human health: The urgent need for a new concept in agriculture. *Frontiers in Public Health, 4*, 148. doi:10.3389/fpubh.2016.00148 PMID:27486573

Nikolinakos, N. T. (2023). EU Policy and Legal Framework for Artificial Intelligence, Robotics and Related Technologies-The AI Act. *Law, Governance and Technology Series, 53.*

Njoku, J. N., Nwakanma, C. I., Amaizu, G. C., & Kim, D. S. (2023). Prospects and challenges of Metaverse application in data-driven intelligent transportation systems. *IET Intelligent Transport Systems, 17*(1), 1–21. doi:10.1049/itr2.12252

Nour, B., Sharif, K., Li, F., Biswas, S., Moungla, H., Guizani, M., & Wang, Y. (2019). A survey of Internet of Things communication using ICN: A use case perspective. *Computer Communications, 142–143,* 95-123. doi:10.1016/j.comcom.2019.05.010

Nykyri, M., Kuisma, M., Kärkkäinen, T. J., Hallikas, J., Jäppinen, J., Korpinen, K., & Silventoinen, P. (2019). IoT Demonstration Platform for Education and Research. *In Proceedings of the 2019 IEEE 17th International Conference on Industrial Informatics (INDIN)* (Vol. 1, pp. 1155–1162). IEEE. 10.1109/INDIN41052.2019.8972280

Olawumi, T. O., Chan, D. W., Ojo, S., & Yam, M. C. (2022). Automating the modular construction process: A review of digital technologies and future directions with blockchain technology. *Journal of Building Engineering*, *46*, 103720. doi:10.1016/j.jobe.2021.103720

Omitaomu, O. A., & Niu, H. (2021). Artificial intelligence techniques in Smart grid: A survey. *Smart Cities*, *4*(2), 548–568. doi:10.3390martcities4020029

Omrani, E., Khoshnevisan, B., & Shamshirband, S. (2014). Potential of redial basis function- based support vector regression for apple disease detection. Measurement, 55.

Omrčen, L., Leventić, H., Romić, K., & Galić, I. (2021, September). Integration of Blockchain and AI in EHR sharing: A survey. In *2021 International Symposium ELMAR* (pp. 155-160). IEEE. 10.1109/ELMAR52657.2021.9550953

Orchi, H., Sadik, M., & Khaldoun, M. (2021). On using artificial intelligence and the internet of things for crop disease detection: A contemporary survey. *Agriculture*, *12*(1), 9. doi:10.3390/agriculture12010009

Oriol, P., Piol, S. R., Joakim, E., & Thiemo, V. (2015). BSD Based Elliptic Curve Cryptography for the Open Internet of Things. *7th International Conference on New Technologies, Mobility and Security (NTMS'15)*, Paris.

Otoum, S., Al Ridhawi, I., & Mouftah, H. T. (2021). Preventing and controlling epidemics through blockchain-assisted ai-enabled networks. *IEEE Network*, *35*(3), 34–41. doi:10.1109/MNET.011.2000628

Özcan-Top, Ö. (2023). Improving Medical Supply Chain Disruption Management with the Blockchain Technology. In *Supply Chain Risk and Disruption Management: Latest Tools, Techniques and Management Approaches* (pp. 217–230). Springer Nature Singapore. doi:10.1007/978-981-99-2629-9_11

Paidi, V., Fleyeh, H., Håkansson, J., & Nyberg, R. G. (2018). Smart Parking Sensors, Technologies and Applications for Open Parking Lots: A Review. *IET Intelligent Transport Systems*, *12*(8), 735–741. doi:10.1049/iet-its.2017.0406

Palanikkumar, D., & Patwari, N. (2023). *Symmetrized Feature Selection with Stacked Generalization based Machine Learning Algorithm for the Early Diagnosis of Chronic Diseases.* 2023 5th International Conference on Smart Systems and Inventive Technology (ICSSIT), Tirunelveli, India.

Panda, S. K., & Satapathy, S. C. (2021). Drug traceability and transparency in medical supply chain using blockchain for easing the process and creating trust between stakeholders and consumers. *Personal and Ubiquitous Computing*, 1–17. doi:10.100700779-021-01588-3

Pandey, A. A., Fernandez, T. F., Bansal, R., & Tyagi, A. K. (2022). Maintaining Scalability in Blockchain. In A. Abraham, N. Gandhi, T. Hanne, T. P. Hong, T. Nogueira Rios, & W. Ding (Eds.), *Intelligent Systems Design and Applications. ISDA 2021. Lecture Notes in Networks and Systems* (Vol. 418). Springer. doi:10.1007/978-3-030-96308-8_4

Pandian, A., Kanchanadevi, K., Rajalakshmi, N. R., & Arulkumaran, G. (2022). An Improved Deep Residual Convolutional Neural Network for Plant Leaf Disease Detection. *Computational Intelligence and Neuroscience*, 2022. doi:10.1155/2022/5102290 PMID:36156945

Pantelic, J., Nazarian, N., Miller, C., Meggers, F., Jason, K. W. L., & Licina, D. (2022). Transformational IoT Sensing for Air Pollution and Thermal Exposures. *Frontiers in Built Environment*, *8*, 236. doi:10.3389/fbuil.2022.971523

Park, A., & Li, H. (2021). The effect of blockchain technology on supply chain sustainability performances. *Sustainability (Basel)*, *13*(4), 1726. doi:10.3390u13041726

Parthasarathy, A. R. (2022). Analysis of the Trust and Resilience of Consumer and Industrial Internet of Things (IoT) Systems in the Indian Context. Khanna, K., Estrela, V.V., Rodrigues, J.J.P.C., (eds.) Cyber Security and Digital Forensics, Lecture Notes on Data Engineering and Communications Technologies. Springer, Singapore. doi:10.1007/978-981-16-3961-6_21

Parul, G., Sahoo, A., & Sharma, T. (2020). Internet of Things: Applications, Security and Privacy: A survey. *Materials Today: Proceedings*, 1–8.

Pathak, N., & Patil, P. R. (2023, March). Securing VANET Communication Using Block Chain: A Review. In *2023 International Conference on Communication System, Computing and IT Applications (CSCITA)* (pp. 215-218). IEEE 10.1109/CSCITA55725.2023.10105098

Paula, E. L. (2016). Deep learning anomaly detection as support fraud investigation in brazilian exports and anti-money laundering. *2016 15th ieee international conference on machine learning and applications (icmla)*. IEEE. 10.1109/ICMLA.2016.0172

Pawar, A. B., & Ghumbre, S. (2016). A survey on IoT applications, security challenges and counter measures. *International Conference on Computing, Analytics and Security Trends (CAST)*, Pune, India. 10.1109/CAST.2016.7914983

Pazikadin, A. R., Rifai, D., Ali, K., Malik, M. Z., Abdalla, A. N., & Faraj, M. A. (2020). Solar Irradiance Measurement Instrumentation and Power Solar Generation Forecasting Based on Artificial Neural Networks (ANN): A Review of Five Years Research Trend. *Science of The Total Environment, 715*. doi:10.1016/j.scitotenv.2020.136848

Pedro, S. M., Nam, T., Brandon, C., Behnam, D., & Yuhong, L. (2018). Analyzing The Resource Utilization of AES Encryption on IoT Devices. *Asia Pacific Signal And Information Processing Association Annual Summit And Conference*, Hawaii, USA.

Peng, C. (2020). Blockchain for vehicular Internet of Things: Recent advances and open issues. *Sensors, 20*.

Peng, X. (2016). A novel authentication protocol for vehicle network. IEEE International Conference on Systems and Informatics. IEEE. 10.1109/ICSAI.2016.7811036

Peng, Y., & Wang, Y. (2019). Real-Time Forest Smoke Detection using Hand-Designed Features and Deep Learning. *Computers and Electronics in Agriculture*, *167*, 105029. doi:10.1016/j.compag.2019.105029

Pereira, F. G., Vassallo, R. F., & Salles, E. O. T. (2013). Human-robot interaction and cooperation through people detection and gesture recognition. *J. Control Automat. Elect. Syst.*, *24*(3), 187–198. doi:10.100740313-013-0040-3

Perilla, F. S., Villanueva, G. R., Cacanindin, N. M., & Palaoag, T. D. (2018). Fire Safety and Alert System Using Arduino Sensors with IoT Integration, *7th International Conference on Software and Computer Applications (ICSCA)*. ACM. 10.1145/3185089.3185121

Pettersson Ruiz, E., & Angelis, J. (2022). Combating money laundering with machine learning–applicability of supervised-learning algorithms at cryptocurrency exchanges. *Journal of Money Laundering Control*, *25*(4), 766–778. doi:10.1108/JMLC-09-2021-0106

Pflanzner, T., & Kertész, A. (2018). A Taxonomy and Survey of IoT Cloud Applications. EAI Endorsed Transactions on Internet of Things, 3(12). doi:10.4108/eai.6-4-2018.154391

Pilozzi, A., & Huang, X. (2020). Overcoming Alzheimer's disease stigma by leveraging artificial intelligence and blockchain technologies. *Brain Sciences*, *10*(3), 183. doi:10.3390/brainsci10030183 PMID:32210011

Połap, D., Srivastava, G., Jolfaei, A., & Parizi, R. M. (2020, July). Blockchain technology and neural networks for the internet of medical things. In IEEE INFOCOM 2020-IEEE conference on computer communications workshops (INFOCOM WKSHPS) (pp. 508-513). IEEE. doi:10.1109/INFOCOMWKSHPS50562.2020.9162735

Poornima, M. (2021). *Security and privacy in the internet of things: computational intelligent techniques-based approaches, Recent Trends in Computational Intelligence Enabled Research*. Academic Press.

Porkodi, S., & Kesavaraja, D. (2021). Healthcare robots enabled with IoT and artificial intelligence for elderly patients. *AI and IoT-Based Intelligent Automation in Robotics*, (pp. 87-108).

Prasad, K., Srinivas, V., & De, T. (2023). Wi-Fi Router Signal Coverage Position Prediction System using Machine Learning Algorithms. *2023 International Conference on Sustainable Computing and Smart Systems (ICSCSS)*, Coimbatore, India.

Prasad, S., & Hegde, N. (2022). Information Security and Privacy in Smart Cities, Smart Agriculture, Industry 4.0, Smart Medicine, and Smart Healthcare. *IoT Based Control Networks and Intelligent Systems: Proceedings of 3rd ICICNIS 2022*. IEEE.

Puri, V. (2020). A vital role of blockchain technology toward Internet of vehicles. In *Handbook of Research on Blockchain Technology* (pp. 407–416). Academic Press. doi:10.1016/B978-0-12-819816-2.00016-2

Puthal, D., Nepal, S., Ranjan, R., Chen, J., & Mohanty, S. P. (2019). The Blockchain as a Decentralized Security Framework [Guest Editorial]. *IEEE Cloud Computing*, *6*(1), 16–21.

Pyingkodi, M., Thenmozhi, K., Nanthini, K., Karthikeyan, M., & Suresh Palarimath, V. (2022). Sensor Based Smart Agriculture with IoT Technologies: A Review. *International Conference on Computer Communication and Informatics (ICCCI)*, Coimbatore, India. 10.1109/ICCCI54379.2022.9741001

Qiu, T., Zhao, A., & Ma, R. (2018). A task-efficient sink node based on embedded multi-core SoC for Internet of Things. *Future Generation Computer Systems, 82*. doi:10.1016/j.future.2016.12.024

Queralta, J. P., Qingqing, L., Zou, Z., & Westerlund, T. (2020, April). Enhancing autonomy with blockchain and multi-access edge computing in distributed robotic systems. In *2020 Fifth International Conference on Fog and Mobile Edge Computing (FMEC)* (pp. 180-187). IEEE. 10.1109/FMEC49853.2020.9144809

R.Preethi, S.Priyanka, U.Priyanka, A.Sheela, (2015), Efficient knowledge based system for leaf disease detection and classifcation, *International Journal of Advance Research In Science And Engineering, 4*(1).

Raffik, R., Sathya, R. R., Vaishali, V., & Balavedhaa, S. (2023, June). Industry 5.0: Enhancing Human-Robot Collaboration through Collaborative Robots–A Review. In *2023 2nd International Conference on Advancements in Electrical, Electronics, Communication, Computing and Automation (ICAECA)* (pp. 1-6). IEEE.

Rahman, A., Sara, U., Kundu, D., Islam, S., Islam, M. J., Hasan, M., & Nasir, M. K. (2020). *Distb-sdoindustry: Enhancing security in industry 4.0 services based on distributed blockchain through software defined networking-iot enabled architecture.* arXiv preprint arXiv:2012.10011.

Rahman, M. A., Asyhari, A. T., Obaidat, M. S., Kurniawan, I. F., Mukta, M. Y., & Vijayakumar, P. (2021). IoT-Enabled Light Intensity-Controlled Seamless Highway Lighting System. *IEEE Systems Journal, 15*(1), 46–55. doi:10.1109/JSYST.2020.2975592

Raiter, O. (2021). Applying supervised machine learning algorithms for fraud detection in anti-money laundering. *Journal of Modern Issues in Business Research, 1*(1), 14–26.

Rane, S., Shah, P., & Sekhar, R. (2022, August). Survey of technologies for industry 4.0. In *2022 6th International Conference on Computing, Communication, Control And Automation (ICCUBEA* (pp. 1-6). IEEE.

Ranjan, K. K. (2023). *Industrial Automated Multipurpose Robot Using WIFI.* 2023 4th International Conference for Emerging Technology (INCET), Belgaum, India.

Ranjan, K. K. (2023). *Wireless and Solar-Powered Multipurpose Robot for Industrial Automation: A Sustainable Solution.* 2023 8th International Conference on Communication and Electronics Systems (ICCES), Coimbatore, India.

Rao, K. P. N., & Manvi, S. (2023). Survey on Electronic Health Record Management Using Amalgamation of Artificial Intelligence and Blockchain Technologies. *Acta Informatica Pragensia, 12*(1), 179–199. doi:10.18267/j.aip.194

Rathee, G., Ahmad, F., Sandhu, R., Kerrache, C. A., & Azad, M. A. (2021). On designing and implementing a secure blockchain-based hybrid framework for Industrial Internet-of-Things. *Information Processing & Management, 58*, 102526.

Rathee, G., Kerrache, C. A., & Ferrag, M. A. (2022). A Blockchain-Based Intrusion Detection System Using Viterbi Algorithm and Indirect Trust for IIoT Systems. *Journal of Sensor and Actuator Networks, 11*(71).

Ray, P. P. (2023). *ChatGPT: A comprehensive review on background, applications, key challenges, bias, ethics, limitations and future scope.* Internet of Things and Cyber-Physical Systems.

Raza, S., Wallgren, L., & Voigt, T. (2017). Edge Computing for Industry 4.0: Scenarios and Challenges. *2017 IEEE 37th International Conference on Distributed Computing Systems (ICDCS),* (pp. 1451-1456). IEEE.

Reddy, A. (2021). A Review on IoT Enabling Technologies and Back-End Data-Sharing Model. *Journal of Electronics, Computer Networking and Applied Mathematics(JECNAM), 01*(1), 41–47. doi:10.55529/jecnam11.41.47

Rehman, Z. U., Khan, M. A., Ahmed, F., Damaševičius, R., Naqvi, S. R., Nisar, W., & Javed, K. (2021). Recognizing apple leaf diseases using a novel parallel real-time processing framework based on MASK RCNN and transfer learning: An application for smart agriculture. *IET Image Processing, 15*(10), 2157–2168. doi:10.1049/ipr2.12183

Revathi, P., & .Hema Latha, M. (2012). Classification Of Cotton Leaf Spot Diseases Using Image Processing Edge Detection Techniques. *International Journal of Applied Research.*

Revathy, M., Ramya, S., Sathiyavathi, R., Bharathi, B., & Anu, V. M. (2018). Automation of Street Light for Smart City. *IEEE International Conference on Communication and Signal Processing.* IEEE. 10.1109/ICCSP.2017.8286503

Rind, Y. M., Raza, M. H., Zubair, M., Mehmood, M. Q., & Massoud, Y. (1974). Smart Energy Meters For Smart Grids, An Internet of Things Perspective. *Energies, 16*(4), 2023. doi:10.3390/en16041974

Ripka, P. (2010). Electric Current Sensors: A Review. *Measurement Science & Technology, 21*(11), 1–32. doi:10.1088/0957-0233/21/11/112001

Rivas, A. (2020). Faults in Smart Grid Systems: Monitoring, Detection and Classification. *Electric Power Systems Research, 189.* 10.1016/j.epsr.2020.106602n

Roehrs, A., da Costa, C. A., & da Rosa Righi, R. (2017). OmniPHR: A distributed architecture model to integrate personal health records. *Journal of Biomedical Informatics, 71*, 70–81.

Rowan, S., Clear, M., Gerla, M., Huggard, M., & Mc Goldrick, C. (2017). *Securing Vehicle to Vehicle Communications using Blockchain through Visible Light and Acoustic Side-Channels.* Cryptography and Security.

Saeed, H., Malik, H., Bashir, U., Ahmad, A., Riaz, S., Ilyas, M., Bukhari, W. A., & Khan, M. I. A. (2022). Blockchain technology in healthcare: A systematic review. *PLoS One, 17*(4), e0266462. doi:10.1371/journal.pone.0266462 PMID:35404955

Salah, K., Rehman, M. H. U., Nizamuddin, N., & Al-Fuqaha, A. (2019). Blockchain for AI: Review and open research challenges. *IEEE Access : Practical Innovations, Open Solutions, 7*, 10127–10149. doi:10.1109/ACCESS.2018.2890507

Salama, R., Al-Turjman, F., & Culmone, R. (2023, March). AI-Powered Drone to Address Smart City Security Issues. In *International Conference on Advanced Information Networking and Applications* (pp. 292-300). Cham: Springer International Publishing.

Saldamli, G., & Razavi, A. (2020, December). Swarm robotics meets blockchain to deploy surveillance missions. In *2020 32nd International Conference on Microelectronics (ICM)* (pp. 1-6). IEEE. 10.1109/ICM50269.2020.9331824

Samani, H. H. G. A., & Wahaishi, A. (2015). Privacy in Internet of Things: A Model and Protection Framework, *6th International Conference on Ambient Systems, Networks and Technologies*. Science Direct. 10.1016/j.procs.2015.05.046

Samaniego, M., & Yang, B. (2020). A Survey on Blockchain Technology for Networking and Communications: Challenges and Opportunities. *IEEE Communications Surveys and Tutorials*, *22*(1), 713–734.

Samantaray, R. & Nayana Hegde, N. (2022), Smart Cart Powered by Artificial Intelligence. *2022 3rd International Conference on Communication, Computing and Industry 4.0* (C2I4). IEEE.

Samiksha, S., & Vinay, C. (2016). Analysis of AES encryption with ECC. *International Conference on Engineering Science And Management*, Goa.

Samimi, M. H., & Ilkhechit, H. D. (2020). Survey of Different Sensors Employed for the Power Transformer Monitoring. *IET Science, Measurement & Technology*, *14*(1), 1–8. doi:10.1049/iet-smt.2019.0103

Samsung Corp. (2015). *Privacy Policy—Smart TV Supplement.* Samsung Corp.

Saritha, K., Kurni, M., Madhavi, K., & Nagadevi, D. (2021). Integration of Artificial Intelligence and the Internet of Things with Blockchain Technology. In *Proceedings of the 2nd International Conference on Computational and Bio Engineering: CBE 2020* (pp. 449-457). Springer Singapore. 10.1007/978-981-16-1941-0_45

Savary, S., Ficke, A., Aubertot, J.-N., & Hollier, C. (2012). Crop losses due to diseases and their implications for global food production losses and food security. *Food Security*, *4*(4), 519–537. Advance online publication. doi:10.100712571-012-0200-5

Schmidt, B., Borrison, R., Cohen, A., Dix, M., Gärtler, M., Hollender, M., & Siddharthan, S. (2018, October). Industrial virtual assistants: Challenges and opportunities. In *Proceedings of the 2018 ACM International Joint Conference and 2018 International Symposium on Pervasive and Ubiquitous Computing and Wearable Computers* (pp. 794-801).

Sethi P. & Sarangi S. R. (2017). Internet of Things: Architectures, Protocols, and Applications. *Journal of Electrical and Computer Engineering*. . doi:10.1155/2017/9324035

Shaheen, S. (2005). *Smart Parking Management Field Test: A Bay Area Rapid Transit(Bart) District Parking Demonstration*. Institute of Transportation Studies.

Shah, K., Chadotra, S., Tanwar, S., Gupta, R., & Kumar, N. (2022). Blockchain for IoV in 6G environment: Review solutions and challenges. *Cluster Computing*, *25*(3), 1927–1955. doi:10.100710586-021-03492-0

Shahnaz, A., Qamar, U., & Khalid, A. (2019). Using blockchain for electronic health records. *IEEE Access, 7,* 147782–147795.

Shankar, S., Varadam, D., Agrawal, H. & E, N. (2022). Blockchain for IoT and Big Data Applications: A Comprehensive Survey on Security Issues. In M. Niranjanamurthy, S. Peng, E. Naresh, S. Jayasimha & V. Balas (Ed.), *Advances in Industry 4.0: Concepts and Applications* (pp. 65-86). De Gruyter. doi:10.1515/9783110725490-004

Sharma, A., Rani, R. P., Abdulrahman, I. S., Sreeram, A., & Shukla, S. K. (2023, May). Resource Allocation Technology Mobile Edge Computing System Using Blockchain Technology. In *2023 3rd International Conference on Advance Computing and Innovative Technologies in Engineering (ICACITE)* (pp. 632-636). IEEE. 10.1109/ICACITE57410.2023.10183102

Sharma, N., Shamkuwar, M., & Singh, I. (2019). The History, Present and Future With IoT. *Internet of Things and Big Data Analytics For Smart Generation, 154*, 27–51. doi:10.1007/978-3-030-04203-5_3

Sharma, V., Tan, T. G., Singh, S., & Sharma, P. K. (2021). Optimal and Privacy-Aware Resource Management in Artificial Intelligence of Things Using Osmotic Computing. *IEEE Transactions on Industrial Informatics, 18*(5), 3377–3386. doi:10.1109/TII.2021.3102471

Sheth, H. S. K. (2022). Deep Learning, Blockchain based Multi-layered Authentication and Security Architectures. *2022 International Conference on Applied Artificial Intelligence and Computing (ICAAIC),* (pp. 476-485). IEEE. 10.1109/ICAAIC53929.2022.9793179

Shi, S., He, D., Li, L., Kumar, N., Khan, M. K., & Choo, K. K. R. (2020). Applications of blockchain in ensuring the security and privacy of electronic health record systems: A survey. *Computer Security, 97*, 101966.

Shinde, R., Patil, S., Kotecha, K., & Ruikar, K. (2021). Blockchain for securing ai applications and open innovations. *Journal of Open Innovation, 7*(3), 189. doi:10.3390/joitmc7030189

Shivaputra, S. (2016). Pomegranate Leaf Disease Detection Using Support Vector Machine. *International Journal Of Engineering And Computer Science, 5*(6).

Shivashankar, M. S. (2017). Implementation of swarm intelligence in obstacle avoidance. in *RTEICT 2017 - 2nd IEEE International Conference on Recent Trends in Electronics, Information and Communication Technology*. IEEE.

Shrimali, B., & Patel, H. B. (2022). Blockchain state-of-the-art: Architecture, use cases, consensus, challenges and opportunities. *Journal of King Saud University. Computer and Information Sciences, 34*(9), 6793–6807. doi:10.1016/j.jksuci.2021.08.005

Siano, P. (2014). Demand response and smart grids—A survey. *Renewable & Sustainable Energy Reviews, 30*, 461–478. doi:10.1016/j.rser.2013.10.022

Siddharth, M. (2021). Issues and Challenges (Privacy, Security, and Trust) in Blockchain-Based Applications, In: Opportunities and Challenges for Blockchain Technology in Autonomous Vehicles. IGI Global. doi:10.4018/978-1-7998-3295-9.ch012

Singh, G., Banga, V. K., & Yingthawornsuk, T. (2022). *Artificial Intelligence and Industrial Robot.* 2022 16th International Conference on Signal-Image Technology & Internet-Based Systems (SITIS), Dijon, France. 10.1109/SITIS57111.2022.00098

Singh, M., & Kim, S. (2018). Branch based blockchain technology in intelligent vehicle. *Computer Networks, 145*, 219–231. doi:10.1016/j.comnet.2018.08.016

Singh, P., Masud, M., Hossain, M. S., & Kaur, A. (2021). Blockchain and homomorphic encryption-based privacy-preserving data aggregation model in smart grid. Computers &amp. *Computers & Electrical Engineering, 93*, 107209. doi:10.1016/j.compeleceng.2021.107209

Singh, P., & Singh, N. (2020). Blockchain with IoT and AI: A review of agriculture and healthcare. *International Journal of Applied Evolutionary Computation, 11*(4), 13–27. doi:10.4018/IJAEC.2020100102

Singh, S. K., Rathore, S., & Park, J. H. (2020). Blockiotintelligence: A blockchain-enabled intelligent IoT architecture with artificial intelligence. *Future Generation Computer Systems, 110*, 721–743. doi:10.1016/j.future.2019.09.002

Siyal, A. A., Junejo, A. Z., Zawish, M., Ahmed, K., Khalil, A., & Soursou, G. (2019). Applications of blockchain technology in medicine and healthcare: Challenges and future perspectives. *Cryptography, 3*(1), 3. doi:10.3390/cryptography3010003

Soltani, R., Nguyen, U., Yang, Y., Faghani, M., Yagoub, A., & An, A. (2016). A new algorithm for money laundering detection based on structural similarity. *2016 IEEE 7th Annual Ubiquitous Computing, Electronics & Mobile Communication Conference (UEMCON).* IEEE.

Soltanisehat, L., Alizadeh, R., Hao, H., & Choo, K. K. R. (2020). Technical, temporal, and spatial research challenges and opportunities in blockchain-based healthcare: A systematic literature review. *IEEE Transactions on Engineering Management, 70*(1), 353–368. doi:10.1109/TEM.2020.3013507

Soofastaei, A. (2021). In A. Soofastaei (Ed.), *Introductory Chapter: Virtual Assistants. Virtual Assistant* (pp. 1–9). IntechOpen.

Soofastaei, A. (2021). Introductory Chapter: Virtual Assistants. In A. Soofastaei (Ed.), *Virtual Assistant* (pp. 1–9). IntechOpen. doi:10.5772/intechopen.100248

Sorin, T. (2023). *Innovative Approach for the Implementation of Blockchain in Command and Control Systems of Critical Infrastructure Networks.*

Sousa, M. J., Dal Mas, F., Gonçalves, S. P., & Calandra, D. (2022). AI and blockchain as new triggers in the education arena. *European Journal of Investigation in Health, Psychology and Education, 12*(4), 445–447. doi:10.3390/ejihpe12040032 PMID:35447750

Srivastava, S., Singh, S. V., Singh, R. B., & Shukla, H. K. (2021). Digital Transformation of Healthcare: A blockchain study. *International Journal of Innovative Science, Engineering & Technology, 8*(5).

Srivastava, S., Pant, M., Jauhar, S. K., & Nagar, A. K. (2022). Analyzing the Prospects of Blockchain in Healthcare Industry. *Computational and Mathematical Methods in Medicine, 2022*, 2022. doi:10.1155/2022/3727389 PMID:36506597

Stadtler, H. (2014). Supply chain management: An overview. *Supply chain management and advanced planning: Concepts, models, software, and case studies*, 3-28.

Stolojescu-Crisan, C., Crisan, C., & Butunoi, B. P. (2021). An IoT-Based Smart Home Automation System. *Sensors (Basel), 21*(11), 3784. doi:10.339021113784 PMID:34070717

Sukmana, H. T., Farisi, M. G., & Khairani, D. (2015). Prototype Utilization of PIR Motion Sensor for Real Time Surveillance System and Webenabled Lamp Automation. *IEEE Asia Pacific Conference on Wireless and Mobile (APWiMob).* IEEE.

Sulaiman, A., Nagu, B., Kaur, G., Karuppaiah, P., Alshahrani, H., Reshan, M. S., AlYami, S., & Shaikh, A. (2023). Artificial Intelligence-based secured Power Grid Protocol for Smart City. *Sensors (Basel), 23*(19), 8016. doi:10.339023198016 PMID:37836846

Sumithra, M. G., & Saranya, N. (2021). Particle Swarm Optimization (PSO) with fuzzy c means (PSO-FCM)–based segmentation and machine learning classifier for leaf diseases prediction. *Concurrency and Computation, 33*(3), e5312. doi:10.1002/cpe.5312

Sun, Y., Peng, M., & Mao, S. (2018). Deep reinforcement learning-based mode selection and resource management for green fog radio access networks. *IEEE Internet of Things Journal, 6*(2), 1960–1971. doi:10.1109/JIOT.2018.2871020

Suresh, S., Bhavya, J., Sakshi, S., Varun, K., & Debarshi, G. (2016). Home Monitoring and Security system. *International Conference on ICT in Business Industry & Government (ICTBIG),* (pp. 1–5). IEEE. 10.1109/ICTBIG.2016.7892665

Suryadevara, N. K., & Mukhopadhyay, S. C. (2018). Wireless Sensor Network Based Industrial Monitoring Applications: A Comprehensive Survey. *IEEE Communications Surveys and Tutorials*, *20*(2), 1403–1427.

Tabassum, N., Devanagavi, G. D., & Biradar, R. C. (2017). Clock synchronization in wireless sensor networks using least common multiple. *AEÜ. International Journal of Electronics and Communications*, *2017*(82), 446–457. doi:10.1016/j.aeue.2017.10.014

Tabassum, N., Devanagavi, G. D., Biradar, R. C., & Lazarescu, M. T. (2020). Comparative node selection-based localization technique for wireless sensor networks: A bilateration approach. *International Journal of Communication Systems*, *33*(15), e4559. doi:10.1002/dac.4559

Tabassum, N., Geetha, D. D., & Biradar, R. C. (2022).. . *Joint Position Estimation and Synchronization of Clocks in WSN Lecture Notes in Networks and Systems*, *2022*(235), 409–418.

Tabassum, Z., & Shastry, B. S. (2022a). Peak power management of residential building using demand side management strategies. *International Journal of Health Sciences*, 8978–8997. doi:10.53730/ijhs.v6nS2.7333

Tabassum, Z., & Shastry, B. S. C. (2022). Short term load forecasting of residential and commercial consumers of Karnataka Electricity Board using CFNN. *International Journal of Electrical and Electronics Research*, *10*(2), 347–352. doi:10.37391/ijeer.100247

Tagde, P., Tagde, S., Bhattacharya, T., Tagde, P., Chopra, H., Akter, R., Kaushik, D., & Rahman, M. H. (2021). Blockchain and artificial intelligence technology in e-Health. *Environmental Science and Pollution Research International*, *28*(38), 52810–52831. doi:10.100711356-021-16223-0 PMID:34476701

Taherdoost, H. (2022). Blockchain technology and artificial intelligence together: A critical review on applications. *Applied Sciences (Basel, Switzerland)*, *12*(24), 12948. doi:10.3390/app122412948

Tahiliani, V., & Dizalwar, M. (2018). Green IoT Systems: An Energy Efficient Perspective. *Eleventh International Conference on Contemporary Computing*, Noida, India. 10.1109/IC3.2018.8530550

Tailor, R. K., Pareek, R., & Khang, A. (2022). Robot process automation in blockchain. *The data-driven blockchain ecosystem: Fundamentals, applications, and emerging technologies*, (pp. 149-164).

Taloba, A. I., Elhadad, A., Rayan, A., Abd El-Aziz, R. M., Salem, M., Alzahrani, A. A., Alharithi, F. S., & Park, C. (2023). A blockchain-based hybrid platform for multimedia data processing in IoT-Healthcare. *Alexandria Engineering Journal*, *65*, 263–274. doi:10.1016/j.aej.2022.09.031

Tandon, R., Verma, A., & Gupta, P. K. (2022, November). Blockchain enabled vehicular networks: A review. In *2022 5th International Conference on Multimedia, Signal Processing and Communication Technologies (IMPACT)* (pp. 1-6). IEEE.

Tanguturi, R. (2022*).* Improved Secure Communication Mechanism for IoT Platform Devices. *2022 International Conference on Smart Generation Computing, Communication and Networking (SMART GENCON)*, Bangalore, India.

Tantawi, K. H., Sokolov, A., & Tantawi, O. (2019). *Advances in Industrial Robotics: From Industry 3.0 Automation to Industry 4.0 Collaboration.* 2019 4th Technology Innovation Management and Engineering Science International Conference (TIMES-iCON), Bangkok, Thailand. 10.1109/TIMES-iCON47539.2019.9024658

Tanwar, S., Parekh, K., & Evans, R. (2020). Blockchain-based electronic healthcare record system for healthcare 4.0 applications. *Journal of Information Security and Applications*, *50*, 102407.

Tapscott, D., & Tapscott, A. (2016). *Blockchain revolution: how the technology behind bitcoin is changing money, business, and the world.* Penguin.

Tariq, A. (2018). Internet of Things: A Comprehensive Study of Security Issues and Defense Mechanisms. *IEEE Access : Practical Innovations, Open Solutions, 7*, 11020–11028.

Tasgaonkar, P. P., Garg, R. D., & Garg, P. K. (2020). Vehicle Detection and Traffic Estimation with Sensors Technologies for Intelligent Transportation Systems. *Sensing and Imaging, 21*(1), 29. doi:10.100711220-020-00295-2

Tauseef, M., Kounte, M. R., Nalband, A. H., & Ahmed, M. R. (2023). Exploring the Joint Potential of Blockchain and AI for Securing Internet of Things. *International Journal of Advanced Computer Science and Applications, 14*(4). doi:10.14569/IJACSA.2023.0140498

Tellaeche, I. (2015). Human robot interaction in industrial robotics. Examples from research centers to industry. *2015 IEEE 20th Conference on Emerging Technologies & Factory Automation (ETFA)*. IEEE.

Tertychnyi, P., Godgildieva, M., Dumas, M., & Ollikainen, M. (2022). Timeaware and interpretable predictive monitoring system for anti-money laundering. *Machine Learning with Applications, 8*, 100306. doi:10.1016/j.mlwa.2022.100306

Thanki, R., & Joshi, P. (2023). Advanced Technologies for Industrial Applications. In *Advanced Technologies for Industrial Applications* (pp. 73–95). Springer International Publishing. doi:10.1007/978-3-031-33238-8_6

Tibrewal, I., Srivastava, M., & Tyagi, A. K. (2022). Blockchain Technology for Securing Cyber-Infrastructure and Internet of Things Networks. In A. K. Tyagi, A. Abraham, & A. Kaklauskas (Eds.), *Intelligent Interactive Multimedia Systems for e-Healthcare Applications*. Springer. doi:10.1007/978-981-16-6542-4_17

Tripathi, G., Ahad, M. A., & Sathiyanarayanan, M. (2019). The Role of Blockchain in Internet of Vehicles (IoV): Issues, Challenges and Opportunities. *2019 International Conference on contemporary Computing and Informatics (IC3I)*. IEEE. 10.1109/IC3I46837.2019.9055613

Trivedi, D. M., & Raval, T. J. (2018). Improving Communication Security in IoT using Hybrid Encryption Decryption. *International Journal of Advance Research and Innovative Ideas in Education, 4*(2), 4175–4182.

Tyagi, A. & Kumar, S. (2022). Blockchain Technology for Securing Internet of Vehicle: Issues and Challenges. *2022 International Conference on Computer Communication and Informatics (ICCCI)*, (pp. 1-6). IEEE. 10.1109/ICCCI54379.2022.9740856

Tyagi, A. (2021a). Applications of Blockchain Technologies in Digital Forensic and Threat Hunting. Recent Trends in Blockchain for Information Systems Security and Privacy. CRC Press.

Tyagi, A. (2021b). Analysis of Security and Privacy Aspects of Blockchain Technologies from Smart Era' Perspective: The Challenges and a Way Forward. In Recent Trends in Blockchain for Information Systems Security and Privacy. CRC Press.

Tyagi, A. (2021c, October). AARIN: Affordable, Accurate, Reliable and INnovative Mechanism to Protect a Medical Cyber-Physical System using Blockchain Technology. *IJIN, 2*, 175–183.

Tyagi, A. (2022a). Preserving Privacy using Distributed Ledger Technology in Intelligent Transportation System. In *Proceedings of the 2022 Fourteenth International Conference on Contemporary Computing (IC3-2022)*. Association for Computing Machinery. 10.1145/3549206.3549306

Tyagi, A. (2023). Decentralized everything: Practical use of blockchain technology in future applications. In Rajiv Pandey, Sam Goundar, Shahnaz Fatima, (eds.) Distributed Computing to Blockchain. Academic Press. doi:10.1016/B978-0-323-96146-2.00010-3

Tyagi, A. K., Fernandez, T. F., & Aswathy, S. U. (2020). Blockchain and Aadhaar based Electronic Voting System. *2020 4th International Conference on Electronics, Communication and Aerospace Technology (ICECA), Coimbatore*, (pp. 498-504). IEEE. 10.1109/ICECA49313.2020.9297655

Tyagi, A. K. (2022b). SecVT: Securing the Vehicles of Tomorrow Using Blockchain Technology. In A. A. Sk, T. Turki, T. K. Ghosh, S. Joardar, & S. Barman (Eds.), *Artificial Intelligence. ISAI 2022. Communications in Computer and Information Science* (Vol. 1695). Springer. doi:10.1109/ICCCI54379.2022.9740965

Tyagi, A. K., Chandrasekaran, S., & Sreenath, N. (2022). Blockchain Technology:– A New Technology for Creating Distributed and Trusted Computing Environment. *2022 International Conference on Applied Artificial Intelligence and Computing (ICAAIC)*, (pp. 1348-1354). IEEE. 10.1109/ICAAIC53929.2022.9792702

Tyagi, S., & Bhatnagar, P. (2023). Roboethics: A Bibliometric Review [2005–2021]. *International Journal of Business Analytics & Intelligence (IJBAI), 11*(1).

Ullo, S. L., & Sinha, G. R. (2021). Advances in IoT and Smart Sensors For Remote Sensing And Agriculture Applications. *Remote Sensing (Basel), 13*(13), 2585. doi:10.3390/rs13132585

Uma, S. (2023). Blockchain and AI: Disruptive Digital Technologies in Designing the Potential Growth of Healthcare Industries. In *AI and Blockchain in Healthcare* (pp. 137–150). Springer Nature Singapore. doi:10.1007/978-981-99-0377-1_9

van Dijk, J. (2020). *State-of-the-art of Blockchain Technology in Intelligent Transportation Systems: Literature review.* TU Delft Repositories.

Varsha, J. & Tyagi, A. (2022). Security Optimization of Resource-Constrained Internet of Healthcare Things (IoHT) Devices Using Lightweight Cryptography. Information Security Practices for the Internet of Things, 5G, and Next-Generation Wireless Networks. IGI Global. doi:10.4018/978-1-6684-3921-0.ch009

Varsha, R. (2020, January 1). 'Deep Learning Based Blockchain Solution for Preserving Privacy in Future Vehicles'. *International Journal of Hybrid Intelligent Systems, 16*(4), 223–236.

Vasavi, P., Punitha, A., & Rao, T. V. N. (2022). Crop leaf disease detection and classification using machine learning and deep learning algorithms by visual symptoms: A review. *Iranian Journal of Electrical and Computer Engineering, 12*(2), 2079. doi:10.11591/ijece.v12i2.pp2079-2086

Vellido, A. (2019). Societal issues concerning the application of artificial intelligence in medicine. *Kidney Diseases, 5*(1), 11–17. doi:10.1159/000492428 PMID:30815459

Venkatakiran, S., Ashreetha, B., & Reddy, N. S. (2023). Implementation of a Machine Learning-based Model for Cardiovascular Disease Post Exposure prophylaxis. *2023 International Conference for Advancement in Technology (ICONAT)*, Goa, India.

Verma. (2022). Blockchain for Industry 5.0: Vision, Opportunities, Key Enablers, and Future Directions. *IEEE Access, 10*, 69160-69199.

Verma, A., Bhattacharya, P., Madhani, N., Trivedi, C., Bhushan, B., Tanwar, S., Sharma, G., Bokoro, P. N., & Sharma, R. (2022). Blockchain for industry 5.0: Vision, opportunities, key enablers, and future directions. *IEEE Access : Practical Innovations, Open Solutions, 10*, 69160–69199. doi:10.1109/ACCESS.2022.3186892

Vignesh Saravanan, K., Jothi Thilaga, P., Kavipriya, S., & Vijayalakshmi, K. (2023). Data Protection and Security Enhancement in Cyber-Physical Systems Using AI and Blockchain. In *AI Models for Blockchain-Based Intelligent Networks in IoT Systems: Concepts, Methodologies, Tools, and Applications* (pp. 285–325). Springer International Publishing. doi:10.1007/978-3-031-31952-5_13

Vijayalakshmi, S., Savita, Gayathri, S. P., & Janarthanan, S. (2021). Blockchain security for artificial intelligence-based clinical decision support tool. *Internet of things, artificial intelligence and blockchain technology*, 209-240.

Vishnu, R., & Krishna Prakash, N. (2022). Mobile application-based virtual assistant using deep learning. In *Soft Computing and Signal Processing: Proceedings of 3rd ICSCSP 2020,* (pp. 609-617). Springer Singapore. 10.1007/978-981-16-1249-7_57

Volkova, A., Patil, A. D., Javadi, S. A., & de Meer, H. (2022). Accountability challenges of AI in Smart Grid Services. *Proceedings of the Thirteenth ACM International Conference on Future Energy Systems*. ACM. 10.1145/3538637.3539636

Vrontis, D., Christofi, M., Pereira, V., Tarba, S., Makrides, A., & Trichina, E. (2022). Artificial intelligence, robotics, advanced technologies and human resource management: A systematic review. *International Journal of Human Resource Management*, *33*(6), 1237–1266. doi:10.1080/09585192.2020.1871398

Vyas, S., Shabaz, M., Pandit, P., Parvathy, L. R., & Ofori, I. (2022). Integration of AI and blockchain technology in healthcare and agriculture. *Journal of Food Quality*, 2022.

Wang, Q., Li, R., & Zhan, L. (2021). Blockchain technology in the energy sector: From basic research to real world applications. *Computer Science Review, 39*, 100362.

Wang, C. (2012). An IBE Based Security Scheme on Internet of Things. *2nd International Conference on Cloud Computing and Intelligence Systems*, China.

Wang, C., Daneshmand, M., Dohler, M., Mao, X., Hu, R. Q., & Wang, H. (2013). Guest editorial - special issue on Internet of things (IoT): Architecture, protocols and services. *IEEE Sensors Journal*, *13*(10), 3505–3510. doi:10.1109/JSEN.2013.2274906

Wang, C., Jiang, C., Wang, J., Shen, S., Guo, S., & Zhang, P. (2022). Blockchain-Aided Network Resource Orchestration in Intelligent Internet of Things. *IEEE Internet of Things Journal*, *10*(7), 6151–6163. doi:10.1109/JIOT.2022.3222911

Wang, D. Q., Feng, L. Y., Ye, J. G., Zou, J. G., & Zheng, Y. F. (2023). Accelerating the integration of ChatGPT and other large-scale AI models into biomedical research and healthcare. *MedComm - Future Medicine*, *2*(2), e43. doi:10.1002/mef2.43

Wang, Q., Zheng, J., Xu, H., Xu, B., & Chen, R. (2017). Roadside Magnetic Sensor System for Vehicle Detection in Urban Environments. *IEEE Transactions on Intelligent Transportation Systems*, *19*(5), 1365–1374. doi:10.1109/TITS.2017.2723908

Wang, Y., Kumar, N., Singh, R., & Alowaisheq, E. (2018). Secure Data Sharing Framework for Industrial IoT Using Blockchain and Attribute-Based Encryption. *IEEE Transactions on Industrial Informatics*, *14*(5), 2038–2046.

Watson, I. (2015). Design and Implementation of an automatic street light Control System,. *International Journal of Emerging Technology and Advanced Engineering*, *5*(3), 336–340.

Weizenbaum, J. (1976). Computer power and human reason: From judgment to calculation.

Wenhua, Z., Qamar, F., Abdali, T. A. N., Hassan, R., Jafri, S. T. A., & Nguyen, Q. N. (2023). Blockchain technology: Security issues, healthcare applications, challenges and future trends. *Electronics (Basel)*, *12*(3), 546. doi:10.3390/electronics12030546

Wetula A., Bień A., & Parekh M. (2021). New Sensor for Medium and High-Voltage Measurement. *Energies, 14*(15). doi:10.3390/en14154654

Wu, M., Lu, T.-J., Ling, F.-Y., Sun, J., & Du, H.-Y. (2010). Research on the architecture of Internet of Things. *3rd International Conference on Advanced Computer Theory and Engineering(ICACTE)*, Chengdu. 10.1109/ICACTE.2010.5579493

Xia, Q., Sifah, E. B., Smahi, A., Amofa, S., & Zhang, X. (2017). BBDS: Blockchain-based data sharing for electronic medical records in cloud environments. *Information, 8,* 44.

Xiao, P., Salleh, M. I., Zaidan, B. B., & Xuelan, Y. (2023). Research on risk assessment of Blockchain-driven Supply Chain finance: A systematic review. *Computers & Industrial Engineering, 176*, 108990. doi:10.1016/j.cie.2023.108990

Xie, L., Ding, Y., Yang, H., & Wang, X. (2019). Blockchain-Based Secure and Trustworthy Internet of Things in SDN-Enabled 5G-VANETs. *IEEE Access : Practical Innovations, Open Solutions, 7*, 56656–56666. doi:10.1109/ACCESS.2019.2913682

Xie, Q., Tan, D., Zhu, T., Zhang, Q., Xiao, S., Wang, J., & Yi, P. (2019, May). Chatbot application on cryptocurrency. In *2019 IEEE Conference on Computational Intelligence for Financial Engineering & Economics (CIFEr)* (pp. 1-8). IEEE.

Xu, J., Gu, B., & Tian, G. (2022). Review of agricultural IoT technology. *Artificial Intelligence in Agriculture, 6.* doi:10.1016/j.aiia.2022.01.001

Xu, H., Klaine, P. V., Onireti, O., Cao, B., Imran, M., & Zhang, L. (2020). Blockchain-enabled resource management and sharing for 6G communications. *Digital Communications and Networks*, *6*(3), 261–269. doi:10.1016/j.dcan.2020.06.002

Xu, L. D., He, W., & Li, S. (2018). Internet of Things in Industries: A Survey. *IEEE Transactions on Industrial Informatics*, *10*(4), 2233–2243. doi:10.1109/TII.2014.2300753

Yang, Y., & Wu, M. (2020, July). Supervised and unsupervised learning for fraud and money laundering detection using behavior measuring distance. In *2020 IEEE 18th International Conference on Industrial Informatics (INDIN)* (*Vol. 1*, pp. 446-451). IEEE. 10.1109/INDIN45582.2020.9442099

Yang, Y., Li, X., Qamar, N., Liu, P., Ke, W., Shen, B., & Liu, Z. (2018). Mindshare: A novel hybrid cloud for medical resource sharing among autonomous healthcare providers. *IEEE Access, 6,* 46949–46961.

Yang, D., Rao, K., Xu, B., & Sheng, W. (2015). PIR Sensors Deployment with the Accessible Priority in Smart Home Using Genetic Algorithm. *International Journal of Distributed Sensor Networks*, *11*(11), 1–10. doi:10.1155/2015/146270

Yang, H., & Lee, H. (2019). Understanding user behavior of virtual personal assistant devices. *Information Systems and e-Business Management*, *17*(1), 65–87. doi:10.100710257-018-0375-1

Yang, Q., Zhao, Y., Huang, H., Xiong, Z., Kang, J., & Zheng, Z. (2022). Fusing blockchain and AI with metaverse: A survey. *IEEE Open Journal of the Computer Society*, *3*, 122–136. doi:10.1109/OJCS.2022.3188249

Yang, Z., Wu, S., Ning, H., & Lu, R. (2019). A Blockchain-Based Access Control Framework for Industrial IoT. *IEEE Transactions on Industrial Informatics*, *15*(1), 471–480.

Yaqoob, I., Salah, K., Jayaraman, R., & Al-Hammadi, Y. (2021). Blockchain for healthcare data management: Opportunities, challenges, and future recommendations. *Neural Computing & Applications*, 1–16.

Ylianttila, M., & Vasilakos, A. V. (2017). Security in Industrial IoT Networks: Approaches and Challenges. *IEEE Network*, *31*(3), 82–88.

Yoon, M. Y., Cha, B., & Kim, J. C. (2013). Recent trends in studies on botanical fungicides in agriculture. *The Plant Pathology Journal*, *29*(1), 1–9. doi:10.5423/PPJ.RW.05.2012.0072 PMID:25288923

Younes, Z., Alhamrouni, I., Mekhilef, S., & Khan, M. R. (2021). Blockchain applications and challenges in smart grid. *2021 IEEE Conference on Energy Conversion (CENCON)*. IEEE. 10.1109/CENCON51869.2021.9627297

Youssef, B., Bouchra, F., & Brahim, O. (2023). State of the Art Literature on Anti-money Laundering Using Machine Learning and Deep Learning Techniques. *The International Conference on Artificial Intelligence and Computer Vision.* Cham: Springer Nature Switzerland. 10.1007/978-3-031-27762-7_8

Yuan, Y., & Wang, F.-Y. (2016). Towards blockchain-based intelligent transportation systems. *IEEE International Conference on Intelligent Transportation Systems.* IEEE.

Zaman, H. U., & Rafiunnisa, A. M. (2018). A User-Friendly Low-Cost Mobile App Based Home Appliance Control And Circuit Breaker, *2nd International Conference on Computing Methodologies and Communication (ICCMC)*, (pp. 203–208). IEEE.

Zhamshid, V., Nusratilloevich, Y. A., Shahida, K., & Azamat, K. (2023). Rise of the Machines: The Legal Implications of Robotics and Automation for the Digital Workforce. *International Journal of Cyber Law*, *1*(4).

Zhang, P., Walker, M. A., White, J., Schmidt, D. C., & Lenz, G. (2017). Metrics for assessing blockchain-based healthcare decentralized apps. In *Proceedings of the 2017 IEEE 19th International Conference on e-Health Networking, Applications and Services (Healthcom)* (pp. 1–4). IEEE. 10.1109/HealthCom.2017.8210842

Zhang, H., Xiao, Y., Bu, S., Niyato, D., Yu, F. R., & Han, Z. (2017). Computing resource allocation in three-tier IoT fog networks: A joint optimization approach combining Stackelberg game and matching. *IEEE Internet of Things Journal*, *4*(5), 1204–1215. doi:10.1109/JIOT.2017.2688925

Zhang, Q., & Fitzek, F. H. P. (2015). Mission Critical IoT Communication in 5G. In V. Atanasovski & A. Leon-Garcia (Eds.), *Future Access Enablers for Ubiquitous and Intelligent Infrastructures, FABULOUS 2015, Lecture Notes of the Institute for Computer Sciences, Social Informatics and Telecommunications Engineering* (Vol. 159). Springer. doi:10.1007/978-3-319-27072-2_5

Zhang, R., Xue, R., & Liu, L. (2019). Security and privacy on blockchain. *ACM Computing Surveys*, *52*(3), 1–34. doi:10.1145/3316481

Zhang, Y., & Trubey, P. (2019). Machine learning and sampling scheme: An empirical study of money laundering detection. *Computational Economics*, *54*(3), 1043–1063. doi:10.100710614-018-9864-z

Zhang, Z., Song, X., Liu, L., Yin, J., Wang, Y., & Lan, D. (2021). Recent advances in blockchain and artificial intelligence integration: Feasibility analysis, research issues, applications, challenges, and future work. *Security and Communication Networks*, *2021*, 1–15. doi:10.1155/2021/9991535

Zhao, G., Si, X., Wang, J., Long, X., & Hu, T. (2011). A novel mutual authentication scheme for Internet of Things. *IEEE International Conference on Modelling, Identification and Control.* IEEE.

Zhavoronkov, A., Vanhaelen, Q., & Oprea, T. I. (2020). Will artificial intelligence for drug discovery impact clinical pharmacology? *Clinical Pharmacology and Therapeutics*, *107*(4), 780–785. doi:10.1002/cpt.1795 PMID:31957003

Zheng, P., Su, B., Jiang, Z., Yang, C., Chen, J., & Wu, J. (2023, July). Exploring Heterogeneous Decentralized Markets in DeFi and NFT on Ethereum Blockchain. In *2023 IEEE 10th International Conference on Cyber Security and Cloud Computing (CSCloud)/2023 IEEE 9th International Conference on Edge Computing and Scalable Cloud (EdgeCom)* (pp. 259-267). IEEE. 10.1109/CSCloud-EdgeCom58631.2023.00052

Zhong, C. L., Zhu, Z., & Huang, R.-G. (2015). Study on the IoT Architecture and Gateway Technology, *14th International Symposium on Distributed Computing and Applications for Business Engineering and Science (DCABES)*, Guiyang, China. 10.1109/DCABES.2015.56

Zhu, X., Mukhopadhyay, S. K., & Kurata, H. (2012). A Review of RFID Technology and Its Managerial Applications in Different Industries. *Journal of Engineering and Technology Management*, *29*(1), 152–167. doi:10.1016/j.jengtecman.2011.09.011

Zorzi, M., Gluhak, A., Lange, S., & Bassi, A. (2010). From Today's Intranet Of Things To A Future Internet of Things: A Wireless-And Mobility-Related View. *IEEE Wireless Communications*, *17*(6), 44–51. doi:10.1109/MWC.2010.5675777

zWave. (n.d.). *Z-Wave compatible products.* zWave. https://www.z-wave.com/shop-z-wave-smart-home-products

# About the Contributors

**Rajashekhar Biradar** serves as Vice Cancellor in the School of Electronics and Communication Engineering at REVA University.

**Geetha Devanagavi** Serves as a Professor in the School of Computing and Information Technology at REVA University.

**Nikhath Tabassum** completed her Ph.D. from REVA University in 2021. She has published in reputed journals and conferences. Her research area includes Wireless Sensor Networks, Artificial Intelligence and Machine Learning.

**Nayana Hegde** did her Bachelors in Electronics and Communication from Karnataka University, India in year 2001 and Masters in Digital Communication and Networking from VTU, India in 2014. She did her PhD in Electronics and Communication and Engineering from REVA University Bangalore, India in year 2022. She is currently Assistant Professor in School of Electronics and Communication Engineering, REVA University Bangalore, India. Before that she worked as Lecturer in Impact College of Engineering Bangalore from 2011-2012. Before that she worked for Infosys Technologies Bangalore from 2003 to 2009. She was Graduate Engineer Trainee at NTTF from 2002-2003. Her area of research is: wireless communication, vehicular network.

**Mihai Lazarescu** serves as an Associate Professor in the Department of Electronics and Telecommunications (DET) at Politecnico di Torino.

***

**Sarthak Agrawal** is a Final year B.Tech CSE student in M.S. Ramaiah University of Applied Sciences. Sarthak's primary focus revolves around acquiring knowledge in AI/ML, Blockchain and Data Science.

**Abdul Azeez** (Senior Member, IEEE) received his Bachelor of Engineering and M.Tech Degrees in Electronics and Communication Engineering and Digital Electronics and Communication Engineering from Visvesvaraya Technological University (VTU), Belgaum, India in 2005 and 2007 respectively, and the Ph. D Degree in Wireless Communication Engineering from Visvesvaraya Technological University (VTU), Belgaum, India. He has been associated with HKBK college of Engineering where he is currently working as Associate Professor in Department of Electronics and Communication Engineering. He is an active IEEE Member in IEEE COMSOC society, IEEE ITS Society, IEEE Computer Society, IEEE Signal Processing Society, IEEE Photonics Society. He is serving as a treasurer for IEEE ITS Society Bangalore Section. He is pioneer in starting IEEE ITS Student Branch and IEEE Computer Society student branch in the college. His research interests includes Wireless Communication, Signal Processing, Information Theory and Coding, Evolutionary algorithms. He has served as a Reviewer for various IEEE conferences. He has published several research papers in refereed and international conferences.

**Aryan Bharadwaj** is a Final year B.Tech CSE student in M.S. Ramaiah University of Applied Sciences. Aryan's primary focus revolves around acquiring knowledge in Blockchain, Data Science technologies and exploring the intersection of AI/ML and digital healthcare with AI.

**Sachin Chaudhary** completed his Graduation from MJPRU, and Post Graduation from AKTU, Moradabad, U.P. Currently Pursuing his Ph.D. in Computer Science and Engineering from Govt. Recognized University. Presently, he is working as an Assistant Professor in the Department of Computer Science and Applications, IIMT University, Meerut, U.P, India. He has been awarded as Excellence in teaching award 2019. He is the reviewer member of some reputed journals. He has published several book chapters and research papers of national and international reputed journals.

**Shraddha Dayananda** is a final year student pursuing B.Tech. in Computer Science at M.S Ramaiah University of Applied Sciences. Shraddha's main areas of interest involves in AI, ML, System Design, and Software Development.

**Dankan Gowda V** is currently working as an Assistant Professor in the Department of Electronics and Communication Engineering at BMS Institute of Technology and Management in Bangalore. Previously, he worked as a Research Fellow at ADA DRDO and as a Software Engineer at Robert Bosch. With a total experience of 14 years, including teaching and industry, Dr. Gowda has made significant contributions to both academia and research. He has published over 60 research papers in renowned international journals and conferences. In recognition of his innovative work, Dr. Gowda has been granted six patents, including four from Indian authorities and two international patents. His research interests primarily lie in the fields of IoT and Signal Processing, where he has conducted workshops and handled industry projects. Dr. Gowda is passionate about teaching and strives to create an engaging and effective learning environment for his students. He consistently seeks opportunities to enhance his knowledge and stay updated with the latest advancements in his field.

**Veena K.N.** has received her Ph.D degree in wireless sensor networs which was awarded by Kuvempu University under the guidance of Dr. B.P. Vijaya Kumar, Professor and Head, Dept. of Information Science and Engg., MSRIT, Bangalore. She is working as an Associate Professor in the school of Electronics and Communication Engineering, Reva university.She has had around 50 international journals which are indexed in scopus, google scholor and web of science. She was Qualified for GATE All India level exam in 1999 and joined NITK Surathkal for M.Tech in Digital Electronics and Advanced Communication course. She has had many national and international patents under his belt, out of which few patents are granted. GOOGLE SCHOLOR ID: Veena K.N EMAIL ID: veenakn@reva.edu.in

**Mahabaleshwar S. K.** is presently working as Associate Professor, Electronics and Communication Engineering, Basaveshwar Engineering College, Bagalkote, Karnataka, India. He has experience of 21 years in teaching and research. His research interests are Vehicular Ad hoc Networks, Wireless Networks, and Internet of Things. He has published 02 Patents, 59 papers in national and international conferences, 37 papers in national and international journals, and 07 books/books chapters. He is a reviewer and programme committee member for many journals (published by IEEE, Springer, Elsevier, etc.) and international conferences, respectively. He received "Seed Money to Young Scientist for Research" from VGST Karnataka in 2015. Under his guidance, totally 07 PhD research scholars have received their degrees.

**Shridevi Jeevan Kamble** is currently working as Assistant Professor, REVA University, Bengaluru, India. She has 12 years of teaching experience and Administration and University exam conduction and 5 years of Research experiences. She has completed bachelors in ECE and Masters in DE and PhD in ECE. She has published publications in various reputed journals and Internal conferences. Here overall publication has reached citation over 110+ citations in google scholar and H-index is 5. Here research interests include Machine Learning, IoT Edge Computing, Internet of Vehicles, Wireless Communication, Block chain Technologies etc. she has also served as reviewer of various peer reviewed International Journals from publishers like IEEE, Springer, Elsevier.

**Manjunath Kounte** is currently working as Professor and Head, Electronics and Computer Engineering, REVA University, Bengaluru, India. He has over 16+ years of Teaching, Research and Administrative experience. He has completed his bachelors in ECE and Masters in Computer Network Engineering from VTU Belagavi and PhD from Jain University, Bengaluru. Dr. Manjunath Kounte is Senior Member, IEEE and member of various international professional bodies such as IETE, ISTE. He has over published over 20+ Indexed Journal papers, 25+ International Conference papers, 1 book and 3 book chapters. His overall publications have reached decent citations with over 500+ citations in google scholar and hindex of 12. His research interests include Machine Learning, IoT Edge Computing, Internet of Vehicles, Signal Processing, Wireless Communication, Blockchain Technologies etc. He is also heading Machine Learning and Blockchain (MLB) research lab in REVA University. He has also served as reviewer of various peer reviewed International Journals from publishers like IEEE, John Wiley, Springer, Elsevier, MDPI etc.

**Bhupendra Kumar** completed his Graduation and Post Graduation from Chaudhary Charan Singh University, Meerut, U.P. and Ph.D. in Computer Science and Engineering from Mewar University, Hapur. Presently, he is working as a Professor in the Department of Computer Science and Applications, IIMT University, Meerut, U.P. He has been a huge teaching experience of 19 years. He is the reviewer member of some reputed journals. He has published several book chapters and research papers of national and international reputed journals.

**Pullela SVVSR Kumar** completed his Ph.D. in Computer Science & Engineering from Acharya Nagarjuna University, Guntur, Andhra Pradesh, India in the area of Image Processing and Data Mining. He is having more than 23 years of experience in teaching. He is currently working as Professor, Department of Computer Science & Engineering, Aditya College of Engineering, Surampalem. His major areas of interest include Image Processing, Data Mining and Big Data.

**Chayalakshmi C. L.**, completed her B.E. in Instrumentation Engineering in the year 1996 from B.V.B. College of Engineering and Technology, Hubli, M. Tech. in Digital Communication in the year 2006 from Basaveshwar Engineering College, Bagalkot and Ph.D. in the year 2019 from Visvesvaraya Technological University, Belagavi. She has teaching experience of 25 years. Her research interests are Internet of Things, sensors, Embedded Systems, Industrial Instrumentation and Signal conditioning. She has two project grants from AICTE and VGST. She has published many papers in National/International journals and conferences and has won two best paper awards. She has translated four NPTEL courses in Kannada language. Presently she is working as an Associate Professor in Electrical and Electronics Engineering, Basveshwar Engineering College, Bagalkote, Karnataka, India.

**Soumya Ranjan Mishra** is presently working as an Assistant Professor (II) at the School of Computer Engineering, Kalinga Institute of Industrial Technology (Deemed to be University), Bhubaneswar Odisha, India. He has teaching and research experience of more than 10+ years and Industry experience of 2+ years in IT Companies. He did his doctoral degree Ph.D. in the field of Computer vision and machine learning from the National Institute of Technology, (NIT) Durgapur, WB, India. His research interest includes Image Processing, Intelligent Systems, Human-Computer Interaction, Computer Vision and similar innovative areas. His research contribution includes various proceedings/books which include ASIC Springer series, and publications in reputed conferences, book chapters, and journals indexed in SCI/ESCI.

**Hitesh Mohapatra** received the B.E. degree in Information Technology from Biju Patnaik University of Technology (BPUT), Odisha in 2006, and the MTech. Degree in CSE from Odisha University of Technology and Research (OUTR) University, Odisha in 2009. He received his Ph.D. in Computer Science & Engineering in 2021 from Veer Surendra Sai University of Technology (VSSUT), Burla, India. He has contributed 30 SCI and Scopus indexed research papers, 21 international/national conferences and books on Software Engineering and C Programming respectively. He has 15 years of teaching experience both in industry and academia. He has served the research community with various capacities like session chair, technical chair, keynote speaker etc. His research interests include wireless sensor network, smart city, smart grid and smart water. Currently he is working as Associate professor at School of Computer Engineering, KIIT DU, Bhubaneswar, and Odisha.

**Nidhi N P** is a second-year student pursuing BTech in Computer Science Engineering at M.S Ramaiah University of Applied Sciences. Nidhi's main areas of interest involve AI/ML, System Design, and Software Development.

**Ashwath Narayan** is a Final year student of B.Tech AI/ML at M.S. Ramaiah University of Applied Sciences. His main areas of interest lie in gaining knowledge about Data Science and AI/ML

**Tushar Narendra Kumar** is a final year student pursuing his B.Tech in Computer Science Engineering at PES University. His main areas of interest are Machine Learning, Artificial Intelligence and Data Analytics.

**Sahana. P. Shankar** completed her B.E in Information Science and Engineering and MTech in Software Engineering from Ramaiah Institute of Technology, Bangalore. She is a third rank holder in B.E and a first rank holder with gold medal in MTech. She has worked with Unisys as an Associate Engineer for three years. She holds a US Patent in the field so software testing. She has worked with RIT as a Teaching Assistant for 6 months under the TEQIP-II scholarship program. She has published papers in various international journals and conferences. She has also authored several book chapters. She is an ISTQB certified software tester. She is currently working with Ramaiah University of Applied Sciences, Bangalore as an Assistant Professor in the Department of Computer Science and Engineering since 2018.

**Rajneesh Panwar** Graduated and Post Graduated in Mathematics and Computer Application from Ch. Charan Singh University, Meerut (U.P.) and received his M. Tech. in Computer Science from Shobhit University, Meerut. Presently, he is working as an Assistant Professor in the School of Computer Science and Application IIMT University, Meerut, U.P. He qualifies GATE 2021 and UGC-NET June 2020 and December 2020. He has published several book chapters and research papers of national and international repute.

**Sanjana Prasad** (Member, IEEE) received the B.E and M.E degree in Electronics and Communication Engineering and Communication Systems from Anna University, Chennai in the year 2013 and 2015 respectively. In 2020, she received the Ph.D. degree in Information & Communication Engineering from Anna University, Chennai. She has full time research experience in the field of Wireless Communications and teaching experience at PSG College of Technology, Coimbatore. Her Ph.D thesis title is "Investigations on techniques to reduce Peak to Average power ratio reduction in OFDM systems". Dr. Sanjana Prasad is an accomplished researcher and an academician with a strong academic background and diverse expertise brings a wealth of knowledge as an Associate Professor in the Department of ECE at HKBK College of Engineering, Bangalore. She has around 20 publications, including international conferences and journals, 3 Patents published. She is a reviewer for many international journals such as IET Communications, International Journal of Communication systems, IEEE Access, International Journal of Engineering Research and Technology, International Journal of Antennas and Propagation and Wireless Personal Communications. She is also a member of professional bodies such as IEEE, IAENG and IETE. Known for her mentoring skills and guiding UG students towards innovative ideas and projects, takes part in research-related programmes conducted in India and abroad, thereby supports the impact of research in the organization. Her research areas include Wireless Communications, Optimization techniques, 4G/5G/B5G Systems, Machine learning, Artificial intelligence, IoT and Blockchain Technology.

**Rajani S. Pujar** is presently working as Assistant Professor, Electronics and Communication Engineering, Basaveshwar Engineering College, Bagalkote, Karnataka, India. She has experience of 20 years in teaching and 10 years research. Her research interests are Wireless Networks, Signal Processing and Internet of Things. She has published many papers in journals and conferences.

**Deepashree Rajendraprasad** is a passionate health informatics professional with a foundation in clinical dentistry. Her fascination with healthcare processes and a passion for technology has led her to further her education in the United States, where she is currently pursuing her Masters in Health Informatics with a focus on Applied Data Analytics at Boston University. In addition, she is currently employed as Integrated Data & Analytics (Population health) Intern at Corewell Health, USA which is one of the major healthcare industries in USA. She brings a unique blend of healthcare and IT proficiency with skills in Microsoft Office Suite, SQL, Java, Python, R, Tableau, and Power BI. Her specialization includes healthcare data analysis, clinical data management, medical coding, and strict adherence to HIPAA compliance and ICH/GCP guidelines. Additionally, she has a thorough understanding of health information systems, healthcare quality improvement strategies, and health data standards. With hands-on clinical experience, she is adept at addressing real-world healthcare challenges and navigating Medicare, Medicaid, and health insurance claims systems. Her primary goal is to leverage her skills in health informatics and data analytics to improve patient outcomes, enhance healthcare operations, and boost the efficiency of healthcare delivery. She eagerly anticipates the chance to contribute her skills to a challenging health informatics role in the global level.

**Harsha.R.** Currently working as Embedded software design in Teufel Berlin, Germany. I have received my Master degree in VLSI and Embedded systems. Currently persuing my Doctor of philosophy in the school of Electroniccs And Communinaction Engineerin,REVA university. I was working as an Assistant Professor in Dr Ambedkar Institute of Technology Bengaluru.During my tenure as Assistant Professor I have applied for seven national and international patents, out of which few patents are granted and the rest are published. I have also published many international journals which are indexed in scopus, google scholor and web of science. I take this opportunity to thank almighty, my parents, my friends and my guide who are constantly supporting me during my research. REVA University I would thank REVA university for providing me the oppourtunity for persuing my Doctrate of philosophy. The infrastructure in the campus is just excellent which makes any scholor to put in effort for persuing his education. EMAIL ID: harsh4no1@gmail.com GOOGLE SCHOLOR ID: Harsha Raju

**Saujanya Mathanda Sanjaya** completed bachelor of engineering in Electronics and Communication From RITM, Bangalore. She received MSc. in Cybersecurity. Currently pursuing PhD from Munster Technological University, Ireland. Here main interests in research are Cybersecurity, Block Chain and ML.

**Manoj Kumar S B** works as an associate professor in BGS Institute of Technology and has 13.5 Years of experience. He has Published 23 research papers and 2 Book chapters. He is a lifetime member for 4 professional bodies. Under his supervision KSCST 43rd and 44th series SPP projects were selected and received 22 thousand rupees. He filed and published 4 National Patents, Currently he is working as PG Coordinator from Jan 2017 to till date, Project Coordinator, Internship Coordinator, Industrial visit coordinator, Innovation and science club coordinator.

**Rashmi Rani Samantaray** is currently working as an Assistant Professor in HKBK College Of Engineering, Bangalore. She brings with her a wealth of experience of 12 years dedicatedly teaching undergraduate courses. She earned her M.Tech degree in Electronics from BMS College of Engineering (VTU) and is currently pursuing her Ph.D. from BMS College Of Engineering under VTU. Her research is centered around wireless communication, specifically delving into the contemporary landscape of 5G, 6G, and Filter Bank multi-carrier modulation. Alongside her academic focus, Rashmi exhibits a profound interest in Artificial Intelligence (AI) and Machine Learning (ML) techniques. Her valuable contributions to academia extend to publishing and presenting papers in various IEEE conferences and peer reviewed journals. Additionally, Rashmi holds a lifetime membership with the ISTE (Indian Society for Technical Education).

**Tanvi Saxena** is a Final year student pursuing his B.Tech in Computer Science Engineering at Ramaiah Universty of Applied Sciences. Her main areas of interest are Machine Learning, Artificial Intelligence and Data Science.

**Anirudh Shankar** is a Final year student of B.Tech AI/ML at M.S. Ramaiah University of Applied Sciences. Shankar's main areas of interest lie in gaining knowledge about next-gen technologies and AI/ML

**Kewal Krishan Sharma** is a professor in computer sc. in IIMT University, Meerut, U.P, India. He did his Ph.D. in computer network with this he has MCA, MBA and Law degree also. He did variously certification courses also. He has an overall experience of around 33 year in academic, business and industry. He wrote a number of research papers and books.

**Vikas Sharma** completed his Graduation and Post Graduation from Chaudhary Charan Singh University, Meerut, U.P. Currently Pursuing his Ph.D. in Computer Science and Engineering from Govt. Recognized University. Presently, he is working as an Assistant Professor in the Department of Computer Science and Applications, IIMT University, Meerut, U.P. He has been awarded as Excellence in teaching award 2019. He is the reviewer member of some reputed journals. He has published several book chapters and research papers of national and international reputed journals.

**Zahira Tabassum** obtained her Doctorate in Electronics Engineering from Jain University in 2023."Analysis and Design of Demand Side Management in Smart Grids" was the title of her thesis. She is presently employed with the HKBK College of Engineering as an Associate Professor in the ECE Department. Dr. Zahira has 22 years of broad teaching experience and has made major contributions to research and academia. She won the best "Innovative Teaching" award in 2022 thanks to her enthusiasm for teaching and utilisation of a variety of teaching methodologies. She has papers published at IEEE conferences and reputable journals. Her areas of research interest include Smart Grids, Embedded Systems, IoT and Machine Learning. She constantly looks for ways to learn more and keep up with the most recent developments in her areas of expertise.

**Md. Tauseef** completed both his graduation and post-graduation at Visvesvaraya Technological University, Belgaum. With over 8 years of teaching experience, he currently serves as an Assistant Professor at the School of Electronics and Communication Engineering, REVA University in Bengaluru, India. His primary focus at REVA University is conducting research in cutting-edge technologies such as IoT, Blockchain, Artificial Intelligence, and Embedded Systems. His research contributions have been featured in national and international conferences/journals indexed in Scopus and Web of Science. Notably, he holds Indian and Australian patents for groundbreaking inventions and has significantly contributed to the development of innovative products. Beyond academia, he passionately shares his knowledge through workshops and hands-on training sessions for educational institutions and industry professionals. His goal is to inspire and empower others to explore the vast possibilities offered by technology.

**Amit Kumar Tyagi** is working as Assistant Professor, at National Institute of Fashion Technology, 110016, New Delhi, India. Previously he has worked as Assistant Professor (Senior Grade 2), and Senior Researcher at Vellore Institute of Technology (VIT), Chennai Campus, 600127, Chennai, Tamilandu, India for the period of 2019-2022. He received his Ph.D. Degree (Full-Time) in 2018 from Pondicherry

Central University, 605014, Puducherry, India. About his academic experience, he joined the Lord Krishna College of Engineering, Ghaziabad (LKCE) for the periods of 2009-2010, and 2012-2013. He was an Assistant Professor and Head- Research, Lingaya's Vidyapeeth (formerly known as Lingaya's University), Faridabad, Haryana, India for the period of 2018-2019. His supervision experience includes more than 10 Masters' dissertations and one PhD thesis. He has contributed to several projects such as "AARIN" and "P3- Block" to address some of the open issues related to the privacy breaches in Vehicular Applications (such as Parking) and Medical Cyber Physical Systems (MCPS). He has published over 100 papers in refereed high impact journals, conferences and books, and some of his articles awarded best paper awards. Also, he has filed more than 20 patents (Nationally and Internationally) in the area of Deep Learning, Internet of Things, Cyber Physical Systems and Computer Vision. He has edited more than 20 books for IET, Elsevier, Springer, CRC Press, etc. Also, he has authored 3 Books on Internet of Things, Intelligent Transportation Systems, Vehicular Ad-hoc Network with BPB Publication, Springer and IET publisher. He is a Winner of Faculty Research Award for the Year of 2020, 2021 and 2022 (consecutive three years) given by Vellore Institute of Technology, Chennai, India. Recently, he has been awarded the best paper award for a paper titled "A Novel Feature Extractor Based on the Modified Approach of Histogram of Oriented Gradient", in ICCSA 2020, Italy (Europe). His current research focuses on Next Generation Machine Based Communications, Blockchain Technology, Smart and Secure Computing and Privacy. He is a regular member of the ACM, IEEE, MIRLabs, Ramanujan Mathematical Society, Cryptology Research Society, and Universal Scientific Education and Research Network, CSI and ISTE.

**Deepak Varadam** holds a Masters Degree in Microsystems Engineering from Heriot-Watt University, Edinburgh (Scotland), U.K. and P.G.Diploma in ECE from Napier University, U.K. He worked as a Postgraduate Masters Research Scientist under Biomedical Microengineering Group (BMG) at Heriot-Watt University. He is currently working as an Assistant Professor in Faculty of Engineering and Technology, Computer Science and Engineering Department at Ramaiah University of Applied Sciences (RUAS) since 2010. He has been actively participating in delivering corporate trainings, seminars and have publications in national, international journals and conferences. His research interests is in MEMS, Embedded Systems, Software Engineering, Computer Graphics, Blockchain technology, AI, IoT and Big Data. He is currently an ExeCom Member of the IEEE RUAS Student Chapter.

**Deepak Varadam** holds a Masters Degree in Microsystems Engineering from Heriot-Watt University, Edinburgh (Scotland), U.K. and P.G.Diploma in ECE from Napier University, U.K. He worked as a Postgraduate Masters Research Scientist under Biomedical Microengineering Group (BMG) at Heriot-Watt University. He is currently working as an Assistant Professor in Faculty of Engineering and Technology, Computer Science and Engineering Department at Ramaiah University of Applied Sciences (RUAS) since 2010. He has been actively participating in delivering corporate trainings, seminars and have publications in national, international journals and conferences. His research interests is in MEMS, Embedded Systems, Software Engineering, Computer Graphics, AI, Blockchain, and IoT. He is currently an ExeCom Member of the IEEE RUAS Student Chapter.

**Tarun Kumar Vashishth** is an active academician and researcher in the field of computer science with 21 years of experience. He earned Ph.D. Mathematics degree specialized in Operations Research; served several academic positions such as HoD, Dy. Director, Academic Coordinator, Member Secretary of Department Research Committee, Assistant Center superintendent and Head Examiner in university examinations. He is involved in academic development and scholarly activities. He is member of International Association of Engineers, The Society of Digital Information and Wireless Communications, Global Professors Welfare Association, International Association of Academic plus Corporate (IAAC), Computer Science Teachers Association and Internet Society. His research interest includes Cloud Computing, Artificial Intelligence, Machine Learning and Operations Research; published more than 20 research articles with 1 book and 10 book chapters in edited books. He is contributing as member of editorial and reviewers boards in conferences and various computer journals published by CRC Press, Taylor and Francis, Springer, IGI global and other universities.

# Index

## Q



## R



## S



## T



## U



## V



## Z

www.ingramcontent.com/pod-product-compliance
Lightning Source LLC
LaVergne TN
LVHW062112300926
849161LV00010B/2444
*9798369306598*